THIRD EDITION

The Effective Teaching of Language Arts

Donna E. Norton
Texas A&M University

Merrill Publishing Company
A Bell & Howell Information Company
Columbus Toronto London Melbourne

To my husband, Verland, and my children, Saundra and Bradley,
for their constant support, immense understanding, and insightful viewpoints

Cover Art: Leslie Beaber

Published by Merrill Publishing Company
A Bell & Howell Information Company
Columbus, Ohio 43216

This book was set in Souvenir

Administrative Editor: Jeff Johnston
Developmental Editors: Amy Macionis and Linda Scharp
Production Coordinator: Linda Bayma
Art Coordinator: Gilda Edwards
Cover Designer: Cathy Watterson
Text Designer: Anne Daly
Photo Editor: Gail Meese

Photo Credits: All photos copyrighted by individuals or companies listed. Andy Brunk/Merrill, pp. 1, 378, 420; Ben Chandler/Merrill, p. 183; Chicago Historical Society and Clarion Book/Ticknor & Fields, a Houghton Mifflin Company, p. 452; Paul Conklin, p. 472; Kevin Fitzsimons/Merrill, pp. 196, 293, 581, 601; Jean Greenwald/Merrill, pp. 6, 24, 59, 72, 125, 140, 173, 221, 267, 307, 330, 363, 368, 439, 505, 507, 531, 561; Bruce Johnson/Merrill, p. 587; John McNamara, pp. 115, 529, 541; Bradley Norton, p. 383; Mike Penney, p. 463, Charles Quinlan, pp. 236, 519; Michael Siluk, pp. 17, 191; David S. Strickler/Strix Pix, pp. 39, 53, 67, 161, 253; Larry Thurston, p. 484; Dan Unkefer/Merrill, pp. 133, 320, 343, 427; Cynda Williams/Merrill, pp. 210, 214; Gale Zucker, p. 9.

Library of Congress Catalog Card Number: 88–63824
International Standard Book Number: 0–675–20649–9
Printed in the United States of America
   2   3   4   5   6   7   8   9—92   91   90   89

Preface

The effective teaching of language arts requires a commitment to excellence on the part of the classroom teacher. The excellence cannot be attained without a thorough knowledge of the language arts and an understanding of methods that will develop children's language arts skills. This is an awesome task because the language arts include the diverse skills of speaking, listening, reading, and writing.

Current educational practice emphasizes the role of children's literature in all phases of the language arts curriculum. Consequently, this edition of the text includes strong emphasis on children's literature. There are discussions about and instructional approaches for using children's literature in chapters on language and cognitive development, oral language development, listening, grammar and usage, composition, literature, reading, and multicultural education. By referring to the combined information in chapter 10, "Literature" and chapter 11, "Reading and Literature," students will be able to develop a literature-based language arts and reading program. Methods for integrating language arts through literature are illustrated in a unit of study: "A Literary Approach to the Study of Biography."

This edition includes two chapters on writing. Chapter 8, "Composition: The Writing Process," emphasizes the theoretical basis for processing and the development of composition skills through a process approach to writing.

Chapter 9, "Composition: Expressive, Poetic, and Expository Writing," explores the three types of writing developed in the language arts curriculum. The integration of the language arts is illustrated in this chapter through a unit of study: "A Composition Approach to the Study of Biography." By studying the two units on biography, students can identify numerous ways that literature may provide a focus for and enhance the integration of the language arts in a literature-based program.

Each chapter in this text contains features geared to the development of effective language arts instruction. Major topics are outlined at the beginning of each chapter. Chapter objectives provide an overview of content. This overview gives students a preview of materials to be covered and stresses the major learnings to be mastered in the chapter.

Diagnostic procedures are suggested for many areas of the language arts. Because teachers are accountable to the children they teach, the administrators of the school, and the parents, they must obtain accurate knowledge about children's strengths, weaknesses, and interests so that they can be most effective in language arts instruction. Informal assessment techniques that the classroom teacher can easily use and interpret are stressed.

Diagnosis is worthless, however, without research-based knowledge about learning environments, motivation, objectives, instructional procedures, grouping, and teacher effectiveness. Effective language arts teachers cannot ignore these important elements. Examples of lesson plans, instructional units, and learning centers allow the student to visualize both the methods of teaching suggested and the theory behind them. These lessons have been used with children and frequently include examples of children's written and oral products.

In order to experience planning lessons, evaluating language arts skills, or teaching a lesson, reinforcement activities are included throughout each chapter and at the conclusion of each chapter. All of the reinforcement activities have been used with college language arts classes or inservice presentations with experienced teachers. The professor and the students may choose as many of these reinforcement activities as they wish to develop the objectives of the chapter and to explore effective teaching strategies.

Effective teaching of language arts cannot occur if special considerations are not given to the linguistically different child or to multicultural education. A separate chapter on the linguistically different child stresses research the classroom teacher can use to improve instruction, effective techniques for diagnosing language arts needs, and factors influencing performance. The section on multicultural education emphasizes evaluating, selecting, and using multicultural literature in the classroom.

Teachers must meet the instructional needs of all children. Mainstreaming children with special needs requires knowledge of children's characteristics and effective instructional approaches. Consequently, a majority of chapters include charts that identify characteristics of learning disabled students and mentally handicapped students. These charts also identify specific language arts methods that have been used successfully.

Finally, chapter 15 discusses classroom organization as it influences effective instruction. The goal of this chapter is the effective management of the classroom through grouping, flexible room arrangements, adequate instructional time, appropriate assignments, and blocking of teacher-directed and individually completed activities.

The methods development of a language arts text would be impossible without the enthusiasm, critical evaluation, and suggestions of many individuals. My special appreciation is extended to the many language arts students and teachers who have tried specific language arts methods in their classrooms, shared stimulating experiences, and criticized sections of this text. Outstanding teachers have shared their experiences with me by suggesting lesson examples, units, and learning centers that have been used successfully in the classroom. Both undergraduate and graduate language arts classes read the text to ensure understanding of the content. The enthusiasm and sincerity of these students are appreciated.

My sincere appreciation is also extended to Elizabeth Antley, University of Arizona; Thomas Devine, University of Lowell; Barbara Erwin, Angelo State University; George Hess, Kennesaw College; Walter Prentice, University of Wisconsin-Superior; and Carolyn Reeves-Kazelskis, University of Southern Mississippi, who made many helpful suggestions as they reviewed the manuscript. My thanks go also to executive editor Jeff Johnston, developmental editors Amy Macionis and Linda Scharp, and production editor Linda Bayma.

The goal of this text is to provide support, motivation, and knowledge for the classroom teacher who is responsible for the effective teaching of all children. This goal will be achieved if we can instill in teachers a commitment for excellence.

Contents

Chapter One

WHAT ARE THE LANGUAGE ARTS?
HOW TO APPROACH THE LANGUAGE ARTS
Chapter Objectives ■ *Diagnostic Procedures* ■
Reinforcement Activities ■ *Research* ■
Developmental Instruction ■ *The Literature
Connection* ■ *Linguistically Different Children and
Multicultural Education* ■ *Summaries*

CURRENT ISSUES AND CONCERNS
AFFECTING LANGUAGE ARTS TEACHERS
SUMMARY

*After completing this introduction to the language
arts, you will be able to:*

1. *Describe the contents and importance of the
   language arts curriculum.*
2. *Describe how language arts is presented in
   this textbook.*
3. *Identify current issues that may affect language
   arts instruction.*

Introduction to the Language Arts

*L*anguage is the most important form of human communication. Language communication includes speaking, writing, listening, and reading; as such, it is the most important, as well as the most exciting, part of the elementary curriculum. Its importance is reflected in estimates that the average person listens to the equivalent of a book each day; talks the equivalent of a book each week; reads the equivalent of a book in a month; and writes the equivalent of a book each year.

How do children develop these important language arts communication skills? It is your responsibility as a language arts teacher to provide ability to communicate effectively. The purpose of this textbook is to provide you with effective methods for developing language arts skills for all children.

WHAT ARE THE LANGUAGE ARTS?

A brief overview of the language arts curriculum reinforces the suggestion that language arts is both important and exciting. When we enter a stimulating elementary school where learning is taking place, we can easily see how the teachers use language arts in all areas of the curriculum.

We have mentioned that oral language is a vital form of human communication; the effective teacher provides many opportunities for oral language development and does not assume that oral language skills cannot be improved. Good oral questioning strategies guide children to focus their attention on an initial question, extend their information and understanding, clarify facts, and raise their comprehension levels. Exciting discussions take place in all subject areas when teachers use effective oral questioning strategies that encourage children to develop their individual thought processes. Discussions are also extended through such techniques as buzz sessions, round table and panel discussions, and debates. Original drama is an extremely exciting method for developing creative thinking and oral language skills. In a stimulating elementary school, we can visit classrooms where children are presenting puppet plays, doing improvisations, using creative play activities, pantomiming characters and objects, role playing experiences, and doing choral poetry reading. Children in such an environment are involved in all aspects of oral language, and the teachers know how to increase their students' participation in oral language as well as how to evaluate and improve it.

The teachers in this environment recognize that effective oral language instruction requires an audience, and that this audience needs to refine its listening skills. Research shows that listening skills can be improved through instruction, although because of the multiple definition of listening, both evaluation and instruction are complicated. Listening includes hearing; auditory discrimination; literal, interpretational, critical, and evaluational comprehension; and appreciation. The effective language arts teacher recognizes the importance of each aspect of listening. Audiometers are used to screen children who have a suspected hearing loss; auditory discrimination tests identify students who may have problems with auditory perception; informal tests identify children who may have problems with achievement due to inattentive listening; and listening comprehension tests are used to evaluate the students' ability to restate, interpret, and critically evaluate what is heard.

The teacher uses the information obtained through evaluation to provide exciting and effective learning activities. To develop auditory awareness, young children listen to and describe sounds they hear in their environment. They listen to music, rhymes, and limericks to become aware of the lovely sounds in our language. To improve auditory discrimination, they listen to and describe likenesses and differences in sounds, words, and rhymes. Many of these discrimination activities take the format of games. To develop attentive listening skills, teachers help children set purposes for listening, provide opportunities for children to give as well as listen to directions, and provide instruction that requires children to listen for specific purposes. Many stimulating methods are used to help children develop listening comprehension. In one class, children might listen to a taped conversation between two famous people in history for the purpose of identifying the individuals. Another class might listen to the beginning of a story, then predict its outcome. Students can listen to commercials and critically evaluate the propaganda techniques they hear, or they may develop appreciative listening skills by listening to poetry, music, or literature for pure enjoyment.

Teachers of an effective language arts curriculum provide many opportunities for children to develop their writing skills. Folders of each child's writing are kept so student and teacher can evaluate growth in writing. The teacher realizes environment is critical in developing writing skills. She stimulates, reacts to ideas, confers with the writer, listens to and reads the writing, and provides opportunities for sharing writing, rather than merely assigning projects and providing group instruction. Stimulation is essential. Before children write poetry, they may go out into a field or the woods to experience nature; before they write a creative story they may listen to music or discuss a thought-provoking item or picture. Children in this environment experience the development of a composition by identifying their audience and purpose, deciding on the subject, and organizing their ideas. They also write a great deal—biographies, classroom newspapers, and creative stories.

The effective language arts curriculum also includes the more mechanical aspects of written and oral language. Young children experience readiness activities for handwriting. The first-grade child is usually instructed in manuscript printing; in about third grade, the child receives instruction in switching from manuscript to cursive writing. The numerous writing activities provided in the classroom offer children many opportunities to use and improve their handwriting skills. Because the study of modern grammar stresses that children should understand their language, the effective language arts teacher uses activities such as sentence-pattern exploration, sentence combining, sentence expansion, and sentence transformation to allow children to explore their language and discover how it works. Children also examine appropriate levels of usage. They learn that different levels of usage are appropriate for different audiences and purposes; instruction is flexible and increases the levels of usage available to the child. The effective teacher assesses children's spelling-ability levels and provides instruction at the appropriate level for each child.

Literature is a dynamic part of the effective language arts classroom. Books are everywhere. The teacher reads to the students and allows them many opportunities to read for enjoyment and to discuss the books they read with other children. The teacher models behavior and helps children through the decision-making process as they approach various literary elements in literature. If we enter a classroom where literature is important, we will probably see a colorful library corner with attractive and comfortable places for children to read. Children share their books through creative oral, written, and art ideas. Most important, the teacher reads and tells stories to the students. During the storytelling, the students are probably sitting in front of the teacher in excited anticipation; they know storytelling is a wonderful experience. Literature is used to stimulate research and interest in social studies and science subjects. A class studying the westward expansion, for example, may be adding to its knowledge of frontier life by reading the Little House books by Laura Ingalls Wilder. A class studying science may learn more about a famous scientist by reading a biography or autobiography. The teacher in this environment has also evaluated the students' interests in literature as well as their ability to read and uses this information to help children find stimulating books they can read and enjoy.

Tchudi (1986) maintained that literature and reading materials are excellent sources for extending the curriculum into what he referred to as multicultural reading, multidisciplinary reading, and multimedia reading. Multicultural reading includes infusion of Black and ethnic literature into all language arts classes. Multidisciplinary reading emphasizes that literature from a wide variety of discipline areas and genres is appropriate in language arts classes. Likewise, quality literature is appropriate in other disciplines such as science, mathematics, history, and social studies. Multimedia reading includes television and other media as integral parts of the curriculum.

Language arts instruction that utilizes the various media can be highly motivating for many children. Television, rather than contributing to the decline of the language arts, can be used by the creative teacher to motivate reading, to develop discussion skills, and to develop critical thinking. Commercial films and filmstrips add visual interpretations to books, stimulate creative writing, and stimulate oral expression. A visit to this effective language arts class might find the students actively involved in making their own film, providing a musical background for it, and showing the finished product to an appreciative audience. Another class might be developing critical evaluation skills by studying propaganda techniques used in newspaper, television, and radio commercials. They can use the newspaper to distinguish fact from opinion in writing, or write their own newspaper ads, news stories, or special features. A class-produced newspaper, television show, radio program, or film is an exciting culminating activity for a study of media.

The subjects covered in the language arts curriculum are obviously important. The effective language arts teacher must be able to provide instruction in each of these subject areas—oral language, listening, written composition, handwriting, grammar, usage, literature, reading, media, and reference skills—to a wide variety of children. This text is designed to prepare the prospective teacher for this exciting work.

HOW TO APPROACH THE LANGUAGE ARTS

This text includes several features to help the preservice and inservice teacher become an effective language arts teacher. Let us look at each feature to see how it can improve understanding of and instruction in the language arts.

Chapter Objectives

Each chapter includes a list of objectives for that section. This list provides an overview of the chapter, outlines the topics covered in it, and provides a means of evaluating the learning that occurs by completing the chapter. A glance at all the chapter objectives also provides a review of the language arts subjects covered in the text.

Diagnostic Procedures

Effective language arts teaching requires that children be taught in small groups, large groups, or individually, according to their needs. This requirement would be an impossible task without adequate knowledge of each child's strengths and weaknesses. For this reason, many chapters describe both formal and informal diagnostic procedures for assessing children's abilities in that particular area of language arts.

Informal assessment techniques that the classroom teacher can easily use are especially valuable in language arts. Because many of the standardized tests given in elementary schools do not adequately assess language arts skills, the teacher cannot use the information gained from these tests to improve instruction. Assessment of oral language is an example of this problem. Group-administered, standardized tests do not have an oral language component. If oral language is a vital part of language arts instruction, then we must use assessment techniques that allow for observation and evaluation of each child's total language apparatus under many different circumstances.

Checklists for immediate use by the teacher are included in the text. Examples of these checklists include an individual student profile to accompany oral language evaluation; an auditory inventory; a listening comprehension inventory; a handwriting analysis checklist; a writing evaluation checklist; a usage, punctuation, and capitalization inventory; an interest inventory for literature; and a library reference skills checklist.

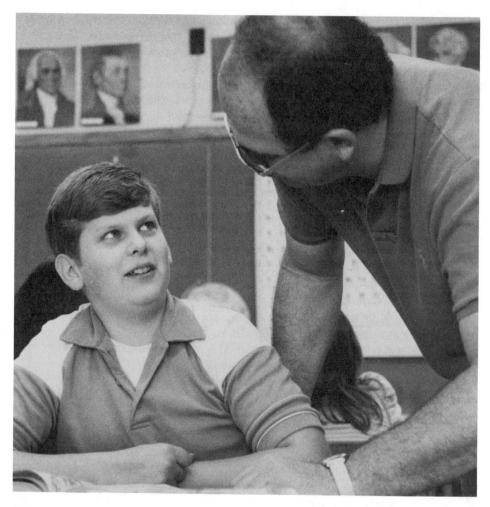

Effective language arts teachers understand children's individual needs.

The teacher must also know the ability level of each child so that valuable instructional time will not be wasted teaching skills the child has already mastered. Spelling is one example. Without assessment of spelling levels, a child may be receiving instruction on words he already knows, which would be boring for the child and a waste of valuable instructional time. It is equally wasteful to provide spelling instruction at a level so difficult that it frustrates the child.

Reinforcement Activities

As we all know from experience, we learn and retain knowledge by becoming actively involved in our own education. This is particularly true in an educational methods course such as language arts. We thoroughly understand assessment techniques only after we have used them with children or adults. Likewise, the best way to develop skills in teaching various language arts subjects is to design lessons and teach them, either to children or to peers. For this reason, reinforcement activities are included throughout each chapter, so that you can immediately apply a language arts assessment or instructional technique.

Research

Teachers of language arts are fortunate that a great deal of research has been done in most areas of the language arts. This is important because sound instructional practices are based on research. This text includes a review of the research that provides a rationale for suggestions for diagnosis and instruction. This research may also be used as a source for further study in that area of language arts. Research findings have changed some of the instructional emphases in language arts and have explored the most effective role for the teacher. Writing research is an example of this change: Researchers such as Flower and Hayes (1978), Graves (1981), Applebee (1984), Britton et al. (1979), and Hillocks (1986) either developed foundations for effective instruction or identified effective practices that improve students' abilities to respond during different phases of the writing process. Likewise, research in modeling and implications from schema theory have improved instruction in literature and have reinforced the strong connections between literature and reading.

Research from cognitive psychology is another example of this change. Cognitive mapping or webbing strategies now enhance all areas of language arts. Research has shown that use of students' creative and critical thinking skills increases their interactions and understandings in all content areas. In 1986 the National Council of Teachers of English Committee on Classroom Practices in Teaching English identified the development of students' creative and critical thinking skills as the current major concern for English language arts teachers. Undoubtedly, this will be an area of research concern for the next decade.

Developmental Instruction

After diagnosis, the next step is to provide the most effective language arts instruction for each child. This instruction encompasses the learning environment, motivation, objectives, instructional procedures, and teacher effectiveness. None of these elements can be ignored in the effort to maintain an effective language arts classroom. The development of writers in the classroom exemplifies the interrelationship of these essential elements as students proceed from a motivation for writing, to the consideration of a topic, to a tentative approach for writing, to gathering materials, to organizing their ideas and materials, to writing their first drafts, to reviewing and rereading their drafts, to revising and making changes in the content, to proofing and editing their works, and to sharing their works with the intended audiences. These various prewriting, writing, and postwriting activities encourage interactions between students as well as between students and teachers. This text explores these instructional interactions in a chapter on the writing process and in a chapter on developing the expressive, the poetic, and the expository writer.

Researchers such as Roehler and Duffy (1984) and Gordon (1985) developed instructional approaches that place the teacher in an active learning role with the students and that show students how the teacher approaches thought processing before they are expected to perform similar tasks. Effective modeling approaches proceed from identifying skill requirements, to examples showing that skill, to identifying why the skill is important to students, to identifying text samples, to developing questions, to considering the answers to the questions, to citing the evidence that supports the answers, and to thinking through the reasoning process used to acquire the answers. As the modeling progresses, the approach proceeds from total teacher modeling, to gradual student interaction, to total student responses. Modeling, including oral language, grammar, and literature, is demonstrated in several chapters in this text.

Examples of lesson plans, instructional units, and learning centers are included to help you visualize instructional procedures and to give you ideas for developing similar lessons. Examples of lesson plans include the development of language experience chart stories, expressive writing, vocabulary webs, manuscript writing, sentence transformations, usage, and modeling various literary elements and cohesive devices. Examples of units include a puppetry project, a proofreading unit, a folktale unit, a composition approach to biography, a literary approach to biography, a filmmaking project, an advertising unit for critical evaluation of propaganda techniques, and a library reference unit. Examples of learning centers include a listening learning center and a newspaper learning center. These lesson plans, units, and learning centers have all been used effectively with children.

Where feasible, examples of elementary-students' products from a lesson are included. The examples, many of which are in creative poetic writing, range from first grade through middle school, and illustrate the results of instruction.

Developmental instruction is not accomplished in isolation. Any language arts topic requires the interaction and integration of numerous areas. For example, a series of language arts lessons based on folktales may require the development of critical and appreciative listening, the analysis of literary elements, the comparison of folklore types, the creation of a play or oral storytelling based on folktales, and the creative writing of folktales that follow similar forms. Consequently, these lessons require the integration of listening, literature, reading, oral language, writing, and social studies. In addition, folklore references may be selected from the school media center, and a film or filmstrip may be developed. The integration of such language arts subjects and various content areas are emphasized throughout the text.

The Literature Connection

The role of primary literature sources in the language arts and reading curriculum is receiving renewed interest. There are two types of considerations. First are the concerns for using literature to motivate students to read and to appreciate literature. Second are the concerns for instructional approaches that encourage understanding of various genres, story structures, literature elements, and reading improvement.

Effective language arts teachers use literature to encourage children's oral language development.

Two chapters within this text emphasize the dual role of pleasure and understanding. The literature chapter focuses on the selection of literature from various genres, the identification of students' interests that can relate to literature, the stimulation of interest and appreciation of literature, and the methodologies that help students understand particular genres and literary elements within books. The literature and reading chapter emphasizes additional methodologies that may be used in the reading program.

To help you increase the use of primary sources within all areas of language arts, chapters on oral language, listening, writing, and multicultural education include strong literature components.

Linguistically Different Children and Multicultural Education

Many children in our schools do not speak and read the language of the classroom and the teacher, and, as a result, these children are frequently unsuccessful in our schools. The chapter on linguistically different children stresses research the classroom teacher can use to improve instruction, effective techniques for diagnosing language arts needs of linguistically different children, and factors influencing performance. Programs that have been successful with speakers of Black English and with Mexican-American children are described. This chapter also places special stress on the evaluation of literature for and about linguistically different children.

Summaries

A summary at the end of each chapter reviews the most important information presented in the chapter. As with the chapter objectives, you can read all the summaries before you read the text to obtain a quick overview of the information covered. You should also read the summary after completing each chapter to reinforce the important concepts that have been covered.

REINFORCEMENT ACTIVITY

Preview this language arts text. Read the objectives at the beginning of each chapter and the summaries at the end of each chapter. Locate the sections containing diagnostic procedures; briefly glance over the informal checklists. Read the reinforcement activities; they provide another overall view of the information covered in the text. Look for research pertaining to a language arts subject. Find an example of a lesson plan, unit, or learning center that illustrates developmental instruction. In class, or with a group of teachers, discuss the scope and importance of the subject matter covered in the language arts.

CURRENT ISSUES AND CONCERNS AFFECTING LANGUAGE ARTS TEACHERS

Issues, concerns, and mandates surround education and language arts instruction in particular. Many of the current developments can be classified under either control of our profession from the outside or improvement of our profession from within. College students, college professors, and classroom teachers are affected by these influences.

Control of the profession from the outside includes legislative mandates, public opinion responses to studies on education, and censorship attempts by various interest groups. For example, states such as Illinois and Texas mandated language arts objectives that must be taught and assessed. In 1985 Illinois Public Act 84-126 amended the School Code of Illinois to include specific goals and objectives for reading, listening, writing, oral communications, literature, and language functions. Likewise, the essential elements mandated by Texas law provide detailed objectives for language arts teaching and assessment. Many states across the country either have similar mandates or are in the process of developing or considering such mandates.

Responses of school districts to these mandates reflect both negative and positive reactions. Negative responses include tight controls, narrow definitions of what can be taught in language arts, minimal individualization, and teaching to the assessment instruments. In schools where such practices are common, teachers lose control over curricular development. In contrast, positive responses include using the mandates to identify inservice needs and knowledge areas that should be improved through varied types of instruction. This positive response views mandates as minimal requirements. Within this attitude educators search for ways to improve instruction based on sound theory and practice rather than allowing the mandates to dictate curricular decisions. Methodologies that have been developed to show teachers how to both improve language arts instruction and meet legislative mandates are included throughout this text.

Public responses to national studies, evaluative reports, and critical statements affect language arts instruction at all levels of education and cause local school districts and colleges of education to evaluate their programs. Newspapers and television and radio broadcasts across the nation highlight such reports and frequently analyze the efforts of local school districts to overcome any cited deficiencies. For example, results from a recent study by the National Endowment for the Humanities, a report requested by the U.S. Congress, were highlighted on September 1, 1987 by the CBS News. The same study was reported as the lead front page headline in *The Dallas Morning News*. The headline read "Teens Know Little About America's Past, Report Says." The secondary headline stated "Public schools faulted for failing to teach literary classics, history" (Holloway, 1987, p. 1A). The news story reported that a majority of seventeen-year-olds do not know when the signing of the Constitution occurred or when the Civil War took place, do not recognize works of authors such as Nathaniel Hawthorne and Herman Melville, and do not know literary classics and history because "Our system of elementary and secondary education stresses skills rather than knowledge" (p. 1A). After reporting these national findings the news article continued with interviews with local educators and descriptions of programs such as magnet schools and proposed new programs designed to increase knowledge of literature and history.

Other national studies indicate declining standards in our school systems and consequently raise public concern. Reasons hypothesized for this lack of excellence include (1) nonrigorous curriculum, especially in the high schools; (2) curriculum that teaches skills rather than content; (3) too much time and money spent on nonacademic subjects such as sports; (4) teachers who are poorly qualified and inadequately trained; (5) low salaries and nonprofessional educational environments that either do not attract the most qualified personnel or do not retain them after a few years of experience; (6) school days/years that are too short to provide excellent education; and (7) lack of support from parents and communities.

A third type of control from outside forces includes organized group efforts to censor children's literature and school textbooks. Certain groups believe that books that they suspect are capable of subverting children's religious, social, or political beliefs should be censored and that teachers should not be allowed to use the materials in the classroom. According to a survey by People for the American Way, censorship attempts "increased by 20 percent during the 1986–87 school year and by 168 percent in the last five years" (Wiessler, 1987, p. 1). A report by Dronka (1987) indicated that organized groups are increasing their sponsorships of censorship. According to Dronka, 17 percent of the censorship incidents in 1982–83 were linked to organizations. By 1985–86, 43 percent were associated with organized group efforts. Litigation efforts over textbooks in Tennessee and Alabama demonstrate the power of censorship efforts. Although the appeals court decisions in both Tennessee and Alabama upheld the rights of the school districts to use the books, we can be certain that censorship attempts will continue and will influence language arts instruction.

The forces concerned with improving our profession from within include resolutions by professional societies, studies conducted by respected language arts authorities, studies and recommendations by university groups, and criticisms expressed by our peers in college and university English departments.

Issues related to both control and improvement of the curriculum are found in the resolutions adopted by the National Council of Teachers of English (Maxwell and Allen, 1986). For example, NCTE (1986) responded to legislative controls with this resolution:

> RESOLVED, that the National Council of Teachers of English affirm that as professional practitioners English language arts teachers are best qualified to decide what constitutes informed practice and curriculum content; that NCTE urge legislative bodies and the agencies that regulate education to directly involve professional language arts organizations in the development of all legislation, regulations, and guidelines governing English language arts practice and curriculum; and that NCTE oppose the imposition by mandate of curriculum and practice that have not been developed with the involvement of professional language arts teachers. (pp. 101–2)

NCTE resolutions geared toward instructional improvement emphasize the integration of an adequate and accurate account of racial and ethnic minorities and their contributions to American history and literature; recommend the discontinuance of isolated grammar and usage exercises not supported by theory and research; and endorse the inclusion of meaningful listening, speaking, reading, and writing activities at all levels of language arts instruction.

Studies by Durkin (1986) exemplify research in which educators analyze classroom practices, classroom textbooks, and college-level methodology textbooks. Durkin's studies question the quality and quantity of comprehension instruction provided in both public school classes and college methods courses. After evaluating the quality of comprehension instruction in college-level reading methodology textbooks, Durkin concluded that the specific suggestions for teaching comprehension are brief and meager, that suggestions for assessing comprehension are falsely identified as methods for teaching comprehension rather than testing comprehension, that textbooks underemphasize the need for text-based comprehension instruction, and that instructors of methods courses need to supplement the textbooks with specific instructions in how to teach comprehension. This textbook includes methodologies for teaching comprehension through listening (chapter 4), through literature (chapters 10 and 11), and through media (chapter 12).

Colleges of education are becoming leaders within the forces that seek improvements in education. For example, research universities across the nation have joined the Holmes group, a group of educational institutions whose goal is the study and the improvement of education. As a result of research and surveys conducted by colleges of education, innovative programs are being developed and evaluated. Some colleges of education are implementing and evaluating five- and six-year programs in teacher education. These extended programs for teacher education frequently include a core curriculum in liberal arts and sciences, an academic specialization in a specific teaching field, a series of professional education courses, an internship, and a probationary induction period during which both the colleges of education and the public schools provide guidance for beginning teachers.

The concepts of teacher–scholar and levels of professional practice are extending the sequence for teacher education and the collaboration between colleges and public schools. Teacher–scholars are involved in research to improve educational practice and to enhance professionalism. Different categories for career professionals are identifying different job descriptions and different levels of expertise. Public schools are beginning to identify highly qualified professional teachers and are providing different contractual arrangements and job opportunities for these educators. Shanker (1987) described the lead teacher proposal, a concept inspired by the Carnegie Report, that is being implemented in Rochester, New York with the collaboration of the University of Rochester. Shanker stated, "The Rochester contract gives us a glimpse of a new kind of school . . . one with high standards for teaching, attractive salaries and conditions, and real professional responsibilities and decision-making authority for teachers" (p. 7E).

The final forces that are trying to improve the profession from within include the positions of college educators such as Hirsch (1987) and Bloom (1987) who argued that elementary, secondary, and college educators are not emphasizing the rich content on which an education should be based. The response of the public to this argument is reflected in the fact that both Hirsch's *Cultural Literacy: What Every American Needs to Know* and Bloom's *Closing of the American Mind* are on the 1987 list of best sellers published by *The New York Times*. When interviewed by Brinkley (1987) on ABC News, Hirsch stated that the popularity of his book resulted because "People are realizing the negative thesis. Kids cannot read, write, speak, and think as well as they should." Likewise, as part of the same interview, Bloom declared "Parents are concerned about their children." Hirsch's book, which attacks the content of elementary and secondary education, will probably have more impact on language arts instruction than Bloom's book, which attacks college liberal arts instruction. Throughout his text Hirsch argues for a strong content base in language arts instruction emphasizing literature, history, science, and geography. In such an educational environmental envisioned by Hirsch, skills would not be separated from content. How to balance skills and content will undoubtedly be an important issue in language arts instruction. This textbook emphasizes various content on which language arts instruction can be based.

REINFORCEMENT ACTIVITY

Read and discuss an article or a book on an issue that might affect the teaching of language arts. You may wish to discuss one of the articles or books listed below:

- *Banfield, Beryle, and Wilson, Geraldine L. "The Black Experience Through White Eyes—The Same Old Story Again." In* The Black American in Books for Children: Readings in Racism, *edited by Donnarae MacCann and Gloria Woodard, pp. 192–207. Metuchen, N.J.: The Scarecrow Press, 1985.*
- *Donelson, Ken. "Almost 13 Years of Book Protests—Now What?"* School Library Journal *31 (March 1985): 93–98.*
- *"Education Secretary Bennett's Suggested Reading List for Elementary-School Pupils."* The Chronicle of Higher Education *(September 14, 1988): B3.*
- *Goddard, Connie Heaton. "Interview: Putting Reading Research Into Practice." (Interview with Richard Anderson).* Instructor *(October, 1988): 8–10.*
- *Hirsch, E. D. Jr.* Cultural Literacy: What Every American Needs to Know. *Boston: Houghton Mifflin Co., 1987.*
- *Roser, Nancy. "Research Currents: Relinking Literature and Literacy."* Language Arts *64 (January 1987): 90–97.*

SUMMARY

Effective teaching of language arts includes assessment of all areas of the language arts. The language arts curriculum includes instruction that fosters development of oral language, listening, handwriting, creative writing, written composition, grammar, usage, and spelling. This curriculum also allows children to develop an appreciation for literature and other media.

The language arts classroom must accommodate children who demonstrate various needs and ability levels. In addition to children who acquire the various language arts skills without undue difficulties, the classroom must also serve children who require slower-paced instruction and those who demonstrate various learning disabilities.

Control of the profession from the outside includes legislative mandates, public opinion responses to studies on education, and censorship attempts by various interest groups. The forces concerned with improving our profession from within include resolutions by professional societies, studies and recommendations by university groups, and criticisms expressed by our peers in college and university English departments.

BIBLIOGRAPHY

Applebee, Arthur N. "Writing and Reasoning." *Review of Educational Research* 54 (Winter 1984): 577–96.

Bloom, Allan. *The Closing of the American Mind.* New York: Simon & Schuster, 1987.

Brinkley, David. "This Week With David Brinkley." ABC News, Sept. 6, 1987.

Britton, James; Burgess, Tony; Martin, Nancy; McLeod, Alex; and Rosen, Harold. *The Development of Writing Abilities (11–18).* Schools Council Research Studies, London: Macmillan, 1979.

Dronka, Pamela. "Forums for Curriculum Critics Settle Some Disputes: Clash Persists on Students' Thinking About Controversy." *Update* (March 1987): 1, 6, 7.

Durkin, Dolores. "Reading Methodology Textbooks: Are They Helping Teachers Teach Comprehension?" *The Reading Teacher* 39 (January 1986): 410–17.

Flower, Linda, and Hayes, John. "The Dynamics of Composing: Making Plans and Juggling Constraints." In *Cognitive Processes in Writing,* edited by Lee W. Gregg and Erwin R. Steinberg. Hillsdale, N.J.: Lawrence Erlbaum Associates, 1978.

Gordon, Christine J. "Modeling Inference Awareness Across the Curriculum." *Journal of Reading* 28 (February 1985): 444–47.

Graves, Donald. *A Case Study Observing the Development of Primary Children's Composing, Spelling, and Motor Behaviors During the Writing Process.* Final Report, NIE Grant No. G-78-0174. Durham, N.H.: University of New Hampshire, 1981.

Hillocks, George. *Research on Written Composition: New Directions for Teaching.* Urbana, Ill: National Conference on Research in English, 1986.

Hirsch, E. D., Jr. *Cultural Literacy: What Every American Needs to Know.* Boston: Houghton Mifflin Co., 1987.

Holloway, Karel. "Teens Know Little About America's Past, Report Says." *The Dallas Morning News.* August 31, 1987, pp. 1A, 6A.

Maxwell, John C., and Allen, Diane. "NCTE to You." *Language Arts* 63 (January 1986): 99–106.

NCTE Committee on Classroom Practices in Teaching English. *Activities to Promote Critical Thinking.* Urbana, Ill: National Council of Teachers of English, 1986.

Roehler, Laura, and Duffy, G. "Direct Explanation of Comprehension Processes." In *Comprehension Instruction,* edited by Gerald G. Duffy, Laura R. Roehler, and Jana Mason, 265–80. New York: Longman, 1984.

Shanker, Albert. "Where We Stand." *The New York Times.* August 23, 1987, p. 7E.

Tchudi, Stephen N. "Reading and Writing As Liberal Arts." *Convergencies: Transactions in Reading and Writing,* edited by Bruce T. Peterson, 246–59. Urbana, Ill.: National Council of Teachers of English, 1986.

Wiessler, Judy. "Book Censorship Attempts Are Soaring, Group's Survey Says." *Houston Chronicle.* August 28, 1987, p. 8, Section 1.

Chapter Two

THEORIES OF LANGUAGE ACQUISITION
The Behaviorist Theory ▪ The Genetic Theory
▪ The Sociocultural Theory

SEQUENCE OF DEVELOPMENT
Early Language Development ▪ Language
Development in the Elementary Grades
▪ Linguistic Investigations ▪ Psycholinguistic
Investigations

DEVELOPING COGNITIVE SKILLS
Basic Operations Associated with Thinking
▪ Semantic Mapping and Vocabulary Development

SUMMARY

After completing this chapter on language and
cognitive development, you will be able to:

1. Describe the behaviorist's theory of language
   acquisition and analyze instructional
   outcomes according to a behaviorist's view
   of language acquisition.
2. Describe a genetic theory of language acquisition and analyze instructional outcomes
   according to a geneticist's view of language
   acquisition.
3. Describe a sociocultural theory of language
   acquisition and evaluate the impact of sociocultural research on language acquisition
   theory.
4. Relate a child-development-oriented philosophy to an instructional program.
5. Compare the role of the teacher and the
   role of the student in a program with a behaviorist perspective and in a program with
   a child-development perspective.
6. Describe the sequences of language development in the preschool and early-elementary-
   age child.
7. Understand that all children go through approximately the same stages of language development, but that the rate of development
   varies from child to child, and relate this understanding to instructional practice.
8. Describe some implications for language
   study developed from linguistic investigations.
9. Describe some implications for language
   study developed from psycholinguistic investigations.
10. Understand the importance of cognitive de-
    velopment and describe some instructional
    activities that stimulate cognitive
    development.

Language and
Cognitive Development

*I*f someone asks you to name the most important means of communication, what will you answer? Most of you will probably respond, "Language." Language is the key to human communication. We can use written language to convey information, or to read the vast accumulation of knowledge found in books.

The spoken and written language that separates us from the animal kingdom is made up of sounds and symbols that are grouped into words. Although words are the fundamental unit of language, they do not contain complete meaning until they are placed within the structure of the sentence. For example, if you see the word *wind,* can you even pronounce the word correctly without knowing the sentence context in which it is to be placed? This isolated word might refer to a breeze, as in the sentence, "The wind blew the kite up into the sky." In contrast, the same word may mean an action, as in "You must wind the clock." These identical spellings have two pronunciations, two meanings, and are two different parts of speech, depending on the words that form the surrounding sentence.

Every language has sounds, or *phonemes,* that combine to form words; the words are put into certain sequences that comprise a grammatical form that is structured within the language (syntax). Because the way this language is put together is fundamental to every aspect of the language arts, we investigate some questions about and implications of language acquisition and development before proceeding to other aspects of the language arts curriculum.

THEORIES OF LANGUAGE ACQUISITION

How do children acquire language? This question has fascinated people since the early days of recorded history. As early as 600 B.C., Herodotus reported that the Egyptian King Psammetichus I tried to answer this question by having two children actually raised in a speechless environment. Although modern experimenters dispute both the scientific and ethical status of such early research, they are still trying to answer the same question.

If we look at more recent studies in language, we see that the last fifty years have produced many changes in the way language is studied. Early studies often took the form of diary investigations, in which investigators listened to young children and recorded their language as the children progressed through the various stages of language development. The 1930s and 1940s produced a movement calling for scientific rigor in research. During this time, earlier studies by parents were considered suspect, because it was assumed that parents would show bias in their recordings.

Concern for rigorous scientific experimentation led to language studies with large groups of children. These studies described properties of children's speech such as the average length of utterance, parts of speech used, and the number of different words spoken by a child during a specific period of time. A study by Madorah Smith (1933) was one of the earliest to conclude that children's language was systematic and rule-governed.

In the 1950s, language researchers started to investigate what children know about language. This research ignored study of large numbers of children, and stressed systematic study of fewer children over longer periods of time. This type of research has been summarized by Ervin-Tripp (1966). The studies concluded that children learn an underlying linguistic system, rather than all of the sounds, words, and possible sentences in a language. The studies also implied that children's language is systematic and rule-governed, and inspired linguists to start other research to discover more about this underlying system of rules.

In the late fifties and the sixties, researchers sought to establish a description of the rule systems that could account for children's use of sentences. Language investigators of the sixties and seventies were developing theories about how children acquired language. Research in the seventies and eighties became more sophisticated as researchers used computer analysis and electronic recording to systematically investigate language. Language acquisition investigators in the seventies and eighties also considered the social context in which children's language develops. Researchers such as Lawton (1968) from England and Poole (1976) from Australia investigated the linguistic strategies of working-class and middle-class subjects and added to the increasing body of research in language development. Models of language acquisition were developed by proponents of such diverse learning theories as behaviorism, genetics, and sociocultural theory. We will examine each of these theories in some detail.

The Behaviorist Theory

The emergence of the behaviorists' theory of education is believed to have begun with B. F. Skinner's article, "The Science of Learning and the Art of Teaching," published in 1954. Prior to this time, much of the behaviorist's technology was considered a laboratory subject. Although the early behaviorists, represented by Pavlov, worked primarily with conditioning behavior in laboratory settings, modern behaviorists have placed major emphasis on the application of laboratory implications to human learning and educational settings. Skinner saw a strong parallel between his activities in the laboratory and practices that he felt would improve education. Thus, we now hear of and use such concepts as imitation, reinforcement, successive approximations, and shaping.

The concepts of imitation, reinforcement, successive approximations, and shaping are also used to explain language acquisition and development (Staats, 1964, 1968). The behaviorist believes that the young child learns language through the process of imitating the language of other speakers in the environment, and that language is not instinctive in the young child. Language learning is explained in terms of reinforcement of the imitated language. Consequently, language that is positively reinforced is learned, whereas language that is negatively reinforced is not learned. If the behavior, in this case language, increases in frequency, the behaviorist believes the behavior is positively reinforced.

We can understand this concept better if we view it in terms described by Jenkins and Palermo (1964). They believe the babbling of an infant becomes infant speech because parents or other adults reinforce those speech sounds that are close to sounding like adult speech. The reinforcement may be in the form of paying attention to the child, talking to the child, responding with a smile, providing food, or holding the child when the child produces a sound resembling adult speech. This happy consequence of selected behavior is referred to as positive reinforcement. The behaviorist believes all human behavior is controlled by such reinforcement principles. In contrast, babblings that do not resemble adult speech may be ignored, and are thus

REINFORCEMENT ACTIVITY

The DISTAR (Direct Instruction Systems for Teaching Arithmetic and Reading, Science Research Associates, 1968) program was developed at the Institute for Research on Exceptional Children at the University of Illinois. The major authors of the materials are Bereiter and Engelmann. The DISTAR materials have been used with disadvantaged preschool children in order to increase language facility. The following description of the DISTAR program is based on a review by Aukerman (1984) and observation by this author. What components of the program correspond with the behaviorist theory of learning? Why do you believe a behaviorist would formulate a highly structured program?

The learning approach of the DISTAR program includes the following points:

1. *A concentrated, structured, no-nonsense program is necessary to bring pre-schoolers' performance up to a level that will permit them to succeed in school.*
2. *Culturally disadvantaged children are generally nonverbal when spoken to in a normal classroom manner. Strenuous intervention and direct teaching are necessary.*
3. *One must determine the specific educational deficit and structure the sequenced teaching of skills to overcome those deficits. Objectives must be stated in terms of desired facts, skills, or behaviors.*
4. *The learning process must be teacher-dominated because the teacher is responsible for the children's learning.*
5. *Direct instruction will result in learnings that can be tested. It can therefore be demonstrated that learning has taken place.*

negatively reinforced. These speech sounds would not be acquired as infant speech. Negative reinforcement does not produce a pleasurable consequence for the child.

The behaviorist believes that as the child grows older, her language behavior is modified as she is rewarded for successive approximations of both adult pronunciation and grammar. This modification of behavior, which is believed to occur when a set of responses such as a phrase or sentence is brought under the control of a positive reinforcer, is called "shaping" by the behaviorist. The child is rewarded when a listener answers questions produced by the child, or when the child makes a request that is granted. If the child wants an ice cream cone, she will have to ask for it. If she asks for the ice cream cone in a way that is understood by adults and in the appropriate location, she may be rewarded with this choice food. Eachus (1971) contended that teaching is a continuous process of just such shaping behavior. Shaping is considered one of the most important features of behaviorist technology.

The development of the DISTAR Language Program is a result of behaviorist theories of language. If you look at the program, you will see how both imitation and reinforcement are used to develop language. A fast-paced verbal instruction program that uses similar techniques to teach reading and arithmetic is described by Engelmann (1974).

Creative teachers are told not to resort to language-arts approaches because direct instruction does not provide for lesson embellishment; this would, according to the authors of DISTAR, break the proper sequence of instruction.

Instruction includes patterning in sequence in order to train children to pay attention. All of the instruction demands that the children pay close attention and respond correctly. For example, during one lesson, this author observed that the children sat directly in front of the teacher so that the teacher could focus on each child. The pace was quite rapid as the teacher asked every child to do exactly what she did. This teacher did some patterned clapping activities, then had the students repeat exactly the same patterns. The group of five-year-olds worked on sentence patterns by answering questions posed by the teacher. The teacher showed a picture of a ball on a table and asked the children, "Where is the ball?"; the children replied in unison, "The ball is on the table." This activity continued with a number of different pictures and more sentence responses in unison from the children. The teacher responded with praise for correct answers, using such phrases as "You did it right," and "That was good remembering." She also provided a reward of raisins for hard work. The half-hour of language instruction was very rapid, required close attention by every child, and required identical responses by every child. (Some child development and early childhood educators have shown concern about these aspects of the approach.)

The DISTAR program is a highly structured program in both methods and materials that appeals to teachers who feel the need for considerable structure and for step-by-step procedures. The program differs greatly from the enrichment approach found in many nursery school and kindergarten classrooms.

The Genetic Theory

While behaviorists look upon the task of language learning as a form of behavior that can be explained by the conditioning processes of imitation, reinforcement, and shaping, another group of researchers believe that children possess innate or instinctive language ability. According to this theory, language learning is largely instinctive rather than imitative. All children are thus believed to be born with the ability to use language: they start with a language of their own and amend their language to conform to adult language.

Researchers such as Carol Chomsky (1969) argue that imitation theories of language acquisition do not explain how children create sentences they have never heard. This viewpoint was reinforced by language acquisition studies in the sixties and early seventies. For example, a study by Menyuk (1963) found that preschool children develop their own model of grammar that differs from the adult language in their environments. Additional insights into the role of imitation and innate ability are provided by Ervin-Tripp's (1964) study. She asked children to repeat sentences such as "Mr. Miller will try." Young children who were speaking at the two-word level responded with "Miller try" rather than the longer adult sentence. Ervin-Tripp concluded that children imitate only those language structures that have already appeared in their speech.

A rule-learning theory began to emerge as researchers investigated children's grammatical structures. Researchers such as Slobin and Welsh (1973) and Cazden (1972) added to the knowledge about children's syntactic development. Slobin and Welsh asked two- and three-year-old children to repeat sentences. They found that children repeat sentences according to their own levels of grammatical development. Likewise, Cazden concluded that children's language is rule-governed. Consequently, children process language data around them, draw rules that they test, and revise rules based on feedback they receive. In this way children's speech gradually approaches adult speech.

The Sociocultural Theory

Language acquisition is probably neither totally genetic nor totally behavioristic. The interactive reality of language acquisition is stressed in the sociocultural theory. While many researchers in the sixties and seventies concluded that language is an innate capability triggered by the presence of language in the environment, other researchers in both America and England (Wells, 1979, 1981) considered the importance of the social environment in which language is acquired and the interaction that takes place between children and adults. Bruner described a sociocultural viewpoint of language acquisition: Language is "encountered in a highly orderly interaction with the mother, who takes a crucial role in arranging the linguistic encounters of the child. What has emerged is a theory of mother-infant interaction in language acquisition—called the fine tuning theory—that sees language mastery as involving the mother as much as it does the child" (1978, p. 44).

Support for a social context for language acquisition frequently relies on cases in which children do not interact with adults. For example, Moskowitz (1978) described a boy with normal hearing who is raised by deaf parents. Although the child listened to television daily, by the age of three he could neither speak nor understand English. He was, however, fluent in his parents' sign language. Moskowitz concluded, "It appears that in order to learn a language a child must also be able to interact with real people in that language" (p. 94).

Snow (1977) and Lindfors (1987) suggested that adults play an important role in this socialization process. Snow studied the interactions between infants and their mothers. She found that mothers adjust their speech to meet the needs of their children's conversational skills. Likewise, Lindfors found that adults adjust their language to fit the oral language skills of their children. Consequently, early conversations include shorter sentences characterized by considerable contextual support, repetitions, and exaggerated intonational patterns. The goal of these early conversations is apparently a meaningful exchange of language.

Conversational skills apparently develop quite early in infants. Shugar (1978) found that the interchange of conversation appears before very young children are able to speak in two-word utterances. She maintained that the initial burden for oral interaction rests upon the adult who identifies the child's meaning and builds a conversation upon that meaning. During this interaction, socialization also occurs.

Language acquisition research has changed since the sixties. Fox, following a review of language research, stated, "The current attention of child language researchers has shifted from rate and stages of acquisition to variations of language use. This change has focused upon interactional language and the social settings in which it takes place, a direction which limits the number of subjects involved. The role of interaction in language learning begins practically at birth. It seems that, almost instinctively, mothers recognize the need to establish meaning with their infants through negotiation" (1983, p. 237).

Stages in intellectual development   If you are aware of the teachings of Piaget, you are already familiar with the concept that children's interaction with their physical and social environment is critical to both intellectual and language development. As you may recall, Piaget related intellectual development to several stages of child development. Raven and Salzer (1971) developed some of the instructional implications of Piaget's theory. According to Raven and Salzer, during the sensori-motor period, from birth to about two years, children should manipulate objects and materials to enable them to develop images and to stimulate cognitive growth. During this period, other exploratory activities are considered more important than teaching babies to read. (This was a popular notion in women's periodicals.)

The second stage, the preconceptual period, is divided into two developmental phases. During the first phase, from two to four years, it is important to provide many varied concrete experiences. Sensory and motor activities are considered necessary for developing concepts and complex thinking, both of which are necessary for the development of advanced language art skills. The second phase, or preoperational period, is the intuitive phase from the ages of four through seven years. During this time, an activity-oriented curriculum in which students are able to interact with materials and explore the environment is desirable. Children should have a chance to

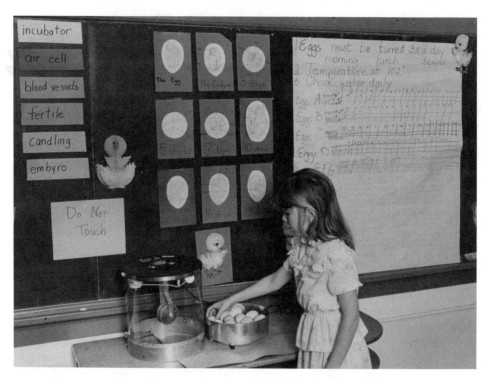

A rich classroom environment encourages language and cognitive development.

manipulate and explore such skills as language and reading books. This theory suggests that children in this stage should not be taught in a program that demands the mental gymnastics necessary for reading or spelling programs based on rules of grapheme-phoneme relationships.

During the third stage, the concrete-operational period from ages seven to eleven years, children have the ability to reason, but can reason adequately only about direct experiences and not about abstractions. Because children will be able to reason about what they read only if it relates closely to direct experience, the environment should enrich that direct experience. During the concrete-operational stage, students should have opportunities to combine sentence and word elements, and associate the elements in different ways. They should be encouraged to transform the word elements and observe the differences produced. Questioning techniques developed from the child's own experience would be used.

During the final stage hypothesized by Piaget, the formal-operational period that begins at eleven or twelve years of age, children have the ability to control formal logic. They can now begin to deal with propositions and hypotheses that are not related to direct experience. During this period, instructional procedures should be structured so that students may analyze various logical relationships. Oral discussion, listening, and writing experiences all allow children opportunities to deal with logic.

Educational strategies formulated from Piaget's beliefs include close observation of children, numerous opportunities for children to manipulate things in the environment and to work individually at tasks of their own choosing, and encouragement of children's oral language development through such activities as arguing and debating.

REINFORCEMENT ACTIVITY

If you believe that language acquisition is inherent within the child and is developmental, rather than totally controlled by the environment, your instructional approach to language will be quite different from a behaviorist's approach. As you read the following recommendations by Anastasiow (1979), analyze what theory of language acquisition and development lies behind these recommendations. In your language arts class, discuss how a theory of language acquisition and development affects not only the type of instruction provided but also the instructional outcomes. Finally, compare the recommendations by Anastasiow with the DISTAR program described earlier in this chapter.

Anastasiow's Suggestions for Developmental Instruction

1. *Play is perceived as an integral part of any preschool or kindergarten program. Toys are the tools with which children think.*
2. *Dramatization of stories, songs, or rhymes is another major activity for any preschool or kindergarten program. Dramatization engages the child in thinking, the most critical activity of all school experiences.*
3. *Dance can be a daily activity. Responding to or moving with the beat of a drum, records, or songs are culturally expressive experiences.*
4. *Play and dramatization allow the child to be physically involved. All humans must be active, particularly the young child; it is through these activities that a child learns.*
5. *Multiple experiences are recommended, including field trips, walks, chart stories, songs, and dictated stories.*
6. *All learning is done by the child; therefore, it is the teacher's role to encourage, nurture, and plan for learning. Any direct attack on a child's language is a direct attack on that child's thinking ability, and can have overwhelmingly negative effects.*

SEQUENCE OF DEVELOPMENT

Linguists have provided us with increasing knowledge about children's language development by asking such questions as: What features of phonological (sound), grammatical, and semantic (meaning) development seem to be universal? Why is one linguistic skill acquired before another? What is the nature of the child's linguistic ability during various stages of development? There appears to be more agreement about the sequence of language development than about language acquisition.

Early Language Development

The first few years of a child's life produce dramatic changes in language ability. The infant begins communicating by crying, then progresses to cooing and babbling. Linguists have used several different terms to label the speech produced by a child during the first year of life. *Infant vocalization, preverbal period,* and *prelinguistic vocalization* have all been used to describe the baby's early speech. Researchers interested in this stage of language development have listened to young children and analyzed what they have heard.

In this early stage, the infant is showing increasing control over vocal production. Fry described this babbling stage:

> During the babbling stage, the child is doing two important things: he is trying out mechanisms that will be needed for speech, combining phonation with articulation and no doubt gaining a certain control of the respiratory system, and he is establishing the circuits by which motor activity and auditory impressions are firmly linked together. He is learning the acoustic effect of making certain movements and finding out how to repeat a movement, how to do it again and again to get more or less the same acoustic results. In one sense he is learning a trick, and the experience lasts him, so to speak, for the rest of his life. (1966, p. 190)

During the first few years of life, children move from vocalizing to meaningful language use. When children have the ability to produce and comprehend meaningful sentences they have never heard before, we say that they have learned the language. Linguists have studied the early speech of children in many different countries and have concluded that children throughout the world have the innate capacity to make the various speech sounds. (As you remember, this was one argument used by advocates of a genetic theory of language acquisition to justify their belief.) Linguists believe that some of these universal sounds are encouraged by the specific language heard, and that others disappear from early language because they are not part of the child's environment. Consequently, the child learns a particular grammar, language, and phonological system. Linguists refer to this process as *internalizing.*

Language development is very rapid in most children. They usually speak their first word at about one year of age. As you know, this is an important occasion; your parents can probably remember your first word and when you said it. These early words probably mean more than the single word implies. For example, the child who says "mama" may mean "Mama come here," "Where is mama?" or "Mama pick me up." Linguists use the term *holophrastic* to describe this speech.

At about eighteen months, children usually begin putting two words together. According to Braine (1978), the number of different two-word combinations increases slowly, then shows a sudden upsurge around the age of twenty-three or twenty-four months. Braine recorded the speech of three children from the ages of eighteen months and reported the cumulative number of different two-word combinations for one child in successive months was 14, 24, 54, 89, 350, 1400, 2500+. Obviously, this is a rapid expansion of speech in such a short time.

The time when children put two words together for speech is also considered important by linguists. According to Slobin (1979), this is the time when a child's active grammar can be investigated. Brown (1973) conducted one of the most extensive longitudinal studies of early language acquisition. In this study, the speech of three children was recorded over a five-year period. Brown found that during approximately the first two years of life speech could be described as "telegraphic." During this stage, the children's speech was made up of content words belonging to the large open classes called nouns, verbs, and adjectives. The speech in this telegraphic period did not utilize function words such as prepositions, articles, auxiliary verbs, or pronouns. When children say such phrases as "pretty flower" or "allgone milk," they are using telegrapic speech.

Brown's research also provides us with valuable insight into the next stage of language acquisition. It is during this second stage that children acquire the ability to use grammatical morphemes, which are considered the smallest significant unit of syntax. These units may be whole words, as in the use of "came" to show the past tense of "come," or the units may be only parts of words, as in adding "s" to a word to form the plural or possessive, as in "balls" or "baby's."

In *A First Language/The Early Stages,* Brown analyzes the results of his longitudinal study and combines them with other early language studies to determine the order in which children acquire fourteen of the grammatical morphemes. This order is apparently quite consistent for most children. Table 2–1 shows this order of language acquisition and provides an example of each grammatical morpheme.

All children appear to go through the same stages of language development, although the rate of development varies from child to child. Even Brown's small sample showed great variance among the three children. For example, one child successfully used six grammatical morphemes by the age of two years, three months, while a second child did not master them until the age of three years, six months. The third child was four years old before reaching an equivalent stage in language development.

Grammatical Morpheme	Example
1. Present progressive -ing	Jimmy eating
2. and 3. in, on	toy in box, Sally sit on bed
4. Plural -s	tables
5. Past irregular	came, went
6. Possessive	Sandy's chair
7. Uncontractible copula (linking verb)	am, is, are, be
8. Articles	the, a
9. Past regular -ed	Billy walked home
10. Third person regular -s	He plays
11. Third person irregular -s	He does
12. Uncontractible auxiliary is	This is going fast
13. Contractible copula (linking verb) = s	Billy's sleepy
14. Contractible auxiliary	He's flying

TABLE 2–1
The order of grammatical morphemes acquired in early stages of language

SOURCE: Roger Brown, *A First Language/The Early Stages.* (Cambridge, Mass.: Harvard University Press, 1973). Copyright © 1973 by the President and Fellows of Harvard College. Reprinted with permission.

The children in any one nursery school, kindergarten, or elementary class are in various stages of language development. Because language is such a vital aspect of education and is basic to the skills of oral communication, listening, writing, and reading, it is necessary for the teacher to understand the development process of language acquisition. Such knowledge is essential for the teacher to diagnose oral language skills or to create an educational environment that will foster development of all the communication skills.

Language Development in the Elementary Grades

The most extensive study of language development in school-age children is a longitudinal study conducted by Loban (1976). In his study, Loban examined the language development of the same group of over 200 children from the age of five to eighteen years. This study has numerous implications for the language arts teacher, because Loban investigated stages of language development, identified differences between students who ranked high in language proficiency and those who ranked low, and stressed the use of taped oral language samples rather than published language tests.

What are the differences between children who ranked high in language proficiency and those who ranked low? First, students who demonstrated high language proficiency excelled in the control of expressed ideas. Both speech and writing showed unity and planning. They spoke freely, fluently, and easily; used a rich variety of vocabulary; and adjusted the pace of their words to their listeners. The high proficiency group also used more words in each oral sentence. The difference in language development was so dramatic that the higher group had reached a level of oral proficiency in first grade that the lower group did not attain until the sixth grade. The lower group's oral communication was characterized by rambling and unpurposeful dialogue that showed a meager vocabulary.

In addition, in the area of writing, the high group was more fluent, used more words per sentence, showed a richer vocabulary, and was superior in using connectors such as "meanwhile" and "unless" in their writing. This group also used more subordination in combining thoughts into complex forms. The high group again showed greater proficiency at a much younger age than their peers; the fourth-grade level of proficiency shown by the high group was not shown by the low group until the tenth grade.

Moreover, those students who were superior with oral language also ranked highest on listening. They were both attentive and creative listeners. Children who were superior in oral language in kindergarten and first grade also excelled in both reading and writing in the sixth grade. In a recent review of the implications of language research, Loban (1986) concluded, "The awesome importance of oral language as a base for success with literacy is crystal clear" (p. 612). Consequently, children need oral language instruction that helps them organize ideas and illustrate complex generalizations. Oral discussion should be a vital part of the elementary school program.

Young children apparently go through similar stages of language development, although the rate of development shows wide variations. This is seemingly true also for school-age children. The school-age child's power over language increases through successive control over forms of language, including the ability to handle pronouns, to use appropriate verb tenses, and to use connectors. There is also steady growth in the average number of words per sentence and the average number of dependent clauses per sentence. Consequently, although language development is well advanced by the time a child enters school, it is far from complete (Lamb, 1977). An earlier study by Carol Chomsky (1969) emphasized the expanding acquisition of syntax in children ages five through ten years. Chomsky found that children in this age group are still acquiring syntactic structures. Language learning does not cease, nor is it limited to an expansion of previously acquired structures.

Although the rate of language development differs, table 2-2 may help you visualize the language acquisition stages common to many children at specific age levels. This table was compiled from studies by Brown (1973), Loban (1976), and a report by Bartel (1986).

REINFORCEMENT ACTIVITY

Listen to the oral language of several preschool children or elementary-age children. How does their oral language development compare to findings of Brown or Loban? Write some specific examples of speech found at a certain age level and share your findings with your classmates.

Linguistic Investigations

Linguists study language in order to learn how it functions. They observe and record how people actually use language to discover not only how it functions today, but how it is evolving. They use their data to draw conclusions and formulate generalizations about the nature of language. Linguists have thoroughly investigated the code systems of language. Because both speaker and listener must know what the code means, linguists have spent a considerable amount of time investigating these code signals. Linguists usually refer to several categories of this code system, including phonology, morphology, syntax, and semantics.

Phonology is the study of the sound system within the language. Each language has a set of sounds that provide meaningful differences within that language. The smallest distinctive unit of sound within the language is called a *phoneme*. The phoneme is a minimal linguistic unit whose replacement can result in a meaning difference. For example, in the words c/a/t and m/a/t, the /c/ and the /m/ sounds are distinct phonemes. The written representations of these sounds are called *graphemes*. There are forty-four sounds in the English language that communicate such meaning. Because there are only twenty-six graphemes in English, some children have trouble relating the correct phoneme to the grapheme, especially in reading and spelling. In an effort to limit the sound/symbol inconsistencies, some instructional systems present only consistent spellings in the beginning materials; these materials present sentences such as, "The cat sat on a bat." Another attempt to overcome inconsistencies is the

TABLE 2–2

A general overview of language development

Age	General Language Characteristics
3 months	The young child starts with all possible language sounds and gradually eliminates those sounds that are not used around her.
1 year	Many children are speaking single words (e.g., "mama"). Infants use single words to express entire sentences. Complex meanings may underlie single words.
18 months	Many children are using two- or three-word phrases (e.g., "see baby"). Children are developing their own language rule systems. Children may have a vocabulary of about 300 words.
2–3 years	Children use such grammatical morphemes as plural suffix /s/, auxiliary verb "is," and past irregular. Simple and compound sentences are used. Understands tense and numerical concepts such as "many" and "few." A vocabulary of about 900 words is used.
3–4 years	The verb past tense appears, but children may overgeneralize the "ed" and "s" markers. Negative transformation appears. Children understand numerical concepts such as "one," "two," and "three." Speech is becoming more complex, with more adjectives, adverbs, pronouns, and prepositions. Vocabulary is about 1,500 words.
4–5 years	Language is more abstract and most basic rules of language are mastered. Children produce grammatically correct sentences. Vocabularies include approximately 2,500 words.
5–6 years	Most children use complex sentences quite frequently. They use correct pronouns and verbs in the present and past tense. The average number of words per oral sentence is 6.8. It has been estimated that the child understands approximately 6,000 words.
6–7 years	Children are speaking complex sentences that use adjectival clauses, and conditional clauses beginning with "if" are beginning to appear. Language is becoming more symbolic. Children begin to read and write and understand concepts of time and seasons. The average sentence length is 7.5 words.
7–8 years	Children use relative pronouns as objects in subordinate adjectival clauses. ("I have a cat which I feed every day.") Subordinate clauses beginning with "when," "if," and "because" appear frequently. The average number of words per oral sentence is 7.6.
8–10 years	Children begin to relate concepts to general ideas through use of such connectors as "meanwhile" and "unless." The subordinating connector "although" is used correctly by 50 percent of the children. Present participle active and perfect participle appear. The aveage number of words in an oral sentence is 9.0.
10–12 years	Children use complex sentences with subordinate clauses of concession introduced by "nevertheless" and "in spite of." The auxiliary verbs "might," "could," and "should" appear frequently. Children have difficulties distinguishing among past, past perfect, and present perfect tenses of the verb. The average number of words in an oral sentence is now 9.5.

i/t/a/ alphabet developed by Sir James Pitman. This alphabet has forty-four graphemes, one for each phoneme. Children learn to read and spell in this alphabet and then go through a transitional period in which standard English spelling is introduced (Fink and Keiserman, 1969). Figure 2–1 illustrates the i/t/a alphabet.

The second code signal studied by linguists is morphology. *Morphology* represents the study of meaning in relation to speech sounds. The smallest meaning-bearing units are called *morphemes*. Thus, "girl" is a morpheme, and it is also the smallest meaningful unit; if any of the graphemes are removed, this combination of letters would not mean "girl." Because "girl" can also stand alone in meaning, it is called a *free morpheme*. If we add an "s" to "girl" to form "girls," we have now represented two smallest units of meaning. We still have the meaning "girl," but the "s" has added the meaning of plurality. Because "s" cannot stand alone and still mean plurality, it is called a *bound morpheme*. Prefixes and suffixes have meanings, and also change the meaning of root words when they are added to the root. Studies of word changes attributable to bound morphemes will help children understand meaning and develop vocabulary. For example, consider the following list of free and bound morphemes. What happens to the meaning of the words when the prefix "un" is added?

FIGURE 2–1
The i/t/a/ alphabet (Reprinted with the permission of Pitman Learning, 6 Davis Drive, Belmont, CA 94002)

bound morpheme	free morpheme
un	happy
un	done
un	desirable
un	domesticated

A third code signal is *syntax*, or the study of how words are put together in a meaningful order to form sentences. This study in school may be referred to as grammar. Grammar refers either to structural grammar, which looks at the grammatical structure of a sentence, or, more recently, to transformational grammar, which looks at how meaning is communicated in sentences. According to many linguists, instruction should emphasize the systematic building of sentences according to certain patterns, rather than the diagramming of sentences found in the study of traditional grammar. Sentence building would also stress the function of words in sentences rather than mere categorization of words as parts of speech. Consequently the term *usage* is often preferred to *grammar*.

The final code usually studied by linguists is semantics. *Semantics* refers to the study of meaning, and is probably the most important code system, as we will see in our discussion of psycholinguistic implications.

The linguist, through the study of code signals, has presented us with several conclusions about linguistic development in children. Lamb (1977) summarized some of these conclusions:

1. Language development, although well advanced by the time a child enters school, is far from complete.
2. Growth occurs in all the areas identified—phonology, morphology, syntax, and semantics.
3. There are probably close corollaries between cognitive growth and linguistic development. (The stages identified by Piaget seem to be significant periods in language development.)
4. Teachers' expectations in terms of reading comprehension, skill in composition, and acquisition of phoneme-grapheme correspondence generalizations should be adjusted to account for the growth still occurring.

Psycholinguistic Investigations

Whereas the linguist studies the various code systems of language, including phonology, morphology, syntax, and semantics, the psycholinguist stresses the interdisciplinary study of psychology and linguistics wherein language behavior is examined. The psycholinguist, therefore, maintains that the various code systems of language cannot be studied independently of one another, nor can instruction be provided in one system without looking at the total language system. Burke (1972) described the psycholinguist's view of the language system as if it were a circle, with meaning in the center, grammar as the next ring of the circle, and the sound and written symbols of language as the outer circle. If you cut a wedge out of this circle, your wedge would include meaning, grammar, and the sound and written symbols of language. Meaning is considered the core of language, with the grammatical structure fused to meaning; the sound and written symbols are vehicles by which meaning is displayed. Burke maintained, "The first successful attempts of language study deal with the systems of language as integral parts of a whole. There is knowledge that language elements, when pulled out of the total language process, do not retain the properties of that process"(p. 27).

As an example of this interrelated concept, let us look at the letter *e*. In isolation, the letter may have several different sounds. It is not until it is put into the context of a word that you know the correct pronunciation. Even when it is put into a sequence of letters, such as in "read," we still cannot provide an accurate definition and pronunciation until it is placed in a sentence context. The pronunciation and meaning of "read" is quite different in "He read the book yesterday" from "Will you read the story tomorrow?" Total meaning is not available until all the systems of language are utilized.

Roach Van Allen (1976) developed an excellent diagram of the interrelationship of these language systems and their relationship to reading, oral skills, environmental influences, and communication skills.

As you can see in figure 2–2, the systems of language are interrelated in this linguistically based concept of instruction. Strand 1 refers to experiencing communication, and includes self-expression through talking, painting, singing, dancing, acting, and writing. Strand 2 includes studying communication in such a way that students will understand how language works, understand sound-symbol relationships, and acquire vocabularies. Strand 3 includes relating communication of others to yourself,

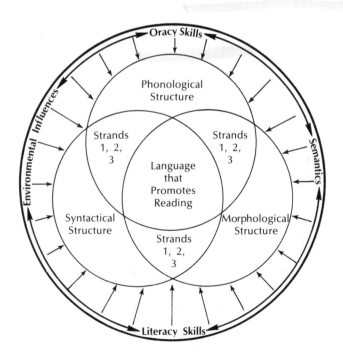

FIGURE 2–2

Psycholinguistic base for language instruction (Roach Van Allen, *Language Experiences in Communication*. Copyright © 1976 by Houghton Mifflin Co. Used with permission.)

and emphasizes the influences of language and ideas of others as the individual reads, sees and hears films, listens to records and music, or enjoys art prints and sculpture. Roach Van Allen uses this diagram as a rationale for employing language-experience approaches with children (see chapter 14 in this text).

Kenneth Goodman has conducted much psycholinguistic research. His theory of instruction is based on psycholinguistic principles, which he has related to reading. According to Goodman (1972):

1. Meaning must always be the immediate, as well as the ultimate, goal of reading.
2. Language systems are interdependent, so language cannot be divided into words for instructional purposes.
3. Children learning to read their native language are competent language users.
4. Children will find it easier to read language that is meaningful and natural to them.
5. Children must learn strategies for predicting, sampling, and selecting information, guessing, confirming or rejecting guesses, correcting, and reprocessing.
6. Special reading strategies are needed for reading special forms of language.
7. The reader must be able to relate his reading experiences to real experiences.

Psycholinguistic theory stresses that students need uninterrupted reading time in order to discover strategies for gaining meaning; they need opportunities to use the meaning and grammatical systems of language to predict an appropriate word. If you teach reading from a psycholinguistic perspective, you do not teach sounds or words in isolation. Instead, you would teach in sentence contexts, so that the child always has the support of the grammatical and meaning cueing systems in language. (We discuss this later in the chapter about reading.)

DEVELOPING COGNITIVE SKILLS

Cognitive psychologists study not only how information is presented to students, but also how the learner processes the information. Within cognitive psychology, learning is considered a process in which learners actively participate in the teacher-learner act. Teaching approaches include teaching students how to think, how to learn, how to remember, and how to motivate themselves. Weinstein and Mayer (1986) identified the following types of learning strategies that are frequently identified under cognitive processes: (1) rehearsal strategies, such as repeating names in an ordered list and underlining pertinent information; (2) elaborative strategies, such as forming mental images, paraphrasing, and creating analogies; (3) organizational strategies, such as grouping, categorizing, and creating diagrams; (4) comprehension monitoring strategies, such as using self-questioning techniques and guide questions to structure reading behavior; and (5) affective strategies, such as reducing external distractions by studying in a quiet place. Terminology that is frequently associated with cognitive processing includes *schema* (an individual's conceptual system or structure for understanding something), *metacognition* (an individual's knowledge of the functions of his or her own mind and the conscious efforts to monitor or control those functions), and *semantic maps* or *webs* (diagrams that help students visualize how words or concepts are related to one another). Schema theory is discussed at greater length in chapter 11. The "Essentials of English" adopted by the National Council of Teachers of English (1983) emphasized the importance of stimulating cognitive development. Child development authorities Mussen, Conger, and Kagan (1984) emphasized the importance of developing cognitive processes necessary for perception. They identified the following essential processes:

1. Perception—the deletion, organization, and interpretation of information.
2. Memory—the storage and retrieval of perceived information.
3. Reasoning—the use of knowledge to make inferences and draw conclusions.
4. Reflection—the evaluation of the quality of ideas and solutions.
5. Insights—the recognition of new relationships between two or more segments of knowledge.

Current research in cognitive processing stresses both the relationship between cognitive processing and achievement and the importance of creating environments and instructional tasks that help children visualize and develop concepts and increase their ability to handle basic operations associated with thinking. Children with learning difficulties often demonstrate deficiencies in cognitive processes, especially at higher cognitive levels required for imagery, verbal processing, and concept formation. Kavale (1980) discovered that children with learning problems frequently experience difficulties understanding concepts related to time, quantity, and space. Torgesen (1979) found deficiencies in memory and the development of memory strategies in students who have reading problems. Finally, Myers (1983) identified three characteristics associated with students who benefit from cognitive processing tasks: lack of fluency, lack of synthesis, and poor development of ideas.

What types of instructional strategies are recommended for developing cognitive processes? Interestingly, silence can be an important strategy. Allowing time to think about a discussion, to formulate questions, and to answer questions is an important teaching strategy that will also increase the cognitive level of both responses and questions. Tobin (1987) reviewed research that extended the average one-second wait time in discussions to between three and five seconds. Tobin reported that such an extension caused teachers to repeat fewer questions, to ask more higher level questions, and to ask more probing questions. Positive changes in student behavior included more and longer responses, increases in the cognitive levels of responses, greater initiation of discussions, more student-to-student interaction, decreases in confusion, and higher achievement. Tobin concluded that "the silence provides teachers with time to think and to formulate and use higher quality discourse that then influences the thinking and responding of students" (p. 87).

Activities that encourage children to organize information into meaningful wholes are suggested by Myers (1983). Myers also recommended semantic mapping strategies that help children visualize concepts and relationships. A similar approach for concept and vocabulary development is recommended by Johnson and Pearson (1984). A curriculum emphasizing cognitive processes and problem-solving tactics increases mental abilities of disadvantaged students (Rand, Tannenbaum, and Feuerstein, 1979). Squire (1982) focused instruction on language tasks that encourage analyzing, reporting, persuading, interpreting, reflecting, imagining, and inventing. Norton (1987) recommended literature-related oral activities that develop observing, comparing, classifying, hypothesizing, organizing, summarizing, applying, and criticizing.

These important cognitive operations should be developed through oral language activities as well as in other language arts areas. Activities that encourage the development of these operations may be based on the various content areas. Literature, history, geography, art, math, and science provide excellent sources. The oral language activities frequently precede or accompany written composition, listening, reading, and literature activities.

Basic Operations Associated with Thinking

Strickland (1977) stated that teachers should capitalize on every opportunity to develop operations associated with thinking. She recommended that teachers focus attention on observing, comparing, classifying, hypothesizing, organizing, summarizing, and applying.

Observing   Language arts teachers have many opportunities to develop observational skills. Teachers of young children may focus children's observations on concrete objects in the school, home, or neighborhood. Children can describe the color, size, shape, and use of objects in their environments.

Colorful illustrations in picture books enhance observational skills as children identify objects and describe the content. Suse MacDonald's *Alphabatics* (1986) entices viewers to follow the illustrations and to describe the changes in each letter of the alphabet as the letter is transformed into an object that represents the letter. For example, A proceeds from a brown letter within a white background, to an angled A floating on blue water, to an upside down A on blue waves, to an ark shape on waves, to an ark filled with animals. Audrey Wood's *King Bidgood's in the Bathtub* (1985) includes humorous illustrations of objects that are not usually found in a bathtub. Older students can search for literary, art, or historical objects in Mitsumasa Anno's *Anno's U.S.A.* (1983) or *Anno's Britain* (1982). The science curriculum can be enhanced through observation and description of Heiderose and Andreas Fischer-Nagel's *Life of the Honeybee* (1986) and Jerome Wexler's *From Spore to Spore: Ferns and How They Grow* (1985). History and architecture are both enhanced when students describe the content in Gian Paolo Ceserani's *Grand Constructions* (1983). The illustrations and text proceed chronologically from Stonehenge to skyscrapers.

Word association activities enhance observational skills, vocabulary development, and writing capabilities. These activities are appropriate for any age level, but they should be changed to meet children's developing language capabilities. Younger children may consider words that create strong emotional feelings. The teacher may ask, "What causes you to feel angry? How do you show your anger? How do other people or even animals display anger?" Similar discussions and observations may invoke students into thinking about feelings of love, happiness, and fear.

A verbal observation activity for older children could follow the format developed by Catherine Daughtery, a headmistress of a primary school in Scotland. In an effort to help children visualize and describe their internal visual pictures associated with emotional experiences, she tells them a short story. For example, to enhance observations about misery, she tells children a story about a puppy lost in the snow after it is taken from its mother. She asks the children to describe orally their visual pictures. Next, they make a list of words, mainly verbs and adjectives, that convey the emotional message of the story. Finally, each child provides one sentence, with carefully selected words, that describes a visual picture. This activity can expand as children first observe and then describe people in the environment, on film, or in pictures.

Daughtery maintained that oral language development is very important if children are to first develop observational skills and imagination and then use these competencies in oral and written compositions. She stated, "Broadly speaking, we must begin by training the children's observation of things which are to them commonplace and very often not worthy of notice. We must then explore the way in which they tell or write the results of their observations. Initially, these will be 'flat' matter-of-fact observations with no attendant aids from emotive language or imagery. The use of emotive language and imagery is not essentially instinctive unless to a particular gifted child, so a great deal of groundwork, including vocabulary building, must be done. The bulk of this work should be, in the initial stages, oral. Free expression and expansion of thought are inhibited by the superfluous process of recording at this level" (personal communication, June 20, 1983).

Comparing   When children describe various attributes of concrete objects, literary selections, or pictures, they can be encouraged to compare those attributes with other objects, literary selections, or pictures that have either similar or very different attributes. For example, attributes of real hats can be observed, described, and compared. Next, children can consider the attributes of hats illustrated in Stan and Janice Berenstain's *Old Hat, New Hat* (1970). Illustrations include hats that are heavy, light, tall, flat, small, and big.

Comparisons may be made between preparing for a feast during two time periods by using Aliki Brandenberg's *A Medieval Feast* (1983) and Joan Anderson's *The First Thanksgiving Feast* (1984). Children may consider clothing, housing, food, preparations, guests at the feast, and purposes for the celebration.

Older students may compare plot developments and characterizations in two stories that have similar person versus self conflicts. For example, in Marion Bauer's *On My Honor* (1986), personal conflict results when a boy disobeys his father, swims in a river, and his best friend drowns. In Paula Fox's *One-Eyed Cat* (1984) personal conflict results when a boy disobeys his father, shoots an air rifle, and wounds a cat.

Different artists' renditions of the same story or folktale provide numerous opportunities for comparisons and critical evaluations. Four 1983 versions of Margery Williams's text, originally published in 1922, provide an excellent example. Ellerman stated, "the appearance of four new and different *Velveteen Rabbits* this spring is surprising and dictates a need for careful study before selecting, while at the same time it offers a stimulating exercise in artistic comparison" (1983, p. 1337). Allen Atkinson's version of *The Velveteen Rabbit* published by Knopf includes over fifty illustrations varying from full-page spreads to tiny decorations that dot the pages. Michael Hague's full- and double-page illustrations for the text published by Holt focus on the boy as well as the rabbit. Ilse Plume's version published by Godine creates a more contemporary feeling. Finally, Tien's version published by Simon and Schuster contains muted colors. Because the text of the four versions is the same, children can consider the impact of color, line, and design. They can also consider the accuracy of the illustrations. The illustrations in two versions show a rabbit with four legs. The text, however, indicates that the Velveteen Rabbit had no hind legs.

Comparison may be made between both text and illustrations in various versions of Grimm's folktales, Charles Perrault's folktales, and Hans Christian Andersen's literary fairy tales. Easy-to-read versions of the classics may be compared with original versions. Made-for-television versions may be compared with original novels, or movie versions may be compared with versions that are written following a movie's release. Oral discussions focusing on such comparisons may follow text reading by teachers or students.

Classifying   Concrete objects in the environment may be described and classified according to shape, color, size, and use. Concept books emphasizing alphabetical order, colors, numbers, sizes, and shapes provide useful sources for oral discussions. For example, Anita Lobel's *On Market Street* (1981), an alphabetical trip through a market street of an earlier time, encourages children to identify, describe characteristics, and add to various objects categorized in groups such as apples, books, clocks, instruments, and toys. Children can classify items that begin with the same sound using Elizabeth Cleaver's *ABC* (1985). Classifications of objects familiar to young children are enhanced through Janet and Allan Ahlberg's *The Baby's Catalogue* (1982). Classifications of seasonally related activities associated with working and playing are possible through illustrations in Anne Rockwell's *First Comes Spring* (1985). Classification of objects according to color may be enhanced using Eric Carle's *My First Book of Colors* (1974) and *The Mixed-Up Chameleon* (1975) and Donald Crews's *Freight Train* (1978). Shape classifications are stimulated using Tana Hoban's *Shapes, Shapes, Shapes* (1986).

Hypothesizing   Literature selections can encourage children to hypothesize about language the author will use, occurrences in descriptive book and chapter titles, and plot development. Rhyming patterns in books for younger children provide powerful language cues. Teachers may read such books as Dr. Seuss's *Hop on Pop* and ask

Classifying activities stimulate children's cognitive development.

children to complete a missing word. Older children may listen for and complete the rhyming elements in "The King of Cats" found in Nancy Willard's *A Visit to William Blake's Inn* (1981). In addition, poems in Jack Prelutsky's *The Random House Book of Poetry for Children* (1983) contain many examples of enjoyable rhyming.

Through repetition, a cumulative tale adds a new incident with each telling until the story is complete. Consequently, tales such as Verna Aardema's *Why Mosquitoes Buzz in People's Ears* (1975), Janet Stevens's *The House That Jack Built* (1985), the "Gingerbread Boy," "Henny Penny," and "The Fat Cat" encourage children to hypothesize about the language and join in with appropriate words.

Descriptive and imaginative book titles such as Judith Viorst's *Alexander and the Terrible, Horrible, No Good, Very Bad Day* (1972), James Stevenson's *The Wish Card Ran Out* (1981), Margot Zemach's *It Could Always Be Worse* (1976), Paula Underwood Spencer's *Who Speaks for Wolf* (1983), and Sid Fleischman's *The Whipping Boy* (1986), encourage hypothesizing about story content and reading to verify the hypothesis. Chapter titles in longer books provide sources for older children.

Organizing   Children frequently have difficulty understanding time concepts, following a sequence of time, and identifying the order in which things happen. Several picture books develop stories according to seasonal changes. Donald Hall's *Ox-Cart Man* (1979) proceeds from fall, through winter, into spring. Helen Craig's book for young children, *Mouse House Months* (1981), traces a tree as it goes through seasonal changes. Both books contain colorful illustrations and details that support the changing seasons.

Informational books for older students frequently follow a chronological order. In Kathryn Lasky's *Sugaring Time* (1983), photographs and text show the procedures used to collect sap from maple trees and process it into maple syrup. William Jaspersohn's *Magazine: Behind the Scenes at Sports Illustrated* (1983) follows the production of a major sports magazine.

Stories with strong sequential plots help children analyze how plots are developed. Flannel board presentations help children identify logical organization in these stories:

1. "The Three Little Pigs"—progessing from flimsiest to strongest building materials.
2. "The Little Red Hen"—following chronological order in the steps for preparing bread.
3. "The Three Billy Goats Gruff"—proceeding according to size from smallest to largest.

Summarizing   Children should have many opportunities to summarize stories heard or read, points made by speakers, and content covered in course work. These summaries may vary as children summarize the most exciting part of a story, the actions of the main character, the funniest part of a story, and the supporting ideas in informational writing.

Applying   Books that encourage children to observe, gather data, experiment, compare, and formulate hypotheses are excellent for cognitive development. For example, Seymour Simon's *The Secret Clocks, Time Senses of Living Things* (1979), encourages children to become involved with the principles of biological time clocks by providing step-by-step directions for experiments with bees, animals, plants, and themselves. Jim Arnosky's *Drawing from Nature* (1982) and *Drawing Life in Motion* (1984) provide detailed drawings that show how to observe nature and how to draw the various animals and natural environments. Oral language discussions and interaction between children and teachers will help stimulate these crucial cognitive capabilities.

Semantic Mapping and Vocabulary Development

Semantic mapping and webbing, a method for graphically and visually displaying relationships among ideas and concepts, emphasizes congnitive processes and encourages problem solving. Semantic mapping encourages higher thought processes, stimulates ideas, and encourages oral interactions among students and teachers as they consider and complete various portions of the web. Semantic mapping procedures increase vocabulary development (Johnson and Pearson, 1984; Toms-Bronowski, 1983); enhance literary discussions that highlight plot development, setting, characterization, and theme (Norton, 1987); improve reading comprehension (McNamara and Norton, 1987; Prater and Terry, 1985); enhance the development of instructional units (Norton, 1982, 1987); stimulate the composition process (Myers and Gray, 1983); encourage interaction and understanding in various content areas (Heimlich and Pittelman, 1986); and encourage the integration of reading, literature, writing, listening, and oral discussion within the language arts curriculum (McNamara and Norton, 1987).

   Each of these objectives for using the semantic mapping strategy uses slightly different procedures. Consequently, an in-depth approach to each procedure is provided as it is developed within the appropriate chapters of this text. General procedures usually begin with a brainstorming activity in which the teacher encourages students to verbalize associations or ideas while the teacher maps the ideas on the chalkboard. The content of the initial map varies according to the teaching and learning objectives and the subject matter to be visualized. The semantic map may be used as an introductory activity to organize and extend previous knowledge or to explore various possibilities, as a discussion activity to visualize relationships and to teach understandings within the lesson, or as a follow-up activity to reinforce learning or even to test understandings. Semantic mapping strategies may be used to help children identify words with similar meanings, expand a precise vocabulary, understand multiple meanings for words, develop concepts, and perceive relationships among words and ideas. The following activities are stimulated through brainstorming and group or class discussion.

Similar meanings and semantic precision   There are numerous words such as *said* and *went* that are overused in speaking and writing. Helping children identify and use words that express precise meanings will improve both speaking and writing activities and stimulate cognitive development. The following procedures were developed in a fourth grade class using the word *went.*

First, the teacher wrote and then read several sentences on the chalkboard. For example:

Jackie *went* to the park to find a lost football.
Mike *went* to the store to spend his birthday money.
Sharon *went* to school to take a spelling test.
Terry *went* to the park to play in the championship soccer game.

Next, the class identified the word *went* and discussed the appropriateness of the word to express the feelings and the actions of the subjects. In an effort to illustrate different meanings, they dramatized each sentence. They also dictated additional sentences in which *went* could be used but might not be appropriate.

The teacher led a brainstorming activity in which he encouraged children to identify numerous words that could be used in place of *went.* The following words illustrate a partial list:

ran	jogged	hurried	strutted
marched	danced	pranced	promenaded
raced	flounced	dawdled	sauntered
ambled	swaggered	rambled	crawled
glided	drifted	skipped	bounced
wandered	staggered	bounded	meandered
strolled	dashed	trotted	hopped
hastened	shuffled	scampered	capered
lurched	stumbled	wobbled	tottered
swayed	moseyed	struggled	limped

In an effort to categorize meanings for *went,* the teacher and class identified various meanings for the word: happy meanings, tired meanings, slow meanings, fast meanings, proud meanings, frightened meanings, and sad meanings. These meanings were used as subheadings in a map or web around the word *went* (see figure 2–3).

FIGURE 2–3
First step in a web designed to categorize the meanings of *went*

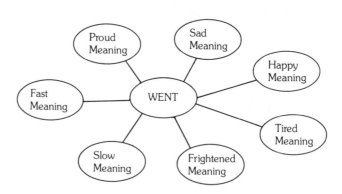

Finally, the class considered the words on their list and placed them around the appropriate meaning or meanings. Additional words were also identified as children considered each meaning classification. The relationships among words that had several meanings were discussed. The semantic map in figure 2–4 represents a partial web developed by the class.

Multiple meanings and different parts of speech   Even simple words may become complex and difficult for children to understand if the words are used in unfamiliar contexts or as previously unused parts of speech. Three meanings for the word *act* as well as two parts of speech are illustrated in the semantic map (see figure 2–5) developed in an upper elementary social studies class.

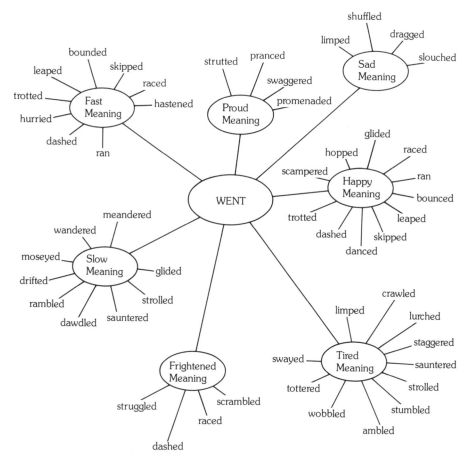

FIGURE 2–4
A completed web showing various meanings for the word *went*

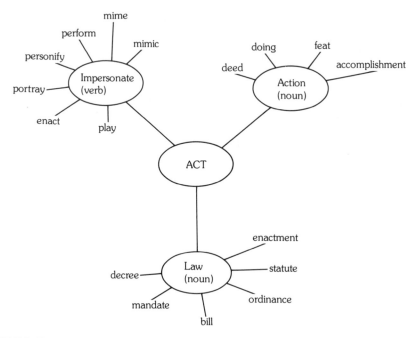

FIGURE 2–5
Web illustrating three meanings for the word *act*

Semantic maps representing concepts Johnson and Pearson recommended developing semantic maps to illustrate class and property links that "capture the character of our intuitive notions about the similarities and differences among concepts" (1984, p. 29). They illustrate the concepts of *dog* and *cat* and some of the related concepts in figure 2–6. (In this semantic map *is a* refers to class and *has, is,* or *does* refer to property.)

Semantic maps and story comprehension Semantic maps encourage understanding a story and relationships among such elements as character traits, setting, and plot development. Figures 4–3 and 4–4 develop semantic maps following a literature listening activity and illustrate the mapping of character traits, setting, and plot development.

SUMMARY

Language is the basis of human communication. In fact, language is uniquely human, and is the only form of communication that allows its user to devise theories about the language system itself. In this chapter, we have learned about three theories of language acquisition. The behaviorist believes that language is learned through imitation and, consequently, relates language acquisition to such concepts as imitation,

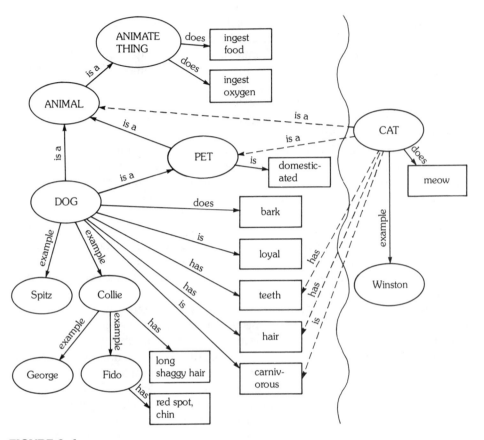

FIGURE 2–6

An incomplete semantic network representation of the concepts *dog* and *cat* and related concepts, using substitutions for class, property, and example relationship terms (Adapted from *Teaching Reading Vocabulary,* 2d ed. by Dale D. Johnson and P. David Pearson. Copyright © 1984 by Holt, Rinehart & Winston. Reprinted by permission of the publisher.)

reinforcement, successive approximations, and shaping. These concepts have been used to teach language to young children, as illustrated in the DISTAR language program.

A contrasting theory of language acquisition, the genetic theory, states that language is not learned through imitation, but is innate or instinctive. Proponents of this theory believe all children are born with the ability to use language. They begin with a language of their own, and amend it to conform to adult language. Linguists give three reasons to support the theory that language ability is biologically inherited: (1) there are many features common to all languages; (2) word patterns that children construct are not imitations of adult speech; and (3) the sequence of language development in children is orderly and systematic and is related to the physical and motor aspects of development.

We also reviewed the concept that a child's interaction with her physical and social environment is critical to both language and intellectual development. This concept is called the sociocultural theory. Pose Lamb suggested that there are probably close corollaries between cognitive growth and language development. Anastasiow's recommendations for developmental instruction were quite different from the principles employed in the DISTAR program. We saw that theories of language and child development influence the focus of language arts programs and both teachers' and children's roles in the classroom.

We have also seen that, although scholars disagree about theories of language acquisition, there is substantial agreement about the sequence of events in language development. Research has concluded that all children appear to go through approximately the same stages of language development, although the rate of development varies from child to child. Longitudinal studies by Brown and Loban have provided educators with extensive knowledge about language development. Although language development is advanced by the time most children enter school, it is far from complete; thus, teacher expectations in written and oral composition as well as in reading comprehension should be adjusted to the child's level of language development.

This chapter also reviewed a linguistic analysis of language and discussed some of its implications for instruction. Linguists investigate the code systems of language, including phonology (the study of the sound system), morphology (the study of meaning in relation to speech sounds), syntax (the study of how words are put together to form sentences), and semantics (the study of meaning).

Knowledge about the interrelationships of the various linguistic code systems has developed through psycholinguistic research. From a psycholinguistic perspective, the various linguistic cues are so closely related that they should not be separated in providing language instruction. Psycholinguists consider meaning the core of language. The grammatical structure is fused to meaning, and the sound and written symbols are vehicles by which meaning is displayed.

Learning environments must stimulate children's cognitive development. Therefore, teachers should provide stimulating activities for students. Activities that encourage children to organize information into meaningful wholes, to visualize concepts and relationships, and to develop problem-solving tactics are important.

ADDITIONAL LANGUAGE AND COGNITIVE DEVELOPMENT ACTIVITIES

1. *Look at several different types of instructional materials used to develop language or reading skills with children. Do the materials fragment the instruction of the linguistic code into phonology, morphology, syntax, and semantics, or do they stress the interrelationship between the linguistic systems? Find some examples of each approach and share them with your class.*

2. *Visit several nursery school or kindergarten classrooms. Observe the environment and the instructional activities. Can you identify a theory of language acquisition supported by the instructional activities?*

Interview a professor of early childhood education development, educational psychology, or special education. What is the professor's view on language acquisition? What type of learning environment and instructional activities does the professor recommend? Share the results of your interview with your language arts class. Do all the professors agree on a theory of language acquisition and learning environments? Can you provide any reasons for differences of opinion?

3. Tape-record a conversation between a very young child and her parents. How does the parent adjust his or her language to the conversational skills of the child? How is meaning achieved?

4. Read a recent journal article or chapter in a text emphasizing language acquisition or development. What is the major emphasis of the author? What theory of language acquisition does the author support? What evidence does the author include? What implications for instruction are included or implied? Examples of articles to read include the following:

Holbrook's "Oral Language: A Neglected Language Art" (1983)
Hall's "Teaching and Language Centered Programs" (1986)
Loban's "Research Currents: The Somewhat Stingy Story of Research into Children's Language" (1986)
Lindfors's "Understanding the Development of Language Structure" (1985)
Olson's "Perspectives: Children's Language and Language Teaching" (1983)
Myers's "Approaches to the Teaching of Composition" (1983)

5. Read an article on cognitive development and identify an instructional application that could be important in the language arts class.

6. Select one of the cognitive operations discussed in the chapter. Develop a lesson plan that would help children observe, compare, classify, hypothesize, organize, summarize, or apply.

7. Plan a lesson that would encourage vocabulary development. With a group of children or your peers, develop a semantic map that identifies words with similar meanings, expands a precise vocabulary, or increases understanding of multiple meanings.

BIBLIOGRAPHY

Allen, Roach Van. *Language Experiences in Communication.* Boston: Houghton Mifflin Co., 1976.

Anastasiow, Nicholas. *Oral Language Expression of Thought.* Newark, Del.: International Reading Association, 1979.

Aukerman, Robert C. *Approaches to Beginning Reading.* 2d ed. New York: Macmillan Co., 1984.

Bartel, Nettie. "Assessing and Remediating Problems in Language Development." In *Teaching Children with Learning and Behavior Problems,* 3d ed., edited by Donald Hammill and Nettie Bartel. Boston: Allyn & Bacon, 1986.

Braine, Martin D. S. "The Ontogeny of English Phrase Structure: The First Phase." In *Readings in*

Language Development, edited by Lois Bloom. New York: John Wiley & Sons, 1978.

Brown, Roger. *A First Language/The Early Stages.* Cambridge, Mass.: Harvard Univ. Press, 1973.

Bruner, Jerome. "Learning the Mother Tongue." *Human Nature* 1 (September 1978): 42–49.

Burke, Carolyn. "The Language Process: Systems or Systematic?" In *Language and Learning to Read,* edited by Richard Hodges and E. Hugh Rudorf. Boston: Houghton Mifflin Co., 1972.

Cazden, Courtney B. *Child Language and Education.* New York: Holt, Rinehart & Winston, 1972.

Chomsky, Carol. *The Acquisition of Syntax in Children from 5 to 10.* Research Monograph 52. Cambridge, Mass.: The M.I.T. Press, 1969.

Chomsky, Noam. "Review of Skinner's Verbal Behavior." *Language* 35 (1959): 26–58.

Eachus, Herbert Todd. *In-Service Training of Teachers as Behavior Modifiers: Review and Analysis.* Washington, D.C.: Sponsoring Agency, Bureau of Educational Personnel Development, Division of Assessment and Coordination, September 1971.

Ellerman, Barbara. "Focus: The Velveteen Rabbit." *Booklist* (June 15, 1983): 1337.

Engelmann, Siegfried. "The Effectiveness of Direct Verbal Instruction on IQ Performance and Achievement in Reading and Arithmetic." In *Control of Human Behavior,* edited by Roger Ulrich. Glenview, Ill.: Scott, Foresman & Co., 1974.

Ervin-Tripp, Susan. "Imitation and Structured Change in Children's Language." In *New Directions in the Study of Language,* edited by E. H. Lenneberg. Cambridge, Mass.: The M.I.T. Press, 1964.

_____. "Language Development." In *Review of Child Development Research,* edited by L. Hoffman and M. Hoffman. New York: Russell Sage Foundation, 1966.

Fink, Rychard, and Keiserman, Patricia. *i.t.a. Teacher Training Workbook and Guide.* New York: Initial Teaching Alphabet Publications, 1969.

Fox, Sharon E. "Research Update: Oral Language Development, Past Studies and Current Directions." *Language Arts* 60 (February 1983): 234–43.

Fry, Denis B. "The Development of the Phonological System in the Normal and the Deaf Child." In

The Genesis of Language: A Psycholinguistic Approach, edited by Frank Smith and George A. Miller. Cambridge, Mass.: The M.I.T. Press, 1966.

Goodman, Kenneth. "The Reading Process: Theory and Practice." In *Language and Learning,* edited by Richard Hodges and E. Hugh Rudorf. Boston: Houghton Mifflin Co., 1972.

Hall, Mary Ann. "Teaching and Language Centered Programs." In *Roles in Literacy Learning: A New Perspective,* edited by Duane R. Tovey and James E. Kerber, 34–41. Newark, Del.: International Reading Association, 1986.

Heimlich, Joan E., and Pittelman, Susan D. *Semantic Mapping: Classroom Applications.* Newark, Del.: International Reading Association, 1986.

Holbrook, Hilary Taylor. "Oral Language: A Neglected Language Art?" *Language Arts* 60 (February 1983): 255–58.

Jenkins, J. J., and Palermo, D. S. "Mediation Processes and the Acquisition of Linquistic Structure." *Monographs of the Society for Research in Child Development* 29 (1964): 141–69.

Johnson, Dale D., and Pearson, P. David. *Teaching Reading Vocabulary* 2d ed. New York: Holt, Rinehart & Winston, 1984.

Kavale, Kenneth A. "The Reasoning Abilities of Normal and Learning Disabled Readers on Measures of Reading Comprehension." *Learning Disability Quarterly* 3 (Fall 1980): 34–45.

Lamb, Pose. *Linguistics in Proper Perspective.* 2d ed. Columbus, Ohio: Merrill Publishing Co., 1977.

Lawton, Denis. *Social Class, Language, and Education.* New York: Schocken Books, 1968.

Lindfors, Judith W. *Children's Language and Learning.* 2d ed. Englewood Cliffs, N.J.: Prentice-Hall, 1987.

_____. "Understanding the Development of Language Structure." In *Observing the Language Learner,* edited by Angela Jaggar and M. Trika Smith-Burke. Urbana, Ill.: National Council of Teachers of English, 1985.

Loban, Walter. *Language Development: Kindergarten through Grade Twelve.* Urbana, Ill.: National Council of Teachers of English, 1976.

_____. "Research Currents: The Somewhat Stingy Story of Research into Children's Language." *Language Arts* 63 (October 1986): 608–16.

McNamara, James, and Norton, Donna E. *An Evaluation of the Multiethnic Reading/Language Arts Program for Low Achieving Elementary and Junior High School Students,* Final Report, College Station, Tex: Texas A&M University, 1987.

Menyuk, Paula. "Syntactic Structures in the Language of Children." *Child Development* 34 (June 1963): 407–22.

Moskowitz, Breyne Arlene. "The Acquisition of Language." *Scientific American,* November 1978: 92–108.

Mussen, Paul Henry; Conger, John Janeway; and Kagan, Jerome. *Child Development and Personality.* 6th ed. New York: Harper & Row, 1984.

Myers, Miles. "Approaches to the Teaching of Composition." In *Theory and Practice in the Teaching of Composition: Processing, Distancing and Modeling,* edited by Miles Myers and James Gray, pp. 3–43. Urbana, Ill.: National Council of Teachers of English, 1983.

_____, and Gray, James. *Theory and Practice in the Teaching of Composition: Processing, Distancing, and Modeling.* Urbana, Ill.: National Council of Teachers of English, 1983.

National Council of Teachers of English. "Forum: Essentials of English." *Language Arts* 60 (February 1983): 244–48.

Norton, Donna E. *Through the Eyes of a Child: An Introduction to Children's Literature.* 2d ed. Columbus, Ohio: Merrill Publishing Co., 1987.

_____. "Understanding and Appreciating Folklore." In *Language Arts Instruction and the Beginning Teacher,* edited by Carl R. Personke and Dale D. Johnson. Englewood Cliffs, N.J.: Prentice-Hall, 1987, pp. 189–98.

_____. "Using A Webbing Process to Develop Children's Literature Units." *Language Arts* 59 (April 1982): 348–56.

Olson, David R. "Perspectives: Children's Language and Language Teaching." *Language Arts* 60 (February 1983): 226–33.

Poole, Millicent E. *Social Class and Language Utilization at the Tertiary Level.* St. Lucia: University of Queensland Press, 1976.

Prater, D. C., and Terry, C. A. *The Effects of a Composing Model on Fifth Grade Students Reading Comprehension.* Paper presented at the American Educational Research Association, 1985 (ERIC Document Reproduction Service No. ED 254 825)

Rand, Yáacou; Tannenbaum, Abraham, J.; and Feuerstein, Reuven. "Effects of Instrumental Enrichment on the Psychoeducational Development of Low-Functioning Adolescents." *Journal of Educational Psychology* 71 (December 1979): 751–63.

Raven, Ronald J., and Salzer, Richard T. "Piaget and Reading Instruction." *Reading Teacher* 24 (April 1971): 630–39.

Shugar, G. W. "Text Analysis as an Approach to the Study of Early Linguistic Operations." In *The Development of Communication,* edited by Catherine Snow and N. Waterson. Chichester, England: John Wiley, 1978.

Slobin, Dan I. *Psycholinguistics.* 2d ed. Glenview, Ill.: Scott, Foresman & Co., 1979.

_____, and Welsh, Charles A. "Elicited Imitation as a Research Tool in Developmental Psycholinguistics." In *Studies of Child Language Development,* edited by Charles A. Ferguson and Daniel I. Slobin. New York: Holt, Rinehart & Winston, 1973.

Smith, Madorah. "Grammatical Errors in the Speech of Preschool Children." *Child Development* 4 (1933): 182–90.

Snow, Catherine E. "The Development of Conversation between Mothers and Babies." *Journal of Child Language* 4 (February 1977): 1–22.

Squire, James R. "The Collision of the Basics Movement with Current Research in Writing and Language." In *The English Curriculum under Fire: What Are the Real Basics?,* edited by George Hillocks, Jr., 29–37. Urbana, Ill.: National Council of Teachers of English, 1982.

Staats, A. W. "A Case in and Strategy for the Extension of Learning Principles to the Problems of Human Behavior." In *Human Learning.,* edited by A. W. Staats. New York: Holt, Rinehart & Winston, 1964.

_____. *Learning, Language, and Cognition.* New York: Holt, Rinehart & Winston, 1968.

Strickland, Dorothy S. "Promoting Language and Concept Development." In *Literature and Young Children,* edited by Bernice Cullinan and Carolyn Carmichael. Urbana, Ill.: National Council of Teachers of English, 1977.

Tobin, Kenneth. "The Role of Wait Time in Higher Cognitive Level Learning." *Review of Educational Research* 57 (Spring 1987) 69–95.

Toms-Bronowski, S. "An Investigation of the Effectiveness of Selected Vocabulary Teaching Strategies with Intermediate Grade Level Students." *Dissertation Abstracts International* 1983, 44, 1405 A (University Microfilms No. 83–16, 238).

Torgesen, Joseph K. "Factors Related to Poor Performance on Memory Tasks in Reading Disabled Children." *Learning Disability Quarterly* 2 (Summer 1979): 17–23.

Weinstein, Claire E., and Mayer, Richard E. "The Teaching of Learning Strategies." In *Handbook of Research and Teaching,* 3d ed., edited by Merlin C. Wittrock, 315–327. New York: Macmillan, 1986.

Wells, Gordon. "Describing Children's Linguistic Development at Home and at School." *British Educational Research Journal* 5 (1979): 75–98.

_____. ed. *Learning through Interaction: The Study of Language Development.* London: Cambridge University Press, 1981.

CHILDREN'S LITERATURE REFERENCES

Aardema, Verna. *Why Mosquitoes Buzz in People's Ears.* New York: Dial Press, 1975.

Ahlberg, Janet and Ahlberg, Allan. *The Baby's Catalogue.* Boston: Little, Brown, & Co., 1982.

Anderson, Joan. *The First Thanksgiving Feast.* Photographs by George Ancona. New York: Clarion, 1984.

Anno, Mitsumasa. *Anno's Britain.* New York: Philomel, 1982.

_____. *Anno's U.S.A.* New York: Philomel, 1983.

Arnosky, Jim. *Drawing from Nature.* New York: Lothrop, Lee & Shepard, 1982.

_____. *Drawing Life in Motion.* New York: Lothrop, Lee & Shepard, 1984.

Bauer, Marion. *On My Honor.* New York: Houghton Mifflin Co., 1986.

Berenstain, Stan and Berenstain, Janice. *Old Hat, New Hat.* New York: Random House, 1970.

Brandenberg, Aliki. *A Medieval Feast.* New York: Crowell, 1983.

Carle, Eric. *The Mixed-Up Chameleon.* New York: Crowell, 1975.

_____. *My First Book of Colors.* New York: Crowell, 1974.

Ceserani, Gian Paolo. *Grand Constructions.* Illustrated by Piero Ventura. New York: Putnam's, 1983.

Cleaver, Elizabeth. *ABC.* New York: Atheneum Pubs., 1985.

Craig, Helen. *Mouse House Months.* New York: Random House, 1981.

Crews, Donald. *Freight Train.* New York: Greenwillow, 1978.

Fischer-Nagel, Heiderose, and Fischer-Nagel, Andreas. *Life of the Honeybee.* Minneapolis: Carolrhoda, 1986.

Fleischman, Sid. *The Whipping Boy.* New York: Greenwillow, 1986.

Fox, Paula. *One-Eyed Cat.* Scarsdale, NY: Bradbury Press, 1984.

Hall, Donald. *Ox-Cart Man.* New York: Viking Press, 1979.

Hoban, Tana. *Shapes, Shapes, Shapes.* New York: Greenwillow, 1986.

Jaspersohn, William. *Magazine: Behind the Scenes at Sports Illustrated.* Boston: Little, Brown & Co., 1983.

Lasky, Kathryn. *Sugaring Time.* Photographs by Christopher G. Knight. New York: Macmillian Co., 1983.

Lobel, Anita. *On Market Street.* New York: Greenwillow, 1981.

MacDonald, Suse. *Alphabatics.* New York: Bradbury Press, 1986.

Prelutsky, Jack, ed. *The Random House Book of Poetry for Children.* New York: Random House, 1983.

Rockwell, Anne. *First Comes Spring.* New York: Crowell, 1985.

Simon, Seymour. *The Secret Clocks, Time Senses of Living Things.* New York: Viking Press, 1979.

Spencer, Paula Underwood. *Who Speaks for Wolf.* Illustrated by Frank Howell. Austin, Tex.: Tribe of Two Press, 1983.

Stevens, Janet. *The House that Jack Built.* New York: Holiday, 1985.

Stevenson, James. *The Wish Card Ran Out!* New York: Greenwillow, 1981.

Viorst, Judith. *Alexander and the Terrible, Horrible, No Good, Very Bad Day.* New York: Atheneum Pubs., 1972.

Wexler, Jerome. *From Spore to Spore: Ferns and How They Grow.* New York: Dodd, Mead & Co., 1985.

Willard, Nancy. *A Visit to William Blake's Inn.* San Diego: Harcourt Brace Jovanovich, 1981.

Williams, Margery. *The Velveteen Rabbit.* Illustrated by William Nicholson. New York: Doubleday & Co., 1922.

Wood, Audrey. *King Bidgood's in the Bathtub.* Illustrated by Don Wood. New York: Harcourt Brace Jovanovich, 1985.

Zemach, Margot. *It Could Always Be Worse.* New York: Farrar, Straus & Giroux, 1976.

Chapter Three

OBJECTIVES OF INSTRUCTION EVALUATION

Using Language Experience Stories ▪ Using Story Retelling to Evaluate Language ▪ Unaided Recall of Story ▪ Obtaining Language Samples from Reluctant Speakers ▪ Evaluating Social Functions of Language

ENVIRONMENT

The Teacher and the Environment ▪ Instructional Practices

DEVELOPING ORAL COMMUNICATION SKILLS

Conversations ▪ Wordless Books ▪ Discussion ▪ Creative Dramatics ▪ Pantomime ▪ Puppetry ▪ Selecting and Making the Puppet for the Creative Drama ▪ A Puppet Project: Helping Freddie Frog—Second-grade Level ▪ Choral Speaking and Reading ▪ Reader's Theatre

MAINSTREAMING AND ORAL LANGUAGE DEVELOPMENT

SUMMARY

After completing the chapter on oral language development, you will be able to:

1. State the objectives of oral language instruction.
2. Design a stimulus for eliciting an oral language sample.
3. Evaluate a child's oral language after administering several informal tests.
4. Describe several instructional practices that help to build an environment in which oral language flourishes.
5. Describe how oral communication skills may be developed through conversation.
6. Describe and demonstrate the sequential development of questioning strategies designed to develop students' higher-level thought processes.
7. Demonstrate the use of pantomime as a basic dramatic technique.
8. Describe the five steps a teacher would use to guide children in dramatization.
9. Develop a puppetry activity.
10. Describe and demonstrate the major types of choral speaking arrangements.
11. Describe the use of oral reading and Reader's Theatre.

Oral Language Development

O   ral language is our chief method of communication. A joint statement prepared by the Early Childhood and Literacy Development Committee of the International Reading Association (1986) emphasized that learning should take place in a supportive environment that permits "children to build upon already existing knowledge of oral and written language" (p. 819) and that encourages children to develop positive attitudes toward themselves and toward language and literacy. In reference to oral language instruction, the committee recommended that instructional activities (1) focus on meaningful language and experiences rather than on isolated skill development, (2) respect the language children bring to school and use that language as a base for language and literacy activities, (3) help children see themselves as people who enjoy exploring language, (4) foster children's affective and cognitive development by encouraging oral communication, (5) encourage children to be active participants in the learning process, and (6) provide opportunities for children to interact with a wide variety of poetry, fiction, and nonfiction.

Loban's (1976) longitudinal study of language development (discussed in chapter 2) points out the importance of oral language and the need for oral language instruction. The children in Loban's study who were superior in oral language in kindergarten and first grade, before they learned to read and write, were also the children who excelled in reading and writing by the time they reached sixth grade.

OBJECTIVES OF INSTRUCTION

Oral language is basic to learning in all disciplines because it is a primary means for communication. Through oral language children develop self-identity and shape their experiences and knowledge. If we agree that oral language instruction is vital, we must also decide on objectives for instruction. The "Essentials of English" adopted by the National Council of Teachers of English (1983) provides excellent objectives in areas of speaking and creative thinking. Both areas are closely related to the development of effective oral communication. The objectives for speaking include:

Students should learn

- to speak clearly and expressively about their ideas and concerns.
- to adapt words and strategies according to varying situations and audiences, from one-to-one conversations to formal, large-group settings.
- to participate productively and harmoniously in both small and large groups.
- to present arguments in orderly and convincing ways.
- to interpret and assess various kinds of communication, including intonation, pause, gesture, and body language that accompany speaking. (p. 246)[1]

The development of thinking skills is essential for all subjects. A well-balanced curriculum must develop children's thinking capacities as well as increase their knowledge. In our world with its rapidly expanding knowledge base, knowing how to think creatively, logically, and critically is essential. The ability to analyze, classify, compare, formulate hypotheses, make inferences, and draw conclusions is crucial for all adults. Early and continuous stimulation is necessary, however, if these important cognitive skills are to develop. The National Council of Teachers of English identifies the following creative, logical, and critical thinking objectives:

In the area of creative thinking, students should learn

- that originality derives from the uniqueness of the individual's perception, not necessarily from an innate talent.
- that inventiveness involves seeing new relationships.
- that creative thinking derives from their ability not only to look, but to see; not only to hear, but to listen; not only to imitate, but to innovate; not only to observe, but to experience the excitement of fresh perception.

In the area of logical thinking, students should learn

- to create hypotheses and predict outcomes.
- to test the validity of an assertion by examining the evidence.
- to understand logical relationships.
- to construct logical sequences and understand the conclusions to which they lead.
- to detect fallacies in reasoning.
- to recognize that "how to think" is different from "what to think."

In the area of critical thinking, students should learn

- to ask questions in order to discover meaning.
- to differentiate between subjective and objective viewpoints; to discriminate between opinion and fact.
- to evaluate the intentions and messages of speakers and writers, especially at attempts to manipulate the language in order to deceive.
- to make judgments based on criteria that can be supported and explained. (p. 247)[2]

---

[1]From the National Council of Teachers of English, "Forum: Essentials of English," *Language Arts* 60 (February 1983): 244–48. Copyright © 1983 by the National Council of Teachers of English. Reprinted by permission of the publisher and author.
[2]Ibid.

EVALUATION

Researchers frequently recommend that teachers use informal means to assess oral language development. Loban (1976), for example, recommended that "because no published tests measure power over the living language, the spoken word, [the teacher should] devise ways of assessing oral language using cassettes, tape recorders, and video tapes" (p. 1). Olson (1983) emphasized, "The only sure evidence of competence is some form of expression, such as speaking or writing, sorting out ideas, making them subject to revision. Evaluation should put the emphasis on what the kid can do" (p. 231). Goodman (1985) stated that evaluation is an ongoing activity and that "informal, naturalistic observation is the most effective way to learn about children's language and their ways of learning" (p. 8). According to Goodman, language observation should evaluate children's ability to adjust language to meet the demands of new situations and new settings. Consequently, during evaluation teachers need to observe children using language to explore concepts in many areas such as art, social studies, math, and physical education.

Otto and Smith (1980) provided a rationale for using the informal approach to evaluate oral language development of disabled students. First, they believed that standardized instruments for diagnosing speaking deficiencies are often based on language standards that are more contrived than real, and that generally do not give any more information than informal assessments. Moreover, they feel a teacher must carefully observe a child's total language development in order to formulate the best instructional approach for that child; this observation should consider the child's effectiveness of communication in terms of purpose, message, and audience. They recommended that all diagnoses of speaking deficiencies be constructed in a highly personal way, and that comparing a disabled student's performance to a standardized scale should be avoided. Of course, in order to do this, the teacher will need enough background in language development to operate from this individual perspective.

In this section we will review some procedures teachers may use to elicit oral language from children for the purpose of diagnosis. Techniques that teachers have found successful include the following: evaluation of language experience stories, storytelling, unaided recall of a story, puppetry, picture discussions using wordless picture books, and social functions of language.

Using Language Experience Stories

Several authors recommend using the dictated story, or language experience, in order to evaluate language informally. (The language experience approach is described in chapter 14.) Anastasiow (1979) recommended that the child first be allowed to draw or paint a picture and then tell the story to the teacher or an aide, who writes it down as it is dictated. The completed story will provide the teacher with information about both the child's language and thought processes.

A more structured use of a language experience approach for evaluating language is recommended by Dixon. She stated: "The ease with which the language experience stories can be obtained and their reflection of the uniqueness of each child, make their use as informal diagnostic-evaluative tools worth further consideration" (1977, p. 501). Dixon distinguished among three types of information that can be obtained from evaluating the language experience. First is the observable behavior, which includes watching as words are written, pacing dictation, pausing at the end of a phrase or sentence, providing an appropriate title, and attempting to read the language experience back to the teacher. Second is the evaluation of global language usage, which includes dictating complete sentences and using a variety of words. Third is refined language usage, which includes counting the number of adjectives, adverbs, prepositional phrases, and embedded sentences.

Dixon's motivation for her language experience story is a puppet from the Peabody Language Development Kit. She allows a child to feel and manipulate the puppet while discussing it with the examiner. The examiner asks questions such as, "What does it look like? What does it feel like? What colors is it? What do you think it is?" The child is then asked to tell a story about the puppet, and a two-minute sample is written. Following the dictation, the child is given an opportunity to "read" the story with the examiner. The language sample is then scored according to the checklist shown in table 3–1.

Dixon suggested that the checklist's weighted score is the most desirable, because it allows equal value within categories. She believed that low ability in the first category may imply the need for more language enrichment activities and exposure to stories before normal reading instruction is started, because a child who does not watch or pace dictation while the words are being written may not be aware that the words being written are the same as those being spoken. Low scores in the second category may indicate that language enrichment and development activities should be a large part of the student's instructional program. Low scores in the final category may imply the need for vocabulary development and language refinement activities as part of the language arts program.

Using Story Retelling to Evaluate Language

Story retelling is a means of evaluating the child's ability to perceive and recall events. Pickert and Chase stated that "early assessment of children's ability to comprehend, organize, and express language, followed by appropriate education based on this assessment, may be the key to student success in school" (1978, p. 528). They recommended the use of story retelling because the student's organization, comprehension, and sentence structures are not biased by prestructured teacher questions.

For this assessment, the teacher tells a short story individually to a student. The student is then asked to retell the story, and his version is tape recorded. A story is selected that has a plot simple enough for the age of the child but too long for memorization. Pickert and Chase felt that story retelling affords the following information:

1. Does the child comprehend the grammatical forms of the sentences and the meaning of the vocabulary? Children who do not understand sentence structures, words, or concepts may give the story a different but plausible ending.

TABLE 3–1
Language experience checklist

I. Observable Behaviors:		
1. Watches when words are written down	____ (usually)	____ (sometimes)	____ (seldom)
2. Paces dictation	____ (usually)	____ (sometimes)	____ (seldom)
3. Pauses at end of phrase or sentence	____ (usually)	____ (sometimes)	____ (seldom)
4. Appropriate title	____ (very)	____ (acceptable)	____ (poor)
5. Attempts to read back	____ (most)	____ (some)	____ (none)
II. Global Language Usage:		
1. Complete sentences	____ (all)	____ (some)	____ (none)
2. Total words	____ (31+)	____ (16–30)	____ (0–15)
3. Total different words	____ (30+)	____ (15–29)	____ (0–14)
III. Refined Language Usage:		
1. Number of adjectives	____ (9+)	____ (4–8)	____ (0–3)
2. Number of adverbs	____ (3+)	____ (1–2)	____ (0)
3. Number of prepositional phrases	____ (4+)	____ (1–3)	____ (0)
4. Number of embedded sentences	____ (2+)	____ (1)	____ (0)
Total	____3X	____2X	____1X
Total points	____	____	____
Grand total _____		

2. Does the child organize information and recall it in a logical manner? Are events left out, is the order confused, or is the task impossible to perform?
3. Does the child express the story in fluent, connected sentences, using correct grammatical forms? If the teachers understand the differences that age, experience, culture, and individual learning styles have on children's performance, they can use retelling to learn more about the children in their classroom and also learn which errors are typical of young children and which may suggest a more severe language disturbance.

Oral language is evaluated as children retell a story.

Unaided Recall of Story

Another evaluative measure is the unaided recall of a story. This technique can be used to assess the child's oral ability in placing materials in logical order; presenting main ideas, important details, and cause-and-effect relationships; analyzing characters; defending a viewpoint; and in using oral vocabulary and language structures. This approach differs from retelling a story because the selection is read by the child rather than told to him. Goodman and Burke (1971) recommended this technique for use with older children who are able to read.

The child first reads a selection that tells a complete story. I often use *Reader's Digest Skill Builders* for this activity, because they have short, interesting stories ranging in reading difficulty from first through eighth grade. The following is an example of a retelling activity I have used with third-grade students. The three students in this example all had high reading ability.

First, each student read a selection from a fifth-grade *Reader's Digest Skill Builder* entitled, "Mustang's Last Stand." After the reading, each child told the story individually without my asking any structured questions. The unaided recall was taped. In addition, each child was asked, "Why did the author write this story for us to read?" and "How did the story make you feel?" After the taping, the unaided-recall stories were transcribed from the tape and analyzed for understanding of characters, logical outline of the story, grasp of main ideas and important details, and level of oral language development. In order to rate the students' ability to present the story in logical organization, an outline of the story was developed and each student's number was placed by the outline if he recalled the item in its proper order. The outline for these three students was as follows:

"Mustang's Last Stand"

Student Number	Characters in Story
1 2 3	Mustangs—One was called ghost
1 2 3	Cowboys
1 2 3	Hunters
1 2	Man who wanted to save the mustangs

Outline of Story

1 3	1.	Mustangs were plentiful in the West in the 1800s.
1 2	2.	Mustangs were trapped, broken for saddle horses, and used by farmers.
1 2 3	3.	Some mustangs were too swift to be caught: The ghost.
1 2 3	4.	Some mustangs when trapped jumped into the mud and suffocated.
2	5.	Ranchers dislike mustangs eating the grass they wanted for their cattle.
1 2 3	6.	Many mustangs were killed from planes for sport.
1 2 3	7.	Not many mustangs are left.
1	8.	Some friends of mustangs have succeeded in a few states in getting laws passed to protect the mustangs. In other states wild horses are still being killed.
1	9.	In stories of the West, cowboys say thunderstorms are the rumbling hooves of mustangs running across pastures in heaven.

Why did the author write this story for us to read?

Student 1: "The author wrote the story in order to tell us what is actually happening in the world to mustangs today. He wants us to help him get a law through all the states that will prohibit the killing of mustangs. If we don't they'll become extinct. Our next generation won't know anything about mustangs."

Student 2: "To make us think about horses. To think about stop killing them. They are starting to be extinct (sort of). He wants us to tell someone to do something about it."

Student 3: "So we would know about the mustangs. There aren't many left."

How did the story make you feel?

Student 1: "I felt sad about all of the mustangs who have been killed. There used to be millions, now there are only about 20,000."

Student 2: "Sort of in between. I like cattle better than horses and horses are eating too much of the grass and there isn't enough for the cattle and then they would die off."

Student 3: "Sad."

As you can see, even these three higher-ability students demonstrate differences in abilities to develop an outline sequentially and to form opinions based on facts.

Students 1 and 2 demonstrated the highest ability in retelling the story, formulating and supporting an opinion with facts, and level of oral vocabulary development. In contrast, student 3 missed several important details, missed the real purpose of the story, and did not substantiate his opinions with facts. These are all skills that can be improved with instruction. Many lower-ability students find this task difficult, and cannot perform the activity without numerous structured questions. A teacher may find that most students in the class will need instruction in these oral language skills.

Obtaining Language Samples from Reluctant Speakers

Puppets are an excellent motivational device for eliciting children's speech, and can be used in several different ways. First, a child who is reluctant to talk to an adult may speak without hesitation to a hand puppet. The hand puppet can carry on a conversation with the child in order to collect a language sample and other information about the child. Puppets of familiar characters are good for this evaluation. In pre-kindergarten language screening, we have used Winnie-the-Pooh puppets, because children respond well to Pooh Bear. I have had great success with first-grade children using a hand puppet of the elephant Dumbo. He has great big ears, which are ideal for listening to children's speech, and he has such a good memory that he can remember everything about the child. He is also a baby elephant, and needs to have many questions answered about the world around him!

A second way to use puppets for evaluation is to have the child manipulate the puppet and tell a story to his teacher or classmates. The teacher evaluates the story according to some of the same criteria used for storytelling or language experiences.

One of the easiest ways to elicit a language sample is to show the child an interesting picture or series of pictures in a wordless book, and encourage the child to tell you about the picture or to tell you a story suggested by the wordless book. Frame your introduction to the picture or book in such a way that you provide the child with a language activity rather than a set of factual questions, because responses to certain questions may require only one or two words and therefore will not provide an accurate impression of the child's language. For example, you could use any of the illustrations in Chris Van Allsburg's *The Mysteries of Harris Burdick* (1984). Choose an illustration and read the accompanying title and the first line of a proposed story. Then ask a child to tell you the story he believes could accompany the illustration, the title, and the sentence.

Evaluating Social Functions of Language

Educators are increasingly interested in developing children's ability to use language effectively during a wide range of functions. These functions range from satisfying basic needs and interacting on social levels, to acquiring knowledge and communicating information. Pinnell (1985) maintained that "teachers need ways of assessing language that will help them to monitor the child's growing ability to use language skillfully in the social milieu" (p. 60), to assess the language environment within the classroom, and to change the environment—if necessary, in order to encourage and to develop a greater variety of language functions.

REINFORCEMENT ACTIVITY

I used the following student profile when demonstrating the kinds of evaluative information that can be obtained from a dictated story. After you have looked at the profile, read the language experience story, "Our Hunting Trip." Then, using the student profile, evaluate the language ability of this student. Finally, collect language samples from a student using two or three different methods described in this section. Evaluate the oral language information that you gained from the samples.

Student Profile to Accompany a Language Experience Evaluation

Name _____   Date _____
Grade _____   Motivation _____
Average number of words per sentence _____

		Yes	Some- times	No
I.	Dictates complete sentences Example: _____	_____	_____	_____
2.	Uses good sequential order	_____	_____	_____
3.	Demonstrates rich vocabulary	_____	_____	_____
4.	Title indicates grasp of main idea	_____	_____	_____
5.	Child is confident during the dictation experience	_____	_____	_____
6.	Dictated story indicates a background of experiences	_____	_____	_____
7.	Sentences show a consistency in noun-verb agreement	_____	_____	_____
8.	Uses pronouns correctly	_____	_____	_____
9.	Coherent arrangement of words	_____	_____	_____
10.	Includes all words needed for under-standing	_____	_____	_____

The following language experience story was dictated by a second-grade boy. Read the story and list some of the information you now have about his language ability:

Our Hunting Trip

Yesterday my friend Brian and I went hunting with my brother's pellet gun. I got two mallard ducks. They were sitting down in our pond. My dog Laddie was with me. He brought the ducks out of the pond. Laddie is my friend, I can really count on him. It was a happy day and it was my eighth birthday.

Pinnell developed a framework for oral language observation based on Halliday's (1973) seven functions for language. These functions include:

1. Instrumental Language—The language used by young children to satisfy needs and desires and to make requests or by older children to persuade.
2. Regulatory Language—The language used to control behavior of others or to get others to do what we want.
3. Interactional Language—The language used to establish and define social relationships. The maintenance language used in group situations that include negotiation, encouragement, and expressions of friendship.
4. Personal Language—The language used to express feelings, opinions, and individuality.
5. Imaginative Language—The language used to create a world of one's own, and to express fantasy through drama, poetry, and stories.
6. Heuristic Language—The language used to explore the environment, to investigate, and to acquire knowledge and understanding; the language of inquiry.
7. Informative Language—The language used to communicate information, to report facts, to synthesize material, and to draw inferences and conclusions.

The observational form shown in table 3–2 was developed by Pinnell (1985). She recommended that the observational form be used for two purposes. In order to assess individual students, the teacher should record oral language samples during different types of formal and informal activities. These samples should be gathered periodically to note progress. Furthermore, for an assessment of the total language environment, the teacher should record class and small-group responses during different activities and at different times of the day.

Results from the observation may be used to analyze the types of language used by individual children as well as the entire class. Providing individual opportunities or developing changes in the total classroom environment may be necessary if it does not stimulate a range of oral language functions and encourage children to respond effectively.

Pinnell recommended that teachers stimulate instrumental language by teaching children how to state requests effectively, by giving help and direction to peers, and by analyzing advertising and propaganda techniques. Regulatory language is enhanced by allowing children to be in charge of groups and by teaching them appropriate regulatory language. Interactional language is improved by creating situations in which students work and plan together in small groups that encourage discussion and interaction. Personal language is increased by encouraging children to share personal thoughts and opinions and by listening to children's conversations during cafeteria and playground duty. Imaginative language is stimulated through creative drama, role-playing, stories that foster imagination, and art. Heuristic language is stimulated by structuring classroom experiences to arouse interest and curiosity, creating problems to solve, and developing projects that require inquiry. Informative language is developed by planning activities that require students to observe objectively and to summarize

TABLE 3–2
Functions of language observation form

Name: _____
(individual, small group, large group observed)

Time: _____
(time of day)

Setting: _____
(physical setting and what happened prior to observation)

Activity: _____
(activity, including topic/subject area)

Language Function	Examples
Instrumental	
Regulatory	
Interactional	
Personal	
Imaginative	
Heuristic	
Informative	

Note: Check each time a language function is heard and/or record examples.

SOURCE: From "Ways to Look at the Functions of Children's Language" by Gay Su Pinnell, 1985, p. 67. In *Observing the Language Learner*, A. Jaggar and M. T. Smith-Burke, editors. Reprinted with permission of Gay Su Pinnell and the International Reading Association.

and draw conclusions, to keep records and draw conclusions, to use questioning techniques, and to use reporting techniques that encourage discussion and feedback.

ENVIRONMENT

It is quite apparent that if oral language is going to develop in the classroom, the classroom must provide many opportunities for the children to participate in oral language. Just as you cannot become an accomplished tennis player without practice and feedback, neither can you become an effective oral communicator without instruction and the opportunity to try the developing skills. King (1985) emphasized that "it is the obligation of the school to extend the opportunities for children to use language for an ever increasing range of purposes—especially to use it to learn" (p. 37).

The Teacher and the Environment

We have already indicated that in order to develop oral language skills, the teacher should know about language acquisition. This is necessary because so much of the diagnosis connected with oral language development relies on informal techniques for eliciting and evaluating speech. Meers (1976) stressed that, in addition to having a knowledge about language acquisition, the teacher needs to build a warm and supportive relationship with the class. Both interaction and encouragement are necessary for oral language development. The teacher also needs to acquire background information about the children. What language competencies do the children already possess? What language skills must be emphasized to improve oral communication? What are the children's interests? Can these interests be used to motivate language?

Instructional Practices

The teacher must also be aware of the types of experiences that will help develop oral communication skills. Loban (1976) was critical of oral language instruction that merely stresses "talk and chatter." He maintained that oral language instruction should focus on thinking and organizing one's ideas. Other insights into the type of instruction that appear useful for developing oral language are provided by Holbrook (1983). She identified an oral language curriculum that responds to children's needs; builds on what they already know about language; provides opportunities for sharing with an audience; encourages oral language development through interaction with the rhythm, repetition, and familiarity of folktales read aloud; and develops oral language abilities through storytelling, questioning, describing, and thinking.

Several educators report that instructional environments stressing language experience activities also improve oral language ability. Hall (1978) stated, "The significance of oral language in language experience programs is immense." Research conducted by Wells (1975) also found that remedial fourth-grade students showed significant gains in oral language following language experience instruction. (The language experience approach is discussed on pages 549–556 of this text.)

Further suggestions about an instructional environment conducive to oral language development are provided by Edelsky (1978). She reviewed research from interaction studies between parents and their children, and noted some instructional implications from the way children learn language in their home environment. First, parents do not sequence or partition oral language (i.e., phonics is not taught one day and syntax the next). Similarly, the classroom should use more whole-language activities, including real projects that stress the use of oral language. Second, the parent interacts with the learner. In order to meet this requirement, the classroom should provide opportunities for the interaction of language users with the teacher, in pairs, in small groups, and in large-group situations. Third, parents use language for some nonlinguistic purposes—for instance, correcting ideas rather than form of language. The classroom, too, should focus on a child's ideas, asking for clarification, elaboration, and justification. The classroom should provide an atmosphere in which children may explore and struggle with ideas. To do so, children need to be involved in planning, decision-making, and problem-solving activities. Fourth, parents talk to children about things the child can see or is actually attending to. Language concepts are developed using concrete items. Thus, the teacher should notice what the child is focusing on and supply language labels for what the child is actually doing. For example, the teacher may orally label concepts when children are sorting, categorizing, and exploring collections of real objects. Finally, parents expect success and delight in progress. The teacher should also become genuinely excited about a child's progress in both language and thinking.

DEVELOPING ORAL COMMUNICATION SKILLS

This chapter considers the development of oral communication skills through such activities as conversations, storytelling accompanying wordless books, discussion including modeling, creative dramatics, pantomime, puppetry, choral reading, and Reader's Theatre. Instructional units that develop various aspects of oral communication have always been extremely popular with both my undergraduate and graduate students. One student titled her oral language unit "Life In The Himalayas—or how a group of intractable 4th- and 5th-graders can be made to clamber up the mountain, converse with the Sherpas, and tame the abominable snowman." You may be wondering how she approached such a formidable goal. She began by introducing the class to various communication modes such as discussion, mime, oral story development, and creative dramatics. This student not only enjoyed her teaching, she also helped the children learn a great deal about oral language.

Conversations

In the early primary grades and especially in kindergarten, a great deal of time is spent developing oral conversation skills. The kindergarten teacher must frequently spend a great deal of time developing a young child's confidence to the point that he feels at ease holding a conversation with either the teacher or fellow classmates.

Children need opportunities to communicate their ideas to each other and to realize that they do have information worth sharing with someone else. Conversation is a more informal activity than discussion, and allows children to move more freely from one topic to another. One method teachers use to stimulate oral conversation is show and tell.

Show and tell   This activity is almost a daily part of many kindergarten and first-grade classrooms. Teachers find that allowing children to talk about something familiar to them both stimulates conversation and builds confidence in speaking. Show and tell also provides a vital link between the home, where the child may feel very confident, and the school, where he may be shy and reserved. Show and tell also

Conversation is enhanced during play experiences at school.

provides listeners for the conversational experience. Children learn to ask appropriate questions and to respond to other children in the group.

During a show and tell experience, children talk about an activity or show an object that they have brought to school. (A circular or semicircular arrangement, of chairs or with children sitting on the floor, allows more interaction than having children sit in rows.) The children voluntarily take turns talking about their object or activity.

The teacher has a definite role during show and tell time. When show and tell is first initiated, the teacher may have to lead some of the conversation by asking the child appropriate questions that provide him with some structure for the conversation and cues to encourage him to elaborate. The teacher's questions also provide a model for appropriate questions that other children may wish to ask a child. During the show and tell experience, the teacher may also need to encourage other children to ask questions. The teacher acts as a model for attentive listening, and prevents other children from interrupting the speaker when necessary. Finally, the teacher offers an enthusiastic response to all the children.

As an example of the role of the teacher during show and tell, one effective kindergarten teacher asked children to bring one of their favorite toys to class. She asked the children to join her in the conversation circle, and the children began to tell about their toys individually. One little girl showed her toys and said, "This is my panda." At that point, the child could not think of anything else to say, but the teacher carefully prompted her with remarks such as, "Oh what a nice panda, we would like to know his name"; "Who gave you the panda?"; "Where does the panda sleep when he's at home?"; "What games does the panda play with you?"; and "Have you ever seen a real panda?"

The teacher may specialize show and tell and its resulting conversation by asking children to bring something they made, something they drew, an item that illustrates a season of the year, and so forth. One of my language arts students was having difficulty encouraging some of her kindergarten children to take part in show and tell until she asked them to bring a picture of their family and tell the rest of the class something special about the family. (The picture could be either a snapshot or a picture drawn by the child.) She was amazed when every child wanted to share. They held their pictures confidently and talked about something very precious to them. Another student teacher asked first-grade children to bring something small enough to fit in a paper sack which could be described. She asked the children not to tell anyone what was in their sacks. The children took turns describing the objects and allowing other children to guess what was in the sack. This activity also assisted in vocabulary development, as the children (frequently encouraged by the teacher) discovered the importance of descriptive words indicating size, shape, color, and use.

After the children are competent in conversation during show and tell, the teacher may divide the group into smaller conversation circles and allow children to lead the show and tell activity.

Telephone conversations Play telephones in the elementary classroom provide children with an opportunity to develop a specific type of conversation. With telephones, children learn about the importance of the quality of their speaking voices;

they learn to take turns during conversation, and how to respond to another speaker. Through role playing, they may learn how to use the telephone for reaching the fire or police departments. During role playing, one child assumes the role of a person who needs assistance and another child becomes the person at the desired location. Children discover the necessity for giving exact directions or details, and learn procedures to use in real emergencies. Other examples of role playing with telephone conversations include calling a grocery store to place an order, calling a friend to ask about his weekend trip, or calling grandmother to wish her happy birthday. I have also used role playing with telephone conversations as an activity for older students, who pretended to be calling prospective employers about a job. We used the activity as part of a functional literacy unit in which the students were also taught how to use the newspaper for job information.

Wordless Books

Books that encourage language growth and stimulate intellectual development are important materials in the elementary classroom. Wordless books—picture books in which the illustrations tell the story without words—provide excellent frameworks for oral storytelling. If the book has an easily identifiable plot, it develops understanding of and interpretation of sequential plot development. Detailed illustrations encourage observational capabilities and descriptive vocabularies. Wordless books that have easily identifiable characters and develop strong story lines are good for young children. Pat Hutchins's *Changes, Changes* (1971), Martha Alexander's *Out! Out! Out!* (1968), and Jan Ormerod's *Sunshine* (1981), have easily followed plots, familiar environments, and problem-solving situations. Mercer Mayer's series of humorous wordless books are especially appealing to primary children. The antics of a boy, a dog, and a frog stimulate oral interpretations in *A Boy, A Dog, and A Frog* (1967); *Frog, Where Are You?* (1969); *A Boy, A Dog, A Frog, and A Friend* (1971); and *Frog Goes to Dinner* (1974).

However, wordless books are not just for younger children. Mitsumasa Anno's rich details and intricate drawings in *Anno's Journey* (1978), *Anno's Italy* (1980), and *Anno's Britain* (1982) encourage older children to follow journeys through European countries and to identify landmarks, sculptures, paintings, and well-known literary characters.

Because wordless books differ so greatly, teachers may consider the following questions when choosing books to promote oral language development (Norton, 1987, p. 160):

1. Is there a sequentially organized plot that provides a framework for children who are just developing their own organizational skills?
2. Is the depth of detail appropriate for the children's age level? (Too much detail will overwhelm younger children, while not enough detail may bore older ones.)
3. Do the children have enough experiential background to understand and interpret the illustrations? Can they interpret the book or would adult interaction be necessary?
4. Is the size of the book appropriate for the purpose? (Larger books are necessary for group sharing.)
5. Is the subject one that will appeal to children?

Two wordless book lessons   The following lessons show how the same wordless book, in this case Emily Arnold McCully's *Picnic* (1984), may be used to enhance oral language development in early elementary and upper grades. *Picnic* has a sequentially developed plot in which a mouse family prepares for a picnic, proceeds to the picnic grounds, unknowingly loses a child on the way, plays games, discovers the child is missing, and searches for the child. The detailed illustrations flash back and forth between the actions of the picnickers and the actions of the lost mouse child.

A lesson that encourages younger children to observe illustrations, describe actions, expand vocabulary, and create sequential narrative includes the following steps and discussions:

1. Look carefully at the first double page. Who is the story about? Look at each of the mice. How would you describe the mice? (Encourage expansion of descriptions to include color, size, emotions, and relationships within the family.)

2. Look at the same pictures. What do you believe the mice are planning to do? Choose one of the characters and explain what you think the mouse is going to do. How will you know if you are right? What should we see on the next few pages?

3. All stories take place in some setting. What is the setting on this first page? Look carefully at the picture. Describe the woods and the house. What setting do you expect for the remainder of the story? Why?

4. Continue an in-depth discussion of each double-page illustration focusing on descriptions of actions, characters, and setting.

5. Retell the sequence of the story.

The same book may be used with older students to enhance the development of language skills and to build an understanding of specific literary skills. The following lessons highlight setting, characterization, and point of view:

1. Setting—Discuss the importance of setting for developing geographic location and for identifying when the story takes place—past, present, or future. Have the students look at the illustrations and create descriptions that provide an accurate geographical location, that set the mood for the story, and that set the time for the story. Have the students compare the settings in which the young mouse is lost and the settings in which the other mice are joyfully picnicking. Ask them to describe one setting so that we know that the setting could cause problems for the character (setting as antagonist) and to describe the other setting so that we know the setting could enhance a happy, carefree mood. Ask the students to consider the importance of the words they use in describing each setting. Help them experiment with the impact of various descriptive adjectives, harsh verbs, and figurative language such as similes, metaphors, and other comparisons.

2. Characterization—Discuss the importance of characterization for developing characters who seem lifelike and who develop throughout the story. Share with the students that authors allow us to understand characters by

describing appearance, feelings, actions, attitudes, strengths, and weaknesses. Have students select one of the characters in the book and describe that character so that we have a better understanding of the character. This activity can be expanded to teach or reinforce the ways that authors reveal character. For example, students can create conversations through dialogue. Conversations between mother and father mouse before and after they discover the missing mouse child or between the mouse and mother before and after the incident reveal family relationships and feelings. Students can tell about the characters through oral narrative. They can describe the thoughts of the characters. The inner thoughts of the lost mouse child reveal both his fears and his self-determination, while the thoughts of brothers and sisters and of mother and father reveal various attitudes toward the lost mouse. Finally, descriptions of actions reveal both the physical and emotional changes in the lost mouse or in the responses in the other mice toward the mouse.

3. Point of View—Discuss the importance of an author developing a story line from a specific point of view. Explore the impact of how a story might change if a different character is telling the story. Have the students choose a character in the story and tell the story from that viewpoint. Examples include, mother mouse, father mouse, the lost mouse, or any of the other children.

This format may be used with any wordless book that contains a story line and includes enough information to develop characterization, setting, and point of view. The activity can be extended to include the development of conflict and the exploration with various literary styles.

Discussion

The discussion skills included in this section have as their goal the development of higher-level thought processes; consequently, the time spent on these skills may be some of the most rewarding in the elementary classroom. According to Kean and Personke, many educators "feel it is in the give-and-take of open discussion that children learn how to express themselves clearly and convincingly, to share ideas, to appreciate others' opinions, and to cooperate in solving problems" (1976, p. 119).

What is the teacher's role in developing effective oral discussion? For oral discussion skills to stimulate cognitive growth, they must be carefully planned. Gordon (1985) emphasized the advantages of using teacher modeling to help students develop an understanding of the processes necessary to understand inferences. Fisher and Lyons (1974) identified two important teacher responsibilities for involving every child in the discussion process. The first responsibility is to develop a pattern of interchange, through questioning techniques and focusing discussion, which engages the children's minds and imaginations. In addition, the teacher manages the environment in such a way that the discussion group is small enough for children to feel comfortable and motivated to share their thoughts. In this section of the chapter we look at teacher modeling, questioning strategies and grouping techniques for discussion.

Discussions about literature help students develop inferences.

Teacher modeling   Teaching children how to become actively involved in and aware of their thought processing is receiving considerable interest. Researchers such as Roehler and Duffy (1984) and Gordon (1985) developed approaches that place the teacher in an active learning role with the students and that show students how the teacher approaches thought processing.

Gordon (1985) used a five-step structure to help students understand important cognitive processes. First, the teacher identifies the skill to be taught, analyzes requirements for effective reasoning within that skill, identifies text samples that require the skill, and identifies questions that stimulate students to use that skill or thought process. Second, the teacher completely models the process for students by reading the text, asking the question, answering the question, citing the evidence supporting the answer, and then exploring verbally the reasoning process used to acquire the answer. Third, the teacher involves the students in citing the evidence. During this third step the teacher reads the text, asks the question, and answers the question; the students cite the evidence for the answer (the interactive process is enhanced if students write brief answers to the various parts prior to the oral discussion); and the teacher leads a teacher-student discussion in which the reasoning process is explored together. Fourth, the teacher involves the students in answering the question. During this fourth step the teacher reads the text and asks the question; the students answer the question; the teacher cites the evidence, and the teacher leads a teacher-student discussion in which the reasoning process is explored together. Fifth, the students take over most of the process. The teacher asks the question, but the students answer the question, cite the evidence supporting the question, and explain the reasoning involved.

These steps may be used effectively with any content area. The following example is selected from literature. It demonstrates how the five modeling steps may be used with a total book to help students understand the inferences associated with characterization in Patricia MacLachlan's Newbery Award winner, *Sarah, Plain and Tall* (1985).

Step I

1. Teacher identifies skill: Inferring the characterization developed by an author of historical fiction.
2. Teacher analyzes requirements: Going beyond the information the author provides in the text; using clues from the text to hypothesize about a character's feelings, actions, relationships, beliefs, values, hopes, fears; and using background knowledge gained from other experiences and stored in memory.
3. Teacher identifies text examples from which characterization may be inferred: ch. 1, p. 5; ch. 2, pp. 11 and 15; ch. 3, pp. 17, 19, and 21; ch. 4, pp. 23, 25, and 27; ch. 5, pp. 28 and 32; ch. 6, p. 37; ch. 8, p. 49; ch. 9, pp. 52 and 54.

Step II

1. Teacher introduces the skill of inferring characterization; discusses how characterization is inferred; provides examples of inferring characterization; and discusses why inferring characterization is important to the student.
2. Teacher introduces text and begins reading chapter 1: Stop on page 5 after the author infers Anna's feelings about Caleb: "He was homely and plain. . . . But these were not the worst of him. Mama died the next morning. That was the worst thing about Caleb" (p. 5).
3. Teacher asks inference question: "What was Anna's attitude toward her brother Caleb at the time of his birth?"
4. Teacher answers the question: "Anna dislikes her brother a great deal. We might even say that she hates him."
5. Teacher cites evidence: "Anna thinks that Caleb is homely, plain, and horrid smelling. Anna refuses to tell anyone what she really thinks about Caleb. Anna associates Caleb with her mother's death."
6. Teacher explores his or her own reasoning process: "The words Anna uses, especially horrid, are often associated with things that we do not like. I know from my own experience that I do not tell someone something that is bad about them. I know from the descriptions of the happy home and the singing that Anna and her mother were very happy and that Anna loved her mother. When she says that her mother's death is the worst thing about Caleb, I believe that she blames him for the death."

Continue reading orally chapters 1 and 2. (After the first example, keep all students actively involved by writing brief notes that answer questions, cite evidence, and explore reasoning). Stop and model the processes needed to understand Anna's feelings when she thinks, "And then the days seemed long and dark like winter days, even though it wasn't winter. And Papa didn't sing" (p. 5); to understand Anna's needs as revealed in the letter on page 11; and to interpret the meaning of "Tell them I sing" on page 15.

Step III

1.   Teacher continues reading chapters 3 and 4, stopping at the appropriate places. Teacher first emphasizes "Caleb slipped his hand into mine as we stood on the porch, watching the road. He was afraid" (p. 17).

2.   Teacher asks question: "Why was Caleb afraid? What does this fear reveal about Caleb's character?"

3.   Teacher answers question: "Caleb was afraid because Sarah was coming. Caleb was afraid he might disappoint Sarah and she would not stay. Caleb was afraid of the unknown. Caleb, a young child, was unsure of himself. Caleb really longed for a mother."

4.   Students cite evidence: "The text tells that Caleb did his chores without talking, he was afraid of little things like if his face was clean or too clean. The text tells that Caleb stood for a long time holding his sister's hand and watching the road. The text tells that he asked many questions about what Anna thought about Sarah and he worried if Sarah would like him."

5.   Teacher-student exploration of reasoning: "Caleb longed for a mother during the first two chapters. His actions now show that he is afraid that Sarah will not like him or that she will not be nice. We know from our own experiences that we may be very quiet if we are afraid. We know from our own experiences that we worry about unknown people and experiences. We know from our own experiences that we go to someone like an older sister if we are frightened. Caleb's actions let us know that he is close to his sister, that he wants a mother, and that he is afraid that his dreams will not come true. Caleb shows the needs of a little boy."

Continue reading chapters 3 and 4 stopping to model the inference about Sarah's strong personality as reflected in her statement on page 19: "The cat will be good in the barn, said Papa. For mice. Sarah smiled, She will be good in the house too"; the inference about Anna's and Sarah's characterization in Anna's statement, "I wished we had a sea of our own" (p. 21); the characterization inferred by the flower picking incident on page 23; the characterization of Papa inferred by the hair cutting incident on page 25; and the inference about Sarah's conflicting emotions in the discussion of seals and singing on page 27.

Step IV

1.   Teacher continues reading chapters 5 and 6 stopping at appropriate places. Teacher first emphasizes the sheep incident by reading all of page 28.

2.   Teacher asks question: "What type of a person is Sarah? Do you believe that she will make a good mother for Caleb and for Anna? Why or why not?"

3.   Students answer questions: "Sarah is a warm, tender person who likes animals and wants to protect Caleb and Anna from a sad experience. Yes

she would make a good mother because she cares about people and about animals. She would also take good care of Caleb and Anna and bring laughter back into the house."

4. Teacher cites evidence: "The text describes Sarah as naming the sheep after her favorite aunts, as smiling when she looks at and talks to the sheep, as crying when a lamb dies; as protecting the lamb from buzzards, and as not letting Caleb and Anna near the dead lamb. Previous descriptions in the text describe Sarah as caring for the children and being interested in what they do and feel."

5. Teacher-student exploration of reasoning: "Sarah's actions during the lamb incident indicate that she cares about animals. We know from our own experiences that we name things we like after favorite people. We know from experience that if we like an animal we cry when it dies. We know from experience that we want a mother who protects us and who demonstrates love."

Continue reading chapters 5 and 6 emphasizing the implications of the haystack incident and the comparison to sand dunes on page 32 and the inferences gained from Anna's dream described on page 37.

Step V

1. Finish reading orally chapters 7, 8, and 9. Stop and ask the questions and allow the students to complete each of the inferencing tasks. Stop after reading the description of a squall and the families' reactions to the storm on page 49.

2. Teacher asks questions: "What do Sarah's actions and her statement, 'We have squalls in Maine too. Just like this. It will be all right, Jacob.' tell you about Sarah's character? What do you believe Sarah has decided to do?"

3. Students answer question: "Sarah is finally satisfied to live on the prairie. She sees that storms in Maine and storms on the prairie are alike. She has changed her opinion about the importance of living near the sea and appears to be happy on the prairie. She may be saying that her life in the new family may sometimes be rough like a storm, but in the end everything will be fine. Sarah is a strong person who can live with hardships. I believe that Sarah has decided to stay on the prairie."

4. Students cite evidence: "The text tells us that she watched the storm for a long time and that she compared what she saw on the prairie to what she had experienced in Maine. This time she touches Papa's shoulder and tells the family that everything is going to be all right. It sounds as if she has finally made up her mind and she will stay on the prairie even though it may sometimes be rough."

5. Teacher-student discussions of reasoning: "We know from other chapters that Sarah was always making comparisons between Maine and the prairie. Some of the comparisons showed a desire to return to Maine. This time she touches the family members and says that everything will be all right. We

know from our own experiences that we may be silent for a long time when we are making important decisions. Incidents in the book show that Sarah is a strong person who is willing to defend her beliefs and the people and things she loves. Sarah now seems to love the family more than she does her former home."

Finish the book, stopping to discuss the inference reflected in the reactions of the children when Sarah rides to town alone on page 52 and the inference drawn from Anna's comparison of Sarah's driving away in the wagon by herself to her mother leaving in a pine box (p. 54).

REINFORCEMENT ACTIVITY

Choose a skill that you can develop through modeling. Choose a book in which students must apply that skill. Identify the requirements for the skill, develop an example showing that skill, identify why it is important to students, identify the text samples, develop the questions, consider the answers, cite the evidence that supports the answers, and think through your own reasoning process used to acquire the answers. Present your modeling lesson to your language arts class. The following lesson plan may be used for this activity:

Modeling for_____

Requirements for understanding:
Example showing this skill:
Why is _____important for students to understand and to use:

Text Examples:	Question:	Answer:	Evidence Cited:	Reasoning Process:

Questioning strategies   Researchers who investigate teacher questions indicate that many of the questions asked of students are at the simplest, literal level of comprehension, and thus make the least demand on reasoning. Although there is certainly nothing wrong with asking literal questions, which demand a type of comprehension that is prerequisite for higher levels of learning, this kind of questioning should not dominate education. Yet when Guszak (1967) analyzed teachers' questions, he found that 70 percent of the questions could be classified as recognition (requiring the student to locate or identify explicit statements) or recall (demanding that the student produce explicit statements from memory), while only 15 percent of the questions asked children to evaluate what they read or heard. Guszak also found that teachers asked high-ability children more questions calling for evaluation, explanation, or conjecture, and asked lower-ability students more recall and recognition questions. It would appear that children who have the greatest need for instruction in reasoning are not receiving that instruction.

Research by Taba (1964) provides a framework for asking questions that will lead to higher-reasoning abilities. In fact, Taba maintained that a teacher's way of asking questions is by far the single most influential teaching art. She believed the teacher's questions define the mental operations that students can perform and determine which points they can explore and which modes of thought they learn. According to Taba, the steps in the cognitive operations are hierarchical; some operations represent a lower level of abstraction, and the mastery of a higher-level task requires the ability to perform tasks at the lower level.

Taba studied the questioning-response pattern of elementary-school social studies classes in the United States. She investigated whether the teacher questions and student responses focused the thought, extended the thought on the same cognitive level, or lifted the thought to a higher level. Taba concluded that only when students had first been involved at the lower levels of cognition could the discussion be lifted and continued at a higher level. The most important conclusion for teachers was that questions must be sequenced from less to more abstract in order to get students to operate at higher thought levels. She also found that she could instruct teachers in this sequential development.

This sequential development had to progress from data gathering (the lowest level), to data processing, and, finally, to abstraction (the highest level). Tables 3–3 and 3–4 present the sequencing in grouping and processing data, in interpreting data and making inferences, and in applying previous knowledge to new situations. Each table includes examples of the steps as they might be used to develop thought processes with elementary children.

As seen in table 3–3, the teacher's questions and class discussion have allowed the students time to explore the data thoroughly before they are asked to provide abstract labels. The activity may be as complicated as the ability level of the children and subject matter. A kindergarten teacher may lead the children through an activity in which they gather data, process, and label objects according to shape, or they may process data about clothing, such as warm-weather clothing, cold-weather clothing, rainy-weather clothing, and so forth. A science class could progress through a similar sequence in order to categorize machines, plant families, or animal families.

In table 3–4, the teacher encourages development of the ability to interpret data and to make inferences and generalizations. Students begin by assembling the data—the specifics on which a generalization will be based. They handle specific items cognitively before they develop the relationships and formulate generalizations. This situation is the opposite of a learning approach in which the teacher tells the class the generalization and has the students apply the generalization. Table 3–4 follows the sequence that might be used to teach the generalization that when the letter *c* is followed by either *e, i,* or *y,* the *c* usually has the soft, or *s* sound.

The sequence of development described in table 3–4 is extremely useful in all subject areas. Scientific principles may be developed through experimentation and

TABLE 3–3

Grouping and processing data

Ultimate Goal: Labeling Categories of Transportation	
Levels of Abstraction	Example
1. Data gathering—differentiating the specific properties of things, listing specific examples	1. Listing, gathering, and describing various ways of travel. Look at concrete objects and discuss their properties (e.g., cars, trucks, airplanes, sailboats, helicopters, submarines, etc.). Teacher questions help children discover properties of each item.
2. Data processing—grouping related items	2. Children group items that are related. Teacher questions and discussion emphasize how these items are alike and how they are different. Children may discover there are different ways to relate items, such as by color, shape, size, or use, and that sometimes one type of arrangement is more useful than another.
3. Abstraction—labeling each category that has been distinguished	3. Teacher questioning and discussion lead to labeling: car, bus, truck—transportation that moves on land; sailboat, submarine—transportation that travels in the water; airplane, helicopter—transportation that travels in the air.

other data gathering before a final generalization is formulated. The meanings of prefixes, suffixes, and vocabulary words can be approached in this way.

The sequence in table 3–5 illustrates the steps used to help students apply previous knowledge to new situations (deduction). The example would be an appropriate activity to use with children after they have read a literature selection. The premise is to have children apply knowledge they have previously developed about the characteristics of a good plot and story (accomplished through inductive-thinking activities) to a new story, to decide whether the story can be judged as excellent literature, and to have them defend their decision.

This sequential questioning and discussion approach helps students formulate opinions and value judgments. For example, one social studies teacher used this technique to have children reach an opinion about the question, "Should the government allow industrial expansion and other types of development in the Everglades? Why or Why not?" The teacher first designed an activity in which the class identified the data required to reach an intelligent judgment. (This is an open-ended question, so there may be no "correct" answer.) The class listed, with the teacher's assistance, the questions they would have to answer before making a decision. For example: "What animals, plants, and people now live in the Ever-

TABLE 3–4

Intepreting data and making inferences

Ultimate Goal: Stating the Ce, Ci, and Cy Generalization	
Levels of Abstraction	Example
1. Data gathering—assembling, describing, and summarizing the data. What do we need to know before we can make a decision?	1. Teacher presents sentences such as the following: "The *ceilings* in the house are eight feet high." "*Cider* is a drink made from apples." "A *cyclone* is a bad windstorm." Teacher helps students read the sentences. Teacher asks for other words that have a beginning sound like ceiling, cider, and cyclone. The class provides other examples and they are placed under appropriate columns. ce ci cy s cement cinder cycle seam cellar city cypress (words that do not fit)
2. Data processing—relating aspects within the data. How is one item like another item? Is there any item that is different? Why does _____ happen?	2. The teacher asks students to look carefully at each word in the row and decide if there is anything similar about all the words in the first row, etc. The teacher asks students to listen carefully as they say each word. Is there anything similar about their sounds?
3. Abstracting—forming generalizations and inferences	3. Students put together, in their words, a generalization incorporating the principles that have been discovered. What generalization could we make about these letters? "When *c* has an *e, i,* or *y* after it, the *c* usually has the sound of *s.*"

glades?" "Are any of the inhabitants endangered species?" "What conditions do the inhabitants need for survival?" "How would different kinds of development affect the Everglades (i.e., housing, airports, plastics industry, nuclear power plant, etc.)?" "How necessary is each kind of development to the economy?" The class investigated these questions, brought all of their data together, and discussed their findings.

Because students were expressing differences of opinion regarding the effects of development, the teacher's next planned activity was to allow the class to debate the issue. Some students supported ecology; some supported development. The class formed several teams for debate, so everyone had an opportunity to participate. Students who were not debating at a particular time acted as audience for the others, and were responsible for comparing different points of view and how well the views were substantiated. The audience was also encouraged to ask questions. Discussion

TABLE 3–5
Applying previous knowledge to new situations—deduction

Ultimate Goal: Making a Decision Whether a New Literature Selection Should be Judged Outstanding Literature	
Levels of Abstraction	Example
1. Data Gathering—assembling related information and establishing the conditions under which students can make predictions. What information do we need? What are the characteristics of the new example? What information do we already have that relates to the new situation?	1. Students gather information and examples about plot and character development, development of theme. author's use of language, and development of an appropriate and realistic setting. The teacher's questions and discussion expand the base of critical data before the students are asked to make a judgment.
2. Data Processing—relating the new situation to the previously stated generalization. How is the new situation like the one in the original generalization? How is it different? Does the generalization still apply?	2. Students compare their information about the story, plot, characterization, theme, language use. and setting with the set of criteria previously developed for evaluating literature. Do the criteria apply to the new story? If they do, why are they still applicable; if they are not, why are our criteria inappropriate?
3. Abstracting—applying the generalization. Supporting the prediction.	3. The children make a final decision as to whether they consider the new story an outstanding literature selection. They defend their decision with explanations.

was lively because each child had previously investigated and discussed the pertinent data. When the final question was asked, students were able to provide answers demonstrating higher levels of thinking.

It is also important in the sequential questioning process to provide students with opportunities to ask appropriate questions. According to Hennings (1975), "Only when the students assume the role of questioner is the teacher certain that the students can carry on the cognitive operations independently." The teacher in the Everglades example allowed the children to ask many appropriate questions. Activities can also be planned with student questioning as the major goal. In this activity, the teacher presents the group with the ultimate question, and the students, individually or in groups, list pertinent questions. Obviously, some students will require considerable assistance with this activity, whereas others will be able to work independently.

Discussion groupings   The second responsibility of the teacher in effective oral discussion is to manage the environment so that discussion can take place. Some discussion activities may take place in a total classroom setting, whereas others may be

*Choose one of these sequential development areas: grouping and processing data; inter-
preting data and making inferences; or applying previous knowledge to new situations.
You may use any part of the curriculum (science, social studies, literature, etc.) from
which to design sequential activities that progress from data gathering to data processing
to abstraction. Outline the activities you will use and examples of questions you might
use at each level. If you are a student teacher or classroom teacher, use your activity
with children and tape the activity for later evaluation. Listen to the tape and evaluate the
effectiveness of your activities, your questions, and the children's interaction. Were the
students able to progress from data gathering to processing to abstraction? Did any steps
need to be improved? If you are not teaching children at this time, share the activities
you have developed with your language arts class, and discuss the probable effectiveness
of each activity.*

more effective in smaller groups. Several discussion techniques may be useful for
stimulating group interaction. During a brainstorming activity, the children rapidly
present as many ideas on a subject as they can think of; all ideas are accepted
uncritically during the brainstorming experience. Students may also add to the ideas
of others. The teacher or a class secretary writes down all the ideas as they are
presented. The chalkboard or overhead projector works well for this, because the class
will need a copy of the ideas generated during the brainstorming. Brainstorming
stimulates creative thought, and can be the first step in a more detailed problem-
solving or creative-writing activity—for example, thinking of all of the ways a class
might raise money for a class trip, or listing all words that can be used in place of
"said." The latter activity resulted in a list of 85 words, and afforded an introduction
to the use of precise words to improve written composition.

A discussion technique similar to brainstorming requires dividing the class into
smaller groups, called "buzz groups." During the buzz session, students have a limited
time in which to generate a number of ideas or to solve a problem. You may wish to
use teacher-directed brainstorming activities before using the smaller-group activity so
students will understand what they are to do in the buzz grouping. I have often used
buzz groups to stimulate ideas among college classes. For example, one class was
assigned to think of as many ideas as they could for characters, situations, and so
forth, that could be used for pantomime. Over 300 ideas were generated in a
ten-minute period. An elementary class thought of all the ways you can prevent your
dog from going under a fence during five-minute buzz sessions. Another group used
buzz sessions to solve such problems as, "You have a poisonous spider in the
terrarium and you want to get rid of the spider without killing the plants," or "Your
mother put the paper boy's money on a high shelf in the kitchen. The paper boy came
while she was out and you could not reach the money." The children were given a few

minutes to think of as many solutions as they could and to talk about the pros and cons of each solution.

Two other discussion formats are effective in the classroom. The round-table discussion includes a moderator and three to eight participants. This group deals with a problem or shares ideas informally. The moderator, who may also be a student, guides the group and assists it in summarizing conclusions. A round-table discussion may also have an audience, because the class can be divided into several small groups, or one group can participate in the discussion while the rest of the class listens. Activities that develop thinking and questioning strategies are prerequisites to effective round-table discussions because skills in asking questions as well as in answering them are necessary for successful interaction in group discussions.

A more formal small-group discussion technique with which you may be familiar is the panel discussion. During a panel discussion, each member of the panel is responsible for a particular aspect of the subject. Each member must be knowledge-able enough in the subject to present information and answer questions raised by the audience. Many of the successful panel discussions we have used with upper-elementary children have also used role playing. For example, one science class pretended to be well-known scientists in the area of solar power, and presented a panel discussion on the advantages and disadvantages of using solar energy. Classroom discussions can be highly rewarding because they motivate children's interest, develop higher-reasoning abilities, and provide opportunities for oral language development.

Creative Dramatics

Probably no other area of language arts stimulates children's imaginations and language abilities as much as the various forms of drama. Children can explore real-life situations through role playing; they can learn control of the body and how to express emotions confidently through pantomime; they can learn how to interpret literature and to develop creative presentations through puppetry. Dramatic interaction provides a purpose for children to use their oral language and develop crucial thought processes. Petty and Jensen maintained that "creative dramatics fosters creativity in language, thinking, uses of the voice, and body movement" (1980, p. 308). Drama allows each child to develop the ability to use language independently and creatively.

Drama for younger children usually deals with imagined characters based on fantasy or on reality. Older children usually incorporate real-life situations into the content of drama activities. These real-life situations often have a sense of struggle and excitement. Children seem to want to imagine themselves experimenting with courage and loyalty and solving difficult problems.

According to Siks (1983), the drama curriculum should be a planned sequence of learning experiences, geared to the children's developmental level, in which they explore and apply the following concepts: (1) relaxation, concentration, and trust; (2) body movement; (3) use of the five senses; (4) imagination; (5) language, voice, and speech; and (6) characterization. The student progresses from these concepts related to the player, to playmaking and considerations of audience. Siks suggested that these concepts be developed in varying time periods. For example, because children seem to learn player skills in short, concentrated experiences, fifteen minutes per day may be devoted to exploring those concepts. However, when the children progress to playmaking, longer periods are needed to explore the player, playmaker, and audience concepts. These longer periods might consist of several thirty-minute periods a week, or a one-hour period weekly.

Pantomime

Several of my students have developed activities with children that utilize pantomime, the creative drama in which an actor plays his part with gestures and actions without the use of words. Pantomime develops a sense of movement and helps children interpret a situation. The following descriptions of pantomime activities are taken from several undergraduate- and graduate-student experiences:

1. The warm-up: Even though the warm-up takes only a few minutes, it is essential to the rest of the session. If the students are not relaxed and comfortable, they will not get the most they can from the period. The warm-up also sets the mood, so the children know what to expect and what structure they must work within. Exercises that stretch and relax muscles are good.

2. The glob: This exercise is a warm-up for the mind. It's an imagination game in which a pretend glob is passed around a circle. Each individual is asked to do something with the glob and to pass it on. This activity will stimulate the imagination when one child perceives the glob as hot, another cold, another as something sticky, and so forth. The activity also tells the instructor something about the abilities and inhibitions of the group.

3. Ball tossing: This exercise is similar in its level to the previous one. Students stand in a circle and throw a pretend ball around. They begin to watch their bodies as they throw, to see what movements the body makes when throwing a ball. They also begin working with focus, because they must watch the path of the pretend ball to know who will catch it. It's a good idea to change the pretend ball's size and weight. You may begin by pretending to throw a volleyball, then a beach ball, then a bowling ball or football.

4. Mirrors: This is a great exercise for cooperation and concentration. Two students face each other with their legs crossed. One is the leader, and the other is told to imitate his movements as if he were a mirror image. They begin with large, slow movements. Later, if they are adept, they may add facial expressions to mimic also.

5. Pantomime games: There are many games to play with pantomime. One of the most enjoyable is to guess the occupation or sport being panto-mimed. Each student performs a short pantomime activity and the others

guess what it is. The first to guess in any pantomime activity performs the next one. Perhaps the instructor's biggest responsibility is to establish a trusting and open atmosphere in which the students do not laugh at or make fun of other students.

6. People machine: This is a common dramatic game. The students construct a working, interacting, and spontaneous machine. One person begins, and the others attach themselves wherever they see a possible position. Machinelike sounds help them to get into the activity and realize these positions. Usually students will attempt an assembly-line machine, but it is best to encourage a conglomerate of interacting parts. For example, if they choose a tractor, four students can be wheels, one the steering wheel, and others the body.

7. Role playing: Role playing works best with a more mature and experienced group. It is fascinating to watch, and the students learn from it. Start by giving them a situation—something they can easily relate to. For example: The older daughter is setting the table, the younger daughter comes in and knocks a plate off the table and breaks it, and the mother blames the older. Explain the situation to each player, including the reasons for their actions. Tell the mother that she is tired from working all day; she thinks the younger daughter is too young to do much work; she feels the older daughter is often lazy and disobedient. Tell the older daughter that she feels she must do all the work while the younger daughter gets away with everything. Tell the younger daughter she is afraid of the mother because she is not in a good mood and does not want to take the blame.

8. Imitation game: This game is based on the game "telephone." The instructor chooses one child to perform a short pantomime with large motions. It should be something that all the students will be familiar with, such as getting ready for bed or making cookies. The other students can't watch when the teacher watches the student for the first time. Then the students come out one at a time; the first one watches, then must try to imitate the first student exactly. While the second student tries, the third one watches, and so forth. Each child gets only one chance to watch, but after he performs the pantomime, may watch all the rest. This is an exercise in observation, but it is also fun to observe the mutations.

9. Slow motion: Any game in slow motion is good. The students learn how they use their bodies, and must concentrate on what they are doing. They might all do a different sport simultaneously, or can form pairs and guess what each other is performing.

10. Pantomime actions and emotions: Ask children to pretend to be a giant striding, a hobbled prisoner, someone pulling a sled, someone drinking something unpleasant. An individual pantomime game may be played by writing an emotion on paper, giving each child a different emotion, and having them act out the emotion while the others guess. Include such emotions as anger, hunger, sadness, happiness, gleefulness, and so forth.

11. Visiting the beach: Students are asked to lie comfortably on their backs. They are told to clear their minds and concentrate on their breathing. The

instructor then tells them to react to what they hear, and takes them to the beach through her voice. The students slowly "go to the beach," seeing things, hearing things, smelling things; the sun gets hot, the sun goes behind clouds, it becomes hot again, the students want to go for a swim. Suddenly rain clouds gather and the teacher's voice brings them quickly back to the cars. The students will react to the teacher's voice if it is sincere and descriptive. You can evaluate their involvement by their movements. Does the breathing slow down in places? Does it quicken in others? Do the children squirm in the sand? Do they shade their eyes from the sun? Do they pretend to swim?

12. Sense memory: This exercise is only for an experienced and trusting group. It is a way to teach emotional expression, and must be handled with skill. The students are asked to relax on the floor, preferably not close to anyone else. They are told first to concentrate on their breathing, focusing their attention on themselves. Again the voice of the instructor leads them into their memories. The students are asked to recall a time when they felt extremely happy or maybe silly. This memory can be any emotion, but it is usually better to start with a happy feeling that is easy to remember. Later you might ask for anger, loneliness, frustration, and so forth, but they must first learn to deal with the emotions and how to rid themselves of undesirable feelings after the exercise is completed. Ask them questions about the experience, such as, Where were you? Who else was there? What were you doing? What did it look like? Were there any sounds? Guide them through the memory with your voice; afterwards, gather them into a group and talk about the experience. Don't ask them what the experience was, because it may be too personal to share; but if they wish to talk about the experience, they may.

13. Narration: Narrate or read a story aloud, giving children time to pantomime each new act and to be each new character. Have the children pantomime all the roles, including the inanimate objects. Folktales such as "The Three Little Pigs" and "The Three Billy Goats Gruff" are appropriate for younger children. Children in third and fourth grades enjoy pantomine scenes from "The Twelve Dancing Princesses" or the Jewish tale "Mazel and Shlimazel." Older children enjoy scenes from Greek and Norse mythology.

14. Object pantomime game: Write names of objects on separate pieces of paper. Have students choose a paper and portray that object, allowing the rest of the class to guess the object. You might include a light bulb, an alarm clock, the sun rising, the wind in a tree, and so forth.

Playmaking—acting out   Experiences that allow children to experiment with movement and to pantomime interpretations of characters and emotions provide a foundation for other dramatic activities, whether it be acting an original play or interpreting a literature selection read by or to the children. During the process of acting out a role, the children develop new insights and are able to identify with the role. If children wish to share their results with another group, more elaborate settings,

REINFORCEMENT
ACTIVITY

You will only feel comfortable with the kind of movement and pantomime activities discussed in this section if you actually try the activities. First, form a group of your peers and experiment with the activities. Take turns leading the activities. After you feel comfortable as part of this group, select several pantomime and movement activities and present them to a group of children. If you are a student teacher, or a classroom teacher working with upper-elementary or middle-school children, you may wish to try some of the exercises in Fran Averett Tanner's book, Basic Drama Projects *(Pocatello, Idaho: Clark Publishing Company, 1979).*

costumes, or masks may be used to define the characters. Young children respond well when they are allowed to take on a character by dressing in appropriate costumes.

According to Corcoran, in a discussion of creative dramatics with kindergarten children, "the lines for such a production may become somewhat finalized after evaluation, but they should be neither written nor memorized" (1976, p. 123). She suggested that prewritten dialogue limits possibilities for thinking through the reactions of the character, and generally inhibits the natural responses of the young child.

This is of course a concern if we are primarily interested in using creative dramatics to develop creative or divergent thinking. To meet these creative goals, Smith (1977) said creative dramatics must meet three criteria. It must (1) be open-ended; (2) develop divergent-thinking abilities; and (3) result in creative products. Both improvisation and dramatization may meet these criteria.

McIntyre (1974) separated improvisation from dramatization by stating that in improvisation, children develop dramatic experiences out of their own ideas, whereas in dramatization, they use the work and ideas of others to expand and develop their own ideas. Improvisation places upon the teacher a guiding rather than a directing responsibility. As an example of improvisation, one group of third-grade children were motivated to act out a play after watching the television production, "It's the Great Pumpkin, Charlie Brown," by Charles Schulz. The following day at school, the students were talking about what would happen if there really were a great pumpkin and if it had visited Linus in the pumpkin patch. The teacher allowed the children an opportunity to discuss what they felt would have happened in those circumstances. There was enough difference of opinion that several groups were formed to further the discussion. These groups met together, then presented their interpretations to the rest of the class.

Many ideas for dramatization come from story selections. McIntyre offered several criteria for selecting a story appropriate for dramatization.

1. The idea should have worth, and the story should be carefully written.
2. The story should involve conflict.

3. There should be action in the development of the plot, and it should be action that can be carried out satisfactorily.
4. The characters should seem real, whether they are human or animal.
5. The situations should call for interesting dialogue.

Numerous literature selections meet these criteria. For example, many of the folktales by the Brothers Grimm develop humorous conflict, easily identifiable characters and plots that encourage dialogue. John Steptoe's *Mufaro's Beautiful Daughters: An African Tale* (1987) develops conflict between the selfish, spoiled daughter and the kind, considerate daughter. Janet Stevens's expanded version of the Aesop fable *The Tortoise and the Hare* (1984) encourages humorous action and dialogue among the various characters who try to help the tortoise or hinder the hare. Arthur Yorinks's *Hey, Al* (1986) involves conflict between a life with struggle and a life of endless leisure. Yorinks's book encourages expanded dialogue as students consider the consequences of each life style.

Even after the story is selected the teacher still has a responsibility to give more guidance than she did in improvisation. In an earlier writing, Siks (1958) suggested five steps the teacher should follow in guiding children in story dramatization:

1. Motivate the children into a strong mood.
2. Present the story, poem, or idea from which the children are to create.
3. Guide children in planning.
4. Guide children in playing.
5. Guide children in evaluating their own work.

One second-grade teacher used the story "Stone Soup" as a literature selection to motivate dramatization. In order to accomplish the first step in dramatization, motivating the children into a strong mood, the teacher placed a large soup kettle and a stone in the front of the classroom. As the children entered the classroom, they asked many questions. The teacher replied with a motivating question: "If you were hungry, could you get a meal out of a stone?"

To meet the second guiding step, the teacher told the story of "Stone Soup" to the class. While telling the story, she used as props the soup kettle, the stone, water, and the numerous soup ingredients the villagers were finally coaxed into bringing out from hiding. The class showed interest during the storytelling, and a desire to become involved in the presentation.

After hearing the story, the children talked about their favorite characters and scenes in the story. The teacher allowed them to improvise the actions of their favorite characters, such as a farmer hiding his barley from the soldiers, a hungry soldier marching into town, or a sly soldier stirring a stone into water to make soup. The class then discussed which characters should be in the play. The story was expanded with additional villagers the class felt should be in the play. They talked about how the mayor of the town, a baker, or a soldier might talk and act. They discussed a sequence of scenes, and decided they would use the soup kettle, the stone, and soup ingredients for props. The only costuming they considered necessary were single items for each person—an apron for a woman peasant, a badge for the mayor, a cardboard rifle for the soldier, a hat for the baker.

The class next acted out the story. The teacher added suggestions by becoming a character in the story. This was necessary at times to bring the story back to the discussed sequence. The story was acted out several times so the children could play different characters.

After acting out the story, the children, guided by the teacher, discussed and evaluated what had happened during the presentation. The teacher guided questions so that the class talked first about what was good about the play; for example, "How did the villagers show that they did not trust the soldiers?" "What did the baker do so you knew he was trying to hide the food?" "How did you know which actor was the leader of the soldiers?" After several positive questions, the children suggested ways they could improve the play another time.

You may wish to investigate the playmaking process further by developing activities that allow the children to explore elements of the play such as plot, character concepts, theme, and language. For example, children can be led to discover that plot provides the framework for the play, that there is a beginning, in which the conflict is introduced, a middle in which there is a struggle, which also moves the action toward a climax, and an ending, which brings a resolution to the conflict. An excellent source for a thorough study of playmaking is chapter 11, "Playmaking," in Geraldine Siks's *Drama With Children* (New York: Harper & Row, 1983). Additional ideas for dramatization based on story texts are described by Kukla (1987) and Booth (1985). Both authors emphasized the role of story drama in expanding imaginations, stimulating feelings, enhancing language, and clarifying concepts.

The audience   As children move from player to playmaker and present dramas to others, they learn that the audience also has a purpose, which is met by listening to, responding to, and enjoying the presentation. Children need experiences being audience

REINFORCEMENT ACTIVITY

Choose a literature selection you consider appropriate for dramatization. Does the selection contain McIntyre's five criteria for dramatizing a literature selection? Describe how you would develop the drama according to these five steps: (1) motivating the children; (2) presenting the story; (3) guiding the planning; (4) guiding the playing; (5) guiding the evaluation.

If you are a student teacher or a classroom teacher, develop your drama activity with a group of children. If you are not teaching, share your drama activity with your language arts class. If you are not teaching, try to visit a class in which drama is part of the curriculum. Try to evaluate the teacher's use of Siks's five guiding steps.

as well as actors. Siks maintained that it is in the role of audience that the child learns the responsibility of both receiving and responding to peers with honest 'feedback' about an activity. In this interactive process, which is extremely affective in nature, the child gains respect and value for the efforts, imagination, and integrity of peers.

Teachers should provide experiences for children to be an audience in both small and larger groups. Teachers can ask the audience to provide responses with questions such as "What did you see that made you believe . . . ?" "How did you feel when . . . ?" "How did you know the players were in . . . ?"

One of the best ways for children to understand the role of the audience is for them to see excellent drama. Trips can be arranged when appropriate children's theater comes to your city, or you can have drama groups or puppeteers give presentations at your school. In addition to being marvelous opportunities for positive audience response, excellent presentations are also motivating for child-produced drama.

Puppetry

What do you visualize when you hear "Punch and Judy" or "The Muppets"? Most of us associate these terms with a form of creative dramatics that has brought pleasure to both children and adults for many centuries. In Europe, puppet theaters are so elaborate that entire operas are performed by marionettes. The Japanese Bunraku puppets perform classical drama, and the puppeteers in Thailand perform in the temple courtyards. How many of you have observed children watching the Muppets on "Sesame Street" or the puppets in the Land of Make Believe on "Mr. Rogers's Neighborhood"? If you have watched children during a live puppet presentation, you know how responsive and attentive they become.

Young children need many experiences for giving their imaginations a chance to develop. Puppetry provides opportunities for a child to develop all the communication skills. When children present a play, they are developing oral language and drama skills; when they write their own plays, they are developing creative writing skills; when they retell literature stories, they are developing interpretive and comprehension skills, as well as other reading skills; when they design sets and puppets, they are learning about art and the theater; and when they become the audience, they are developing appreciative listening skills.

Puppets are used in prekindergarten and kindergarten programs for developing language skills. It is truly exciting to hear a young child who does not normally speak in the classroom respond through puppetry. Puppets are also used in remedial reading and language classes to encourage reading and communication skills with older children. Puppetry has been used effectively by many of my undergraduate and graduate students, and they are always enthusiastic about their results.

Although puppetry is a highly motivating and enjoyable teaching and learning technique, some authorities in children's drama are quite critical of the way puppetry is often used in the classroom. Discussing puppetry, Siks (1983) stressed that "Many projects initiated in the classroom are less than successful because they are never really brought to a satisfactory close. They stop with the making of the puppet. This is like signing a death warrant for a child before it is born. It denies the puppet the right to 'life' and the child the right to a rich and rewarding experience. . . . A puppet is an extension of a human being who seeks another way to communicate. Puppet and child are one and the same. Serve one well and you serve both."

Introducing puppetry to children   Children often respond to a puppet as if it were a real person or animal. You can introduce the children to puppetry by having them see a puppet show, or by presenting several "live" puppets to them. After children see the various types of puppets, allow them to experiment with the puppets before they develop their own first projects. Through experimentation, they will realize the possibilities of using hand puppets, rod puppets, shadow puppets, and marionettes. Also allow children to explore some of the history of puppetry. Even younger children will be able to acquire information from pictures.

An initial puppet project   After children have been introduced to puppetry and have experimented with puppets, they are usually anxious to begin a puppet project. The producer of the Valentinetti Puppeteers, Aurora Valentinetti (1977) made several suggestions for developing pupil projects in the elementary classroom. First, the beginning project should be a simple one that can be completed and given "life" within an hour. A feeling of early accomplishment is particularly desirable for children in the primary grades. Good beginning projects include paper-plate and paper-bag puppets.

Paper plates can be folded in the center, eyes can be partially cut so they stand up, and features can be added to the puppet. This simple paper-plate puppet can be manipulated so it can talk by folding and unfolding the two plate halves.

Another style can be made with unfolded paper plates. For example, an elephant can be made by cutting a round circle for the nose, gluing on cardboard ears, and adding eyes to the plate. If the puppeteer wears a grey stocking on his hand and puts it through the circle on the plate, the puppet can maneuver his trunk and communicate.

Folded Plate Puppet   **Whole Paper Plate Puppet**

Another simple hand puppet is the paper-bag puppet. For this puppet, the child draws features on a paper bag, allowing the mouth opening to fall on the fold of the bag. Features can be made more lifelike by adding hair made from strips of paper, yarn, or felt. A scarecrow from *The Wizard of Oz* might even have real straw for the hair. Pieces of cloth add authenticity to the clothing.

Paper Bag Puppets

These first projects can be brought to life by having the puppeteers improvise short dialogues or present rhymes or riddles using the puppet. After the children share their puppets, they can gather in small groups and improvise short scenes with the puppets.

Extending the puppet project The next step in the puppetry project is to select a play to be presented. Valentinetti recommended that children begin developing puppetry plays by working with literature, because it provides a foundation of plot and character upon which the play can be built. Learning principles of plot and character through the use of familiar stories will also make the transition to original work easier.

When helping children choose appropriate literature for puppetry, Briggs and Wagner's (1979) suggestions are helpful:

1. Stories should have briskly moving action that can be shown through the movements and voices of the puppet characters.
2. Selections should be interesting to children, well-liked, and easily understood.
3. Characters in the stories should present challenging, imaginative subjects that are not too difficult to construct.
4. The number of characters in a scene is determined by the size of the puppet stage; consequently the story should not have more characters than can be accommodated by the restrictions of the puppet theater.

In addition, it is helpful if the story has easily identifiable dialogues or speeches that children can convert into their own words. If sets are to be added to the puppet theater, the story should require only a few simple sets.

Many fairy and folk tales meet the above criteria, and offer enjoyable projects for kindergarten and early-primary students. Some favorites are "Three Billy Goats Gruff," "The Three Bears," "Henny Penny," "Jack and the Beanstalk," "Bremen Town Musicians," "The Three Little Kittens," "Rumpelstiltskin," "Hansel and Gretel," "The Three Little Pigs," and Mother Goose verses. Also useful for puppetry are more contemporary stories such as *Andy and the Lion* (1966), by James Daugherty; *Babar Stories* and *The Story of Babar* (1961) by Jean de Brunhoff; *Winnie the Pooh,* by A. A. Milne (1954); and *Where the Wild Things Are* (1963), by Maurice Sendak.

Stories suitable for the intermediate grades include folk and fairy tales as well as contemporary and historical fiction. Fairy tales such as "Aladdin and the Lamp" make excellent puppet productions. Children's classics such as Stevenson's *Treasure Island* (1911), Pyle's *The Merry Adventures of Robin Hood* (1883), or Kipling's *Just So Stories* (1987) can also be used for puppetry. Charlotte, Wilbur, and the rest of the delightful characters from *Charlotte's Web* (1952) come to life through puppetry. The tall tales of Paul Bunyan and Mike Fink provide plenty of action, and the revolutionary war period can be created through dramatization of *Johnny Tremain* (1943) by Esther Forbes.

To give each child an opportunity to become involved in puppetry, you will have to perform several plays, or have children take turns performing various roles. Children need experience as both actors and audience. You can read the literature selections to the children, or they can read them themselves, depending on the group's reading ability. Allow the group to discuss the story freely, to see how it might be interpreted for the puppet theater. This is a time for building imagination; as plans progress, they can be refined. Prepare an outline of plot, character, and scenes of the play, but keep in mind that improvisation and free interpretation are preferable to memorization.

Selecting and Making the Puppet for the Creative Drama

Many kinds of puppets are simple enough for even young children to make. Some puppets lend themselves better to a certain type of drama; as your students manipulate the various types of puppets, they also discover the limitations and possibilities of each type.

Hand puppets   Hand puppets fit over the hand and become part of the puppeteer. Hand puppets are useful for dialogue, because many of them can be manipulated to appear as if they are speaking. Hand puppets also require less time and skill to make and perform with than the marionette. We have already described two types of hand puppets, the paper-plate puppet and the sack puppet. Other hand puppets include box puppets, sock puppets, cylinder puppets, molded-head puppets, rod puppets, and humanette puppets.

Box puppets   The box puppet is made by cutting the sides of a rectangular box so that it will fold, or by taping two boxes together so that the openings of the boxes face

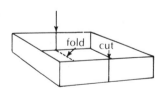

Box Puppet

the puppeteer. Paint a face on the boxes so that half the mouth is on the top box and the lower half of the mouth is on the lower box. Cut features from construction paper, material, or other scraps to give a three-dimensional effect. The puppeteer manipulates the puppet by placing his fingers in the top box and his thumb in the lower box; the puppet talks when the mouth of the box opens and closes.

Sock puppets Make a sock puppet by placing a stocking over your hand and manipulating the hand so the puppet moves and talks. The elephant's trunk is an example of this type of puppet. Make features for the sock puppet by adding buttons or felt for eyes, yarn for hair, and so forth.

Sock Puppet

Cylinder puppets Construct a puppet from a cardboard cylinder by adding facial features to the front of the cylinder. This puppet will be more elaborate and lifelike if you form a papier-mâché head around the top of the cylinder. Paint on features for the face and add paper, yarn, or felt hair. Make clothing from felt or other material. The puppet is manipulated by placing the fingers inside the cylinder.

Cylinder Puppet

Molded-head puppets You can make more lifelike puppets by molding the head around a clay form, a balloon, or a Styrofoam ball. Mold a clay ball or blow a balloon to the desired size. (If you use clay, grease it with vaseline.) Dip paper strips into wallpaper paste and place layers over the head mold. Form the desired features. When the head is dry, pop the balloon, or remove the clay by cutting the head in half. If the latter is done, papier-mâché the two halves together. Cut and sew a cloth costume that will go over the hand, and place the head on top of the costume. Manipulate the puppet by placing a finger inside the head and in each of the costume armholes.

Molded Head Puppet

Rod puppets   These puppets are easy to construct and manipulate. Draw pictures of animals, people, and so forth on stiff paper, or cut pictures from magazines, coloring books, or old story books, and tape or glue them to cardboard. The drawn or pasted character is then attached to the top part of a dowel, straw, or tongue depressor. The child maneuvers the puppet by grasping the lower end of the rod and moving the puppet across the stage so the audience sees only the puppet.

Rod Puppet

Humanette puppets   Humanette puppets are large puppets that do not need a puppet stage. For this puppet, draw a large cardboard shape of a person, animal, or plant. Cut openings in the cardboard figures so the puppeteer's face and arms can be seen and used. The children themselves become the puppets.

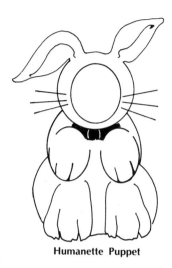

Humanette Puppet

Making the puppet stage   Puppet theaters can be simple or elaborate. You can make a stage by cutting an opening in a very large box, such as a refrigerator box, and decorating it. An instant stage may be designed by turning a rectangular table on its side, or by placing the puppeteers behind a large desk. Open doorways become puppet theaters by placing a length of material (with an appropriate opening cut out)

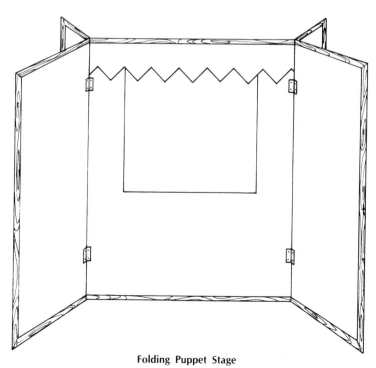

Folding Puppet Stage

across them. I have constructed my own portable puppet theater out of light-weight wood framing covered with burlap. This puppet theater folds for easy storage.

Practice the puppet drama   Each puppet scene should be rehearsed so the children can practice manipulating their puppets while speaking their parts. As already indicated, improvising and interpreting a story freely is better than memorizing a selection. Children will want to work with their puppets so they can develop character voices, and see how the puppet can be given "life." When the puppet drama is presented to an audience, have the puppet exchange some dialogue with the audience. This involves the audience in the play, and provides more opportunities for creative discussion. You can also add sound effects, lighting, and scenery to your puppet production. One of my students even had the children write commercials, which were presented by different puppets between the acts of a formal puppet drama when it was presented for an audience!

When children have learned some of the principles of puppet drama, have them create their own plays. The following example of a creative writing and puppet project was developed by a group of second-grade students. As you read about the project, you will see how puppetry motivated the children and encouraged creative writing, oral communication, reading, and art.

A Puppet Project: Helping Freddie Frog— Second-grade Level

The objectives of this project were:

1. To use a puppet to motivate interest and response.
2. To help children learn to work together—in composing their script, designing a cover for the puppet program, and in designing scenery.
3. To reinforce creative abilities—of singing, drawing, oral expression, and drama.
4. To help children put expression and emotion into their reading.
5. To give children the experience of performing before the class.
6. To develop personal concern in the children for the problem of pollution, by writing a letter to their class, making posters, and by putting on the puppet show for others.

The procedures were as follows: The puppet "Freddie Frog" was made from a folded paper plate. He was colored green, and given bulging eyes made from ping pong balls. The puppet was introduced to the group, and he also delivered the following letter to each child:

Dear _____ ,

My name is Freddie Frog and I hopped over here today to ask you to help me out. You see, I am in two storybooks that only have pictures in them—no words. Can you imagine that? A book that you can't read?

I know how well you can write stories, so I wondered if you could be the author of one of these stories. You could write sentences that tell what happens to me, and also draw pictures so that the other boys and girls in second grade could see what is happening.

First, I want you to look through the books very carefully. Then I will ask you questions to see if you will be able to write the story for me.

Well, what do you think? I hope you can help me. Imagine that! You can be the writer and illustrator of your very own book!

Thanks so much!

Your friend,
Freddie Frog

The children began working on their booklets after looking through two picture books by Mercer Mayer: *A Boy, A Dog and a Frog* (1967), and *Frog, Where Are You?* (1969) The children wrote their stories, copied them into their booklets, and illustrated each page.

After the booklets were finished, Freddie Frog gave each child another letter. Freddie again requested help from the children. Many of his friends and relatives were sick from drinking the polluted water in their pond. Could the children put on a puppet show in which the forest animals would clean up the pond and save their water friends? The following letter was presented by Freddie Frog:

Hi _____ ,

I hopped over here today to see if you can help me. You see, I am very sad today because my wife, Fanny Frog, and my little son, Frankie Frog, are very sick. The water around our log and lily pads is very polluted, so every time we take a drink of water, we get tummy aches.

And do you know what else? My friend Farley Fish can't even swim around in our stream without bumping into pop cans, bottles, and old tires. He is very thin now because he can't eat green plants without swallowing oil and acids in the water. Our other forest friends, Cubby Bear, Danny Dog, Pudgy Porcupine, and Randy Racoon can't even play tag in the water because it is full of garbage.

So do you see how much I need your help? Our pond and stream need to be cleaned up so that everyone will be well again and so we can have fun swimming in clear, clean water. I thought that if you put on a puppet show for the rest of the second-grade boys and girls, they might help stop some of the litterbugs who throw paper and pop bottles on the ground and in the water. Maybe after seeing your puppet show, the children would tell everyone to be helpers, and we could save the lives of the many little fish, turtles and frogs like me, who die because of the dirty water.

I know how well you can write stories and draw pictures, so I think you can put on a very good puppet show. You could make puppets of our forest friends and water friends, and act out a story about how they decide to go swimming and what happens when they see the stream full of trash. What will they do to solve the problem?

Thanks so much for helping Farley Fish, Fanny and Frankie Frog, and all our other forest friends!

Your friend,
Freddie Frog

After reading the letter and discussing the problem, the children listened to a story from the *Ranger Rick* magazine, "How Rick's Rangers Came to Be." This story gave them additional ideas for their puppet show. Children chose a specific character and made up dialogue. This conversation was recorded, and a script typed from the tape. The children made their own puppets out of paper plates. In addition, the children wrote letters about pollution to their classmates, and included some of the letters in the puppet-show program that was distributed to the audience. They worked together in designing a cover for the program, and in designing the background mural for the puppet show. They also made pollution posters, which they showed and explained to the audience after the puppet show. Music was incorporated into the program with the rendition of "A Little Green Frog." Several changes were made in the song so the frog could sing, "He swallowed some water and he said, "I'm sad, 'cause I'm a little sick frog swimming in the water, Glumph! Glumph! Glumph!" The original puppet play, "The Bubbling Pond," was presented to an appreciative audience, and the children enjoyed positive experiences in creative writing, drawing, singing, and creative dramatics.

Choral Speaking and Reading

Choral speaking is the interpretation of poetry or literature by two or more voices speaking as one. Choral speaking and reading allow children to respond to and enjoy rhymes and poetry in a new way. Children discover that speaking voices can be combined as effectively as singing voices in a choir. Young children who are not yet able to read can also enjoy this activity by reciting memorized rhymes and verses. Older children can select anything suitable that is within their reading ability.

Choral speaking and reading synchronize the three language elements of listening, reading, and speaking. This activity helps children develop interpretive skills, and heightens their appreciation of poetry and literature. Choral speaking can also improve children's speech. Teachers of remedial reading and learning disability classes find that choral reading provides an opportunity for repeated reading practice and develops a realization that reading can be fun. Because choral speaking is a group activity, it also builds positive group attitudes and the realization that some activities are better if they are performed as a cooperative effort. This realization can be beneficial to the shy child as well as the aggressive child.

According to McIntyre (1974), the teacher must understand the phases through which children should be guided before attempting this instructional method. McIntyre identified three of these phases; we discuss each phase and some examples of activities.

The first phase is an understanding of rhythm and tempo. Young children are not interested in words or meaning—it is the rhythm or flow of words that delights them. Consequently, rhythm should be explored by allowing children to clap or beat out the rhythm of verses. Children may suggest ways to express rhythm, and should experience fast and slow rhythms as well as happy and sad ones.

During this initial phase, plan activities in which children can sense the rhythm and tempo of music and poetry with their whole bodies. Have them react to different tempos played on the piano or on a record. Rhythm instruments, such as bongo drums or rhythm sticks, will help children appreciate rhythm. My language arts classes often experiment with rhythm and tempo by accompanying a selection with a bongo or stick beat. They realize how a slow tempo creates one meaning, a fast tempo another.

The second phase is understanding the color and quality of the voices available to the choral-speaking choir. Four terms describe the voice presentation of a selection: *inflection* refers to the rise and fall within a phrase; *pitch level* is the change between one phrase and another; *emphasis* is the verbal pointing of the most important word; and *intensity* indicates loudness and softness of the voices.

While you and the children learn about voice color and quality, experiment with these facets of the effective choral chorus. Children learn to be sensitive to inflection, pitch, emphasis, and intensity by listening to and experimenting with simple but exciting materials.

Furthermore, children must understand the different ways choral arrangements can be expressed. Choral speaking is not merely the unison presentation of a poem. In fact, if that were the sole way of making a choral presentation, it would be extremely dull. It is of more value if the children and teacher develop their own arrangements, rather than simply using those suggested by a text. To arrange choral presentations successfully, teacher and students need to know the various alternatives available. We describe five different types of presentation, with an example of each using a well-known Mother Goose rhyme.

Types of Choral Speaking Arrangements

1. The refrain arrangement—In this type of choral speaking, the teacher or a child reads or recites the body of a poem, and the rest of the class responds in unison with the refrain, or chorus. Three poems with refrains are Maurice Sendak's "Pierre: A Cautionary Tale," Stevenson's "The Wind," and Jack Prelutsky's "The Yak." Our example is based on the Mother Goose rhyme, "A Jolly Old Pig."

   Leader: A jolly old pig once lived in a sty,
   And three little piggies had she,
   And she waddled about saying
   Group: "Grumph! grumph! grumph!"
   Leader: While the little ones said
   Group: "Wee! wee!"
   Leader: And she waddled about saying
   Group: "Grumph! grumph! grumph!"
   Leader: While the little ones said
   Group: "Wee! wee!"

2. The line-a-child or line-a-group arrangement—In this arrangement, one child or a group of children read one line, another child or group reads the next line, and a third child or group reads the third line, and so forth. Poems that can be used for line arrangements are Clyde Watson's "One, One," Carl Sandburg's "Arithmetic," and Eleanor Farjeon's "Geography." Our example is "One, Two, Buckle My Shoe."

   Group A: One, two buckle my shoe;
   Group B: Three, four, shut the door;
   Group C: Five, six, pick up sticks;
   Group D: Seven, eight, lay them straight;
   Group E: Nine, ten, a good fat hen.

3. Antiphonal or dialogue arrangements—This choral speaking arrangement involves alternate speaking by two groups. Boys' voices may be balanced against girls' voices, high voices against low voices, and so forth. Poems in which one line asks a question and the next answers it work well for dialogue arrangements. Christina Rossetti's "Who Has Seen the Wind?" A. A. Milne's "Puppy and I," and Rose Pyleman's "Wishes," are dialogue poems. One you may want to try is the Mother Goose rhyme, "Pussy-Cat, Pussy-Cat."

   Group A: Pussy-cat, Pussy-cat, where have you been?
   Group B: I've been to London to visit the Queen.
   Group A: Pussy-cat, Pussy-cat, what did you there?
   Group B: I frightened a little mouse under the chair.

4. The cumulative arrangement—This arrangement, also called the crescendo arrangement, is used when the poem builds up to a climax. One group reads the first line, the first and second groups read the second line, and so forth, until the poem reaches its climax, at which time all the groups read together. Two examples of cumulative poems are Edward Lear's "The Owl and the Pussy-Cat" and James Tippett's "Trains." Our example is "There Was a Crooked Man."

   Group A: There was a crooked man, and
                        he went a crooked mile,
   Groups A,B: And he found a crooked sixpence
                        against a crooked stile;
   Groups A,B,C: He bought a crooked cat, which
                        caught a crooked mouse,
   Groups A,B,C,D: And they all lived together in a
                        little crooked house.

5. The unison arrangement—In the unison arrangement, an entire group or class presents a whole selection together. This kind of presentation can be difficult because it often produces a sing-song effect. Sandburg's "Fog" and Hugo's "Good night" are both suitable for unison arrangements. An example for you to try with a group is "A Big Black Cat," by second-grade authors.

   Whole Group: A big black cat walks down the street,
                     meow, meow, meow.
                  A big black cat with a long black tail,
                     meow, meow, meow.
                  He growls.
                  He spats.
                  He arches his back.
                  The big black cat walks down the street,
                     meow, meow, meow.

Encouraging older children to experiment with the effects of grouping their voices according to light, medium, and dark voices stimulates poetry and literature interpretations. Tanner (1979) recommended experimentation that allows children to discover that light voices can effectively interpret happy, whimsical, or delicate parts; medium voices can add to descriptive and narrative parts; while dark voices can interpret robust, tragic, and heavier material. The following example shows one way that Robert Louis Stevenson's "From a Railway Carriage" (1883) could be interpreted:

(Light)	*Faster than fairies, faster than witches,*
(Medium)	*Bridges and houses, hedges and ditches;*
(Dark)	*And charging along like troops in a battle,*
(Medium)	*All through the meadows the horses and cattle;*
	All of the sights of the hill and the plain
(Dark)	*Fly as thick as driving rain;*
(Light)	*and ever again, in the wink of an eye,*
	Painted stations whistle by.
(Medium)	*Here is a child who clambers and scrambles,*
(Dark)	*All by himself and gathering brambles;*
(Medium)	*Here is a tramp who stands and gazes;*
(Light)	*And here is the green for stringing the daisies!*
(Dark)	*Here is a cart runaway in the road*
	Lumping along with man and load;
(Light)	*And here is a mill and there is a river;*
(All)	*Each a glimpse and gone forever!*

General Guidelines for Choral Speaking and Reading

1. When selecting materials for children who cannot read, choose poems or rhymes that are simple enough to memorize easily.
2. Choose material that will interest children. Young children like nonsense and active words, so you might find it advantageous to begin with a humorous poem. Remember that choral speaking should be fun.
3. Especially for younger children, select poems or rhymes that use refrains. These are easy for nonreaders to memorize and will result in rapid participation from each member of the group.
4. Let the children help select and interpret the poetry. Allow them to experiment with the rhythm and tempo of the poem, improvise the scenes of the selection, and try different voice combinations and various choral arrangements before they decide on the best structural arrangement.
5. Allow children to listen to each other as they try different interpretations within groups.

Reader's Theatre

Reader's Theatre differs from an oral reading selection in that several readers take the parts of the characters in the story or play. Reader's Theatre is not a play with memorized lines, detailed actions, or elaborate stage sets. Instead, it is an oral interpretation of literature read in a dramatic style for an audience who imagines the

REINFORCEMENT ACTIVITY

We have discussed several different arrangements for choral speaking and reading. First, with a group of your peers, practice the various arrangements until you feel confident with the different approaches. Second, select several pieces of poetry, rhymes, or short literary selections that you think are appropriate for choral speaking. Third, instruct a group of children using choral speaking or present the choral speaking or reading activity to your peers.

setting and the action. Consequently, the objectives gained from Reader's Theatre include oral language, listening, reading, literature, and writing. The actors are motivated to read, to think, to enjoy literature, and to express themselves orally. If they develop or adapt their own materials, creative writing is enhanced. Working together fosters teamwork and pride in accomplishment. The audience benefits through improved listening skills, literary enjoyment, and motivation for reading literature. The chosen literary selections provide language models for both performers and audience.

Children's literature provides many excellent sources for Reader's Theatre. Folktales originally told through the oral tradition, picture books designed to be read orally to children, and poetry are good sources for younger elementary children. Realistic stories, plays, and narrative poems are all good for older students. Sloyer (1982) provided criteria for selecting materials. First, the story should be suspenseful with a well-designed plot. It should be an imaginative tale that presents characters in a series of events complicated by problems. The action should turn on a dramatic moment and the ending should be clear and satisfying. The characters should be compelling and understood quickly through the dialogue. The text should have sufficient dialogue or passages that can be changed into dialogue. Although narrative lines should be brief, they may introduce characters and setting or enhance plot development. Finally, repetitive words and phrases produce rhythm patterns enjoyed by children and encourage audience participation.

A few examples of literature appropriate for younger children include Sarah Hayes's *Bad Egg: The True Story of Humpty Dumpty* (1987), Margaret Mahy's *17 Kings and 42 Elephants* (1987), Mem Fox's *Hattie and the Fox* (1987), Dayal Kaur Khalsa's *I Want a Dog* (1987), and Eve Bunting's *Ghost's Hour, Spook's Hour* (1987). More experienced readers can interpret Bill Martin Jr. and John Archambault's *Knots on a Counting Rope* (1987), William Hooks's *Moss Gown* (1987), and folktales such as Grimms's "Rumpelstiltskin" and myths such as "Cupid and Psyche."

Before sharing an oral reading, readers must understand what is read. Readers interpret the materials according to their experience; consequently, they benefit from discussion and/or research prior to the oral presentation. Accelerated students usually enjoy the more in-depth study of a selection, and necessary research allows them to build a background of knowledge about the theater and literary selections. During an instructional sequence, the children discuss theater-related experiences. The teacher emphasizes similarities and differences between Reader's Theatre and other types of theater. Next the students, with the teacher's assistance, select and analyze literature for their presentation. Tanner developed nine points students should use when analyzing a Reader's Theatre selection prior to oral reading:

1. "What do any unfamiliar words mean?" Learn their meanings and correct pronunciation from a dictionary. One word may be the key to understanding a whole selection.
2. "Who is speaking?" Is it the author, a main character, a minor character, etc.?
3. "Who is listening?" Is the selection geared to a general audience, or to a specific listener?
4. "Where and when does the action take place?" Is it in modern times, some time in history, etc.? This question may demand research to ascertain a correct interpretation.
5. "What happens?" What action or plot occurs?
6. "When does the climax occur?" The climax is the most exciting part of the selection. You should identify the exact lines.
7. "What is the basic mood in the selection?" Is it joyfulness, fright, sadness, bitterness, sarcasm, etc.? How is the mood achieved?
8. "What is the theme?" The theme is the basic idea the author is suggesting: it runs underneath the action.
9. "How does this selection keep you in touch with life right now? What in your background gives you appreciation for this literature?"[3]

For oral reading, the student may choose a poem, short story, or portion of a story that both interests and excites him. Next, the student reads the selection silently and responds to its story, mood, and style. The selection should then be studied in order to answer the questions previously listed. Next comes preparation of a brief introduction to the selection, in which the student attracts the audience's attention and provides any background information that will help the audience understand what it is about to hear. The introduction should set a proper mood for the oral reading. Finally, the student should rehearse both the introduction and the selection. The selection should be rehearsed orally until the student is familiar with it. The reader should appear animated in face, voice, and body.

Reader's Theatre allows the audience to enjoy many types of literature, and allows the participants to interpret creatively such diverse scripts as poetry, short stories, radio scripts, essays, and so forth. Tanner (1979) identified two basic principles of Reader's Theatre that the teacher must understand. First, allow the children to explore different ways of presenting the literature. There is no one correct way to perform Reader's Theatre. Use guidelines, but let the material shape the method. Second, help them learn to stimulate the audience intellectually and emotionally by

---

[3]Quoted material from Fran Tanner, *Creative Communication, Projects in Acting, Speaking, Oral Reading* (Pocatello, Idaho: Clark Publishing Co., 1979). Reprinted by permission.

breathing life into the literature. They will be able to do this when they understand the literature and become excited about it.

Henning (1974) described procedures for adopting Reader's Theatre in the classroom. A simplified Reader's Theatre involves three steps. First, the selection is read silently. Each child does the first silent reading individually. The second reading is done orally in a group. Children may choose parts, or the teacher may designate which part each child will read. Children will learn to compromise, because they will not always be able to read their favorite parts. Children also learn to listen during this first oral reading. The third reading is also an oral reading, but this time the play is read and acted out with scripts in hand. The director, who may be the teacher, reads the title, the cast of characters, and descriptions of the setting or action. The players walk through the action as they read their parts; the teacher adds suggestions when a player has problems. There may be a fourth reading in front of an audience. Again, the title, cast of characters, setting, and descriptions of actions are read. There may also be a short musical introduction and simple scenery, but avoid having the play become too elaborate. After several such readings, many children wish to write original plays or adaptations of literature.

REINFORCEMENT ACTIVITY

Choose a selection you feel would be appropriate for an oral reading or a Reader's Theatre production. Answer the nine questions proposed by Tanner about the selection. Try the oral reading or Reader's Theatre activity with a group of children or form a group and do the same activity with a group of your peers.

MAINSTREAMING AND ORAL LANGUAGE DEVELOPMENT

Many of the oral language activities developed in this chapter are excellent for handicapped children who have been mainstreamed into the regular classroom. Table 3–6 identifies specific characteristics of learning disabled children that might interfere with oral language development and lists teaching techniques that are recommended by research studies and/or authorities who work with these children. Table 3–7 provides the same information and guidelines for mentally handicapped children.

SUMMARY

Speech is our chief means of human communication. Linguists emphasize the need for classroom instruction that encourages oral language development and the use of evaluation techniques based on actual language samples. Because most published tests do not assess oral language, we have reviewed informal ways to elicit oral

TABLE 3–6

Mainstreaming and oral language development of learning disabled children

Characteristics of Learning Disabled Children	Teaching Techniques for Language Arts Instruction
1. Problems in language and thought development characterized by poor or repetitive speech, or in comprehending and/or remembering spoken language (Bryan and Bryan, 1986; Lyon and Watson, 1981; Sapir and Wilson, 1978).	1. Teachers' use of praise, flexible verbal behavior, effective questioning strategies, and acceptance of students' feelings are related to academic growth (Smith, 1980). Try creative dramatics (Taylor, 1980). Use the language experience approach to focus children's attention on the purpose of language: to communicate (Hall and Ramig, 1978). Try the neurological impress method in which the teacher sits slightly behind the child and reads directly into his ear while the child either reads along with the teacher or trails slightly in pronunciation (Bader, 1980; Cook, Nolan, and Zanotti, 1980).
2. Demonstrate inability to follow directions (Hallahan and Kauffman, 1985).	2. Enhance the stimulus value of teaching materials; encourage children to rehearse academic tasks verbally; provide explicit, clear verbal directions; use concrete examples (Hallahan and Kauffman, 1985).
3. Demonstrate inability in understanding abstract words, forming abstractions, acquiring and using information or competencies essential to problem solving and reasoning (Fisher, 1980; Johnson and Morasky, 1980; Kavale, 1980; Sapir and Wilson, 1978).	3. Provide instruction in multiple meanings of words, relate unknown words to meaningful experiences, and emphasize vocabulary development within all content areas (Wallace and McLoughlin, 1988). Develop questioning strategies that encourage children to focus their attention on the purpose for an oral discussion or the problem to be resolved, to extend their information on a subject, and to clarify their knowledge by explaining or redefining previous information before they respond to higher-level questions that require inferential information, abstractions, and problem solving (Ruddell, 1978).
4. Demonstrate disability in comprehending information signaled by grammatical morphemes (Fay, Trupin, and Townes, 1981).	4. Emphasize the comprehension information carried by function words, word endings, and word stem changes (McClure, Kalk, and Keenan, 1980). Use pictures to help children build morphological generalizations (Lerner, 1985).
5. Demonstrate deficiencies in cognitive processes especially at higher cognitive levels of imagery, verbal processing, and/or concept formation (Cohen and Plaskon, 1980).	5. Try semantic mapping or webbing strategies to help children visualize concepts and relationships (Cohen and Plaskon, 1980; Johnson and Pearson, 1984; Myers, 1983; Norton, 1977). Increase involvement time in content areas

TABLE 3–6, *continued*

Characteristics of Learning Disabled Children	Teaching Techniques for Language Arts Instruction
	such as science laboratories and decrease textbook-oriented presentations to improve positive attitudes towards class, laboratory, and teachers (Milson, 1979). Develop cognitive processes through all oral language activities accompanying children's literature: observing, comparing, classifying, hypothesizing, organizing, summarizing, applying, and criticizing (Norton, 1987).
6. Demonstrate lack of linguistic sophistication characterized by sentences that are shorter and simpler than sentences developed by age-level peers (Fisher, 1980).	6. Encourage sentence expansion using descriptive words and phrases; transformations of kernel sentences; and meaningful oral language activities such as role playing, show and tell, puppetry, choral speaking, interviewing, and storytelling (Lerner, 1985).
7. Demonstrate difficulties understanding concepts related to time, quantity, and space (Fay, 1981; Kavale, 1982).	7. Provide children with visible, tangible cue systems that illustrate chronological order, quantity, and space relationships (Jordon, 1977).
8. Demonstrate deficiencies in memory and the development of memory strategies (Bryan and Bryan, 1986; Torgesen, 1979).	8. Help children select and organize materials in logical order. Use repetition and frequent reviews. Organize materials to increase chunking of material into already existing memory units. Help children relate what they are learning to what they already know. Encourage children to use or develop mnemonic strategies. Use flannel boards to help them remember cues for a story. Learn nursery rhymes, poems, and finger plays that encourage auditory memory (Lerner, 1985).

language—language experience stories; story retelling; unaided recall of stories; puppetry; picture discussions; and wordless books.

A positive classroom environment is critical to the development of oral language skills. In such an environment, the teacher must (1) have the knowledge necessary for eliciting oral language; (2) be able to evaluate a child's oral language and provide effective instruction; (3) provide an environment conducive to developing oral language, using projects such as puppetry; (4) provide opportunities for children to interact in pairs, small groups, and large groups; and (5) develop critical thinking and reasoning skills through instructional activities that allow children to progress from the concrete to the abstract.

Several instructional approaches were described for improving oral language skills. Conversation, especially through show and tell and telephone activities, was

TABLE 3–7
Mainstreaming and oral language development of mentally handicapped children

Characteristics of Mentally Handicapped Children	Teaching Techniques for Language Arts Instruction
1. Below average language ability when compared with age-level peer group (Gearheart, Weishahn, and Gearheart, 1988).	1. A curriculum emphasizing cognitive processes, problem-solving tactics, and motivation improved mental abilities of twelve- to fifteen-year-old disadvantaged and socially backward students (Rand, Tannenbaum, and Feuerstein, 1979). Alert students to directions that are given in clear, simple language and use vocabulary that is within the understanding of the children (Hart, 1981). Provide many language experiences tied to meaningful situations. Create opportunities for verbal expression, reuse words in numerous contexts, and relate new ideas to concrete rather than abstract ideas (Gearheart, Weishahn, and Gearheart, 1988).
2. Poor self-concept and lower self-esteem due to past failures in academic and social situations (Gearheart, Weishahn, and Gearheart, 1988).	2. Encourage children to discuss issues that are important to them (Hart, 1981). Praise students for small accomplishments and for work well done (Gearheart, Weishahn, and Gearheart, 1988).
3. Below average ability to generalize and see commonalities between similar situations as well as to generalize one set of conditions or rules to another similar situation (Gearheart, Weishahn, and Gearheart, 1988).	3. Integrate mechanical and conceptual skills into the lessons whenever possible and point out how one principle may apply to other academic or social situations (Gearheart, Weishahn, and Gearheart, 1988).

discussed, along with the teacher's role in promoting the child's self-confidence as a speaker.

Three teacher requirements for the development of discussion skills were presented. First, the development of modeling strategies was shown to accompany the oral discussion of characterization in a book. Second, the development of oral interchange through questioning technique was emphasized. Questioning and discussion that progressed from data gathering (the lowest level of abstraction) to data processing, and, finally, to abstraction was recommended. Examples illustrated how the teacher can use question and discussion techniques to help students group and process data, interpret data and make inferences, and apply previous knowledge to new situations. The third teacher requirement for effective classroom discussion is to manage the environment in such a way that children can participate in discussion groups. Brainstorming, buzz groupings, round table discussions, and panel discussions were discussed.

Under the category of creative dramatics, we learned how to develop the student as a player, with activities such as movement and pantomime. Next, we investigated the concepts of improvisation and dramatization. The teacher's role in story dramatization includes motivating children; presenting the story; and guiding children in planning, playing, and evaluation.

Puppetry was suggested as an excellent means for children to utilize all the communication skills, as long as the activity does not end with making the puppet, but goes on to include dramatic technique. To use the method effectively, the teacher must be familiar with the background of puppetry, know how to initiate puppetry, how to select and make puppets, and how to produce the drama.

Choral speaking and Reader's Theatre were recommended because they synthesize three language elements—listening, reading, and speaking. Types of choral speaking arrangements—refrain, line-a-child, antiphonal, cumulative, and unison— were described, with examples of each.

ADDITIONAL ORAL LANGUAGE ACTIVITIES

1. *Identify a group of wordless books appropriate for younger children and another group appropriate for older children. Evaluate the books according to the criteria on page 69. Share one of the books with a child or your language arts class.*

2. *Plan an oral language activity that encourages children to use brainstorming, buzz sessions, round-table discussion, or panel discussion.*

3. *Identify a story appropriate for Reader's Theatre. With a group of your peers or a group of children, adapt the selection for an oral presentation and present it to an audience.*

4. *Develop a file of literature selections that are appropriate for choral speaking arrangements; provide some recommendations for the types of choral arrangements.*

5. *Develop a file of literature selections that are appropriate for puppetry; provide some recommendations for puppetry characters and for types of puppetry construction.*

6. *Develop a dramatization activity that is based on "Narrative Theater" as described by Edmiston, Enciso, and King (1987).*

BIBLIOGRAPHY

Anastasiow, Nicholas, J. *Oral Language: Expression of Thought*. Newark, Del.: International Reading Association, 1979.

Bader, Lois A. *Reading Diagnosis and Remediation in Classroom and Clinic*. New York: Macmillan Co., 1980.

Bohning, Gery. "Show-and-Tell: Assessing Oral Language Abilities." *Reading Horizons* 22 (Fall 1981): 43–48.

Booth, David. "Imaginary Gardens With Real Toads: Reading and Drama in Education." *Theory Into Practice* 24 (1985): 193–98.

Briggs, Nancy E., and Wagner, Joseph A. *Children's Literature through Storytelling and Drama.* Dubuque, Iowa: Wm. C. Brown, 1979.

Bryan, Tanis H., and Bryan, James H. *Understanding Learning Disabilities.* Sherman Oaks, Calif.: Alfred Publishing Co., 1986.

Cohen, Sandra B., and Plaskon, Stephen P. *Language Arts for the Mildly Handicapped.* Columbus, Oh.: Merrill Publishing Co., 1980.

Cook, Jimmie E.; Nolan, Gregory A.; and Zanotti, Robert J. "Treating Auditory Perception Problems: The NIM Helps." *Academic Therapy* 15 (March 1980): 476–81.

Corcoran, Gertrude. *Language Experience for Nursery and Kindergarten Years.* Itasca, Ill.: F. E. Peacock Publishers, 1976.

Dixon, Carol N. "Language Experience Stories as a Diagnostic Tool." *Language Arts* 54 (May 1977): 501–5.

Early Childhood and Literacy Development Committee of the International Reading Association. "Joint Statement on Literacy Development and Pre-First Grade." *The Reading Teacher* 39 (April 1986): 819–21.

Edelsky, Carole. "Teaching Oral Language." *Language Arts* 55 (March 1978): 291–96.

Edmiston, Brian; Enciso, Pat; and King, Martha L. "Empowering Readers and Writers Through Drama: Narrative Theater." *Language Arts* 64 (February 1987): 219–28.

Fay, Gayle; Trupin, Eric; and Townes, Brenda D. "The Young Disabled Reader: Acquisition Strategies and Association Deficits." *Journal of Learning Disabilities* 14 (January 1981): 32–35.

Fisher, Carol J., and Lyons, P. A. "Oral Interaction: Involving Every Child in Discussion." *Elementary English* 51 (November 1974): 1100–1101.

Fisher, Dennis F. "Compensatory Training for Disabled Readers: Research to Practice." *Journal of Reading Disabilities* 18 (March 1980): 25–31.

Fox, Sharon. "Freeing Language in the Classroom." *Language Arts* 53 (September 1976): 612–16.

Gearheart, Bill R.; Weishahn, Mel; and Gearheart, Carol J. *The Exceptional Student in the Regular Classroom,* 4th ed. Columbus, Oh.: Merrill Publishing Co., 1988.

Goodman, Yetta M. "Kidwatching: Observing Children in the Classroom." In *Observing the Language Learner,* edited by Angela Jaggar and M. Trika Smith-Burke. Urbana, Ill.: National Council of Teachers of English, 1985, pp. 9–18.

Goodman, Yetta, and Burke, Carolyn. *Reading Miscue Inventory.* New York: Macmillan Co., 1971.

Gordon, Christine J. "Modeling Inference Awareness Across the Curriculum." *Journal of Reading* 28 (February 1985): 444–47.

Guszak, Frank. "Teacher Questioning and Reading." *The Reading Teacher* 21 (December 1967): 227–34.

Hall, MaryAnne. *The Language Experience Approach for Teaching Reading—A Research Perspective.* Newark, Del.: International Reading Association, 1978.

————, and Ramig, Christopher, J. *Linguistic Foundations for Reading.* Columbus, Oh.: Merrill Publishing Co., 1978.

Hallahan, Daniel P., and Kauffman, James M. *Introduction to Learning Disabilities: A Psycho-Behavioral Approach.* Englewood Cliffs, N.J.: Prentice-Hall, 1985.

Halliday, M. A. K. "The Functional Basis of Language." In *Class, Codes, and Control, Vol. 2, Applied Studies Toward a Sociology of Language,* edited by B. Bernstein. London and Boston: Routledge & Kegan Paul, 1973.

Hart, Verna. *Mainstreaming Children with Special Needs.* New York: Longman, 1981.

Henning, Kathleen. "Drama Reading, an On-Going Classroom Activity at the Elementary School Level." *Elementary English* 51 (January 1974): 48–51.

Hennings, Dorothy. *Mastering Classroom Communication.* Pacific Palisades, Calif.: Goodyear, 1975.

Holbrook, Hilary Taylor. "Oral Language: A Neglected Art?" *Language Arts* 60 (February 1983): 255–58.

Jaggar, Angela and Smith-Burke, M. Trika. *Observing the Language Learner.* Urbana, Ill.: National Council of Teachers of English, 1985.

Johnson, Dale D., and Pearson, P. David. *Teaching Reading Vocabulary,* 2d ed. New York: Holt, Rinehart & Winston, 1984.

Johnson, Stanley, and Morasky, Robert L. *Learning Disabilities,* 2d ed. Boston: Allyn & Bacon, 1980.

Jordon, Dale R. *Dyslexia in the Classroom.* Columbus, Oh.: Merrill Publishing Co., 1977.

Kavale, Kenneth A. "A Comparison of Learning Disabled and Normal Children on the Boehm Test of Basic Concepts." *Journal of Learning Disabilities* 15 (March 1982): 160–61.

_____. "The Reasoning Abilities of Normal and Learning Disabled Readers on Measures of Reading Comprehension." *Learning Disability Quarterly* 3 (Fall 1980): 34–45.

Kean, John M., and Personke, Carl. *The Language Arts Teaching and Learning in the Elementary School.* New York: St. Martin's Press, 1976.

King, Martha L. "Language and Language Learning for Child Watchers." In *Observing the Language Learner,* edited by Angela Jaggar and M. Trika Smith-Burke. Urbana, Ill.: National Council of Teachers of English, 1985, pp. 19–38.

Kukla, Kaile. "David Booth: Drama as a Way of Knowing." *Language Arts* 64 (January 1987): 73–78.

Lamb, Pose. *Linguistics in Proper Perspective,* 2d ed. Columbus, Oh.: Merrill Publishing Co., 1977.

Lerner, Janet W. *Children with Learning Disabilities,* 4th ed. Boston: Houghton Mifflin Co., 1985.

Loban, Walter. *Language Development: Kindergarten through Grade Twelve.* Urbana, Ill.: National Council of Teachers of English, 1976.

Lyon, Reid, and Watson, Bill. "Empirically Derived Subgroups of Learning Disabled Readers: Diagnostic Characteristics." *Journal of Learning Disabilities* 14 (May 1981): 256–61.

McClure, Judith; Kalk, Michael; and Keenan, Verne. "Use of Grammatical Morphemes by Beginning Readers." *Journal of Learning Disabilities* 13 (May 1980): 34–49.

McIntyre, Barbara M. *Creative Drama in the Elementary School.* Itasca, Ill.: F. E. Peacock Publishers, 1974.

Meers, Hilda J. *Helping Our Children Talk.* New York: Longman, 1976.

Milson, James L. "Evaluation of the Effect of Laboratory-Oriented Science Curriculum Materials on the Attitudes of Students with Reading Difficulties." *Science Education* 63 (January 1979): 9–14.

Myers, Miles. "Approaches to the Teaching of Composition." In *Theory and Practice in the Teaching of Composition: Processing, Distancing and Modeling,* edited by Miles Myers and James Gray. Urbana, Ill.: National Council of Teachers of English, 1983, pp. 3–43.

National Council of Teachers of English. "Forum: Essentials of English." *Language Arts* 60 (February 1983): 244–48.

Norton, Donna E. *Through the Eyes of a Child: An Introduction to Children's Literature,* 2d ed. Columbus, Oh.: Merrill Publishing Co., 1987.

_____. "A Web of Interest." *Language Arts* 54 (Nov./Dec. 1977): 928–33.

Olson, David R. "Perspectives: Children's Language and Language Teaching." *Language Arts* 60 (February 1983): 226–33.

Otto, Wayne, and Smith, Richard. *Corrective and Remedial Teaching.* Boston: Houghton Mifflin Co., 1980.

Pearson, P. David, and Johnson, Dale D. *"Teaching Reading Comprehension."* New York: Holt, Rinehart & Winston, 1978.

Petty, Walter T., and Jensen, Julie M. *Developing Children's Language.* Boston: Allyn & Bacon, 1980.

Pickert, Sarah M., and Chase, Martha L. "Story Retelling: An Informal Technique for Evaluating Children's Language." *Reading Teacher* 31 (February 1978): 528–29.

Pinnell, Gay Su. "Ways to Look at the Functions of Children's Language." In *Observing the Language Learner,* edited by Angela Jaggar and M. Trika Smith-Burke. Urbana, Ill.: National Council of Teachers of English, 1985, pp. 57–72.

Rand, Yáacou; Tannenbaum, Abraham, J.; and Feuerstein, Reuven. "Effects of Instrumental Enrichment on the Psychoeducational Development of Low-Functioning Adolescents." *Journal of Educational Psychology* 71 (December 1979): 751–63.

Roehler, Laura, and Duffy, Gerald G. "Direct Explanation of Comprehension Processes." In *Comprehension Instruction,* edited by Gerald G. Duffy, Laura R. Roehler, and Jana Mason. New York: Longman, 1984, pp. 265–80.

Ruddell, Robert B. "Developing Comprehension Abilities: Implications from Research for an Instructional Framework." In *What Research Has to Say about Reading Instruction,* edited by Samuels. Newark, Del.: International Reading Association, 1978, pp. 109–20.

Sapir, Selma, and Wilson, Bernice. *A Professional's Guide to Working with the Learning Disabled Child.* New York: Brunner/Mazel, 1978.

Siks, Geraldine. *Creative Dramatics: An Art for Children.* New York: Harper Brothers, 1958.

_____. *Drama with Children.* New York: Harper & Row, 1983.

Sloyer, Shirlee. *Reader's Theatre: Story Dramatization in the Classroom.* Urbana, Ill.: National Council of Teachers of English, 1982.

Smith, Christine C. "The Relationship between Teacher-Pupil Interaction and Progress of Pupils with Reading Disabilities." *Reading Improvement* 17 (Spring 1980): 53–65.

Smith, James A. *Classroom Organization for the Language Arts.* Itasca, Ill.: F. E. Peacock Publishers, 1977.

Taba, Hilda; Levine, Samuel, and Elzey, Freeman, F. *Thinking in Elementary School Children: Cooperative Research Project Number 1574.* Washington, D.C.: Research Program of the Office of Education, U.S. Department of Health, Education, and Welfare, 1964.

Tanner, Fran Averett. *Creative Communication, Projects in Acting, Speaking, Oral Reading.* Pocatello, Idaho: Clark Publishing Company, 1979.

Taylor, Gail Cohen. "Creative Dramatics for Handicapped Children." *Language Arts* 57 (January 1980): 92–97.

Torgesen, Joseph K. "Factors Related to Poor Performance on Memory Tasks in Reading Disabled Children." *Learning Disability Quarterly* 2 (Summer 1979): 17–23.

Valentinetti, Aurora. "Discovering the World of Puppets." In *Drama with Children,* edited by Geraldine Siks. New York: Harper & Row, 1977.

Wallace, Gerald, and McLoughlin, James A. *Learning Disabilities: Concepts and Characteristics,* 3d ed. Columbus, Oh.: Merrill Publishing Co., 1988.

Wells, Georgia L. "For Blacks, Chicanos, Puerto Ricans and Other Minorities, the Language Experience Approach to Learning." *Adolescence* 10 (Fall 1975): 409–18.

CHILDREN'S LITERATURE REFERENCES

Alexander, Martha. *Out! Out! Out!* New York: Dial Press, 1986.

Anno, Mitsumaso. *Anno's Britain.* New York: Philomel, 1982.

_____. *Anno's Italy.* Ontario, Canada: Collins, 1980.

_____. *Anno's Journey.* New York: Philomel, 1978.

Bunting, Eve. *Ghost's Hour, Spook's Hour.* Illustrated by Donald Clark. New York: Clarion, 1987.

Daugherty, James. *Andy and the Lion.* New York: Viking Press, 1938, 1966.

De Brunhoff, Jean. *The Story of Babar.* New York: Random House, 1933, 1961.

Forbes, Esther. *Johnny Tremain.* Boston: Houghton Mifflin Co., 1943.

Fox, Mem. *Hattie and the Fox.* Illustrated by Patricia Mullins. New York: Bradbury, 1987.

Hayes, Sarah. *Bad Egg: The True Story of Humpty Dumpty.* Illustrated by Charlotte Voake. Boston: Little, Brown & Co., 1987.

Hooks, William H. *Moss Gown.* Illustrated by Donald Carrick. New York: Clarion, 1987.

Hutchins, Pat. *Changes, Changes.* New York: Macmillan Co., 1971.

Khalsa, Dayal Kaur. *I Want a Dog.* New York: Potter, 1987.

Kipling, Rudyard. *Just So Stories*. New York: Macmillan Co., 1902; Viking, 1987.

MacLachlan, Patricia. *Sarah, Plain and Tall*. New York: Harper & Row, 1985.

Mahy, Margaret. *17 Kings and 42 Elephants*. Illustrated by Patricia MacCarthy. New York: Dial Press, 1987.

Martin, Bill Jr. and Archambault, John. *Knots on a Counting Rope*. Illustrated by Ted Rand. New York: Holt, Rinehart & Winston, 1987.

Mayer, Mercer. *A Boy, A Dog, and A Frog*. New York: Dial Press, 1967.

———. *A Boy, A Dog, A Frog, and A Friend*. New York: Dial Press, 1971.

———. *Frog Goes to Dinner*. New York: Dial Press, 1974.

———. *Frog, Where Are You?* New York: Dial Press, 1969.

McCully, Emily Arnold. *Picnic*. New York: Harper & Row, 1984.

Milne, A. A. *Winnie the Pooh*. New York: E. P. Dutton, 1926, 1954.

Ormerod, Jan. *Sunshine*. New York: Lothrop, Lee & Shepard, 1981.

Pyle, Howard. *The Merry Adventures of Robin Hood*. New York: Charles Scribner's Sons, 1946 (1883).

Sendak, Maurice. *Where the Wild Things Are*. New York: Harper & Row, 1963.

Steptoe, John. *Mufaro's Beautiful Daughters: An African Tale*. New York: Lothrop, Lee & Shepard, 1987.

Stevens, Janet. *The Tortoise and the Hare*. New York: Holiday, 1984.

Stevenson, Robert Lewis. *Treasure Island*. New York: Charles Scribner's Sons, 1911.

Van Allsburg, Chris. *The Mysteries of Harris Burdick*. Boston: Houghton Mifflin Co., 1984.

White, E. B. *Charlotte's Web*. New York: Harper & Row, 1952.

Yorinks, Arthur. *Hey, Al*. New York: Farrar, Straus & Giroux, 1986.

Chapter Four

After completing this chapter on listening, you will be able to:

1. Develop a definition of listening that includes all its aspects.
2. Evaluate students' ability to select and to use effective listening strategies.
3. Understand importance of auditory discrimination.
4. Understand the importance of attentive listening, demonstrate several ways to diagnose a child's ability to listen attentively, and plan a series of lessons for improving attentive listening.
5. Evaluate listening comprehension tests, diagnose a student's level of listening comprehension, and develop directed listening for a specific purpose.
6. Develop activities that increase appreciative listening.
7. Understand the importance of motivation and child involvement to listening improvement, and describe classroom conditions and teacher behaviors conducive to good listening.

Listening

C onsidering listening as a subject or a skill that can and should be taught is a fairly recent development in education. The increased interest, both in teaching listening and in research in the field is best illustrated by comparing the number of articles on the subject in two bibliographies published twenty years apart. The listening bibliography published in 1949 contained 30 articles, while 1,332 articles appeared in Duker's 1968 listening bibliography.

Authorities who maintain that listening is a skill deserving to be taught point out that listening is the major language art utilized by both children and adults. An early study by Rankin (1926) reported that children spent 45 percent of their out-of-school time in listening, compared to 30 percent in speaking. Today, in a society where children are constantly bombarded by the mass media, instruction in critical listening is vital in minimizing the development of conformity, misconceptions, prejudices, and stereotypes.

Recent research indicates that elementary children spend over 50 percent of their classroom time listening rather than speaking or reading, but authorities stress that children do not automatically learn the varied listening skills necessary for comprehensive listening. Horowitz and Samuels (1985) found that scores for oral reading and comprehension of both easier and harder texts favored higher-ability sixth-grade readers. In contrast, they found no differences in scores for listening comprehension between sixth graders who were good readers and sixth graders who were poor readers. This finding suggests two possible conclusions. First, all students, even those with higher reading abilities, need instruction in listening comprehension. Second, lower-ability readers gain considerable information through listening.

There is also increasing interest in teaching listening skills to children with reading disabilities. Children with severe reading problems may need to learn the content of the curriculum through audio-visual sources rather than from the printed text. Many teachers, however, have difficulty meeting the listening needs of special education children who are mainstreamed into their classes. Consequently, this chapter con-cludes with characteristics of such children that might interfere with listening devel-opment and teaching techniques that are appropriate for language arts instruction.

This increased concern about listening leads one to ask: Is listening taught in most elementary schools? A study conducted by Landry (1969) reported a serious lack of programs that develop listening skills. Landry also examined elementary textbooks and teacher's manuals; he concluded that there was little evidence to show that children are receiving instruction in listening. Devine (1978) reported that the situation had not changed.

The lack of listening materials in elementary textbooks places a greater responsibility for instructional development on the elementary classroom teacher. In developing an effective listening curriculum, the teacher must understand the components of listening. The teacher must also know how to evaluate a child's listening skills, and how to develop instructional tasks that will improve the child's listening ability.

WHAT IS LISTENING?

You will agree that listening is necessary both for classroom instruction and for effective communication among students and adults. As a teacher, you must also be able to define the listening act in order to develop diagnostic techniques or to provide listening instruction. Defining the listening act is not as easy as it might seem. An educator who wants a definition may refer to several sources; he will find various viewpoints depending on which field he consults. How is it defined by research studies? And, how is it defined by authorities in the field of education?

A review of research studies on listening shows that education authorities do not agree on a definition of listening. Studies may be found that compare listening and reading; experiment with verbal rate of delivery; evaluate attention and achievement; evaluate instruction through taped listening centers; compare auditory discrimination of minority and nonminority students; and evaluate the development of critical listening skills. This is only a partial list of the subjects grouped under the broader term of listening.

Following a review of listening research and theorizing over a fifty-year period, Devine (1978) credited Sara Lundsteen (1979) with the best working definition for listening. Her definition was the result of extensive research analysis. Lundsteen presented a structural definition of listening by asking "What are the parts of listening?" and an operational definition by asking "What does a listener do?" Tutolo (1977) recommended a similar approach for definition and instruction; he suggested a three-part definition that includes (1) acuity/hearing, (2) discrimination, and (3) comprehension—literal, interpretational, critical, and evaluational.

We utilize Tutolo's framework to discuss diagnosis and instruction. This approach does not imply a clear, sequential ordering of listening skills; the parts obviously overlap as one progresses from the hearing of sounds, to auditory perception, to comprehension, and, ultimately, to reaction on complex thinking levels. This chapter describes listening diagnosis and instruction in terms of hearing, auditory perception, attention and concentration, and auditory comprehension.

Hearing

Hearing without distortions is a prerequisite to listening. The term *auditory acuity* is usually applied to hearing, and the lack of this acuity is referred to as deafness.

Auditory Perception

Auditory perception may also be a prerequisite for effective listening. Auditory perception tests evaluate such factors as (1) whether children are able to distinguish one sound from another sound (auditory discrimination); (2) whether children are able to blend sounds together to form words (auditory blending); and (3) whether children are able to hear sounds, remember the sounds, and repeat them in the same sequence they heard them (auditory-sequential-memory). Children with auditory perception problems may have difficulty with reading or spelling approaches that rely heavily on sound and letter relationships.

Attention and Concentration

The listener must be able to focus on speech sounds and select appropriate cues in order to reconstruct the speaker's message. Some students are unable to follow verbal instructions adequately because they are easily distracted by competing noises. This is especially true with certain learning-disabled children.

Auditory Comprehension

Auditory comprehension refers to the listener's highly conscious seeking of meaning from a listening experience. Children may listen for factual or literal understanding, or they may reach the higher level of listening that Lundsteen referred to as "thinking beyond listening." This level of listening includes such skills as classifying information, categorizing, indexing, comparing, defining, predicting, applying, seeing cause-effect relationships, critically evaluating, appreciating, and creative problem solving.

This listening framework will be the basis for listening evaluation and development and remedial approaches to listening instruction.

REINFORCEMENT ACTIVITY

Observe several elementary classrooms. What listening activities do you see? After observing, what do you conclude is the teacher's definition of listening? Ask the observed teachers for their definitions. Do their definitions agree with your observations? Discuss the diagnostic and instructional implications related to the various definitions.

EVALUATION

Children listen under such diverse circumstances and for such different purposes, so it would be extremely difficult to evaluate listening ability through the administration of one test. Listening, like oral language, can be evaluated informally during many of the activities recommended in this chapter. Lundsteen (1979) maintained that evaluation of listening requires live listening situations that examine children during a wide range of activities; these evaluations call for highly specific, imaginative, and even devious ways to evaluate listening behavior.

Auditory acuity, the reception of sound waves at various levels of tone and loudness, is a prerequisite for listening. Consequently, most children are administered hearing tests by nurses or trained teachers. Teachers should, however, observe children informally for any indications of hearing problems. Children who frequently ask for oral repetitions, rub their ears, or show poor pronunciation may have hearing loss and should be referred for further testing and, if necessary, for diagnosis by hearing specialists.

Evaluating Auditory Perception

The most commonly evaluated auditory perception ability within classroom settings is auditory discrimination. Auditory discrimination refers to children's ability to hear differences in letter sounds, words, and nonsense syllables. For example, can children discriminate between "bit" and "bet" when they are spoken by the teacher? Auditory discrimination can be measured be either standardized, norm-referenced tests or by teacher-constructed informal approaches. The Wepman Auditory Discrimination Test (1973) is the best-known standardized test of auditory discrimination. The test is designed for administration to children five through eight years old. In this test, the examiner reads two words to a child, who is positioned so she cannot see the examiner's mouth. The child then tells whether the words she hears are alike or different. This is an individually administered test requiring about five minutes for completion. Another type of auditory discrimination test for young children is found on the reading readiness test administered to many kindergarten and beginning first-grade students.

The activities listed under developing auditory perception skills in this text provide excellent examples for informally evaluating various auditory perception abilities. For example, pages 127 to 128 provide sources for evaluating children's auditory awareness; pages 128 to 129 include auditory discrimination activities related to sounds, words, and rhymes; pages 129 to 130 focus on auditory memory; and pages 130 to 132 emphasize auditory discrimination of beginning, ending, and middle sounds. The informal evaluation of auditory perception takes place in a relaxed, enjoyable atmosphere because many of these activities have a gamelike format. Other sources for developing informal assessment activities include Heilman's *Phonics in Proper Perspective* (1989) and basal reading manuals developed for early elementary grades.

Diagnosing Attentive Listening Ability

Listening is an active process requiring participation on the part of the listener. Poor communication results when students do not pay attention, or when they are thinking about their responses rather than concentrating on what the speaker is saying. Inattentive students may not ask questions when clarification is necessary for understanding. Readers can stop and reread, but most listening situations do not permit listeners to go back and review what was heard. In addition, listening usually takes place in a more public setting; consequently, students may be distracted by noise, mannerisms of the speaker, or actions of other listeners.

What is the relationship between attention and school achievement? A study by Samuels and Turnure (1974) showed a positive relationship between attentiveness and achievement in first grade. Children who demonstrated attentive behavior achieved at a higher level than children who were inattentive. MacGinitie (1976) advanced many possible reasons for a relationship between attentiveness and achievement. For example, students who find the school environment congenial and who are more interested in learning may pay closer attention and, consequently, learn more. Parents who help their children at home may be teaching them to attend during school as well as at home. In contrast, children who make poor initial progress may find the work dull or discouraging, and, consequently, attend less and learn less. MacGinitie said of this cause-and-effect dilemma: "It is hard to know whether attentiveness results in progress or whether progress results in attentiveness" (p. 13). Whatever this relationship may be, it is still the responsibility of the teacher to train students to ignore distractions and to concentrate on the speaker's ideas.

Testing for the ability to pay attention is complicated by the fact that students may attend to a speaker without being able to understand him. Thus, poor communication can be caused either by inadequate attention, or by problems with the student's ability to process ideas. Otto and Smith (1980) suggested that a diagnostic procedure designed to assess attention should not also assess the student's ability to process ideas. These authors prefer informal measures for testing a student's attention.

Informal tests can be given in many different classroom environments, whereas a standardized listening test is usually given under controlled conditions. For this reason, a standardized-listening-test result may not indicate how the student attends in an actual classroom setting. I have also found that students may demonstrate completely different attentive listening responses when tested in an isolated testing room rather than tested informally during a class activity. A student who recently came to the university's Language Arts Laboratory was an excellent example of this problem. During a clinical testing experience, he responded very well to various listening tests. He told me he was trying hard to listen, and that he was concentrating on everything that was spoken. He indicated that this did not happen in school, because, he said, he often sat through a class and could not remember anything that had been covered during the class. This student realized he had two problems: he liked to daydream, and he was easily distracted by noisy fellow students.

Informal tests that assess normal classroom attention, and that also test attention rather than processing, include activities such as following simple directions and repeating concrete ideas presented orally. To find out how well children attend during school situations, they should be assessed in various settings, and should not be informed they are being tested.

In designing an attentive listening task for average achievers, the level of difficulty should be about the same as the student's independent reading level. This may not hold true for remedial students, because listening ability is usually greater than their reading ability.

The ability to follow oral directions explicitly demands attentive listening. The teacher can give oral directions to the class and observe which students are able to complete the task, which students cannot complete the task, and which students ask him to repeat the directions. For example, a kindergarten teacher might say to the class, "Take out your red crayon, your green crayon, and your brown crayon. Pick up your green crayon and show it to me." (Obviously, do not do this activity unless the children can already identify the colors.) An elementary teacher might say, "Put your spelling book and your arithmetic book on top of your desk. Place your spelling book on top of your arithmetic book so I can see the spelling book when I walk around the room."

This type of informal testing can be carried on throughout a normal day, and in the context of different instructional activities. During a social studies class, you might ask a question that has an easy and obvious answer, but that is out of context with what is being discussed. For example, the teacher could ask: "Everyone who can tell me what two plus two is, please raise your hand." During a science class, you could read a short paragraph that presents the simple sequential steps of some logical process, then ask students to summarize the order presented. For example, "Do you often wonder what causes a hard rock to split into several pieces? First, rain falls on the rock. Then the water runs into cracks in the rock. When the weather is cold and the water freezes, the cracks will widen. This process continues until the rock finally splits."

Any of the activities listed later in the section "Developing Attentive Listening" may also be used to test informally a student's attentive listening ability.

Diagnosing Listening Comprehension

As one might expect, listening comprehension shows a strong relationship to verbal intelligence. Spache (1976) reported a correlation of .75 between the listening comprehension score on the Spache Diagnostic Reading Scales and the WISC Verbal I.Q. But, fortunately for educators and for children, Spache also concluded, "This interaction between intelligence and this auditory ability does not imply, as some might assume, that improvement is almost impossible because the performance depends heavily upon the child's intellectual capacity. Both intelligence and auditory comprehension can be improved by careful training. Both are really ways of and skill in thinking, and they can be modified" (p. 63).

Listening comprehension may be diagnosed by standardized listening comprehension tests or by informal inventories that are either group-administered or given individually. Most authorities believe both kinds of assessment are necessary.

Group standardized listening tests   The Sequential Test of Education Progress (STEP) has a listening test for grades four through college. It includes ninety items for the elementary grades, which reportedly measure ability in identifying and interpreting the main idea, remembering details and sequences, and understanding word meanings. The test also has questions related to judging the validity of ideas, distinguishing fact from fancy, and noting contradictions. The STEP test has been criticized because a student must read as well as listen in order to answer the questions. Many items on the test can also be answered without first listening to the oral presentation. This is a problem with many listening tests; to assess listening comprehension accurately, the knowledge should have been gained exclusively from the test source.

The Cooperative Primary Tests provide two assessment forms for grades one through three. In this test, the teacher reads words, sentences, stories, and poems; the child marks correct pictures.

A third group-listening test is the Durrell Listening-Reading Series. This test has three levels comparing reading and listening abilities in grades one through nine. Optional responses are read to the children, so they do not have to read in order to answer the questions.

Brassard (1971) offered help in constructing listening tests that compare listening and reading comprehension in the elementary grades. The following criteria are useful in evaluating listening comprehension tests in order to choose an appropriate one for your students.

1. The test should contain two parts: one to measure vocabulary and one to measure paragraph comprehension.
2. There should be two forms for each test: one for administration as a listening test and the other as a reading comprehension test.
3. Both forms should be the same in length and level of difficulty.
4. The time allowed should be identical for both the listening and reading tests.
5. The test should use a multiple choice format.
6. Pictures should be used to designate the answer choices.
7. Items should include a range of difficulty from below third-grade through sixth-grade levels.
8. Items of varying difficulty should be distributed throughout the test.
9. Directions should be as brief and simple as possible.
10. The listening measurement should be read to the student, not read by the student.

REINFORCEMENT ACTIVITY

Choose a standardized group listening test. Look at the items and how they are administered. What comprehension skills do you believe are being measured? Use Brassard's criteria to judge the test. Could a student answer the test questions without hearing the selections?

Individual listening comprehension tests   One technique for measuring listening comprehension individually is the informal inventory. This inventory may be a published version, or one developed by the teacher. The published inventory usually has three forms, each of which includes paragraphs followed by comprehension questions for preprimer through eighth-grade levels. In order to test listening comprehension, the paragraphs are read to the student. After each paragraph, the student is asked to orally answer factual, vocabulary, and inference questions. The highest level at which a child correctly answers 90 percent of the oral comprehension questions is considered her *independent* level of listening comprehension. The level at which the student is able to answer 75 percent of the comprehension questions correctly is considered her *instructional* listening comprehension level.

An example of a listening comprehension inventory is the *Analytical Reading Inventory* by Woods and Moe (1989). This inventory has three forms, with paragraphs ranging in difficulty from primer through ninth grade. The paragraphs are followed by questions categorized according to main idea, facts, terminology, cause and effect, inferences, and conclusions. The results of these tests have important implications for listening comprehension instruction, because they enable the teacher to analyze which types of questions are causing comprehension problems.

Evaluating Students' Ability to Choose Effective Listening Strategies

The various listening categories identified under our definition of listening imply that students have many different purposes for listening. Likewise, the way they approach listening tasks should change depending on the listening situation and on their purpose for listening. Educators who are interested in process approaches to instruction and in self-monitoring strategies emphasize the need to teach appropriate listening strategies and to teach children to make conscious choices when they approach listening activities.

Tompkins, Friend, and Smith (1987) identified a series of questions that students should ask themselves as they select a listening strategy and then monitor the effectiveness of their choice. The following informal inventory is developed from evaluative questions recommended by Tompkins, Friend, and Smith. Teachers may develop a chart such as the one shown in table 4–1 to help students and to assist teachers in evaluating the effectiveness of listening behaviors and instruction.

Teachers may use the results of this inventory to guide instruction or to help individual children evaluate their own listening strategies. The definitions and descriptions of these listening strategies are developed later in this chapter and in other chapters within this text.

DEVELOPING LISTENING SKILLS

Listening is an active rather than passive activity, and, as a result, students must participate in their own instructional improvement. Listening skills need to be developed through a variety of activities in which children can see the consequences of their listening. Can they see that they have enjoyment when they listen to a story, a good television show, or a musical selection? Can they see that they enjoy playing a game when they have understood the directions? Can they see that they understand a subject better if they have interacted with the class discussion? Can they see that they

TABLE 4–1
Evaluating students' ability to select and to use effective listening strategies

I. Before listening the student asks and answers the following questions:
    1. What is the speaker's purpose?
    2. What is my purpose for listening?
    3. What am I going to do with what I listen to?
    4. Will I need to take notes?
    5. Which strategies could I use?
        a. Imaging?
        b. Categorizing?
        c. Self-questioning?
        d. Discovering the organizational plan?
        e. Note-taking?
        f. Clues from the speaker?
    6. Which one [or ones] will I select?
II. During listening the student asks and answers the following questions:
    1. Is my strategy still working?
    2. Am I putting information into groups?
    3. Is the speaker giving me clues about the organization of the message?
    4. Is the speaker giving me nonverbal cues such as gestures and varied facial expressions?
    5. Is the speaker's voice—pitch, speed, pauses, and repetitions—giving me other clues?
III. After listening the student asks and answers the following questions:
    1. Do I have questions for the speaker?
    2. Was any part of the message unclear?
    3. Are my notes complete?
    4. Did I make a good strategy choice? Why or why not?

Source: Numbered questions from "Listening Strategies for the Language Arts" (p. 39) by G.E. Tompkins, M. Friend, and P.L. Smith, 1987. In *Language Arts Instruction and the Beginning Teacher,* edited by Carl Personke and Dale Johnson, Englewood Cliffs, N.J. : Prentice-Hall. Reprinted by permission.

may lose some freedom of choice if they are not able to evaluate critically and act on what they hear?

A student's listening skills can be improved through specific instruction. Listening may also be taught as part of many content areas, such as reading, literature, music, social studies, and science. For improvement to occur, the teacher must help students develop their own effective listening goals. If students do not see the necessity for and benefits of improved listening, they will not be motivated toward better listening, and will not reach the desired level of competence.

Some of my students have been disappointed in their listening lessons when they have not included their students in the initial planning. For example, one student asked third graders to listen to a story record, then asked them specific questions about the content. She had not adequately prepared the class for the listening activity, and was disappointed in the results. She realized her problem and approached the next

This teacher is developing a purpose for listening before sharing a book.

lesson quite differently. This time, she asked children to list the various listening activities they performed in a day. Then they listened to several quite different activities, including specific oral directions for a worksheet, a short story for pleasure, and a short oral presentation on brushing their teeth. The class then discussed the purposes for listening in each situation. Students included the consequences of good and poor listening for each activity. Finally, they set up guidelines to utilize in improving their own listening ability, including the following:

1. I will get ready for listening by getting rid of distractions.
2. I will know my purpose for listening.
3. I will concentrate on the listening activity.
4. I will expect to get meaning from the listening activity.
5. I will try to see in my mind what I hear.
6. If I am listening to a speaker, I will ask myself:
    a. Did I know his purpose for speaking?
    b. Did I know my purpose for listening?
    c. Did the speaker back up his ideas?
    d. Did I ask intelligent questions?
    e. Can I retell in my own words what the speaker said?

Guidelines developed by students are stronger instructional motivators than guidelines developed by the teacher. The teacher in the above example learned that students in the class could now judge their own behavior based on their own recommendations.

Setting Classroom Conditions for Listening

The teaching environment is crucial to some types of listening instruction. Some listening skills demand few distracting noises and sights. For example, the teacher of a group of open-classroom first graders found that he had to find a quiet and less visually distracting environment for both testing and teaching the auditory discrimination skills prerequisite to an introduction of medial vowels. This teacher used movable bookshelves and screens to create a small area in the classroom where there would be fewer distractions. Another teacher in an open classroom moved to the gym to present the rhythms and movements of a listening experience.

Some students may need special seating arrangements based on results of a hearing acuity screening. Students with hearing loss may need to sit close to the speaker, teacher, or other oral source.

Research by Lundsteen (cited in Porter, 1976) mentioned other necessary teacher behaviors for developing effective listeners. Lundsteen listed the characteristics of an effective listening teacher:

1. Has ability to diagnose, select, and appraise an appropriate sequence of skills for individual learners.
2. Is able to select and develop listening materials at the appropriate levels of difficulty and for the appropriate purposes.
3. Knows why listening instruction is important and guides children toward this realization.
4. Carefully observes children for verbal and non-verbal hints to the boundaries of each learner's performance and competence.
5. Avoids both needless repetition of instruction which encourages careless listening and the presentation of too much listening material at one time.
6. Encourages children to expect meaning from the vocabulary they hear.
7. Pinpoints a focus for listening by providing advanced organizers that provide the students with a sense of direction.
8. Uses challenging vocabulary and syntax that causes children to reach ahead of their boundaries.
9. Uses pacing and other strategies to help children attain concepts through inductive guidance.
10. Uses words as concisely as possible in order to provoke thought and then waits for responses, respecting silences, if necessary.
11. Uses open-ended questioning techniques that extend comprehension from facts to higher cognitive levels.
12. Helps children achieve a sense of direction and uses reinforcement to strengthen desired objectives.
13. Helps children transfer listening ability to new contexts by developing a positive attitude, using meaningful reinforced learning, helping children discriminate similar

aspects in two listening contexts, and provides new instances for generalized transfer.[1]

REINFORCEMENT
ACTIVITY

Introduce a listening-improvement unit to a group of children or to your peers. Develop some listening activities that will help the group understand the various purposes for and benefits of listening. Help the students establish guidelines for their own listening improvement.

Development of Auditory Perception Skills

Considerable auditory discrimination is included in most reading readiness programs. If auditory discrimination is a prerequisite skill for phonics, then, according to Harris (1985), the most effective time to teach the prerequisite skill is immediately before the corresponding discrimination is to be used in printed words. Thus, the teacher must evaluate the sequence of the instructional program as well as assess a child's ability to perform the required tasks. Spache (1976) said no research evidence indicates the exact training sequence to follow, but he did offer guidelines for a logical development of auditory skills. Spache recommended that this sequence begin with auditory awareness, continue with auditory discrimination of sounds and rhymes, auditory memory, auditory discrimination of beginning, ending, and middle sounds, and end with auditory synthesis (blending). We will examine auditory perception skills according to this logical developmental sequence.

Auditory awareness   Auditory awareness is a fundamental skill requiring that a child hear and discriminate among sounds in the classroom, the home, and the rest of her environment. An instructional activity might consist of listening to and identifying sounds heard in a barnyard, on a street, at the airport, and so forth. One kindergarten teacher handled such an activity successfully by having the class walk around inside and then outside the school building. The children stopped periodically to close their eyes and listen to the sounds around them. They discovered, among other things, the difference between kitchen and playground sounds. After several experiences with sound awareness, the teacher taped sounds from different locations. The children listened to the tapes, identified the sounds, and classified them according to their locations.

Auditory awareness can also be developed through the use of a sound box. The teacher places several objects that have a characteristic sound on a box or tray. She might select a bell for ringing, a ball for bouncing, water for pouring, an egg beater for

---

[1]From Sara W. Lundsteen, "Research Review and Suggested Directions: Teaching Listening Skills to Children in the Elementary School, 1966–71," *Language Arts* 53 (March 1976). Copyright © 1976 by the National Council of Teachers of English. Reprinted by permission of the publisher and the author.

mixing, a whistle for blowing, and so forth. Using a screen to hide the objects, the teacher or a student sounds each object. The other students listen carefully and identify the various sounds. Children can also bring in objects from home with which to try to mystify their classmates.

Other auditory awareness activities include imitation of animal sounds or other imitation sound games in which a student or the teacher presents a sound and the class tries to copy it. Rhythm instruments and the piano are also good props for sound awareness activities. Children can identify the sounds of rhythm sticks, blocks, triangles, and tambourines. They can imitate a pattern of sounds with these instruments, or can move—rapidly or slowly, sadly or joyfully—according to the rhythm of the instruments.

Listening to music, rhymes, and limericks helps children become aware of the lovely sounds in our language. Older children can also listen to old radio broadcasts, available on records, and try to identify and duplicate the sound effects. Several of my college students have had children listen to radio broadcasts as a prerequisite to producing a story including sound effects.

Auditory discrimination of sounds, words, and rhymes   Auditory discrimination of likenesses and differences in sounds, words, and rhymes requires a finer auditory perception than auditory awareness. The teacher might provide opportunities for children to decide whether two musical sounds are the same or different. For example, he can play two notes on the piano and have the students say whether the notes are the same or different. They can compare loudness and softness as well as high or low pitch.

Auditory discrimination activities can easily become games. The teacher or child can clap two patterns that may be either alike or different. After listening to the patterns, the children tell whether the two patterns were the same or different.

If we consider auditory discrimination a reading readiness skill, then activities should include those in which the children discriminate between likenesses and differences in words. Two words, such as "red" and "roll," can be carefully pronounced for the children to indicate whether they are the same or different. One way to conduct activities like this is to have children form a circle and close their eyes to listen carefully to the two words. If the words are the same, as, for example, "pet" and "pet," have them raise their hands; if the words are different ("mother" and "father"), tell them not to raise their hands.

Rhyming elements in words, nursery rhymes, jingles, and stories provide many opportunities to develop auditory discrimination and an awareness of the pleasure to be gained from word sounds. Dr. Seuss books are especially good for this activity. The teacher might read *The Cat in the Hat*, then reread it with the rhyming words deleted, to be supplied by the class. Well-known nursery rhymes may also be utilized for this activity. For other rhyming activities, you can do the following:

1. Prepare two sentences in which the last words both rhyme. Read the sentences, omitting the final word, and ask the class to supply the missing rhyming word. (Several answers may be correct).

I just built a boat.
I hope it will_____. (float)
The big black cat
Wore a funny yellow_____. (hat)
The robin built a nest
It was the very_____. (best)

2. Use a series of pictures which have several rhyming elements, such as sack, bat, cat, track, hat, black. Have the pictures in random order and ask the students to put the rhyming pictures together. You can use a flannel board or bulletin board for this activity.

3. Read a sentence and ask the class to supply as many words as they can that rhyme with the final word of the sentence.
    We all went to the store with *Bill*. (will, pill, kill, mill, hill, fill, still, Jill, shrill)
    On her head, Mary wore a red *cap*. (lap, map, gap, strap, trap, tap, clap, flap)
    The baker made a chocolate *cake*. (lake, bake, rake, snake, fake, brake, shake)

Many interesting poems for older children contain lovely rhyming elements. "The Table and The Chair" by Edward Lear describes the adventures of a table and a chair when they decide to take a walk. The poem has such rhyming combinations as heat-feet, walk-talk, air-chair, table-able, down-town, sound-round, and leetle-beetle. Another comical rhyming poem is "The Plaint of the Camel" by Charles Edward Carryl. In this poem, the poor camel complains in rhyme about his food, his housing, his work loads, and his shape. Some of the rhyming elements are feed-seed, crunch-lunch, able-stable, enclosed-exposed, noodles-poodles, treated-heated, and lumpy-bumpy-jumpy. Because many of the rhyming words in this poem contain long and short vowels, this poem may also be used for a listening activity connected with medial vowels. The following books contain rhyming elements or repeated phrases and are excellent for auditory discrimination:

- Eve Merriam's *Halloween ABC* (1987)
- Arnold Lobel's *The Random House Book of Mother Goose* (1986)
- Jack Prelutsky's *Read Aloud Rhymes for the Very Young* (1986)
- Jane Yolen's *The Three Bears Rhyme Book* (1987)
- Nadine Westcott's *Peanut Butter and Jelly: A Play Rhyme* (1987)

Auditory memory   Auditory memory activities include experiences in following and imitating sound, number, and sentence squares. Auditory memory activities can be as simple as repeating a short, clapped rhythm pattern, or as difficult as repeating complex directions. For elementary children, many auditory memory activities can take the form of games. For example, the teacher may play a rhythm on the piano or with rhythm sticks and have the class imitate the rhythm. Rhythms can also be clapped or rapped by the teacher or a child, then imitated by another child or by the class.

Children enjoy circle repetition games in which one child starts and each child has to repeat the same thing and add an item of her own. For example, the first child might say, "I'm going on a trip and I'm going to take an apple." The second child would say, "I'm going on a trip and I'm going to take an apple and a_____." This game may be continued until it is impossible for anyone to remember the sequence of items. Adapt the game to have all items belong to a certain classification, such as food, toys, clothing; follow in alphabetical order, such as apple, book, cat; begin with the sound of *b*, such as book, baby, bottle; or contain the same vowel sound, such as table, rake, pail, shape.

Games of giving simple directions also provide practice in auditory memory. Simple two-part directions, such as "Jennifer, turn the light on and then go to your seat," can be presented by the teacher or other children. Billy might be asked to "Bounce the ball two times and then clap your hands once."

A trunk game also provides practice in following simple directions. For this activity, put a number of items, such as a ball, ruler, book, bat, picture, cap, game, toy, glass, and so forth, in a box or trunk. The child who is the leader names several items she would like from the box, and calls on another child to bring the items to her. If the child responds correctly, she becomes the leader and makes the next request.

Auditory discrimination of beginning, ending, and middle sounds Auditory discrimination of beginning and ending sounds is usually related to the ability to discriminate consonant sounds, whereas the auditory discrimination of middle sounds is usually associated with vowel sounds. Kindergarten and first-grade teachers often associate beginning sounds with the names of students in the class, or with common objects in the classroom. One successful first-grade teacher designed a sound board for her classroom. For example, if the letter *m* was being studied, children brought to school concrete examples of items beginning with the sound of *m*. These items were discussed and grouped on or under a bulletin board. Children also drew items to represent the sounds, such as mittens and monkeys.

You can also use listening activities in which children listen for and identify a beginning sound that is repeated frequently in a story. I have used such an activity as a writing-phonics review activity for second and third graders, and for an auditory discrimination activity for first graders. While teaching a second/third-grade language arts class, I reviewed beginning consonant blends with the children and asked them to write stories that utilized a number of words containing the specific blend being reviewed. These stories were written by the second and third graders and then read to the first graders. The first graders closed their eyes, listened carefully, and raised their hands whenever they heard the specific sound in the story. The older children were motivated to write because they knew they had an audience, and the first graders were interested in listening to stories written by children they knew. Figure 4–1 shows an example of a "gr" story written by a third-grade girl. As you can see, the inhabitants of Grouchtown also live in garbage cans.

Auditory discrimination also includes the ability to tell whether two words end with the same or different sounds. Rhyming activities required a child to listen to a group of sounds; now you are asking children to discriminate between the ending consonants. This is usually harder to hear than rhyming sounds. Picture activities

FIGURE 4–1
"Gr" story written by a third-grade girl

Granie Comes to Grouchtown

One time in Grouchtown, Grinese (a grateful grouch) got a letter from Granie. The letter said Granie was coming on Grouchday.

On Groucholay Grinese made gravy, grapefruit, grapes, and graham crackers for the feast.

When Granie came she brought a kitten for Grinese and she liked the feast.

similar to those described for beginning sounds are appropriate. For example, pictures of a mop, tub, cap, top, rug, and cup could be shown to the class. Ask the children to place all the pictures that end like "map" on a flannel board.

For another word-ending auditory discrimination activity, pronounce pairs of words and ask the children to tell whether the words end with the same sound or different sounds. Examples include:

cot–cat	dump–clap	trunk–scat	Sam–tan
dad–tub	mail–will	flag–snug	jump–wag

Finally, auditory discrimination includes a child's ability to discriminate between medial sounds. This may be the most difficult auditory discrimination task for many students. The necessity for this skill depends on the nature of their spelling and reading programs, and on the sequence of skill development in each program. Some programs demand that the child master this skill at the beginning of first grade, whereas other programs do not require it until second grade.

Examples of useful words for medial auditory discrimination include:

let–pen	bug–bag	box–got
lift–pig	tap–cat	hug–but
hot–rat	plan–plane	read–red

Developing Attentive Listening

Many of my students in college language arts classes have elected to develop and teach units on listening improvement. They are often motivated to do this by their first observation in a language arts practicum. As one college junior said, "My project actually had its beginning during our first session in the classroom, when I discovered a general problem among my fourth graders. Why was the teacher constantly scolding the students for not following directions? Why were the students asking for instructions, messages, and so forth, to be repeated? Why were so many mistakes being made on those assignments and activities taxing the students' aural perception skills? It was quite obvious that any communication bridges between the teacher and students were gradually weakening. The majority of the students simply were not listening to what was taking place around them. It was then that I decided to take action." This perceptive young educator designed a creative program that highly motivated his fourth graders to improve their attentive listening skills. Portions of his unit, Listening Fourth Graders, are included in this chapter.

Other students have discovered the importance of giving concise directions. As a first-grade student teacher described her experience, "Much of the work I did in developing listening skills involved their abilities to follow directions. This was also good practice for me, because one of my weakest areas is in giving concise and ordered directions. We all profited from this experience." Another student discovered her inability to provide accurate directions when she taught a listening lesson requiring third graders to make a bird using the art of origami (Japanese paper folding). She concluded, "In giving the verbal directions for bird folding, I suddenly realized how much clearer and more precise my own instructions need to be. Had I thought of this beforehand, and been more adequately prepared, I think the children might have felt less frustration." As these students so accurately note, children should not be blamed for all attentive listening problems. Often both students and teacher need improvement.

As reported by Samuels and Turnure (1974), attentive listening behavior is particularly important because of its relation to higher school achievement. We have also pointed out that factors such as motivation, auditory acuity, auditory discrimination, lack of purpose, and problems with thought processing influence the development of attentive listening. Teacher behavior is also influential.

Although research has clearly defined the relationship between attention and achievement, it has not been as clear in identifying a methodology that effectively develops attentive listeners. Otto and Smith (1980) suggested the following four factors as causes of inattentiveness: (1) poor motivation to hear a speaker's message;

These children are developing their attentive listening skills through the use of the listening game "Simon Says."

(2) too much teacher talk; (3) excessive distractions; and (4) lack of mental set for anticipating the speaker's message. Eliminating these four factors should, obviously, improve attentive listening behavior.

Implementing the suggestions under "Setting Classroom Conditions for Listening" (p. 126) will eliminate many distractions in the classroom. A review of the teacher behaviors essential for developing effective listeners will show you how closely these behaviors also relate to developing attentive listening. Involving students in setting instructional listening goals will also help overcome some of the problems of poor motivation.

Developing a mental set for purposeful listening may be more difficult than eliminating outside distractions. If students are not able to form a mental set for anticipating a message, they will have difficulty preparing for the type of listening the message demands. Students who listen to complicated directions with the same mental set they use for listening to a musical recording will undoubtedly have difficulty attending in situations requiring strict accuracy of comprehension.

A number of methods may be used to help students develop the attentive listening required for directions. The school day lends itself to this type of instruction, because directions are an essential part of all content areas. Some teachers tape a series of directions that children must follow explicitly. Taping has two advantages: first, it does not encourage the students to ask that directions be repeated before they have been completely given; and second, taping reduces excessive teacher talk. A math teacher who is reviewing shapes might, for example, tape the following directions: "Draw a red triangle near the top of your paper. Draw a blue rectangle under the triangle. Draw a green square on the bottom of your paper." After the class has completed the directions, the tape may be replayed to verify how well the directions were followed. Similar activities could be designed around any content directions, and can be as complicated as is warranted by the students' ability level.

Games reinforce the teaching of attentive listening to directions. An example of such a game is "Listen, Start, and Stop." To play the game, children form a long line, leaving plenty of room for movement. The leader gives a clear direction, such as "hop on your right foot; walk using small steps; pat your head," and so forth. Students continue one action until the leader tells them to stop or blows a whistle. Children must stop immediately and turn toward the leader to wait for the next direction. Children who fail to follow directions or to stop immediately are eliminated from the game. The game "Simon Says" is another good listening game. Other listening games may be found near the end of our discussion of attentive listening.

A student's ability to give clear directions and to follow oral directions should also be evaluated. Pretending that a classmate is a stranger who needs directions to a specific location (principal's office, gym, etc.), the remainder of the students prepare what they believe are explicit oral directions. A class member follows the directions to see if they are indeed adequate. Both listening and oral directions are thus evaluated for adequacy.

Here are some other activities for developing the ability to follow oral directions:

1. Have students write directions for a specific activity such as setting the table or washing a dog. Ask students to read their directions without identifying the activity. Classmates listen carefully and guess what the activity is as soon as they think they have enough clues.
2. Provide oral directions for making a simple object, such as a kite, and have the class make the object without telling them what the object will be.
3. Demonstrate the use of a camera, then ask a child to follow the exact directions to take a picture. This is especially good with instant cameras that provide an immediate picture.

Following directions may be more effective if students number the steps of the oral directions, then repeat the number of steps involved and relate each step to its corresponding number. In addition, each step should relate to its purpose and to the overall purpose for the directions. Helping students map out listening strategies and providing guideposts prior to listening instruction is called "applying advanced organizers." This strategy may be effective with all students, but is especially appropriate for students with learning problems.

Here is an example of an attentive listening unit, the objectives for which would include these points:

1. Following a motivational listening activity, the students will be able to realize the importance of attentive listening and develop their own guidelines for good listening.
2. Students will improve their listening skills through the use of listening games and other creative language arts activities.
3. Creative thinking will be developed through a variety of language arts activities, including oral communications, creative writing, and attentive listening activities.

The following procedures were used to implement the objectives of the unit. First, the fourth-grade students were asked to keep track of the approximate time they spent in one day on the following ways of communicating: speaking, listening, writing, and reading. The next day they discussed their findings. They found that a great deal of time was spent in listening. The teacher presented several types of listening activities, such as following directions, listening to a record, and listening to a story, in order to answer specific questions. Students listed the requirements for each type of listening and developed their own guidelines for listening behavior. (They found they were not very attentive listeners.)

Listening games and exercises proved very motivating, and also served as excellent warm-up drills and challenges to fill free time before the first class period, lunch, and so forth. The listening warm-ups covered three broad areas: reception of sounds, comprehension of listening and understanding the purposes for listening, and reaction to the ideas expressed. The procedures were sequential in nature, starting from simple games and drills and progressing to more challenging activities. The listening games included the following:

1. Telephone Game—The purpose of this game is to send a secret message around the room, to test a student's ability to listen attentively and to speak clearly. One student starts the message, whispering it to the next student; the last person's response is matched with the beginning message.
2. Recipe Game—The purpose of this game is to strengthen the students' ability to listen to and recite a sequential list of items. At first, students read recipes; later, they write their own.
3. Direction Game—When a student reads a task, the remaining students test their ability to follow directions carefully. Students are not allowed to see the written directions.
4. Morning Sounds—Students are assigned a specific time period and told to record all the sounds they hear during this period.
5. Noise Makers—This activity uses a variety of sounds to test the students' aural perception abilities and their use of imagination. Students close their eyes and try to identify various noises. They use creative thinking to choose noises that will stump their classmates.

Many activities to stimulate creative thinking and listening were used, including role playing, which was motivated by large action pictures of events common to most of the students, or hypothetical situations, in which students listened to a taped discussion, then role played their reactions. Other students listened to the role playing and discussed their reactions and the reasons for them. Students wrote and taped many of the situations. Tapes of country sounds and city sounds gave them opportunities to listen attentively, categorize the sounds, and specifically identify each sound. Students listened to the beginning of a story and created an oral or written ending to the story. The recipes game introduced the creative writing of new recipes. They wrote their own "creative recipes" and presented them orally to the class. Other students listened attentively in order to be able to repeat the recipe. Finally, students listened to TV commercials and tried to summarize what they heard. (Using TV commercials to develop evaluative listening skills is described in the multimedia chapter in this text.) After they listened to TV commercials, they wrote their own commercials and presented them to classmates, who, in turn, had to listen attentively in order to summarize what they heard.

Following the completion of this listening unit, students listened to the same types of listening experiences that were used to evaluate listening needs at the beginning of the unit. A majority of the students improved in all types of listening. Informal observations showed that students required less repetition of directions and demonstrated greater accuracy in listening situations. In addition, speaking skills improved.

When you teach a similar listening unit, remember to involve the students in setting their own goals. Remember also that listening improvement is a continuous activity.

Developing attentive listening with students who have learning problems   Students with learning problems require more assistance in listening for a specific purpose than children with no learning handicap. Work by Ausubel (1960) stressed the use of advanced organizers in preparation for the listening material in order to give students a sense of direction. (The use of advanced organizers is effective in listening for directions, as well as in listening for other specific information for analytical purposes.)

An advanced organizer for use with listening for directions might stress listening for key words in the directions, such as "first," "second," "third," "finally," and so forth. The advanced organizer helps the students understand the purpose of the directions, the reason for each step, and the relation of each step to the set of directions.

One technique I have found extremely useful is to ask students with learning problems to close their eyes and visualize themselves performing each step of a set of directions. You may want to try this activity with a pupil-motivated task, such as leaving the building when a fire alarm sounds. The purpose of an orderly, quick exit are easy to understand, because the students are so vitally involved. Ask your students to close their eyes and visualize themselves doing each step of the following: "First, when you hear the fire bell, walk quickly to the classroom door. Second, form a line and turn to the right, walking close to the hall wall. Third, walk straight ahead and go out the double doors in the front of the building. Finally, line up outside the building next to the flagpole so I can count to make sure you are all safe." Children should listen to the advanced organizers signaling sequence—"first," "second," "third," "finally"—and should try to visualize themselves performing each step as you give the directions. Ask them to describe what they see and hear. After they have described each step, have them try to visualize the whole sequence. They can then physically perform the activity, comparing their visualizations with what they actually do. After doing this activity, some children have claimed they could even hear the bell and smell smoke!

Duker (1971) declared that probably no group in school has as much to gain from knowing how to listen well as children who have problems keeping pace with the normal curriculum. If these children are treated as if their only difficulty in language arts is reading, and if it is taken for granted that they can listen without further instruction, they will undoubtedly suffer. Duker concluded, "Such an approach is fallacious as much more often than not these children are suffering from a handicap in all communicative language arts skills. The fact that they perhaps listen better than they read does not justify the conclusion that they therefore listen well" (p. 217).

Listening Comprehension Activities

Following a review of listening comprehension research, Pearson and Fielding (1982) drew the following conclusions about teaching listening comprehension:

1. Direct teaching of listening strategies appears to help children become more conscious of their listening habits than more incidental approaches.
2. Listening training in the same skills typically taught during reading comprehension tends to improve listening comprehension (e.g., main idea, inference, sequence).
3. Listening comprehension is enhanced by active verbal responses on the part of students during and following listening.
4. Listening to literature tends to improve listening comprehension.
5. Instruction directed toward writing or reading comprehension may also improve listening comprehension.

These conclusions offer the language arts teacher numerous recommendations for improving listening comprehension. We consider listening comprehension activities that teach listening strategies that are similar to reading comprehension and require active verbal responses, and activities developed around listening to literature. Listening activities are excellent ways to enrich the content areas. Carefully selected activities teach or reinforce content as well as enhance listening capabilities.

Teaching listening strategies   Tompkins, Friend, and Smith (1987) differentiated between a practice approach to listening instruction and a strategies approach to listening instruction. In the practice approach, which frequently dominates classroom instruction, students listen to an oral presentation and then answer questions about the presentation. This approach assumes that students know what to do when they are given the listening assignment. Unfortunately, this may not be true. In contrast, the strategies approach teaches students strategies they can use during a listening experience to make the message they receive clearer and easier to remember. The strategies approach helps students develop a repertoire of possible listening strategies, teaches students to make conscious choices when selecting appropriate strategies for specific listening situations, and teaches students to monitor the effectiveness of their choices. It is assumed under a listening strategy approach that listening practice follows rather than precedes or replaces strategy instruction.

The listening strategies recommended by Tompkins, Friend, and Smith include imaging (teaching students to make pictures in their minds and to visualize details and descriptions) categorizing (teaching students to group or to cluster information); self-questioning (teaching students to monitor their own understanding by asking themselves questions such as "Could I explain this to someone else?"); discovering the organizational plan (teaching students to recognize and to use patterns such as comparison-contrast, problem-solution, cause-effect, description, and time-order when they are trying to comprehend and to remember oral presentations); note-taking (teaching students to match their purposes for listening with the types of notes they will take); and getting clues from the speaker (teaching students to recognize clues such as gestures, facial expressions, and written information on chalkboards that speakers use to convey important messages).

Consider how these listening strategies are similar to the cognitive processing strategies discussed in chapter 2, to many of the instructional approaches discussed in this chapter, and to the instructional strategies developed in other chapters in this text. The strategies approach also benefits from the modeling approach described in chapter 3. Within the modeling approach, teachers demonstrate what they expect students to do during listening instruction by discussing the requirements and purposes for each listening strategy, by providing examples of that listening strategy, by sharing their own thought processes as they approach and accomplish the listening task, and by exploring the reasoning process used during that listening task.

Listening for main ideas   Listening for main ideas can be incorporated into any content area. One successful method of instruction is to have children identify their purpose for listening. During this activity, the teacher reads short stories or paragraphs aloud and then asks the children to supply a good title for the selection. The best titles usually contain the main idea of the selection. For example, a social studies teacher might read the following paragraph:

> Mount Everest in Tibet is the highest spot on earth and is a challenge to mountain climbers. From 1922 until 1952 many climbers tried to reach the top of Everest but all failed to reach the dangerous summit. In 1953 a British expedition tried again to reach the summit. Inch by inch the group struggled over the ice-packed surface. The air was so thin and cold that the climbers finally had to put oxygen tanks on their backs in order to breathe at such heights. Finally, on May 29, 1953, two members of the group, Edmund Hillary and Tenzing Norkey, reached the summit. At last, Everest had been conquered.

After the oral presentation, students should offer their ideas for the best title. The teacher can write the suggested titles on the chalkboard, and the students can discuss and defend their appropriateness and completeness. Use simpler paragraphs for early-elementary classes.

Two other methods are useful in helping students listen for the main idea:

1. Before they hear a paragraph, have students listen to oral subheadings. Ask them to turn the subheadings into questions they will be able to answer after listening to the selection. For example, a science subheading, "The Shape of Molecules," could be changed to the question, "What is the shape of molecules?" This procedure helps students develop a purpose for hearing specific information.
2. Have children listen to a story, and stop periodically to have them summarize the main ideas of the selection up to that point.

Listening for important details   Students often have trouble differentiating between important and insignificant details. For example, one student listening to a description about Henry Ford's invention of the Model T could remember only that the car was black. Although this was an interesting detail, it was not one of the most significant details of the selection. Instructing children to listen for important details is closely related to listening for main ideas, because important details support the main ideas.

Children can listen for supporting information as well as for the main idea in paragraphs such as that describing the dangerous conquest of Mt. Everest. Science books provide excellent sources for such material.

Having students briefly outline an orally presented selection will also help students listen for both main ideas and supporting details. It is often helpful to write the main idea or headings in the form of questions that can be answered by the selection's important details. For example, a group of third graders listened to an introduction to the study of plants. The selection was called "Plants Are Important to People." The group first discussed the title and their purpose for listening to the selection. Next, the teacher asked for suggestions on how to turn the title into a question that would guide their selective listening, and wrote the question on the chalkboard. Then she read the selection to the class. Finally, the class developed an outline that answered their purpose question.

     I. Why Are Plants Important to People?
        A. Plants give people oxygen to breathe
        B. Plants give people food to eat
        C. Plants like cotton give people clothes to wear
        D. Trees are plants that give people houses
        E. Plants decay and give people coal to burn

This activity helped students define their purpose for listening and organize the presentation logically. Here are some other activities for helping children listen for important details:

1. Present a selection orally and have students classify the details according to whether they are important, helpful, or unnecessary.

Students discover purposes and strategies associated with listening.

2. Read a descriptive selection to your class. Find a selection that mentions such characteristics as size, color, texture, number, and so forth. Have the class listen carefully for descriptive details, then draw a picture that accurately represents the details. Wordless books such as Pat Hutchins's *Changes, Changes* (1971) and Tana Hoban's *Round & Round & Round* (1983) can provide source material for descriptive text developed by the teacher. Informational books such as Sylvia Johnson's *Potatoes* (1984) and Patricia Lauber's *Volcano: The Eruption and Healing of Mount St. Helens* (1986) provide detailed descriptions of science-related topics for older students.

3. Have students write detailed descriptions of characters in familiar stories. Have them read their descriptions to the class without naming the character. Classmates can try to identify the character and indicate which details were most helpful to them. Similar folktales from different cultures provide interesting sources for this activity. For example, students could describe the German Cinderella in a Grimm's version, the Vietnamese Cinderella from Ann Nolan Clark's *In the Land of Small Dragon* (1979), and the Zuni version "Poor Turkey Girl" found in Virginia Haviland's *North American Legends* (1979). In these examples, cultural information adds to the descriptions.

4. While one child reads a detailed description, another draws the description without knowing the identity of the final product. The partners then compare the picture with the original description.

5. Tape imaginary or real conversations between two famous people in history, two people with different occupations, or two school employees. Have students listen carefully for details that suggest the identities of the individuals.

Listening for sequences   We have already presented the concept of using advanced organizers and listening for sequentially related words, such as "first," "second," and so forth when listening to directions. Students should also listen for key words signifying sequences or relationships in other listening situations. Point out to your students that both speakers and writers use terms to help the speaker as well as the listener organize the selection. Have students listen to oral presentations in which speakers use number words to organize their presentations. For example, students may listen to the following oral presentation to discover the usefulness of certain signal words:

> A moth that you see flying from plant to plant in the garden has not always been able to fly. In fact, there are four different stages in the life of that moth. The first stage in the life cycle of the moth is an egg. During the second stage the caterpillar hatches from the egg and starts to eat plants. After the caterpillar has been eating your garden plants for several weeks, he spins a cocoon of silk all around himself. During this third stage in the moth's cycle, he sleeps while his body actually changes. Finally, the cocoon breaks open and out flies the moth.

Which words in the previous paragraph signal a sequence of events that actually took place? Help students discover the usefulness of the purpose statement, "there are four cycles in a moth's life," and that each cycle is identified with signal words—"first," "second," "third," and "finally."

Other useful sequential signal words and phrases are "to begin with," "at the same time," "before," "after," "previously," "my next point," and "my final point." Signal words that emphasize a speaker's conclusion also deserve attention—"in conclusion," "in summation," "thus," "consequently," "therefore," and "hence."

Time lines are exceptionally appropriate devices to help students understand the sequences of orally presented selections. A group of fourth graders, interested in the study of aviation, developed the following time line shown in figure 4–2 after seeing and listening to several films and stories about aviation. This kind of time line can also be presented before a listening activity as a visual advanced organizer for students who need more help than usual.

The teacher can use other activities to help students listen for sequence:

1. Cut up a story into paragraphs. Paste each paragraph on a piece of cardboard. Read the selection in proper sequence and have students place the mixed paragraphs into the right order.
2. Divide a sequentially developed paragraph into sentences. Paste the sentences on cardboard. Have students listen to the paragraph and place the sentences into proper sequence.
3. Comic strips work well for developing sequential skills. Have students listen while you read aloud a comic strip, then have them put individual panels into the right order.
4. A flannel board provides an excellent opportunity for children to listen to a story, hear its sequential organization, then retell the story in appropriate sequence. This method of presentation is described in the chapter on literature.

1783 —World's first recorded balloon flight—passengers included a rooster, a duck, and a sheep
1900 —Zeppelin developed in Germany
1903 —Wright brothers flew at Kitty Hawk
1927 —Lindberg flew the *Spirit of St. Louis* across the Atlantic from New York to Paris
1932 —Amelia Earhart became the first woman to fly across the Atlantic
1937 —Hindenberg burned and crashed
1945 —Fighter planes used in World War II could go 450 miles per hour
1950 —American Air Force used mostly jets during Korean War
1970s—Space travel
1980s—Space shuttles

FIGURE 4–2
Aviation time line

Listening to predict outcomes   When children are given opportunities to predict what happens next in a story or speech, they extend their ability to understand sequential relationships. Predicting also requires them to become involved in listening to the story, so they can find out whether or not their predictions were accurate. One useful technique is to have students listen to a book's chapter titles and predict what will be in each chapter before it is read to them.

For example, before reading *Runaway Slave—The Story of Harriet Tubman,* I asked third-grade children to listen to the title of each chapter and write a brief prediction about what information they thought would appear in the chapter. Writing involves each student in making his own prediction. After reading and discussing the title so everyone understood the time in which the book takes place, they reviewed the concept of "slave." (This was not the first story they had heard about that period in history.) Each chapter title was then read aloud, allowing enough time between titles for each student to write a brief prediction. After finishing their predictions, the class discussed them and gave reasons for their predictions. The chapter headings include: Who Was Harriet Tubman?; At Miss Susan's; Follow the North Star; Trouble; I'm Going to Leave You; Harriet Tubman, Conductor; Nighttime, Daytime; Go On—Or Die; A Sad Christmas; and The War Years. After hearing each chapter, the class reviewed the accuracy of their predictions. This activity was also a highly motivating listening activity because each child listened carefully to see if her predictions were correct.

Here are other activities for predicting outcomes:

1. Present a selection such as Mildred Pitts Walter's *Brother to the Wind* (1985) or Arthur Yorink's *Hey, Al* (1986) orally and stop at an interesting point in the story. Have students tell or write how they think the story will end.
2. Orally present short descriptions of stories such as Hans Christian Andersen's *The Emperor's New Clothes* (1982), Michael Bond's *A Bear Called Paddington* (1960), and E. B. White's *Charlotte's Web* (1952), Caroline Arnold's *Saving the Peregrine Falcon* (1985), Jean Fritz's *Make Way for Sam Houston* (1986), and Colleen Bare's *Guinea Pigs Don't Read Books* (1985). Ask the children to predict whether the story will be fact or fantasy, and explain their decisions.
3. Films are also useful for predicting outcomes. Show a film until you reach an interesting point, then have students finish the story before they see the remainder of the film. Remember that many of their endings may be as good or as logical as the actual film or story ending.

One learning center for listening was created to find activities that would help develop listening for awareness, enjoyment, and for specific purposes. The criteria for a listening learning center should include the following requirements:

1. The activities should provide practice in developing one or more of the listening skills stated in the purpose.
2. The students need to be able to work independently, either alone or in small groups. The listening material must be made available on tape, and directions must be specific.

3. The activities should provide enough enjoyment to motivate without teacher assistance.
4. The activities should use a variety of skills and methods for responding.
5. The center should include activities geared to several ability levels.
6. The activities should be separate, and not require completion in a certain sequence.
7. The activities should be related to other school subjects and to the listening needs of the individual child.

Prepare the students by discussing the purposes for the activities in the center. Teach the students how to use the equipment in the center, and teach at least one activity to the whole class before students use the center individually. Instruct the students to prepare themselves mentally to concentrate on listening, the way athletes concentrate on their sports. Instruct students on how to get rid of distractions, to close their eyes, form a picture in their minds, and pretend they and the voice or sound are all alone in the room.

The specific learning center activities, all on tape, included:

1. A descriptive paragraph—Students drew the picture described, then listened again, and corrected or added to their pictures in another color.
2. A story—Students divided their paper into four squares and drew four pictures to show the sequence of story events.
3. The story and poem, "The King's Breakfast"—Students numbered in sequence printed sentences that described the events in the story.
4. Mystery sounds (ball bouncing, blocks hitting together, water pouring, etc.)—Students attempted to identify each sound.
5. Classmates' voices—Students identified each voice.
6. Verbal descriptions of animals and their characteristics—Children listened to the tape and matched the sounds to the appropriate picture.
7. Vivid verbal descriptions—Students colored a picture after listening to a description.
8. Poems that paint mental pictures such as "Foul Shot," "Steam Shovel," and "Catalogue" all found in Stephen Dunning, Edward Lueders, and Hugh Smith's anthology *Reflections on a Gift of Watermelon Pickle . . . and other Modern Verse* (1967)—Students drew the picture they saw in their minds.
9. Pictures of a city street, a farm, a zoo, and a seashore—Students viewed the pictures and listed the sounds they would hear if they were in the location of one of the pictures.
10. A funny short story with dialogue between characters such as Steven Kellogg's *Tallyho, Pinkerton!* (1982)—Children listened to the tape and drew the sequences in cartoon style.
11. Several short stories—Children wrote a title for each selection.

After using the learning center, the teachers who developed it concluded that working on listening is good training for the teacher as well as for the pupils. They learned that all activities should be carefully planned so students can understand the purpose, order, content, and vocabulary of the exercises. They also indicated that students should be told what to listen for. The teachers learned that the best listening activities demand extensive student involvement.

LITERATURE-RELATED LISTENING ACTIVITIES

Literature is an excellent source for listening materials that encourage active verbal responses, develop understanding of story structure, foster literacy appreciation, and teach reading. Language development, cognitive development, and enjoyment are also stimulated by activities involving children during a story experience. Books such as Arthur Geisert's *Pigs from A to Z* (1986) encourage children to interact with the text during a listening experience.

Young children may become acquainted with story structure and language patterns through interaction with stories that have repetitive language patterns such as "The Three Billy Goats Gruff" and/or repetitive or cumulative story events such as "The Gingerbread Boy." Teachers frequently call such books predictable because the repetitive patterns encourage children to hypothesize about words that follow or events that happen next in the story. When using predictable books with children, present a series of activities during which children listen to a story and select the most useful phonemic, syntactic, and semantic information; predict the most probable language or happening; and confirm if their prediction was correct.

McCracken and McCracken (1986) described a procedure in which they use simple stories to teach beginning reading to kindergarten and first-grade students. The stories they choose are either predictable because of the text or predictable because the children have previously memorized the text. During this type of activity the teacher first encourages children to memorize a story or the repetitive parts of a story by repeatedly reading the story to children and encouraging them to join in during the reading. Pictures are used to help children follow the story sequence. Next, children are introduced to print by either the teacher printing the repetitive parts of the text on the chalkboard or placing accompanying word cards in a pocket chart. During this phase of the reading, the teacher points to the words as the class chants the lines. Next, the children match the words on the word chart with a second identical set of word cards. If the words form a refrain, the teacher either places the whole refrain on the chart and the children match the words by placing identical words on top of the cards or the teacher creates the first line of the refrain and the children reproduce the next lines. During this experience children have many opportunities to 'read' the text. Next, the story is placed on phrase cards in the pocket chart. Children match these phrases to pictures representing appropriate parts of the story content or appropriate characters and their actions. These pictures are placed in appropriate places on the pocket chart. This type of activity continues with many opportunities for individual and group practice until children can rebuild the entire story. By this time many of the children can read the completed story.

Tompkins and Weber (1983) provided a five-step teaching strategy that directs children's attention to repetitive and predictable features of a book. First, the teacher reads orally the title, shows the cover illustration, and asks children what they believe the story is about. Second, the teacher reads through the first set of repetitions and into the point where the second set of repetitions begins. Now the teacher formulates questions that ask children to predict what will happen next or what a character may say. Third, the teacher asks children to explain why they made their predictions. Fourth, the teacher reads the next set of repetitive patterns to allow children to confirm or reject their predictions. Finally, the teacher continues reading the selection as steps two, three, and four are repeated.

The following books contain predictable elements:

- Verna Aardema's *Bringing the Rain to Kapiti Plain* (1981)
- Maggie Duff's *Rum Pum Pum* (1978)
- Mem Fox's *Hattie and the Fox* (1987)
- Paul Galdone's *The Greedy Old Fat Man* (1983)
- Pat Hutchins's *Rosie's Walk* (1968)
- John Ivimey's *Three Blind Mice* (1987)
- Maurice Sendak's *Where the Wild Things Are* (1963)
- Janet Stevens's *The House that Jack Built* (1985)

Story Structure

Traditional tales with easily identifiable conflicts, characters, settings, and plot developments are excellent sources for developing children's understanding of story structure and story elements. The nature of the oral tradition made it imperative that listeners be brought quickly into the action and identify with the characters. Consequently, tales such as "The Little Red Hen," "Puss in Boots," "Jack the Giant Killer," "Three Billy Goats Gruff," and "The Fisherman and His Wife" allow children to be immersed into the major conflict within the first few sentences, understand the nature and importance of the building conflict, identify the characters and character traits, identify the setting, and recognize major features in plot development. For example, the conflict in Paul Galdone's *The Little Red Hen* is between laziness and industriousness. The first sentence introduces the animals who live together. The second sentence introduces the conflict; that is, the cat, dog, and mouse are lazy. Next the industrious hen is introduced. The remainder of the story develops conflict between lazy and industrious animals. The conflict is quickly resolved when the hen eats her own baking.

Drawing semantic maps, or webs, illustrating characters, setting, and plot development helps children visualize the relationships among these important elements. After listening to a story, the teacher and children may develop a web and discuss characterization, setting, and plot development.

Two examples of webs developed around traditional tales illustrate a simpler and a more complex folk tale. The first, "Three Billy Goats Gruff," introduces characters and plot development in a chronological order proceeding from littlest to biggest (see figure 4–3). Notice how the web illustrates the close relationship between characteristics of the characters and plot development.

FIGURE 4–3
Semantic web of "Three Billy Goats Gruff"

The second example, Grimm's "The Fisherman and His Wife," is more complex and more appropriate for older children (See figure 4–4). The characterization changes as the tasks change, the setting becomes more turbulent as the demands of the characters increase, and the plot development relates directly to changes in characterization and changes in setting. Developing a web following a listening activity vividly illustrates these relationships.

Developing Appreciation through Listening

We sometimes become so engrossed in teaching specific skills that we forget that enjoyment is one of the main purposes of listening. Appreciative listening includes listening to poetry, music, literature, the theater, or sounds in nature. In fact, the development of appreciative listening skills can establish lifelong pleasure while it motivates the utilization of other language arts skills. Durkin (1966) found that young children whose parents read to them at home were also high achievers in reading at school. This listening experience apparently illustrates to the child the benefits of learning to read enjoyable books.

Various taxonomies offer the teacher some guidelines for developing appreciative listening with literature. Barrett's (1972) taxonomy of comprehension, for example, stresses the development of appreciative skills. Although this taxonomy was developed in relation to reading, it also has application for listening. According to Barrett, appreciation includes the student's awareness of literary techniques, forms, styles, and structures an author uses to stimulate emotional responses. Barrett identified four tasks that fall under this appreciation component: (1) emotional

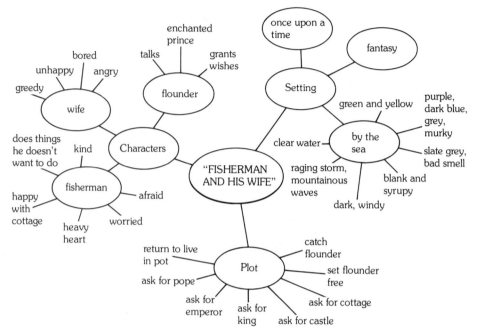

FIGURE 4–4
Semantic map for a book report

response to plot or theme; (2) identification with characters and incidents; (3) reactions to the author's use of language; and (4) reactions to the author's use of imagery.

If we apply this taxonomy to a listening situation, the emotional response task would require students to listen to a selection and determine what the author did to develop the plot or theme to elicit certain responses in the listener. Why did the listener feel excitement, happiness, or sadness? Emotional response is an individual matter, so we certainly will not all respond in the same way to the same selection. As they listen to selections, children may respond through pantomime or other creative dramatics, then discuss how and why a selection made them feel a certain emotion.

Students can also record their favorite stories or poems, then let the listeners discover the qualities in the stories and poems that have emotional appeal for the listener. One creative fourth-grade teacher asked her students to interview their parents; students asked their parents to identify selections from poetry or stories they remembered from their childhood. Then the teacher presented the most popular selections orally. The fourth graders listened to the selections and discussed their universal appeal.

The task for identification with characters and incidents on the appreciation taxonomy requires students to become aware of literary techniques that prompt them to sympathize with, empathize with, or reject a character. Students listen to a selection to discover how the author presents a character, then the teacher asks what descriptive words influenced how they felt about the characters. Did the author use terms such as *heroic, shiftless, lazy, beautiful, strong,* and so forth? How does the author have other people describe the character? Does the character tell you his inner feelings? How do other people in the story react to the character? How does the character overcome adversity?

For the third task, reaction to the author's use of language, students must pay attention to how the writer uses such devices as figures of speech, similes, and metaphors. Authors frequently use similes and metaphors to create new relationships between unrelated things. A *simile* is an expressed comparison of two different things or ideas; a *metaphor* is a implied comparison. For example, in *Secret of the Andes* (1952) by Ann Nolan Clark, the author uses a number of similes. Listen to two examples and visualize how you would respond to these descriptions: "The boy's thoughts were whirling like the foaming rapids on the far side of the valley" and "He bore the proud look of the giant condor circling a cliff nest on a mountain crest." Students may search for similar devices and share them with the class. Besides similes and metaphors, authors often use sentence length and structure to denote pace. Authors may change language patterns to help create the action of the story. Short, staccato sentences impart a feeling of mounting excitement; long, lazy sentences produce a feeling of relaxation. Robert McCloskey used this technique in *Time of Wonder.*

> in the afternoon, when the tide is out, they build a castle out of the rocks and driftwood below the spot where they had belly-whopped and dog-paddled during the morning . . . (p. 24)

> Suddenly the wind whips the water into sharp, choppy waves. It tears off the sharp tops and slashes them in ribbons of smoky spray. And the rain comes slamming down. The wind comes in stronger and stronger gusts. A branch snaps from a tree. (p. 44)[2]

You can read similar examples to children and see how their reactions change as sentence patterns vary.

The final appreciation-taxonomy task, imagery, requires students to react to an author's ability to paint word pictures. Can they close their eyes and visualize the setting, a character, or a feeling? For example, when you listen to *Charlotte's Web,* can you visualize the barn that has such a peaceful feeling, as if nothing bad could ever happen again? My own college classes are introduced to this technique by listening to and acting out stories that include vivid descriptions. One of my favorites, which I have used with both elementary and college students, is "Night in a Haunted Castle" (in Drumbeats, Field Educational Publications, 1970, pp. 75–78). This is the story of a boy who likes to read ghost stories; he is overjoyed when he and his parents spend the summer in a real English castle. The story describes the castle in detail, with its armour and winding staircases. The climax of the story takes place during the boy's final night of sleeping in the castle. During the night, he hears chains and dragging footsteps, sees white ghostlike figures, and feels the bed shaking. When I read the selection, students stand and close their eyes, in order to see, hear, and perform the accompanying

---

[2]From Robert McCloskey, *Time of Wonder* (New York: The Viking Press, 1957). Reprinted by permission.

actions. Everyone usually becomes visually involved with the selection, and each is able to retell what he sees, hears, feels, and sometimes, even, smells.

Sharing poetry with children is an excellent means for developing appreciative listening. Children's responses to Shel Silverstein's *Where the Sidewalk Ends* (1974) emphasize the importance of poetry. Poet José Garcia Villa (Cowen, 1983) maintained that children should be encouraged to enjoy poems for sound values, to hear the rhythm of language, and to be aware of the magic in a poetry line. While listening to poetry children can sense and feel the meaning rather than explicitly state the meaning. Villa recommended that children listen to nursery rhymes and poems by Emily Dickinson, Robert Frost, e. e. cummings, Nathalia Crane, Lewis Carroll, and Edward Lear. For older children Villa also recommended poems by John Donne, Gerard Manley Hopkins, Dylan Thomas, Wilfred Owen, Elinor Wylie, and Marianne Moore. Villa believed these poets will help children discover the magic in poetry. He stressed "the magic to be found in poetry rather than its meaning which is what most educational programs and texts emphasize. Only after the older student or budding poet considers the magical elements inherent in such techniques as the importance of a vivid first line; the musicality, tension, and movement of language; the intricacies of versification; the richness of metaphor; the necessity of economy; the element of surprise; and the implosive drama of the last line, will the neophyte begin to understand the power and magnificence of poetry" (p. 84).

Teachers who want to encourage children to expand the imaginative quality of their minds and appreciate the various elements in poetry may share poems emphasizing such elements as rhythm, rhyme, sound patterns, repetitions, and figurative language. For example, David McCord uses rhythm in "The Pickety Fence" (*Far and Few: Rhymes of the Never Was and Always Is*) to suggest the sounds a stick makes as a child walks by the fence. Likewise, Robert Louis Stevenson relies on rhythm to suggest the dash and rattle of a train as it crosses the countryside in "From a Railway Carriage" (*A Child's Garden of Verses*).

The auditory pleasure of rhyming words is heard in Zilpha Snyder's "Poem to Mud" (*Today Is Saturday,* 1969) and Edward Lear's "The Owl and the Pussy Cat" (*The Nonsense Books,* 1946). Alliteration, the repetition of initial consonants to create sound patterns, creates strong sound patterns in Jack Prelutsky's "The Yak" (*A Gopher in the Garden,* 1966). Repetition in poetry may emphasize words, phrases, lines, or verses. Lewis Carroll's repetitive patterns in "Beautiful Soup" (*Alice's Adventures in Wonderland,* 1984) seem to recreate the marvelous sounds of rich, hot soup ladeled into a spoon and placed noisily into the mouth. Poets use figurative language to encourage their audiences to see things in new ways, to clarify, and to add vividness. Kaye Starbird's imagery in "The Wind" (*The Covered Bridge House and Other Poems,* 1979) depicts a wind that is sneaky, tricky, and full of life. The elementary classroom should provide many opportunities for listening for pleasure.

Evaluating What Is Heard

The final category of listening comprehension usually mentioned by authorities is the ability to evaluate what is heard. Because this subject lends itself so well to listening to television and radio presentations, and to listening to and reading newspaper advertisements and editorials, evaluative listening is discussed in the chapter about Multimedia.

REINFORCEMENT ACTIVITY

We have discussed a number of methods, suggestions, and illustrative units for teaching attentive listening, directed listening for a specific purpose, and appreciative listening. Choose one of these topics and develop a lesson for fostering listening. If you are a student teacher or a classroom teacher, use your lesson with children. Allow the children to participate in developing their goals for the lesson and in evaluating their success. Share the activities with your language arts class. If you are not now teaching children, practice your listening lesson with a group of your peers.

You may also choose to develop a learning center that would be used to teach some aspects of listening. Form a committee with a group of your classmates to determine objectives for the learning center and design appropriate activities for it. Use the learning center with children in a classroom, or demonstrate the use of the learning center to a group of your peers.

MAINSTREAMING AND LISTENING

Many of the activities recommended in this chapter are good for children with special needs. In an effort to identify specific characteristics of children who might have difficulty during listening activities, we have developed two charts that list characteristics that may be found in learning disabled (table 4–2) or mentally handicapped children (table 4–3). The techniques for language arts instruction are identified from research and authorities who work with mainstreamed children.

SUMMARY

We have defined listening to encompass auditory acuity, auditory discrimination, and various levels of listening comprehension, including attentive listening, listening for a specific purpose, and appreciative listening. We have discovered that listening is an important and complicated skill.

A complete profile of children's listening abilities requires understanding of and evaluation of several listening components. Diagnosing listening skills may require tests for auditory acuity, because children cannot comprehend oral communication if they cannot hear; children who appear to hear well may still have weak auditory perception skills. Diagnosis of auditory perception skills may be critical for children who have difficulty blending sounds in a reading or spelling approach that relies on phonic analysis. A positive relationship has been found between attentiveness and

TABLE 4-2

Mainstreaming and listening development for learning disabled children

Characteristics of Learning Disabled Children	Teaching Techniques for Language Arts Instruction
1. May demonstrate difficulty attending to task-relevant information and may be hyperactive, impulsive, and distractible (Bryan and Bryan, 1986; Hallahan and Kauffman, 1986; Samuels and Edwall, 1981; Sapir and Wilson, 1978; Swanson, 1980).	1. Use considerable positive reinforcement. (O'Connor, Stuck, and Wyne, 1979). Help children attend by providing a private space for learning; maintain a structured, well-organized environment; help them focus on the task; provide immediate reinforcement; give several short assignments rather than one long assignment; let children reduce stress by standing; provide quiet opportunities during the school day; and help children experience success (Stoodt, 1981).
2. Auditory modality may be inferior to visual modality. Deficiencies in ability to blend sounds into words and difficulties in auditory closure related to problems in word analysis (Harber, 1979, 1980; Richardson, Di Benedetto, Christ, Press, and Winsberg, 1978; Swanson, 1980).	2. Provide practice in analyzing syllables and short words into phonemes and then blending phonemes into syllables and words (Williams, 1980). Focus on the order of sounds and letters in words, multisensory presentation of stimuli, considerable repetition, and varied materials (Wallace and Kauffman, 1986).
3. Demonstrate difficulties comprehending main ideas in listening materials (Wong, 1979, 1980).	3. Precede a listening experience with questions focusing on the purpose for listening (Wong, 1979, 1980). Teach children to set purposes for listening; give them training in critical thinking through texts such as social studies and propaganda materials (Spache, 1981).

listening; consequently, diagnosis of attentive listening skills may be crucial for future achievement. A profile of listening ability also includes how effectively children comprehend what is heard in a wide range of circumstances. Both standardized and informal tests may be used for these diagnoses. Some of the teacher-made informal tests provide the best information for designing effective instructional programs.

Students must participate in their own instructional improvement because listening is an active rather than passive activity. Children must help set goals for listening instruction. We have also found that classroom conditions and teacher behaviors affect listening comprehension. Methods were introduced for developing the auditory perception skills necessary to catch differences in sounds and words, the attentive listening skills necessary to follow directions, and listening for specific purposes, such as identifying main ideas and sequences, and predicting outcomes. In addition, listening for appreciation is a skill that should be included in instructional planning.

Additional activities offer you opportunities to refine your teaching skills and to visualize theory and methods. The units and lessons in listening have all been taught to children, and will help you conceptualize the listening curriculum. Listening, like the other communication skills, relates to the total language arts program. It cannot be

TABLE 4–3

Mainstreaming and listening development for mentally handicapped children

Characteristics of Mentally Handicapped Children	Teaching Techniques for Language Arts Instruction
1. May demonstrate short attention span due to failure in past efforts, to inappropriate academic expectations, or to auditory or visual distractions in the classroom (Gearheart, Weishahn, and Gearheart, 1988).	1. Demonstrate consistency in classroom management techniques and model desired behavior (Hart, 1981). It is easier to maintain attention and to recall later those things that are concrete rather than abstract, familiar rather than remote, and simple rather than complex. Use language and ideas that are as concrete, familiar, and simple as possible. Eliminate visual and auditory distractions that might decrease students' ability to attend (Gearheart, Weishahn, and Gearheart, 1988).
2. Below average ability to generalize, conceptualize, and perform other cognitive-related tasks (Gearheart, Weishahn, and Gearheart, 1988).	2. Provide instruction that encourages cognitive development and improvement of listening comprehension by encouraging children to respond to inferential questions and rephrase materials that are heard (Zetlin and Gallimore, 1980). Help children set their own purpose for listening by accompanying listening activities with a Directed Reading-Thinking Activity in which children learn to make predictions based on a few cues, either pictorial or textual, then listen to confirm or disprove the prediction (Stauffer, Abrams, and Piluski, 1978). Read materials to children that are similar to those used for reading comprehension. Develop listening for main ideas, details, sequence of events, following directions, making inferences, drawing conclusions, and critical evaluation (Lerner, 1985).
3. Demonstrate problems in auditory discrimination (Cook, Nolan, and Zanotti, 1980).	3. Try supplementing instruction with the Neurological Impress Method in which the teacher and student read aloud in unison (Cook, Nolan, and Zanotti, 1980). Encourage children's sensitivity to sounds by listening with their eyes closed to environmental sounds, recorded sounds, and other teacher-made sounds. Encourage auditory attending to sounds by listening to or repeating sound patterns. Encourage discrimination of sounds by discriminating loud and soft sounds, high and low sounds, and near and far sounds. Encourage awareness of letter sounds by asking children to identify words that begin with the same initial consonant or consonant blend, identify words that have the same endings or vowel sounds, and identify words that rhyme (Lerner, 1985).

taught in isolation; it must be taught in classroom environments similar to those in which it will be used. Listening must also be taught continually, because it is part of everyone's experience from birth through adulthood.

ADDITIONAL LISTENING ACTIVITIES

1. *Compile a list of children's books or poems and rhymes that contain rhyming elements or other interesting sound patterns such as alliteration.*
2. *Develop an advanced organizer that encourages children to listen for a specific purpose such as following directions, paraphrasing the main idea of a selection, or identifying the sequential order in a paragraph or longer selection.*
3. *Create an auditory awareness tape that explores sounds such as nature sounds, animal sounds, street sounds, home sounds, or school sounds. Share the tape with a group of children. Ask them to identify the sounds, classify them according to location, and list other sounds that could be found in the same category.*
4. *With a group of children, develop a set of guidelines that will help them improve their listening ability. Ask them to identify different listening activities that occur in the day, their purpose for listening, and their requirements for that specific listening activity.*

Listening Activity	Purpose	Requirements
Directions for assignments	*Obtain accurate, detailed information that I can follow*	*Attention given to all information*
Music	*Entertainment and appreciation*	*Relaxed environment*

5. *Compile a list of predictable books that rely on repetitive language patterns and/or repetitive story events. Develop a series of questions that directs children's attention to the repetitive or predictable features of a book. Share the book and questions with children.*
6. *Select a story or content area materials that would allow children to practice gaining meaning from a listening experience (e.g., summarizing, clarifying, predicting, and verifying). Identify appropriate places in the story where children should summarize what they hear, places where they might benefit from a clarifying discussion of ideas or content, places where they can predict what might happen next, and locations where predictions can be verified. Share the listening materials with a group of your peers. Evaluate the effectiveness of the listening experience.*
7. *Compile a card file of poems that allow children to hear the rhythm of language, that have rich metaphors, or that have elements of surprise that delight children.*
8. *Read a folktale to a group of children. Map or web the characters, setting, and plot development.*

9. *Evaluate a Listening Learning Center: The listening skills we have discussed—listening for the main idea, details, sequence, and predicting outcomes—may also be reinforced by using a learning center for listening. The listening center previously described, developed by a group of student teachers, was used in both third and fourth grades. Look carefully at each component. What listening skill is being reinforced by each? How would you introduce the center to children? How would you assist children in using the center? Do you believe the purposes of the center were met? Would you add anything else to the center?*

BIBLIOGRAPHY

Ausubel, D. P. "The Use of Advanced Organizers in the Learning and Retention of Meaningful Verbal Material." *Journal of Educational Psychology* 51 (1960): 267–72.

Barrett, Thomas C. "Taxonomy of Reading Comprehension." *Reading 360 Monograph.* Lexington, Mass.: Ginn & Co., 1972.

Baumann, James. "Listening Activities for the Language Arts." In *Language Arts Instruction and the Beginning Teacher,* edited by Carl Personke and Dale Johnson. Englewood Cliffs, N.J.: Prentice-Hall, 1987.

Brassard, Mary Butler. "Planning Test Construction." In *Teaching Listening in the Elementary School,* edited by Sam Duker. Metuchen: N.J.: The Scarecrow Press, 1971.

Bryan, Tanis H., and Bryan, James H. *Understanding Learning Disabilities,* 3rd ed. Mountain View, Calif.: Mayfield Pub., 1986.

Cook, Jimmie E.; Nolan, Gregory A.; and Zanotti, Robert J. "Treating Auditory Perception Problems: The NIM Helps." *Academic Therapy* 15 (March 1980): 473–81.

Cowen, John E. "Conversations with Poet Jose Garcia Villa on Teaching Poetry to Children." In *Teaching Reading through the Arts,* edited by John E. Cowen. Newark, Del.: International Reading Association, 1983, pp. 78–87.

Devine, Thomas G. *Listening Skills Schoolwide: Activities and Programs.* Urbana, Ill.: ERIC Clearinghouse on Reading and Communication Skills, 1982.

Devine, Thomas G. "Listening: What Do We Know after Fifty Years of Research and Theorizing." *Journal of Reading* 21 (January 1978): 296–304.

Duker, Sam. *Listening Bibliography.* Metuchen, N.J.: The Scarecrow Press, 1968.

_____. *Teaching Listening in the Elementary School: Readings.* Metuchen, N.J.: The Scarecrow Press, 1971.

Durkin, Dolores. *Children Who Read Early.* New York: Teachers College Press, 1966.

Furness, Edna Lue. "Proportion, Purpose, and Process in Listening." In *Teaching Listening in the Elementary School,* edited by Sam Duker. Metuchen, N.J.: The Scarecrow Press, 1971.

Gearheart, Bill R.; Weishahn, Mel; and Gearheart, Carol J. *The Exceptional Student in the Regular Classroom,* 4th ed. Columbus, Oh.: Merrill Publishing Co., 1988.

Hallahan, Daniel P., and Kauffman, James M. *Exceptional Children: Introduction to Special Education.* Englewood Cliffs, N.J.: Prentice-Hall, 1986.

Harber, Jean R. "Auditory Perception and Reading: Another Look." *Learning Disability Quarterly 3* (Summer 1980): 19–29.

_____. "Differentiating LD and Normal Children: The Utility of Selected Perceptual and Perceptual-Motor Tests." *Learning Disability Quarterly 2* (Spring 1979): 70–75.

Harris, Albert J., and Sipay, Edward R. *How to Increase Reading Ability.* 8th ed. New York: Longman, 1985.

Hart, Verna. *Mainstreaming Children with Special Needs.* New York: Longman, 1981.

Heilman, Arthur W. *Phonics in Proper Perspective,* 6th ed. Columbus, Oh.: Merrill Publishing Co., 1989.

Horowitz, R. and Samuels, S. Jay. "Reading and Listening to Expository Text." *Journal of Reading Behavior* 17 (1985): 185–98.

Landry, Donald L. "The Neglect of Listening." *Elementary English* 46 (1969): 599–605.

Lerner, Janet W. *Children with Learning Disabilities,* 4th ed. Boston: Houghton Mifflin Co., 1985.

Levin, J. R., and Pressley, M. "Improving Children's Prose Comprehension: Selected Strategies That Seem to Succeed." In *Children's Prose Comprehension: Research and Practice,* edited by C. Santa and B. Hayes. Newark, Del.: International Reading Association, 1981.

Lundsteen, Sara W. *Listening: Its Impact on Reading and the Other Language Arts.* Urbana, Ill.: National Council of Teachers of English, 1979.

MacGinitie, Walter H. "Research Suggestions from the Literature Search." *Reading Research Quarterly* 11 (1975–1976): 7–35.

McCracken, Robert A. and McCracken, Marlene J. *Stories, Songs, and Poetry to Teach Reading and Writing: Literacy Through Language.* Chicago: American Library Association, 1986.

Neville, Mary H. "Effect of Reading Method on the Development of Auditory Memory Span." *The Reading Teacher* 22 (October 1968): 30–35.

O'Connor, Peter D.; Stuck, Gary B.; and Wyne, Marvin D. "Effects of a Short-Term Intervention Resource-Room Program on Task Orientation and Achievement." *The Journal of Special Education* 13 (Winter 1979): 375–85.

Otto, Wayne, and Smith, Richard. *Corrective and Remedial Teaching.* Boston: Houghton Mifflin Co., 1980.

Pearson, P. David, and Fielding, Linda. "Research Update: Listening Comprehension." *Language Arts* 59 (September 1982): 617–29.

Porter, Jane. "Research Report." *Language Arts* 53 (March 1976): 341–51.

Rankin, Paul. *The Measurement of the Ability to Understand Spoken Language.* Doctoral dissertation, University of Michigan, 1926.

Richardson, Ellis; Di Benedetto, Barbara; Christ, Adolph; Press, Mark; and Winsberg, Bertrand G. "An Assessment of Two Methods for Remediating

Reading Deficiencies." *Reading Improvement* 15 (Summer 1978): 82–95.

Samuels, S. Jay, and Edwall, Glenace. "The Role of Attention in Reading with Implications for the Learning Disabled Student." *Journal of Learning Disabilities* 14 (June/July 1981): 353–61, 368.

Samuels, S. Jay, and Turnure, James E. "Attention and Reading Achievement in First-Grade Boys and Girls." *Journal of Educational Psychology* 66 (February 1974): 29–32.

Sapir, Selma, and Wilson, Bernice. *A Professional's Guide to Working with the Learning-Disabled Child.* New York: Brunner/Mazel, 1978.

Spache, George. *Diagnosing and Correcting Reading Disabilities.* Boston: Allyn & Bacon, 1981.

_____. *Investigating the Issues of Reading Disabilities.* Boston: Allyn & Bacon, 1976.

Stauffer, Russel G.; Abrams, Jules C.; and Piluski, John J. *Diagnosis, Correction, and Prevention of Reading Disability.* New York: Harper & Row, 1978.

Stoodt, Barbara D. *Reading Instruction.* Boston: Houghton Mifflin Co., 1981.

Swanson, H. Lee. "Auditory and Visual Vigilance in Normal and Learning Disabled Readers." *Learning Disability Quarterly* 3 (Spring 1980): 71–78.

Tompkins, Gail E.; Friend, Marilyn; and Smith, Patricia L. "Listening Strategies for the Language Arts." In *Language Arts Instruction and the Beginning Teacher,* edited by Carl Personke and Dale Johnson. Englewood Cliffs, N.J.: Prentice-Hall, 1987.

Tompkins, Gail E., and Weber, Mary Beth. "What Will Happen Next? Using Predictable Books with Young Children." *The Reading Teacher* 36 (February 1983): 498–502.

Tutolo, Daniel J. "A Cognitive Approach to Teaching Listening." *Language Arts* 54 (March 1977): 262–65.

Vernon, M. D. *Reading and Its Difficulties: A Psychological Study.* Cambridge: Cambridge University Press, 1971.

Wallace, Gerald, and Kauffman, James A. *Teaching Students With Learning and Behavior Problems.* Columbus, Oh.: Merrill Publishing Co., 1986.

Weidner, M.J. *A Study of the Effects of Teacher Oral Reading of Children's Literature on the Listening and Reading of Grade Four Students.* Doctoral dissertation, Boston University, 1976.

Wepman, Joseph M. *The Auditory Discrimination Test.* Chicago: Language Research Associates, 1973.

Wepman, Joseph M., and Morency, Anne. *The Auditory Memory Span Test.* Chicago: Language Research Associates, 1973.

Williams, Joanna P. "Teaching Decoding with an Emphasis on Phoneme Analysis and Phoneme Blending." *Journal of Educational Psychology* 72 (February 1980): 1–15.

Wong, Bernice Y. "Activating the Inactive Learner: Use of Questions/Prompts to Enhance Comprehension and Retention of Implied Information in Learning Disabled Children." *Learning Disability Quarterly* 3 (Winter 1980): 29–37.

_____. "Increasing Retention of Main Ideas Through Questioning Strategies." *Learning Disability Quarterly* 2 (Spring 1979): 42–47.

Woods, Mary Lynn, and Moe, Alden, J. *Analytical Reading Inventory,* 4th ed. Columbus, Oh.: Merrill Publishing Co., 1989.

Zetlin, Andrea G., and Gallimore, Ronald. "A Cognitive Skills Training Program for Moderately Retarded Learners." *Education and Training of the Mentally Retarded* 15 (April 1980): 121–31.

CHILDREN'S LITERATURE REFERENCES

Aardema, Verna. *Bringing the Rain to Kapiti Plain.* New York: Dial Press, 1981.

Andersen, Hans Christian. *The Emperor's New Clothes.* Retold by Anne Rockwell. New York: Harper & Row, 1982.

Andrews, Jan. *Very Last First Time.* Illustrated by Ian Wallace. New York: Atheneum Pubs., 1986.

Arnold, Caroline. *Saving the Peregrine Falcon.* Photographs by Richard R. Hewett. Minneapolis: Carolrhoda, 1985.

Bare, Colleen. *Guinea Pigs Don't Read Books.* New York: Dodd, Mead & Co., 1985.

Bond, Michael. *A Bear Called Paddington.* Illustrated by Peggy Fortnum. Boston: Houghton Mifflin Co., 1960.

Carroll, Lewis. *Alice's Adventures in Wonderland.* London: Macmillan Co., 1866, New York: Alfred A. Knopf, 1984.

Clark, Ann Nolan. *In the Land of Small Dragons.* Illustrated by Tony Chen. New York: Viking Press, 1979.

_____. *Secret of the Andes.* New York: Viking, 1952.

Duff, Maggie. *Rum Pum Pum.* New York: Macmillan Co., 1978.

Dunning, Stephen; Lueders, Edward; and Smith, Hugh. *Reflections on a Gift of Watermelon Pickle . . . and other Modern Verse.* New York: Lothrop, Lee & Shepard, 1967.

Fritz, Jean. *Make Way for Sam Houston.* Illustrated by Elise Primavera. New York: G. P. Putnam's, Sons, 1986.

Fox, Mem. *Hattie and the Fox.* Illustrated by Patricia Mullins. New York: Bradbury, 1987.

Galdone, Paul. *The Greedy Old Fat Man.* Boston: Houghton Mifflin Co., 1983.

_____. *The Little Red Hen.* Minneapolis: Seabury, 1973.

Geisert, Arthur. *Pigs from A to Z.* Boston: Houghton Mifflin Co., 1986.

Haviland, Virginia. *North American Legends.* Illustrated by Ann Strugnell. New York: Philomel, 1979.

Hoban, Tana. *Round & Round & Round.* New York: Greenwillow, 1983.

Hutchins, Pat. *Changes, Changes.* New York: Macmillan, 1971.

_____. *Rosie's Walk.* New York: Macmillan Co., 1968.

Ivimey, John. *Three Blind Mice.* Illustrated by Paul Galdone, New York: Clarion, 1987.

Johnson, Sylvia. *Potatoes.* Minneapolis: Lerner, 1984.

Kellogg, Steven. *Tallyho, Pinkerton!* New York: Dial Press, 1982.

Lauber, Patricia. *Volcano: The Eruption and Healing of Mount St. Helens.* New York: Bradbury, 1986.

Lear, Edward. *The Complete Nonsense Book.* New York: Dodd, Mead & Co., 1946 (Original 1846, 1871).

Lobel, Arnold. *The Random House Book of Mother Goose.* New York: Random House, 1986.

McCloskey, Robert. *Time of Wonder.* New York: Viking, 1957.

Merriam, Eve. *Halloween ABC.* Illustrations by Lane Smith. New York: Macmillan Co., 1987.

Prelutsky, Jack. *A Gopher in the Garden.* New York: Macmillan Co., 1966.

_____. *Read–Aloud Rhymes for the Very Young.* Selected by Jack Prelutsky and illustrated by Marc Brown. New York: Alfred A. Knopf, 1986.

Rockwell, Anne. *The Old Woman and Her Pig, and Ten Other Stories.* New York: Thomas Y. Crowell Co., 1979.

Sendak, Maurice. *Where the Wild Things Are.* New York: Harper & Row, 1963.

Silverstein, Shel. *Where the Sidewalk Ends.* New York: Harper & Row, 1974.

Snyder, Zilpha. *Today is Saturday.* New York: Atheneum Pubs., 1969.

Starbird, Kaye. *The Covered Bridge House and Other Poems.* New York: Four Winds, 1979.

Stevens, Janet. *The House that Jack Built.* New York: Holiday, 1985.

Walter, Mildred Pitts. *Brother to the Wind.* Illustrated by Diane and Leo Dillon. New York: Lothrop, Lee & Shepard, 1985.

Westcott, Nadine. *Peanut Butter and Jelly: A Play Rhyme.* New York: E. P. Dutton, 1987.

White, E. B. *Charlotte's Web.* Illustrated by Garth Williams. New York: Harper & Row, 1952.

Yorinks, Arthur. *Hey, Al.* Illustrated by Richard Egielski. New York: Farrar, Straus and Giroux, 1986.

Yolen, Jane. *The Three Bears Rhyme Book.* Illustrated by Jane Dyer. San Diego: Harcourt Brace Jovanovich, 1987.

Chapter Five

ISSUES IN HANDWRITING INSTRUCTION
Context of Handwriting

EVALUATING HANDWRITING SKILLS
Assessing Handwriting

HANDWRITING INSTRUCTION
Readiness ■ *Manuscript Printing* ■ *Cursive Handwriting*

TEACHING LEFT-HANDED WRITERS
MAINSTREAMING AND HANDWRITING
SUMMARY

After completing this chapter on handwriting, you will be able to:

1. State some of the issues related to handwriting instruction.

2. Identify some of the factors necessary for readiness in formal writing instruction. Describe some informal means of assessing readiness.

3. Describe and use a technique for collecting handwriting samples under the three conditions of normal, best, and fastest writing.

4. Evaluate handwriting samples for letter formation, slant, spacing, and line quality.

5. Identify the manuscript errors made most frequently by first-grade students.

6. Describe some instructional activities that enhance a child's perception of the need for and use of handwriting, and that develop fine motor skills and readiness for letter formations.

7. Demonstrate the formation of manuscript and cursive letters.

8. Develop and demonstrate a lesson that teaches manuscript printing to beginning writers.

9. Develop and demonstrate a lesson that introduces students to cursive writing.

10. Describe a kinesthetic approach for teaching handwriting to learning-disabled students.

Handwriting

andwriting instruction that emphasizes ornate, consistent penmanship for every student is no longer included in modern curricula. Modern handwriting instruction, although it does not develop beautiful flourishes, does stress legibility—well-formed letters, consistent slant, and proper spacing of letters and words. Whereas writing was once almost an art form, modern handwriting is instead considered a tool for personal communication. There is more leeway for developing a personal handwriting, although it cannot be so personal that only the writer is able to read it.

Handwriting is receiving renewed attention because of the drive to improve composition skills at all educational levels. Even universities are demonstrating a concern with handwriting. Shaughnessy's (1977) study of compositions of entering freshmen in New York's public university system concluded that poor handwriting is a factor in producing poor essays.

ISSUES IN HANDWRITING INSTRUCTION

Before we discuss either evaluation or instruction, it is helpful to discuss some of the issues associated with handwriting. It becomes evident that handwriting instruction gives rise to numerous unanswered questions. As Kean and Personke (1976) noted: "A bibliography of handwriting studies between 1890 and 1960 lists 1,754 reported studies. A review of these and subsequent studies is discouraging to a conscientious educator who is unconcerned with which form of r is used but who seeks to learn how time is best spent helping children develop legible handwriting." Sheldon's (1978) review of language arts research cites only three studies related to handwriting instruction. Handwriting is taught an average of fifteen to twenty minutes per day in most elementary schools; consequently, this lack of concrete knowledge is indeed depressing.

Context of Handwriting

Handwriting should be taught within the broader context of composition and communication, rather than as the subject of isolated drills. According to Graves (1978), "Too much of current handwriting instruction has lost its 'toolness.' It exists as an end in itself. Handwriting has become the main event, composition the side show" (p. 394).

One reason instruction may emphasize penmanship in isolation, and out of the context of composition, is the concrete nature of handwriting. It is a basic skill for which samples can be collected, measured, and compared over a period of time. It may also be easier for the teacher to evaluate handwriting rather than other aspects of a composition. Handwriting skills, or the lack of them, are highly visible. Many language arts authorities fear that if handwriting practice is a predominately isolated drill, or if handwriting is given more importance than the developmental process of composing, children may think handwriting is an end in itself, instead of a means to self-expression and communication.

But not all language arts authorities believe handwriting should be viewed in the broader context of composition. Enstrom (1969) advocated treating handwriting as a separate subject. He maintained that "to build automatic skill in handwriting there must be separate, daily learning sessions. To combine lessons on handwriting with lessons in other subjects is not unlike trying to learn to play the violin while learning, at the same time, the history of the invention of the oboe."

Steven Fairchild (1987) reviewed handwriting research and identified the following practices that lead to effective handwriting instruction:

1. Knowing the purpose for the activity increases participation of the students.
2. Visual demonstrations by teachers and other students help students see what is to be practiced.
3. Students who can verbalize a task before performing it show greater improvement.
4. Practice sessions need to be spaced.
5. Analyzing the motor task to be practiced and then identifying small, teachable steps permits students to master the activity.
6. Reinforcement encourages continued practice.
7. Giving students knowledge of the results of their practice and teaching them to evaluate their own progress enhances instruction.

Without question, handwriting is an important tool, because problems with handwriting may generate problems with composition. This chapter presents ways to teach both manuscript and cursive writing. Although there will be times that require some isolated drill (for example, when a first-grade teacher begins manuscript instruction), my own view is that handwriting should be approached as a tool for meaningful communication, not isolated from other concepts that deal with writing.

We also discuss other instructional issues: tools for instruction; whether to use lined or unlined paper; measurement scales; procedures; and transition from manuscript to cursive.

EVALUATING HANDWRITING SKILLS

Most of the children you teach will probably experience some difficulties when first learning to print. It is quite common for the beginner to reverse letter formations, writing *b* instead of *d*, to omit letters from words, or to space poorly. If these errors are still apparent after a reasonable period of instruction, however, the child probably has a handwriting problem that requires special attention. Identifying children who are unable to write with normal legibility and speed is the goal of diagnostic handwriting procedures.

An individualized diagnostic approach is the most effective (Tagatz, Otto, Klausmeier, Goodwin, and Cook, 1968). Unfortunately, a review of handwriting research by Askov, Otto, and Askov (1970) revealed that only one-fifth of the nation's elementary schools attempt to individualize handwriting instruction. As a result of mainstreaming children with various learning disabilities, the teacher now bears responsibility for at least the partial instruction of many children who were previously in self-contained special education classrooms. Because these children often demonstrate handwriting difficulties, many teachers need extra assistance in both diagnosis and instruction of children with handwriting problems. The wide range of abilities in any classroom increases the need for both individualized diagnosis and individualized instruction.

Evaluation of handwriting skills may be considered at two levels. First, one must ask, is the child ready for normal handwriting instruction? And second, does the child demonstrate specific difficulties in penmanship?

Readiness for handwriting requires an interest in and a desire to write, the development of adequate visual acuity and fine motor skills, an understanding of left-to-right progression, and an understanding of the concept of language. These readiness skills can be evaluated through simple teacher observation. The child's interest or lack of interest in writing is usually quite apparent. Does the child ask how to write his name or other words? Does the child pretend he is writing even if he does not know how? Durkin (1966) found that children who read prior to entering school also showed an early interest in learning to print. The learning sequence of these early printers moved from (1) scribbling and drawing, to (2) copying objects and letters of the alphabet, to (3) questions about spelling, and to (4) the ability to read.

Because both writing and reading progress from left to right, it seems logical that understanding this sequence will benefit handwriting instruction. The teacher can also observe the child to see whether, when he tries to read a book, he progresses from left to right. Does the child begin at the left side of the paper when writing his name, or does he write from right to left? If you place two dots on chalkboard or paper, does the child connect the dots in a left-to-right sequence?

Observation during artwork, bead stringing, paper cutting, or holding a pencil will give you a great deal of information about the child's fine-motor coordination. Samples of older children's writing also indicate whether control is smooth and even or uncoordinated. Observation will also indicate the child's hand preference.

The majority of handwriting evaluation tests are thus informal. Lewis and Lewis (1964) did find that one standardized test, the Draw-A-Man test of Mental Maturity, was an excellent predictor of success in beginning writing. The Draw-A-Man test requires a child to draw three pictures, one of a man, one of a woman, and one of the child. The *Goodenough-Harris Drawing Test Manual* by Dale B. Harris (Harcourt, Brace, 1963) contains explicit directions for administering the test, as well as a scoring guide for the drawn picture.

Another evaluation measure is the way the child relates to the space on the paper (de Ajuriaguerra and Auzias, 1975). The authors say teachers need to look at handwriting in the broader language framework as well as in the framework of motor analysis. They believe the simultaneous production of symbols and their meanings enables a child to relate himself to the space on a sheet of paper. The child, in both drawing and scribbling, explores the paper by first using indiscriminate actions that show increased purpose and good use of space. De Ajuriaguerra and Auzias believe the child's use of space can be used as a diagnostic tool. Graves (1978) developed six questions, based on the research of de Ajuriaguerra and Auzias, for use in evaluating how the child explores space in artwork, construction, movement, and writing, and in determining the level of development of the skills necessary for writing.

First, how does the child change in the use of the thumb and forefinger grip? The child who demonstrates poor grip when using crayons in artwork will have difficulty controlling the pencil in writing.

Second, how does the continuousness of the child's writing change? When the child first writes, his movements show stops and starts because he is unfamiliar with what is he doing. The observer wants to find out if the child continues to show erratic movements in artwork or in handwriting. These erratic movements suggest underdeveloped fine motor skills.

Third, how does the position of the elbow and the stability of the body axis change? When a child begins to draw or write, he exhibits a great deal of body and elbow motion. If this motion is not reduced, the child will have difficulty with the small-motor demands of writing.

Fourth, how does the position of the writing surface change in relation to the child's midline? When the child first faces paper, he faces it in a straight position. Gradually, the child comes to understand the relationship of the paper to the body activity.

Fifth, how does the child's distribution of strength change? In writing, large muscles must be suppressed so small muscles can gain control. The teacher can observe changes from light to heavy lines on the child's paper as well as changes in body position.

Sixth, how does the child's use of writing space change? Look at the angle in which letters are chained together. If the child does not have good fine-motor control, the angles will not be consistent nor the spaces even. If the child does not understand the concept of language, he will also show a tendency to run words and sentences together. Graves believed any of these factors can contribute to slowness in handwriting, which can, in turn, affect quality, content, and the child's concept of writing.

REINFORCEMENT ACTIVITY

(1) Observe children in a nursery school or kindergarten. Using informal assessment techniques, describe some of the relevant characteristics of children you feel are ready for writing and children who may not be ready for handwriting. (2) Collect samples of several children's results from the Draw-A-Man test and art projects. Use the test manual to evaluate the Draw-A-Man test, and use Graves's six questions to evaluate the children's use of space. Compare samples from children who appear to be ready for writing with those who do not.

Assessing Handwriting

There are few standardized tests for assessing handwriting. The teacher will have to rely either on informal sampling procedures or on loosely standardized writing scales to evaluate a child's handwriting skills.

For the evaluation of older students' handwriting, Otto and Smith (1980) recommended gathering handwriting samples under three conditions: usual writing, best writing, and fastest writing. Testing in these different situations is important, because legibility may be acceptable when a child works slowly and methodically, but may deteriorate when the child has to write rapidly to keep up with the demands of various school assignments.

Otto and Smith suggested the following procedures for gathering handwriting samples:

1. Use a sentence containing most of the letters of the alphabet; for example, "The quick brown fox jumps over the lazy dog." Write the sentence on the chalkboard (in manuscript or cursive, depending on grade level) and have the students read it several times to become familiar with it.
2. Ask the students to write the sentence five times. Tell them to write as they normally do. Allow them to write for three minutes, then stop the activity.
3. Provide a period of relaxation, then tell the students to write the sentence in their very best handwriting; they may take all the time they need. Ask them to write the sentence three times, with no time restrictions.
4. Provide a second period of relaxation, then tell the students you want to see how fast they are able to write. Allow them three minutes to write the sentence as many times as they can.
5. Compare each child's three writing samples. Use either informal evaluations or an evaluation scale.

Evaluation scales   The advantages of using published handwriting scales to measure the relative quality of different children's handwriting is debatable. In 1963, Greene and Petty recommended the use of handwriting scales to improve accuracy in appraising handwriting quality. In 1973, however, Petty said the use of handwriting scales should be limited, because children may attempt to copy the handwriting on the scale and, consequently, lose their individuality in writing. Teacher-produced evaluations are recommended by Askov, Otto, and Askov (1970) and Petty (1973). They felt an internalized scale, developed through training and experience, will probably be most useful in the classroom.

Handwriting evaluation demands some experience; consequently, beginning teachers may find the *Evaluation Scales for Expressional Growth through Handwriting*, published by the Zaner-Bloser Company (1969), helpful. The scales contain five examples each of manuscript printing for grades one and two. These examples represent high, good, medium, fair, and poor quality. Five examples are also provided of cursive writing for each grade, three through nine. My students have found that looking at the various degrees of legibility helps them evaluate formation, slant, and spacing of letters.

Informal assessment   The most important factor in handwriting legibility is letter formation. Other factors that affect legibility include spacing between letters and words, consistency or inconsistency in the slant of the letters, and evenness of writing pressure. The teacher can assess these factors informally.

To check letter formation, Burns and Broman (1983) recommended using a card with a circular hole cut in the center. This hole, a little larger than the size of a letter, should be moved along the line of writing to expose one letter at a time, allowing the teacher or the child to identify poorly formed or illegible letters.

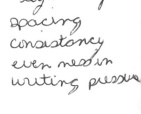

Letter formation is easily observable when children use lined paper, or when lines are drawn across the top of the letters to see if they are uniform. In the example on the left, the second-grade child has formed each letter according to the way he was taught in first grade. The example on the right represents a second-grade child who is having considerable difficulty with letter formation.

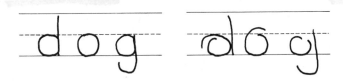

Legibility also relates to the slant of letters. To evaluate consistency in the slant, the teacher or child can draw straight lines through the letters of the words. The slant of the lines quickly indicates whether the letters are slanted consistently. Posture is the most important factor affecting the slant of handwriting, so the teacher should observe the child while he is writing. An example of consistent slant would look like this:

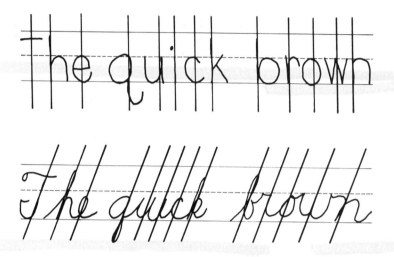

An example of inconsistent slant would look like this:

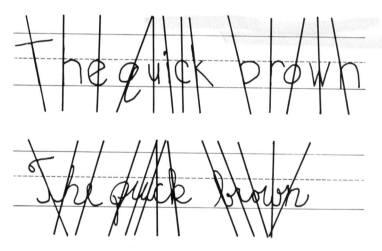

Spacing, in both manuscript and cursive writing, also affects legibility. Too much spacing between letters can slow the speed of reading. Obviously, if there are no spaces between words, the material will be hard to read. Some children place their letters so close together that the letters are difficult to decipher, which can affect assessment of spelling ability. The following example illustrates legible spacing:

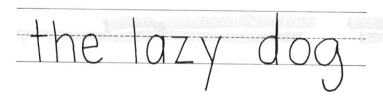

This is an example of poor spacing:

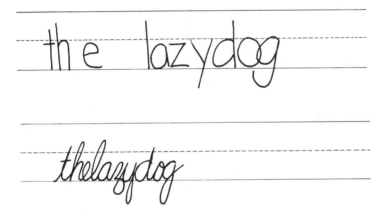

When you assess line quality, you must evaluate the thickness, lightness, or darkness of the lines. If the child has full control of the fine motor skills, the lines will usually show consistent pressure. In contrast, if the child does not have control of these motor skills, there may be a noticeable difference in line quality. An example of uneven line quality is this:

Informal evaluation will provide the teacher with a great deal of diagnostic information, and give her a basis for planning to meet individual handwriting needs. The teacher should keep folders of writing examples for each child, so that both teacher and child are aware of progress. It is helpful for the teacher to develop a progress chart for each child, so that various aspects of handwriting can be evaluated over a period of time. Figure 5–1 illustrates the type of information that may be kept on the progress chart.

Name _____

Grade _____

	Sept.	Jan.	May
1. Formation of letters:			
a. All correctly formed	____	____	____
b. Examples of incorrect letters	____	____	____
2. Slant of letters:			
a. Consistent slant	____	____	____
b. Irregular slant	____	____	____
c. Too much back slant	____	____	____
3. Posture while writing: (Posture of hand, body, paper)			
a. Correct	____	____	____
b. Incorrect	____	____	____
4. Spacing within words:			
a. Spaces too wide	____	____	____
b. Crowded letters	____	____	____
c. Irregular	____	____	____
5. Spacing between words:			
a. Spaces too wide	____	____	____
b. Spaces too narrow	____	____	____
c. Irregular	____	____	____
6. Speed of writing:			
a. Appropriate for grade	____	____	____
b. Too slow	____	____	____
c. Too fast	____	____	____
7. Overall appearance of writing:			
a. Neat	____	____	____
b. Poor appearance due to ____	____	____	____

(Place checks, examples, and comments in appropriate spaces.)

FIGURE 5–1
Handwriting analysis

Research by Lewis and Lewis (1964) provides some helpful information to the first-grade teacher. They analyzed both the number and types of errors made by first-grade children. The teacher can see from this information what types of errors are common, which letters may be easier to learn, and which are more difficult. Here are some conclusions from this first-grade study.

1. The most errors were made in letters that go below the line—q, g, p, y, and j.
2. Errors were frequent in letters in which curves and vertical lines merge—J, U, f, h, j, m, n, r, u.
3. The letters more likely to be reversed were—n, d, q, and y.

4. The most frequent types of errors include improper size, incorrect relationship of parts, and incorrect relationship to line.
5. The easiest letters with fewest errors were—H, O, o, L, l.

Diagnostic procedures can help individualize handwriting instruction so that
valuable time won't be wasted in ineffective group drills. Informal diagnosis provides
opportunities for constant analysis so that errors can be corrected before they become
reinforced through continued practice.

REINFORCEMENT ACTIVITY

*Collect samples of handwriting from several children using the procedures for gathering a
normal handwriting sample, the best handwriting sample, and the fastest handwriting sample. Evaluate the samples using the described informal assessment techniques. After you
have evaluated the samples, place the results on a handwriting analysis chart. Include recommendations for instruction or remediation.*

HANDWRITING INSTRUCTION

Handwriting instruction has three phases: first, the readiness for handwriting that is
developed at home, in kindergarten, and early first grade; the formal teaching of
manuscript writing, beginning usually in first grade; and, third, the formal teaching of
cursive writing, beginning about third grade.

Readiness

Early-childhood educators contend that the readiness phase of handwriting is as
important as the readiness phase in reading. Many activities in nursery school,
kindergarten, and early first grade contribute to a child's readiness for handwriting. As
Graves (1978) suggested, drawing and writing development are more closely related
than is often realized. Both are attempts at representation. At first—at two or three
years of age—there is no difference between them. Later, the child differentiates by
drawing a picture and by making continuous scribbles across the pages to represent
writing. When a child begins to try to write, he is indicating a perception of the need
for handwriting. The child who does not feel this need has unusual difficulties with all
aspects of writing. Drawing and early writing experience also give a child opportunities
to explore the space provided by the paper.

Kindergarten programs that include reading to children and language experience activities in which children dictate group or individual chart stories help children
see the relationship between oral language, printed stories, and handwriting. If
children do not understand the concepts of "word" or "sentence," handwriting will be

meaningless; in fact, they may regard handwriting as an almost impossible task, in which they are expected to make monotonously neat, properly spaced circles on lined paper, rather than as a tool for personal and permanent communication. I remember one five-year-old kindergarten child who had her first experience dictating a sentence about her artwork. After the teacher had printed the short sentence under the picture and had read the sentence back to the child, the child responded with, "Oh, is that what those lines are for?" She did not realize the use or need for handwriting before this experience. Wordless books, such as those recommended on page 69 (oral language chapter), provide enjoyable instructional sources for language experience stories. Teachers should develop many readiness activities during which they allow children to see that the purpose for handwriting is composition and communication, not isolated drill.

Drawing, painting, and cutting all help develop fine-motor skills. Wright and Allen (1975) listed several other exercises to encourage readiness for handwriting. First are activities to help children develop a left-to-right orientation. The folk dances Looby-Loo and Hokey Pokey nurture mastery of bodily concepts related to left and right. Other activities might include drawing cars on a road that goes from left to right, or actually "driving" small cars and trucks on the road in the classroom or playground.

The second group of readiness activities helps children with letter formations. In analyzing the movements needed to make the manuscript alphabet, you will notice they are primarily circles and lines in vertical, horizontal, or angled positions. Children can practice making these strokes on large paper or on the chalkboard. Kindergarten and first-grade teachers often have children make creative pictures from elements such as these:

One first-grade teacher, working with students who demonstrated under-developed fine-motor control, had them make delightful stick figures, animals, and snowmen, using the basic shapes that would later be necessary for formal manuscript writing. They thus experimented with space and line dimensions while enjoying an art activity. Tompkins (1980) recommended using basic penmanship strokes to create stories and pictures. She developed activities around the language game "Let's Go on a Bear Hunt," instructions for making a jack-o-lantern, and a snowy day story.

After children have had opportunities to experiment with shapes and lines on unlined paper, teachers often introduce them to the use of lined paper by having them draw the fundamental shapes between the appropriate lines. For example:

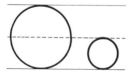   

Printing the child's name provides an opportunity to practice manuscript in the most meaningful context.

Manuscript Printing

Manuscript writing, the name given to printing, is taught to first- and second-graders in most American schools. Manuscript rather than cursive writing is considered preferable for beginners because the manuscript symbols are similar to reading symbols.

Although manuscript is taught in most early-elementary grades, there is some disagreement about what size pencils and what type of paper is most conducive to instruction. According to Askov, Otto, and Askov (1970), the research in this area is too scanty to permit any conclusions. One research article reported that children prefer adult pencils to the larger-sized beginner pencils. Many educators recommend allowing each child to use whatever size pencil is most comfortable because the larger pencils have not been proven better for beginning writers.

There is also a mild controversy over what type of paper to use for beginning instruction. Alice Yardley (1973), an English educator, pointed out that unlined paper offers a surface free of restrictions when a child's early attempts to write are immature and unformed. Consequently, English children spend several years working with unlined paper. In contrast, most American schools use paper that progresses from wider to narrower spaces, even for early instruction.

There are several published instructional programs in the area of handwriting. If the school you are now or will be teaching in has approved one of these programs, you need to learn the formation of the letters for that program. Children are very observant, and will inform even the student teacher if he is not making the letters the same way they have been taught.

Two manuscript alphabets taught in many schools are shown in figures 5–2 and 5–3. The first example is published by the Zaner-Bloser Company. The second example is the D'Nealian Handwriting model published by Scott, Foresman. Compare the two examples for letter shape and strokes.

REINFORCEMENT ACTIVITY

Before you proceed with specific instructional techniques for teaching manuscript, practice printing the manuscript alphabets shown in figures 5–2 and 5–3 or the manuscript alphabet that is used in the schools in your area. Practice the alphabet on paper and on chalkboard until you feel comfortable with it.

Introducing manuscript   Readiness activities help the child understand the techniques that contribute to legible handwriting. Manuscript instruction should involve correct sitting posture, correct letter formation, proper spacing of letters, and uniform slanting of letters.

Several language arts authorities stress the teacher's role in helping children analyze a letter before writing it. Beatrice Furner (1969) recommended a perceptual-motor approach for teaching letter formation. This approach allows children to identify

FIGURE 5–2
Manuscript alphabet (Used with permission from *Handwriting: Basic Skills and Application.*
Copyright 1987. Zaner-Bloser, Inc., Columbus, Ohio.)

the writing position, letter formation, alignment, and spacing of letters before they actually print the letter. Roger Goodson (1968) said the teacher should help children analyze a letter by encouraging them to visualize the letter; study its characteristics (beginning stroke, ending stroke, strokes peculiar to the letter, height of the letter, height of various parts of the letter, and width of the letter); and by checking the quality of the strokes. The following lesson plan utilizes this letter-analysis approach. The objective is to enable the students to analyze and form the manuscript letter "a." The teaching procedures are as follows:

1. Make lines on the chalkboard similar to the lines on manuscript paper. (A musical staff marker with chalk placed in the top, middle, and bottom holders is useful.)
2. Ask the children to watch carefully as you make the letter "a." Ask them questions such as: How am I making the letter "a"? What kinds of strokes

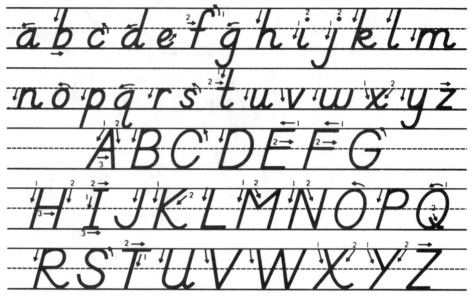

FIGURE 5–3
D'Nealian manuscript alphabet (*D'Nealian ® Handwriting* by Donald Thurber. Copyright ©
1987 by Scott, Foresman & Co. Reprinted by permission.)

do you see? (Circle stroke and straight-line stroke.) You may discuss the fact
that the class has practiced these strokes.

3. Now emphasize the beginning stroke for the letter "a." Ask the class to
    watch as you make the circle portion of the letter. Ask them questions such
    as: Where does the circle begin? (Just below the dotted or middle line.) In
    what direction does the stroke go? (The circle goes to the left.) A mark and
    an arrow will help students visualize this formation.

4. Have the students watch you make the letter "a" again. This time, ask them
    to watch the straight-line formation. Ask: Where does the straight line begin?

(The line begins on the dotted or middle line.) Where does the straight line touch the circle? (On the right side of the circle.) Where does the straight line end? (On the bottom line.)

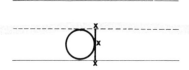

5. Have the students watch you again so they can describe the height of the letter. Ask: How tall is the circle part of the letter? (The circle is one space high; it touches the middle line and bottom lines, but it does not go over either line.) How tall is the straight-line part of the letter? (The straight line is one space high; it touches the middle line and bottom line, but it does not go over either line.)

6. Ask the children to describe how the letter is formed. Have them write the letter independently, describing the process to themselves.

After introducing the letter "a," other letters in the circle group are taught in a similar manner. Children should be allowed to practice their handwriting through a functional approach as soon as possible. Instead of merely copying model sentences, they should practice writing their names, addresses, diaries, invitations, letters, captions for artwork on pictures, weather charts, autobiographies and so forth. In all early writing activities, the teacher needs to remember that writing is a difficult and slow process of many children, so early handwriting practice should not overtire them.

In addition, when children are first learning to write, they need a great deal of supervision so they will not reinforce incorrect letter formations. Individual observation is the best way to help children after they have been taught the fundamental formations. It would be a waste of time for every child to repeatedly practice letter formations already mastered.

REINFORCEMENT ACTIVITY

Develop a lesson plan for teaching one of the letters of the manuscript alphabet. Use your lesson plan with beginning writers, or try the lesson plan with a small group of your peers.

Cursive Handwriting

Most elementary schools require children to change from manuscript printing to cursive handwriting some time between the end of second grade and the middle of third grade; the transition time usually begins in early third grade. Although cursive may be more acceptable for both social and business uses, research has not shown any great advantage in changing from manuscript to cursive writing. Cursive writing is, however, a more personal tool than printing, is usually faster, and does not require the conformity of printing.

Introduction to cursive   Cursive handwriting is not a new skill that must be memorized in the slow, meticulous way that manuscript is often developed. Children have by now had several years of practice with printed language and can make comparisons and contrasts between manuscript and cursive. When Furner (1969) introduced her students to cursive writing, she used two approaches. First, she organized readiness activities in which the students were guided in recognizing and reading cursive letters. They named the letters, read their own names in cursive, and so forth. Next, they examined samples of cursive and its formation, and stated the likenesses and differences between manuscript and cursive.

The teacher can refer to a cursive writing chart to demonstrate likenesses and differences between manuscript and cursive. The cursive handwriting alphabets published by Zaner-Bloser or Scott, Foresman are used in most schools (see figures 5–4 and 5–5).

FIGURE 5–4
Cursive handwriting alphabet (Used with permission from *Handwriting: Basic Skills and Application.* Copyright 1987. Zaner-Bloser, Inc., Columbus, Ohio.)

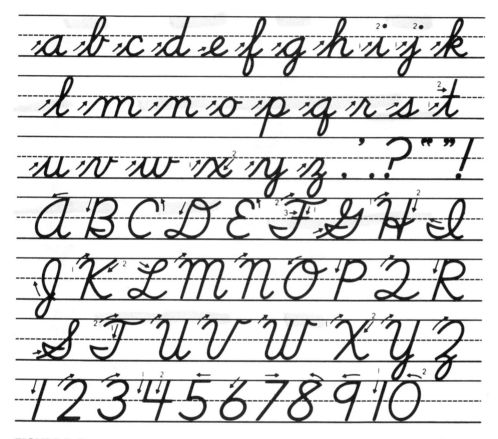

FIGURE 5–5
D'Nealian cursive alphabet (*D'Nealian ® Handwriting* by Donald Thurber. Copyright © 1987 by Scott, Foresman & Co. Reprinted by permission.)

Following observation and discussion, the students and teacher can list some of the similarities and differences between manuscript and cursive writing. They can list letters that are similar, as well as similarities in direction of strokes, size, and whether letters touch the baseline or extend below the baseline. These are some of the differences to note:

1. The letters are joined in cursive, and there are no spaces until the end of the word.

lad vs. *bad*

2. Cursive writing has a slant, whereas manuscript writing is straight.

3. The letters *f* and *z* go below the baseline in cursive, whereas they do not in manuscript.

4. When a word is written in cursive, the *t* is crossed and the *i* and *j* are dotted after the word is completed, rather than after the letter is completed, as in manuscript.

5. The position of the paper may be more slanted for cursive writing.

Continuing instruction   Burns and Broman (1983) recommended teaching at the same time those letters in words that have similar strokes. This procedure would progress from the manuscript to the slanted cursive version of the word in the following steps:

1. manuscript

2. dotted manuscript

3. nonslant cursive

4. cursive slant

Burns and Broman recommended continuing cursive writing instruction with words containing letters that are similar in both manuscript and cursive alphabets, such as a, d, g, h, i, l, m, n, o, p, q, t, u, y, and the capital letters B, C, K, L, O, P, R, and U.

Next, teach the letters that are dissimilar, such as f, k, r, s, and z. Handwriting instruction should use practical types of writing; the children can practice their spelling words, letter writing, and other useful assignments. At this level, instruction needs to be individualized so that each student will have extra practice with letter formations and combinations that he finds difficult.

REINFORCEMENT
ACTIVITY

1. *Develop a lesson or series of lessons to demonstrate the transition from manuscript to cursive handwriting. If you are teaching elementary students at the appropriate age, use your lesson plan with them. If you are not teaching, demonstrate your activity to a group of your peers.*

2. *Investigate the way cursive writing is taught in one of the commercial sets of handwriting materials. Look at suggestions for transition to and development of the cursive alphabet. What skills are stressed? In what order are the alphabet letters presented? Is the instruction in isolation, or are functional, practice activities suggested? How much time is recommended for handwriting instruction?*

TEACHING LEFT-HANDED WRITERS

Teachers today do not try to change left-handed writers to right-handed writers, as has frequently been the practice. Although left-handed writers are allowed to retain their hand preference, they do require some specific instruction in handwriting. Roger Goodson (1968) recommended the following procedures when working with left-handed writers:

1. Have children assume positions that are the *opposite* of those expected of right-handed children, with the added caution that the position should be one that is most comfortable to them.

2. The position of the paper is reversed for the *left-handed* writer (it may be less slanted and more toward a straight up-and-down position for the left-hander). The direction of the pencil, too, is reversed. The grip is basically the same as that for right-handed students.

3. Seat left-handed children so the light comes over their left shoulder.

4. Provide lots of writing on the chalkboard. (It is nearly impossible to use the upside-down hand style on the board.)

5. Furnish left-handed people with pencils that have slightly harder lead than those used by right-handers. Harder lead will not smear as easily, giving less reason for twisting the wrist into the upside-down position. The point should not be so sharp that it will dig into the paper.

6. Encourage left-handed children to grasp their writing instrument at least an inch-and-a-half from the point. If he grips the instrument at this distance,

the child can see over or around his hand. You can put a rubber band around the pencil at the point where he should grasp it.

7. Children should have individual model charts written, if possible, by a left-handed teacher or community worker.

8. At the primary levels, left-handed writers may be seated together near the front of the room where you can give them constant encouragement. After their handwriting habits are formed and no further adjustments are needed, this seating can be discontinued.

9. Encourage left-handed children to develop a writing slant that is natural and "feels good."

10. Avoid any attention that may lead to self-consciousness.

11. Avoid forcing older left-handed pupils to write a certain way. If an awkward position cannot be easily corrected, it is sometimes best to let the child continue.

MAINSTREAMING AND HANDWRITING

The mainstreamed child frequently exhibits problems in handwriting as well as in other areas. The complexity of these problems is stressed by Otto and Smith (1980), who stated that difficulties in handwriting frequently coexist with difficulties in reading and spelling. Both reading and spelling may be influenced by inadequate visual memory, inability or lack of readiness to handle visual symbols, or difficulties with sound-symbol relationships.

Research indicates that a kinesthetic (sensations of position; movement of the body) approach to handwriting is best for certain learning-disabled students (Askov, Otto, and Askov, 1970). Learning-disabled students may require extensive pencil tracing activities, as well as tracing raised letters with the finger. The advantages of adding a tracing component to handwriting instruction for these children are also implied in research conducted by Birch and Leffort (1967). These researchers investigated the ability of five- through eleven-year-olds to draw geometric forms by tracing, connecting dots, and by freehand copying. The tracing activity was the easiest for younger children, and also produced the highest accuracy score. Connecting the dots was the next easiest task, whereas freehand drawing was the most difficult, and resulted in the greatest number of errors.

Tracing was found to be the easiest, and because it provides a kinesthetic component for handwriting, a tracing activity could be added to the beginning analysis approach before learning-disabled children are required to form each letter independently.

If adding a kinesthetic activity is still not enough guidance for a student who is having difficulty, Hammill and Bartel (1975) recommended that handwriting be taught with a procedure that encompasses four developmental levels. Reger, Schroeder, and Uschold (1968) also recommended this procedure. In this four-level approach, teachers would begin with the first level, and move on to the next level only when the student has mastered the skills at the earlier level.

Tracing a raised letter develops a kinesthetic feel for the letter.

The first level introduces handwriting movements to the student. Each of the movements is discussed by the teacher and demonstrated on the chalkboard. For example, the teacher demonstrates the straight-stick movement in manuscript, and says, "We put the chalk on the board and then go down." The teacher demonstrates the movement on the board; then, the student makes the same movement at the board while looking at the teacher's model but not at his own hands. The student says "down" when the movement goes down, or "up" when the movement goes up. The student practices the movement, verbalizing the directions, for several days. Then the auditory clue is removed, and the student makes the movement while looking at the model, but not at the hands. The student then follows the same procedures with crayon, on newsprint, or paper. Figure 5–6 illustrates suggested movement.

FIGURE 5–6

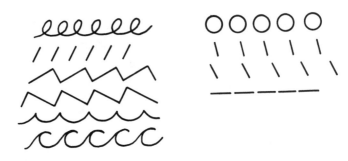

The second level requires the student to repeat the same procedures as level one, but this time he is allowed to look at the paper as the movements are formed. At this level, attention should be paid to correct posture and slant of paper. Lines may be made on the desk with masking tape, to guide the student in placing the paper. It may also be helpful to line the paper with three different colors. You might use the colors of traffic lights, with green at the top, yellow in the center, and red at the bottom. This makes it easy for the child to verbalize directions such as, "Put my pencil on the green line, draw it straight down through the yellow line, and stop on the red line."

During the third-level activities the student moves on to letter formation at the chalkboard, using the procedures described in level 1 for movements. After using the chalkboard, he goes to one-inch lined paper, then to regular lined paper. The sequence for this fourth level is as follows:

1. Name the letter—For example, "The name of this letter is b."
2. Discuss the form of the letter while the child looks at it. "How many shapes are there in b?" (A straight stick, and a circle or ball.)
3. Make the letter for the student to see. You can use a transparency, and make each part with a different color.
4. Show the student the directions for each movement in the formation of the letter, and discuss the directions with him.
5. Develop a kinesthetic feel for the letter form by tracing a raised, script letter with the finger. You can make raised letters by placing a few drops of food coloring in a glue dispenser and writing the word in glue on a strip of tagboard. A kinesthetic method recommended by Goodson (1974) uses a sandpaper board. In this method, 9×11-inch pieces of coarse sandpaper are mounted on a piece of board. The student places a sheet of paper over the sandpaper board, and uses a crayon to write the troublesome letter correctly. He removes the paper and traces the letter with his finger. I have also made alphabets using trays filled with plaster of Paris. While the plaster is still damp, write the letters with a blunt pencil or stick in cursive or manuscript alphabet. The student can use this tray to trace the letters with a pencil or crayon. The tray is reusable and quite inexpensive.
6. The teacher places a model of the letter on each student's paper. (Students with learning problems may have difficulty copying from the chalkboard.)

The teacher and student again discuss the letter form and direction. The student makes the letter while looking at the model instead of at his hand.

7. The teacher helps each student compare the letter with the model. (Close supervision is necessary.)

8. The student or teacher may use auditory clues to help with letter formation. If necessary, the teacher can also hold the student's hand to help with the letter formation.

9. Have the student write the letter on the chalkboard without looking at the model.

10. Have the student write the letter on paper without a model and without watching his hand.

11. Have the student write the letter on paper while watching the paper.

The teacher can use the same procedures to introduce the writing movements in cursive. These begin with chalkboard activities and progress to newsprint and paper. The following movements are suggested for levels 1 and 2:

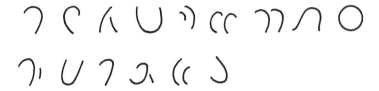

As you can see, this is a complicated approach to handwriting instruction, and it would not be necessary or desirable to use it with every student in the class. It is designed for those who demonstrate difficulties in learning to write by more conventional approaches. The steps can be used for individual instruction, or for small-group instruction of six or seven pupils. It would be too difficult to supervise the movements of larger groups in this detailed approach.

The teacher can also create transparencies. One set might have the manuscript alphabet, with the different parts of the letter shown in different colors: two-space-high straight lines, as in b, d, h, k, l, and t, could be printed in blue; complete circles, as in a, o, b, d, g, q, in orange, and so forth. These transparencies can also be placed over the student's own letters for checking formation.

There are interrelationships between difficulties in reading and spelling, so many of the techniques suggested in the Oral Language and Spelling chapters (chapters 3 and 6) should also be considered.

Table 5–1 presents characteristics of learning disabled children that might interfere with handwriting development and suggests teaching techniques for handwriting instruction. Table 5–2 presents the same information for mentally handicapped children. The techniques for language arts instruction result from research and authorities who work with mainstreamed children.

TABLE 5–1
Mainstreaming and handwriting for learning disabled children

Characteristics of Learning Disabled Children	Teaching Techniques for Language Arts Instruction
1. Demonstrate deficits in visual perception, visual motor integration, and fine motor–skills (Harber, 1979; Lyon and Watson, 1981).	1. Trace on acetate (Spache, 1976). Use four developmental levels suggested by Hammill and Bartel (1975). Use directional arrows, color cues, and numbers to help children trace. Cut with scissors, draw outlines of patterns, and try simple paper folding (Lerner, 1976). Biofeedback training may help children relax their writing forearm and control forearm tension voluntarily (Carter and Russell, 1980). Provide handwriting readiness activities that encourage development of fine motor control: coloring, painting, finger painting, chalkboard activities. An electric typewriter is beneficial for children with severe motor difficulties (Lerner, 1976).
2. Orientation confusions characterized by rotations and reversals of letters and words (Frith and Vogel, 1980).	2. Add color cues to letter formations (Bracey and Ward, 1980). Analyze the situation and try to identify specific bases for children's rotation and reversal errors: a. rotation: Point out that the letter may look better rotated, but it is usually written so it looks upside down. Change to a slanted easel so there is a definite top and bottom. b. reversals: Train children in the importance of left–right orientation of symbols (Frith and Vogel, 1980).

SUMMARY

Modern instruction considers handwriting a tool for personal communication. This viewpoint stresses that handwriting should be taught within the broader context of composition rather than as isolated drill.

Research indicates that the most efficient handwriting instruction is one that utilizes an individualized diagnostic approach. Apparently few elementary schools actually use such procedures. Two areas of diagnosis are stressed in this chapter: first,

TABLE 5–2

Mainstreaming and handwriting for mentally handicapped children

Characteristics of Mentally Handicapped Children	Teaching Techniques for Language Arts Instruction
1. Visual discrimination deficits may cause difficulty identifying letters (Adams, Taylor, and Glendenning, 1982).	1. Positive reinforcement during practice increases letter discrimination capability (Adams, Taylor, and Glendenning, 1982).
2. Tendency to have motor coordination handicaps (Gearheart, Weishahn, and Gearheart, 1988).	2. Provide opportunities for development of motor skills that are prerequisites for writing. Include simplified games and specific motor exercises such as hopscotch, bag throwing, and jumping games that promote eye-hand coordination (Gearheart, Weishahn, and Gearheart, 1988)

evaluation of readiness for formal handwriting, which includes the child's use of space; interest in handwriting; left-to-right progression; and fine-motor control. Research by de Ajuriaguerra and Auzias placed a new emphasis on developmental aspects of the child and how these relate to his utilization of writing space. The second area of diagnosis is handwriting assessment. Few standardized tests are available for assessing handwriting; consequently, we described an informal method of gathering handwriting samples under the conditions of normal, best, and fastest writing. Both evaluative scales and informal means of assessing letter formation, slant, spacing, and line quality were described. We concluded that an internalized scale developed through training and experience may eventually be most helpful to the teacher. Research has identified the most frequent types of letter-formation errors.

This chapter discussed handwriting instruction in terms of readiness, manuscript, and cursive instruction. Readiness activities that develop a child's understanding of the need and purpose for writing are considered necessary, as well as activities that develop fine-motor skills and readiness for letter formations.

The beginning teacher must become proficient in the manuscript or cursive alphabet taught in area schools. A letter-analysis approach for manuscript instruction was presented as a method for teaching beginning manuscript printing.

The transition to cursive writing usually takes place some time early in the third grade, although research has not shown any advantage in changing from manuscript to cursive. Research also indicates that the time of transition is not very important. An approach that helps children understand the similarities and differences between manuscript and cursive writing was described.

Since the inception of mainstreaming, regular classroom teachers are providing instruction for many children who have formerly been in self-contained special education classes. Research indicates a kinesthetic approach is effective for teaching handwriting to many learning-disabled students. This chapter described a kinesthetic approach, which begins with teaching the movements of handwriting and extends to the formation of letters. An individualized, diagnostic approach is thus equally important for all students in the classroom, because there may be many different needs and requirements in the class.

ADDITIONAL HANDWRITING ACTIVITIES

1. *Compile a list of readiness activities appropriate for children who have not developed large-motor and fine-motor coordination.*

2. *Develop a creative penmanship activity that reinforces readiness skills related to basic manuscript strokes.*

3. *Go to the library and locate language arts curriculum guides published by several school districts. What are the goals for handwriting instruction? What recommendations are included for instructional strategies? How much time is recommended for instruction at various grade levels? Is handwriting taught in isolated drills, or is the subject emphasized during meaningful writing activities?*

4. *Use ERIC to locate any programs designed to improve handwriting capabilities. Describe the program. Do you believe it would be successful? Why or why not? Report your findings to your language arts class.*

5. *Secure scope and sequence charts from several published handwriting programs. Compare the programs according to expectations at specific levels, pace of instruction, instructional objectives, instructional emphasis, and evaluation.*

6. *Collect handwriting samples from first-, second-, and third-grade children. Analyze the handwriting at each level. What developmental changes are illustrated in the writing samples?*

BIBLIOGRAPHY

Adams, Gary L.; Taylor, Ronald L.; and Glendenning, Nancy J. "Remediation of Letter Reversals." *Perceptual and Motor Skills* 54 (June 1982): 1002.

Askov, Eunice; Otto, Wayne; and Askov, Warren. "A Decade of Research in Handwriting." *Journal of Educational Research* 64 (November 1970): 100–111.

Birch, H. G., and Lefford, A. "Visual Discrimination, Inter-sensory Integration, and Voluntary Motor Control." *Monographs of the Society for Research in Child Development* 32 (1967): 2.

Bracey, Susan A., and Ward, Jefflyn. "Dark, Dark Went the Bog: Instructional Interventions for Remediating b and d Reversals." *Reading Improvement* 17 (Summer 1980): 104–12.

Burns, Paul C. *Assessment and Correction of Language Arts Difficulties*. Columbus, Oh.: Merrill Publishing Co., 1980.

_____and Broman, Betty L. *The Language Arts in Childhood Education*. 5th ed. Chicago: Rand McNally & Co., 1983.

Carter, John L., and Russell, Harold. "Biofeedback and Academic Attainment in LD Children." *Academic Therapy* 15 (March 1980): 483–86.

Collette, Martha A. "Dyslexia and Classic Pathogenic Signs." *Perceptual and Motor Skills* 48 (June 1979): 1055–62.

de Ajuriaguerra, J., and Auzias, M. "Preconditions for the Development of Writing in the Child." In

Foundations of Language Development, vol. 2, edited by Eric and Elizabeth Lenneberg. New York: Academic Press, 1975.

Durkin, Dolores. *Children Who Read Early.* New York: Teachers College Press, 1966.

Enstrom, E. A. "Those Questions on Handwriting." *Elementary School Journal* 59 (March 1969): 44–47.

Fairchild, Steven H. "Handwriting as a Language Art." In *Language Arts Instruction and Beginning Teaching,* edited by Carl Personke and Dale Johnson. Englewood Cliffs, N.J.: Prentice-Hall, 1987.

Frith, Uta, and Vogel, Juliet M. *Some Perceptual Prerequisites for Reading.* Newark, Del.: International Reading Association, 1980.

Furner, Beatrice A. "Recommended Instructional Procedures in a Method Emphasizing the Perceptual-Motor Nature of Learning in Handwriting." *Elementary English* 46 (December 1969): 1021–30.

Gearheart, Bill R.; Weishahn, Mel; and Gearheart, Carol J. *The Exceptional Student in the Regular Classroom,* 4th ed. Columbus, Oh: Merrill Publishing Co., 1988.

Goodson, Roger A. *Handwriting.* Alexandria City Public Schools, Alexandria, Va., 1968.

_____, and Floyd, Barbara J. *Individualizing Instruction in Spelling: A Practical Guide.* Minneapolis: Denison & Co., 1974.

Graves, Donald H. "Research Update—Handwriting Is for Writing." *Language Arts* 55 (March 1978): 393–99.

Greene, Harry A., and Petty, Walter. *Developing Language Skills in the Elementary Schools.* Boston: Allyn & Bacon, 1963.

Hallahan, Daniel P., and Kauffman, James M. *Introduction to Learning Disabilities: A Psycho-Behavioral Approach.* Englewood Cliffs, N.J.: Prentice-Hall, 1976.

Hammill, Donald D., and Bartel, Nettie R. *Teaching Children with Learning and Behavioral Problems.* Boston: Allyn & Bacon, 1975.

Harber, Jean R. "Differentiating LD and Normal Children: The Utility of Selected Perceptual and Perceptual-Motor Tests." *Learning Disabled Quarterly* 2 (Spring 1979): 70–75.

Johnson, Stanley, and Morasky, Robert L. *Learning Disabilities.* 2d ed. Boston: Allyn & Bacon, 1980.

Kean, John M., and Personke, Carl. *The Language Arts.* New York: St. Martin's Press, 1976.

Lerner, Janet W. *Children with Learning Disabilities.* 2d ed. Boston: Houghton Mifflin, 1976.

Lewis, Edward R., and Lewis, Hilda P. "Which Manuscript Letters Are Hard for First Graders?" *Elementary English* 41 (December 1964): 855–58.

Lyon, Reid, and Watson, Bill. "Empirically Derived Subgroups of Learning Disabled Readers: Diagnostic Characteristics." *Journal of Learning Disabilities* 14 (May 1981): 256–61.

Otto, Wayne, and Smith, Richard. *Corrective and Remedial Teaching.* Boston: Houghton Mifflin Co., 1980.

Petty, Walter T.; Petty, Dorothy C.; and Becking, Marjorie F. *Experiences in Language.* Boston: Allyn & Bacon, 1973.

Reger, R.; Schroeder, W.; and Uschold, K. *Special Education: Children with Learning Problems.* New York: Oxford University Press, 1968.

Shaughnessy, Mina P. *Errors and Expectations.* New York: Oxford University Press, 1977.

Sheldon, William D. "A Summary of Research Studies Related to Language Arts in Elementary Education: 1976." *Language Arts* 55 (January 1978): 65–101.

Spache, George. *Investigating the Issues of Reading Disabilities.* Boston: Allyn & Bacon, 1976.

Tagatz, Glenn E; Otto, Wayne; Klausmeier, Herbert J.; Goodwin, William L.; and Cook, Doris M. "Effects of Three Methods of Instruction upon the Handwriting Performance of Third and Fourth Graders." *American Educational Research Journal* 5 (January 1968): 81–90.

Tompkins, Gail E. "Let's Go on a Bear Hunt! A Fresh Approach to Penmanship Drill." *Language Arts* 57 (October 1980): 782–86.

Wright, Jone P., and Allen, Elizabeth G. "Ready to Write!" *Elementary School Journal* 75 (April 1975): 430–35.

Yardley, Alice. *Exploration and Language.* New York: Citation Press, 1973.

Chapter Six

*After completing the chapter on spelling, you will
be able to:*

1. *Define the objectives of the elementary
   spelling program.*
2. *Place students on their proper instructional
   spelling levels.*
3. *Analyze how standardized spelling tests can
   help the teacher evaluate a student's spelling
   skills.*
4. *Use informal spelling inventories to evaluate
   a student's spelling strengths and weaknesses.*
5. *Evaluate a student's application of spelling
   generalizations.*
6. *Understand the two most commonly used
   criteria for selecting spelling words.*
7. *Use the best developmental approach for
   spelling instruction, including the corrected-
   test method and the most efficient self-study
   method.*
8. *Design a weekly spelling program.*
9. *Teach spelling generalizations inductively.*
10. *Understand the step-by-step procedures
    needed to instruct students who require an
    effective remedial approach.*
11. *Develop accelerated spelling activities for the
    gifted student.*
12. *Teach a proofreading spelling unit to an ele-
    mentary class.*

Spelling

“**W**hy Johnny can't write" is becoming as pressing a concern today as was "why Johnny can't read" in the fifties. In fact, an issue of the *Journal of American Education* concentrated on the writing crisis in America. Spelling errors detract from writing effectiveness; the advantages of good spelling ability and the disadvantages of inferior spelling ability justify careful, systematic spelling instruction.

OBJECTIVES

Petty and Jensen maintained that the long-range objectives for spelling programs should be stated in terms of the following attitudes, skills or abilities, and desired habits (1980, pp. 442–43):

Attitudes

Each child should:

1. Recognize the necessity and responsibility for correct spelling in effective communication.
2. Show a desire to spell all words correctly.
3. Believe that spelling correctly is something she can accomplish.

Skills and Abilities

Each child should be able to:

1. Recognize all the letters of the alphabet in capital and lowercase forms in both printed and handwritten materials.
2. Write all the letters of the alphabet in a legible manner in both capital and lowercase forms.
3. Alphabetize words.
4. Hear words accurately as they are spoken.
5. Pronounce words clearly and accurately.
6. See printed words accurately.

7. Group and connect the letters of a word properly.
8. Use punctuation elements that are necessary for spelling.
9. Use a dictionary, including diacritical markings and guide words.
10. Pronounce unfamiliar words properly.
11. Use knowledge of sound and symbol relationships.
12. Use knowledge of orthographic patterns that recur in language.
13. Use the most effective spelling rules (generalizations).
14. Use personally effective procedures in learning to spell new words.

Habits

Each child should habitually:

1. Proofread all writing carefully.
2. Use reliable sources to find the spellings of unknown words.
3. Follow a specific study procedure in learning to spell new words.

Spelling, like handwriting, is not an end in itself. The major objective is effective written communication. The mature speller can write correctly a great number of words without conscious effort. Consequently, if our goal is to develop effective spellers, the spelling program must be formulated on objectives that enhance the student's independent spelling skills. In addition, the spelling program must include many opportunities for writing for an audience.

Formulating objectives such as these to guide the program is the first step in developing or improving the spelling curriculum. It would be extremely difficult to formulate specific objectives, or to design a successful spelling program, without sound diagnosis of students' level of proficiency and strengths and weaknesses in spelling skills. Thus, we first consider preinstructional spelling development, then we consider the spelling curriculum according to assessment and instructional methods.

DEVELOPMENTAL STAGES IN SPELLING

Research discussed in the Language and Cognitive Development chapter identified stages in children's language development. Researchers in the 1970s and 1980s such as Read (1971), Hendersen and Beers (1980), Gentry (1982), Henderson (1985), and Henderson and Templeton (1986) investigated developmental stages through which children proceed as they acquire spelling ability.

Henderson and Templeton (1986) concluded that "developmental theory validates the skills we should teach and provides a rationale for the pacing and maintenance of instruction in a more detailed and clearly stated manner than has been possible before" (p. 314). These researchers used children's inventive spellings to identify the following five stages in developing spelling competence:

1. Stage I—Preliterate stage in which young children spontaneously use symbols to represent words; there is, however, no knowledge of letter–sound correspondence.

2. Stage II—Children, in approximately first grade, begin to spell alphabetically, matching sounds and letters systematically. Their spellings reflect errors such as MAK = make.
3. Stage III—Children, in approximately third and fourth grades, begin to assimilate word knowledge and conventional alternatives for representing sounds, include vowels in every syllable, use familiar spelling patterns, and intersperse standard spelling with phonetic spelling. Their spellings reflect errors such as CREME = cream.
4. Stage IV—Children, in approximately fourth through sixth grades, increase understanding of word patterns. Their spellings reflect errors such as INOCENT = innocent.
5. Stage V—Children understand that words related in meaning are often spelled similarly; children's knowledge of the English orthographic system and the accompanying rules are established; children's knowledge of base and root words are extensively developed.

Henderson and Templeton (1986) provided suggestions for instruction that apply at each of these stages. For example, at stage II, children should begin to examine systematically groups of words that are organized around common features such as short-vowel phonograms, beginning consonant digraphs, and common long-vowel patterns. Short lists of words that are chosen from their reading vocabularies can be used for examination and memorization. At the same time, children should be instructed in letter formation and provided with many opportunities for creative and purposeful writing.

At stage III, Henderson and Templeton maintained that spelling instruction should emphasize basic pattern features of words including reviewing long-vowel patterns taught in stage II (e.g., /a/: bait); introducing common digraphs (e.g., ow, oi); studying r and l influenced vowel patterns (e.g., card, fall); studying relationships between words in compound words; learning functions of homophones in spelling; examining common inflections and how they are joined to base words (e.g., ed, ing, ly); presenting concept of base words that can stand alone after prefixes and suffixes are removed; and studying sounds and meanings of common prefixes and suffixes.

At stage IV, Henderson and Templeton emphasized continued study of common and uncommon vowel patterns; teaching students to scan words systematically, syllable by syllable; examining prefixes and suffixes for meaning rather than sound (e.g., when to use able or ible, ant or ance, ent or ence); exploring homophones; and studying consonant doubling as it applies to a broader range of vocabulary.

At stage V, Henderson and Templeton emphasized studying silent/sounded consonant patterns (e.g., resign-resignation); studying different vowel alternation patterns that are sequenced for study from long-to-short (e.g., sane-sanity), long-to-schwa (e.g., inflame-inflammation), short-to-schwa (e.g., excel-excellent) and predictable sound/spelling alternations (e.g., explain-explanation); and examining contributions of Greek and Latin forms to the spelling/meaning connection in English.

Recent articles related to spelling frequently focus on either developmental stages in children's spelling or on disputes among those who draw implications from children's spontaneous spelling and those who adhere to more traditional approaches. Although the two camps have not resolved their differences, Henderson and Templeton provided the following support for combining implications from both approaches: Today most list words are selected, as they should be, by frequency counts of their occurrence in the language as a whole and in children's language in particular. Examination of basal spelling programs of the past decade for the primary grades will show a remarkable correspondence between the features derived from the stage theory and the scope and sequence plans for word study. . . . What developmental research has achieved is not a radical revision of traditional spelling instruction but a clarification of those things long practiced, and what is important, an extension of word study principles to the middle and upper grades, where they are presently most crucially needed. (p. 314)

DEVELOPING THE SCHOOL SPELLING PROGRAM

In addition to studies conducted by Henderson and Templeton, considerable research in spelling diagnosis, instructional programs, and teaching techniques provide clear guidelines for spelling instruction. Following a review of research spanning more than fifty years, Hillerich (1982) identified the following effective procedures and instructional approaches:

1. Determine the appropriate instructional spelling level for each student.
2. Teach the list of high-frequency words using a research-based approach.
3. Teach generalizations about apostrophes and word endings.
4. Provide experiences with homophones.
5. Teach dictionary skills along with ways of spelling various sounds.
6. Develop a desire to spell correctly when the writing is designed for an audience other than an informal group.
7. Encourage considerable writing to help children develop an understanding that writing is the only reason for learning to spell.
8. Provide children with many meaningful writing experiences that reinforce the automatic spelling of their security lists.

These recommendations are apparent throughout the discussion of assessment and instructional practices.

ASSESSING SPELLING SKILLS

Spelling offers an excellent opportunity for diagnostic teaching. Assessment is necessary to discover at what grade level to begin spelling instruction, and to evaluate each student's specific spelling strengths and weaknesses. We discuss three approaches to diagnostic teaching. The first diagnostic technique provides a method for placing students on appropriate spelling instructional levels; the second stresses formal assessment of spelling skills; and the third diagnostic tool employs informal evaluation of spelling ability.

Dictionary skills are part of the school spelling program.

Diagnostic Techniques

Studies show great variability in the spelling competency of children in any one class. I saw an example of this diverse spelling ability when I assessed the spelling capacity of several sixth-grade classes. It was not uncommon to find students with third-grade spelling ability in the same class with students who were able to spell lists of words appropriate for high-school students.

The teacher must discover each child's appropriate spelling level because there is no economy in teaching children words they can already spell correctly. It is equally important to diagnose the level of the student who is having spelling problems. Students will not make adequate gains if they are expected to spell words of an unrealistically high level of difficulty.

At first, this assessment may seem formidable. The teacher must utilize the most efficient procedure for placing students in proper ability groupings. Hillerich (1977) believed children should receive instruction in a spelling word list of which they can spell approximately 75 percent of the words; children who are not able to correctly spell 50 percent of a graded word list have not mastered the prerequisite spellings taught at lower grade levels.

It is not difficult to place elementary students on appropriate spelling levels. First, the teacher selects approximately twenty-five words from each graded spelling list. Do not choose only the most difficult or the easiest words from that level. Many teachers find the best procedure is to choose every tenth word from a list of spelling words for each grade. Table 6–1 illustrates this procedure, using a graded word list from the *Skills in Spelling* series published by the American Book Company.

After compiling the lists, the teacher administers the tests to the class to determine spelling levels. He would administer the first-grade list to all students. Children who spell more than 75 percent of the words correctly are tested with the second-grade list. Children who spell fewer than 75 percent correctly do not take a further test; the teacher has discovered their level of spelling ability, and will use first-grade instruction for these children.

Students who have mastered the first-grade list are tested with the second-grade list. If they have not mastered this list, they are provided with second-grade instruction. Students who demonstrate mastery of the list are tested with the third-grade list. The teacher tests children on consecutive days until he has placed all of them in their most efficient instructional level. This diagnostic procedure is not overly time consuming, and assures that each child will be instructed at her specific spelling level.

REINFORCEMENT ACTIVITY

Compile a spelling test list, using the graded lists from one of the spelling series. Administer the list to a group of children, and estimate the spelling instructional level for each member of the group.

TABLE 6–1

Word list from the *Skills in Spelling* series

Grade 1	Grade 2	Grade 3	Grade 4
pig	after	Monday	act
pretty	wet	fed	dream
send	win	trip	drove
she	block	week	eight
five	top	name	tip
purple	game	Friday	base
can	five	these	penny
like	boat	sent	cook
there	stay	cent	paid
come	nest	full	all right
not	friend	keep	everywhere
	fall	then	careful
	band	add	everybody
	dinner	bread	country
	girls	where	church
	dress	fight	comb
	there	asked	coal
	when	things	bow
	thanks	tall	busy
	open	use	ocean
	call	with	circus
	our	stopped	hurry
	at	nuts	master
	calling	snowball	lucky
	something	throw	popcorn

SOURCE: From Bremer, Bishop, and Stone, *Skills in Spelling.* (Dallas: American Book Co., 1976). Reprinted by permission of American Book Co.

Informal Assessment

Informal spelling assessment, including teacher-made tests and observations, will give teachers their most useful information for designing a spelling program. Informal evaluation techniques are vital when specific information is required about a child's spelling. Moreover, informal diagnosis is essential for designing a program to remediate spelling difficulties. Linn (1967), writing about diagnosis and remediation of spelling problems, suggested teachers evaluate a child's spelling in relation to the following questions:

1. Does the child recall letter and sound symbols quickly and accurately?
2. Can the child correctly produce the letter and sound symbols on paper?
3. Can the child fuse the sound parts of the words together to form a whole word? E.g., is she able to blend the individual sounds c-a-t into the word *cat?*
4. Does the child reverse letters in sound parts? E.g., does she write *talbe* instead of *table?*

5. Can the child remember words the teacher has written on the board a few minutes after they are erased?
6. Does the child learn words when she *hears* the letter sequence rather than when she *sees* it? The answer to this question provides insight into the instructional approach that may be of most value to the individual child.
7. Does the child block out or not hear sounds? Can she hear the minimal difference between *pin* and *pen?*
8. Can the child write the correct symbol for single sounds when the sounds are dictated orally?

Answers to these questions are of vital concern to the teacher. The student who has mastered these skills will probably be able to learn to spell without serious difficulty, whereas the student who has difficulty producing letters, blending or hearing sounds, and remembering words and letters may need an individually designed spelling program. A remedial program that has proven beneficial for many children is described when we discuss remedial spelling approaches.

Informal spelling assessment can be maximized by systematic study of a child's spelling errors. If a child follows a pattern in misspelling certain words, the teacher can expect her to make the same type of error when spelling similar words. Spache's Spelling Errors Test provides a list of types of errors often reflected in children's spelling.

[handwritten: most common mistaks]

1. A silent letter is omitted—*take-tak*
2. A sounded letter is omitted—*stand-stan*
3. A double letter is omitted—*letter-leter*
4. A single letter is added—*park-parck*
5. Letters are transposed or reversed—*angel-angle*
6. There is a phonetic substitution for the vowel—*green-grean*
7. There is a phonetic substitution for the consonant—*bush-buch*
8. There is a phonetic substitution for the syllable—*flies-flys*
9. There is a phonetic substitution for the word—*hare-hear*
10. There is a nonphonetic substitution for the vowel—*bats-bot*
11. There is a nonphonetic substitution for a consonant—*camper-canper*

The teacher can use this list to analyze test papers and other written work. Various creative writings and written assignments in content areas provide sources for informal spelling assessment.

Besides observing written work, the teacher can also analyze a child's oral spelling and speech performance. How does the child approach oral spelling? Does she sound words one letter at a time, by syllables, or as whole-word units? Are the child's oral errors similar to her written errors? Does she enunciate words carefully? How does the child pronounce the words she spells incorrectly? Does her speech reflect a dialect? Does the dialect interfere with spelling patterns? Is the child able to blend sounds orally into a whole word?

The teacher can compile other useful information by observing the child's study skills in regard to spelling. Does the child apply an effective method of study to spelling-word lists? (An excellent method for self-study is presented under the "procedures" subhead.) A positive attitude toward spelling is essential. Does the child feel spelling is important to her? Does she carefully proofread her papers for spelling errors? Does she use a dictionary to verify spelling questions? Does the child use her new spelling words in other daily assignments?

The teacher can also obtain valuable diagnostic information by questioning children. Ask children to tell you how they spell an unknown word. Question them about their best method for remembering a spelling word. Many children have told me they must carefully write a word several times before they are able to remember it. One learning-disabled child indicated that he traced words with his hands when his

REINFORCEMENT ACTIVITY

The following examples are excerpts from a third-grade student's daily written work. The first selection is an outlining activity following a basal reading lesson. The second example is from a social studies assignment. Study the child's papers and find the spelling errors. Classify the errors according to the Spache spelling-error categories and then according to Henderson and Templeton's stages in developmental spelling defined on pages 193–94. How do the errors compare with Henderson and Templeton's stages? At what stage or stages is this student? What proof do you have for your answer? Do you see any consistent errors? Discuss the implications of these errors for remedial instruction.

I. *Daytime*
   A. *a sailer showd Mark around*
   B. *there was a spider*
   C. *Mark went to play with his frends*
   D. *the boys climed around*
   E. *the sailer toke the spider to a sientest*
II. *Nighttime*
   A. *the tarantual was luse*
   B. *Marks Mother was wered*
   C. *he wated until 9 o'clock with the spider*
   D. *Tim quickly looked at the spider*
   E. *Tin cauld another polece*

Fur Seals

Seals are excellent swimers and they spend lots of time in the water. They have fur all over them.

The mail seals are cauld bulls and femail seals are cauld cows. They have there pups on land and they usualy have one pup but sometimes two.

Seals spend much of there time on land or on floating chuncks of ice.

Fur seals stay at sea for 6 to 8 months.

They are found along the costes of continents in most parts of the world.

A few kinds live in fresh water lakes and cloes to inland seas.

teacher asked him to spell orally. Other children say they benefit from spelling words aloud while writing. One fourth-grade girl explained her method for overcoming a dialect difference between herself and her teacher. She solved her difficulty by whispering each word before writing a dictated word. Children can frequently describe their greatest spelling problems; this information should be a vital part of the informal spelling assessment.

Another useful informal assessment diagnoses the child's ability to apply spelling generalizations. Ernest Horn (1963) recommended teaching rules that apply to a large number of words and have few exceptions. This is a problem, because few spelling generalizations have 100 percent utility. The teacher can administer a cloze test to assess the child's knowledge of the more reliable generalizations. A cloze test for specific spelling generalizations might include the following examples: *most often used*

1. Generalization: Words that end with a silent *e* usually drop the *e* before adding a suffix beginning with a vowel.
   Cloze: Jim was rak_____ the leaves. (ing)

2. Generalization: Words ending in silent *e* retain their *e* when adding a suffix beginning with a consonant.
   Cloze: The president lives in a stat_____ mansion. (ely)

3. Generalization: If a word ends with a consonant plus *y*, change the *y* to *i* before adding suffixes except those beginning with *i*. Do not change the *y* to *i* when adding suffixes to words ending in a vowel plus *y*.
   Cloze: The bab_____ were all crying at once. (ies)
   Cloze: The horse was carr_____ a heavy load. (ying)
   Cloze: The boy scouts were enjo_____ the football game. (ying)

4. Generalization: Double the final consonant in one-syllable words or words accented on the last syllable, ending in a single consonant preceded by a single vowel before adding a suffix beginning with a vowel.
   Cloze: The horse was run_____ a very fast race. (ning)
   Cloze: Sunday is the begin_____ of the week. (ning)

5. Generalization: The letter *q* is always followed by the letter *u* in the English language.
   Cloze: At night it is q_____iet in the house. (u)

6. Generalization: Proper nouns should begin with capital letters.
   Cloze: The dog's name is _____over. (R)

SELECTING SPELLING WORDS

An examination of the word lists in major spelling books illustrates two different viewpoints on the selection of spelling words. According to one judgment, the English language system is irregular; consequently, spelling instruction should emphasize gradual accumulation of the most useful words. In contrast, those who advocate teaching spelling by means of spelling generalizations believe the English language is remarkably regular, and that efficient spelling programs should have greater phonetic emphasis.

Frequency-of-Use Approach

Many spelling programs concur with the philosophy that the English language system is irregular, and that instruction should be based on the words most frequently used. Ernest Horn, whose writings and research span the past fifty years in education, is the most influential proponent of this view. As early as 1919, Horn stated: "One must show that a rule can be easily taught, that it can be remembered, and that it will function in the stress of actual spelling. Evidence seems to cast a doubt on all three of these assumptions." In 1954, he said: "The limited success in attempts to teach pupils to learn and apply even a few spelling rules suggests that we should not be too optimistic about the practicality of teaching the more numerous and complicated rules or principles in phonetics," and in 1957, "There seems no escape from the direct teaching of the large number of common words which do not conform in their spelling to any phonetic or orthographic rule." By 1963, Horn believed that some emphasis on phonics should be included in the spelling program, but he cautioned that "Instruction in phonics should be regarded, however, as an aid to spelling rather than a substitute for the systematic study of the words in the spelling list."

The words Horn believed should comprise the spelling list are those most frequently used by children and adults. In selecting these high-frequency words, Horn recommended that:

1. The easiest words should be taught in the beginning grades and the most difficult words in the advanced grades.
2. Words most frequently used in writing should be taught first.
3. Words commonly used by children in a specific grade should be taught in that grade.
4. Words needed in other content subjects should be taught in the appropriate grades.

Researchers have compiled several spelling lists that can be useful in evaluation and selection of high-frequency spelling words. James Fitzgerald (1951) compiled a Basic Life Spelling Vocabulary consisting of the 2,650 words most frequently written by children and adults; these words constitute approximately 98 percent of all words written. Horn compiled nine studies of written vocabularies; his list included the 4,052 words that appear most frequently in correspondence. The first ten words on Horn's list (I, the, and, to, a, you, of, in, we, for) constitute 25 percent of the words used in various kinds of writing. The first 100 words in the list account for 65 percent of the words written by children and adults. Encouraging mastery of meaningful lists is appropriate because spelling efficiency increases when children are taught the words they need to spell most frequently.

Spelling Approaches with Phonetic Emphasis

Some linguists disagree with Horn's opinion that the English language system is irregular; they maintain that there is a high degree of regularity between the graphemic presentation of a written word and its corresponding sounds. In 1965, Hodges and Rudorf investigated the relationship between phonemes (sounds) and graphemes (printed letters) in over 17,000 words, and found a high degree of regularity between the sounds and the written letters.

Hanna and Hanna (1965) recommended that a spelling program reflect the findings of the Hodges and Rudorf investigation; they believed children should learn to spell words using a phonic approach. In an earlier article, Hanna and Moore (1953) recommended putting aside some time for a concentrated attack on translating sounds into written symbols, during which they would have a child learn phonic patterns inductively. The simple patterns, such as the consonant-vowel-consonant in "cat" would be learned first, and the complex phonic patterns later. Hanna and Moore believed the English words that follow a rarely occurring phonic pattern should be memorized.

Relationship of Spelling and Meaning

Research by Chomsky (1970) suggests that words that mean the same tend to look the same, even though they may have different pronunciations. Thus, the words *nation* and *national* have a similar spelling; the meanings are consistent even though the phonetic pronunciations are different. Likewise, certain spelling patterns are considered consistent because they indicate a specific part of speech. From this perspective, the writing system seems both sensible and systematic.

Zutell (1978) suggested that teachers apply linguistic and psycholinguistic findings by helping children develop their understanding of the writing system. To do so, "teachers must consciously construct environments in which children have the opportunity to systematically examine words and to freely generate, test, and evaluate their own spelling strategies" (p. 847). Such a spelling program would allow children to examine their own writing; provide experiences for reading and vocabulary development; and provide word study activities that allow children to compare, contrast, and categorize words according to root words, word origins, and similarities in structural patterns.

PROCEDURES

The teacher in a developmental spelling program must consider such questions as: how many words should be taught; how many minutes a day should be spent on spelling instruction; what words should be studied; what is the most efficient individual study method; how often should review be included; and how should students receive instruction in spelling generalizations?

Several factors contribute to a decision about how many words to teach during spelling lessons. The teacher must consider the number of words that students can learn efficiently in a specific grade, as well as how much emphasis is placed on spelling in other content areas. Most spelling series contain fewer than 4,000 words. One spelling authority describes this accumulation of words in terms of daily tasks. If 2 new words a day are presented in second and third grades, 3 new words a day in fourth and fifth grades, and 4 new words a day in sixth, seventh, and eighth grades, students can master the spelling of 4,180 words. If the words are chosen according to high frequency, this accumulation would account for 98 percent of the words written by children and adults. It would not seem an impossible task for most students to master the essential spelling lists if spelling is systematically taught. Most authorities believe these spelling words can be learned during concentrated daily spelling instruction periods of approximately fifteen minutes.

The teacher's problem now is deciding what words each student should master. We know it would be inefficient to have students work on words that are either too easy or too difficult, so the program should begin with assessment procedures to ascertain each student's individual spelling instructional level. After assessment, the teacher will be able to divide the class into appropriate groupings. The average elementary class will have, perhaps, four spelling groups. The teacher initiates spelling instruction with the corrected-test method.

Corrected-Test Method

Research by Horn indicates that the single best method for learning to spell is the corrected-test method. Personke (1987) stated that the approach "has been reexamined many times and remains the single best method for learning to spell words. The reason for its success is simple: it provides immediate, positive feedback" (p. 79). In fact, Hillerich (1977) indicated that the use of a pretest, with immediate correction by the child, accounts for about 95 percent of all learning that takes place in spelling. In the corrected-test method, the teacher administers a spelling test before the students study the words. This procedure liberates students from systematic study of words they already know.

First, the teacher administers the test. Immediately afterward, the pupils self-correct their tests; they may self-correct from either their own spelling list or a typed list provided by the teacher, or they can listen to the teacher slowly read the correct spelling. The pupils immediately write the correct spelling of any words they have missed. This method gives students essential positive feedback. If the teacher collects the tests, corrects them, then returns them to the children, they may retain the misspellings rather than the correct spellings. The teacher should monitor the corrected-test procedure carefully to insure the students are checking accurately, and writing their misspelled words correctly.

Classroom teachers have found two successful techniques for administering more than one spelling list to a class. Some teachers test the various ability groupings simultaneously. These teachers read, in turn, one word from each spelling list. For example, the teacher dictates the first-grade word, "pretty," to one group, the

touch each letter

second-grade word, "block," to the next group, the third-grade word, "Friday," to the third group, and the fourth-grade word, "everywhere," to the last group. After each group receives a word, the teacher returns to the initial group's next word and follows the same sequence until all students have been tested on their lists. Other teachers prefer to test one group at a time, which takes longer, but gives the teacher a smaller group to monitor. Self-correction and immediate writing of misspelled words follow both methods of administration. Each child's misspelled words form the list of words she systematically studies for that week.

The Self-Study Method *for good spellers*

The self-corrected test procedure gives students an individualized list of words they need to master. Now, each student must be taught to learn these misspelled words efficiently. Teachers know students do not automatically use the most effective study methods. In fact, proper study methods should be reviewed at the beginning of each grade.

Horn recommended teaching the following self-study procedure:

1. Pronounce the word correctly. Correct pronunciation is an important factor in learning to spell. Enunciate each syllable distinctly while looking closely at the word.
2. Close your eyes and try to recall how the word looks, syllable by syllable. Pronounce the word in a whisper while you try to visualize the word.
3. Open your eyes and make sure you have recalled the word correctly. Repeat steps two or three three times.
4. When you feel you have learned the word, cover it, and write it without looking. Check your written word with the correct, covered word.
5. Repeat this writing procedure until you have correctly spelled the word three times. Check the correct spelling after each writing.

Students who demonstrate more severe spelling problems will need to overlearn their spelling words; they may need more than three correct repetitions of the spelling word. Some of these students may benefit from the more rigorous remedial technique described later in this chapter.

A Weekly Spelling Plan

Use of a corrected test and study plan might result in a Monday through Friday arrangement similar to that recommended by Burns and Broman (1983):

1. On Monday, a pretest is given to each spelling group. The pretest consists of all the words the group is expected to master during the week. Children check their own words, using the corrected-spelling approach. They study each misspelled word immediately.
2. On Tuesday, children continue to study misspelled words. They also work on exercises to expand the meaning of the spelling words. Such exercises might consist of writing phrases with each word, or adding affixes to the

words. For example, affixes to the word *decide* produce *decided, deciding,* and *undecided.*

3. On Wednesday, a midweek test of all the words given on Monday is given again to each spelling group. Children who spell the list correctly on Wednesday are not required to take an additional test. Children who miss words immediately use the same self-study procedure with their words.

4. Thursday's spelling lesson would include enrichment activities that use the words in various situations. Unless children use the spelling words they learn, they will quickly forget them. Some students will need self-study time before the Friday test.

5. On Friday, students who missed spelling words on both Monday and Wednesday take a final test. Words they miss on Friday are added to the student's spelling list for periodic review.

Students can keep charts to help them visualize their individual growth in spelling. They should chart their spelling scores from the Wednesday or Friday tests. The charts will resemble that in figure 6–1.

Horn emphasized periodic review as an essential aspect of an efficient spelling program. The first review should occur within a day or two after the words are taught; the Wednesday test provides this review. The second review should take place during the following month. Missed words should be placed on a special list and reviewed regularly. Finally, words the children find especially difficult during the year should be included on the instructional list for the following grade.

FIGURE 6–1
Individual pupil spelling progress chart

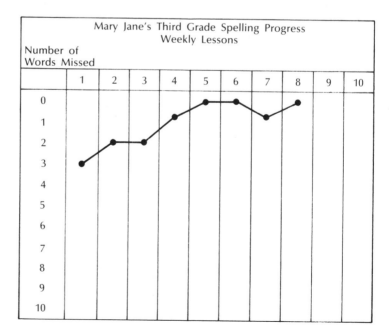

Looking over the developmental spelling program, we find six important factors to remember when planning the spelling curriculum. The initial step is diagnostic testing, to find the most efficient spelling level for each student. Second, the weekly program begins with a pretest, followed immediately by the corrected-test procedure. Third, each child should work only on the words that are difficult for her, and the teacher should show her a definite method for learning these words. Fourth, rigorous reviews will emphasize these more difficult words. Fifth, the pupil should chart her daily, weekly, monthly, and yearly progress. Finally, the students need to apply their spelling words in various contexts.

Inductive Teaching of Spelling Generalizations

Most spelling authorities believe the instructional program should include some spelling generalizations. Disagreement over the teaching of spelling generalizations pertains to the number of generalizations considered reliable; the more reliable generalizations should definitely be included in a developmental spelling program. Students will be better able to apply these generalizations if they learn them inductively.

Using the inductive method, the teacher initiates the learning experience by writing on the board several words that demonstrate the particular generalization. Then, he asks the children to supply more words like those listed, and adds them to the list. Any exceptions to the rule are added to a separate list. The class is then guided to look carefully at the words to discover the generalization involved. After discovering the generalization the children turn to their spelling books or lists to verify their discovery. Inductive teaching tends to be more active and pupil-involved; consequently, it both motivates and provides its own positive reinforcement.

An Inductive Approach for Spelling Generalizations

An excellent third-grade teacher developed this lesson plan. Begin by placing the following words on the board in two columns. Singular forms should be in the left-hand column, plural forms in the right-hand column.

key	keys
day	days
monkey	monkeys
boy	boys
donkey	donkeys
baby	babies
cherry	cherries
lady	ladies
berry	berries
daisy	daisies

Pronounce all the words with the children. Have individual children produce oral sentences utilizing the listed words, then ask the children to add similar words to the list. Put the children's words on the board next to the appropriate list.

Have the class examine the left-hand column and decide on a common element for all of the words (all words end with *y*). Next, have the class look at the right-hand column and determine how the first five words were made plural (*s* was added to each word). Ask how the second five words were made plural (*es* was added and the *y* was changed to *i*). Also ask why the *y* was changed in some words but not in others. The question should lead the children to formulate a generalization similar to the following:

If a word ends in *y*, add *s* to make the plural if a vowel is before the *y*. If a consonant is before the *y*, change the *y* to *i* and add *es* to form the plural.

Students should then test their generalization on new words. These words conform to the above generalizations:

sky	lorry	ferry	penny	play	tray
bay	trolley	body	country	toy	

Finally, ask the class to open their spelling and reading books to find additional words that fit the "*y*" generalization.

(You can use this approach with other reliable generalizations listed on page 201 and with apostrophe generalizations.)

REINFORCEMENT ACTIVITIES

Administer a list of appropriate spelling words to an elementary class. Following the corrected-test activity, instruct the class in the most efficient method for self-study. Develop a chart, with student assistance, to remind the students how to use the study procedures.

Develop a lesson plan that shows the step-by-step procedure for teaching a spelling generalization. If you are a preservice teacher, demonstrate your activity to a group of peers in your college class. If you are an inservice teacher, instruct a group of children with your lesson plan.

REMEDIAL APPROACHES

Slower spellers in the class may need special remedial consideration. Accurate diagnosis of spelling ability levels and specific strengths and weaknesses are especially important for effective remedial spelling instruction. The teacher may need to reduce by half the list of words he will expect remedial spellers to master, because of the more detailed study method that may be necessary.

Fernald Approach

Fernald's multisensory approach has been successful with many disabled spellers (Campbell, 1976). The teacher begins by writing a word on a card in large letters. (Use cursive writing for older children and manuscript for younger students.)

Next, the teacher shows this card to the child and names the word. He asks the child to look carefully at the word and say it with him. The teacher traces the word with two fingers while the child repeats the word. Next, the child traces the word with two fingers while she pronounces each syllable; she does not spell the word. The child repeats this tracing and verbalizing until she feels she can write the word without copying it.

When she feels she is ready, she turns the card face down and writes the word without looking at it. After writing the word, she turns the card over to verify her spelling. If she spells the word correctly, she goes through the writing and verification procedure three more times without copying either the card or her previously written words. If unsuccessful, she goes back to the finger-tracing and verbalizing experience. The child is not permitted to erase part of the word; she must rewrite the whole word if she makes an error. Words learned in this manner are retested the following day.

The Fernald approach can be used individually or with small groups. One teacher who used this approach with severely remedial spellers reported that her pupils increased mastery of their spelling words by 80 percent. They learned fewer words, but used the words they learned in daily work and remembered them.

Reinforcement Games for Remedial Students

The remedial student also needs reinforcing spelling activities. Games offer diversion from the more drill-centered remedial approaches. The Tic-Tac-Toe and the Sports and Spelling games enrich and motivate remedial instruction.

Tic-Tac-Toe   In the Tic-Tac-Toe game, the week's spelling words are printed on one side of tag board cards. One player shuffles the cards while the second player draws a tic-tac-toe grid. The players decide who is to be X and who is to be O. Player X draws the first card, and asks player O to spell the word on the card. If player O spells the word correctly, she places an O on the grid and draws a card for player X to spell. When a player is unable to spell a word, the card is placed on the bottom of the card pile and the player is not allowed to place a marker on the grid. The winner is the player who places three of her markers in a row.

Sports and Spelling   For the Sports and Spelling game, have the students divide into pairs. Give each pair the sports section from the daily newspaper. The children must find as many words from their spelling list as possible. The winners are the team or teams who find the most words.

The sports section of the newspaper provides an opportunity for children to locate sports-related spelling words.

REINFORCEMENT ACTIVITY

Develop a spelling game for reinforcing a weekly spelling list. Play the game with children of the appropriate age. Caution: when you use games to reinforce learning, be sure to supervise the play. Children may circumvent the rules of the game and make up their own. When this happens, the teacher's objective for the game is not realized.

THE ACCELERATED STUDENT

When the teacher uses a diagnostic and pretest approach, accelerated students will not have to waste class periods on words they can already spell. The gifted speller can thus take part in a spelling period without excessive boredom.

Accelerated students may learn their words very rapidly, and will thus have time during the spelling period to increase their vocabularies and work on word lists that suit their particular interests. As an example, one gifted third-grade student was especially interested in astronomy and another in dinosaurs. They learned to spell subject-related words that allowed them to write about their favorite topics. These words are not normally taught in the third-grade spelling curriculum, but these students used them frequently.

Successful spelling programs for gifted spellers are usually individualized. These students should not be punished by having to master a second list of words after they have learned one list. One student who described her spelling experience to me indicated that her spelling teacher always gave her an additional twenty-five spelling words if she spelled the first twenty-five correctly. The second list proved to be punishment for this girl, rather than an award for accurate spelling. The girl solved her problem by intentionally missing words on the first list so she would not have to write an additional list of words. Obviously, the teacher had not placed this child on the correct spelling level and had not used a pretest to determine the words she did not need to study.

The teacher can enrich the spelling curriculum with vocabulary-building activities. Students can find synonyms for overworked words and discover when a sentence's meaning would improve with the use of a more exacting synonym. For example, the word *small* has several different meanings. Students who are investigating synonyms would find three categories of words relating to *small:*

little		*minor*		*small-minded*	
diminutive	tiny	unimportant	trivial	ignoble	narrow
miniature	wee	insignificant	modest	bigoted	petty
mini	dwarfish	inconsiderable	secondary	stingy	limited
Lilliputian		inconsequential	trifling	provincial	intolerant

Students can categorize the various synonyms according to their proper meanings and discuss the context for using each of them. Among other words with multiple meanings and numerous synonyms are *strong, big, rub, rude, nice, real, look, safe* and *fair.*

SPELLING UNIT

Virginia Smith developed the following unit to teach the proofreading skills associated with spelling improvement. She taught the unit to a fourth-grade class. This unit was based on research by Personke and Yee (1971). They concluded that a student must attain an attitude toward spelling that they label "spelling conscience"; students must be instructed in the techniques necessary for proofreading their own compositions efficiently. The dictionary is the primary reference tool for such proofreading.

Smith's unit begins with an informal diagnostic test, followed by proofreading activities that stress dictionary skill, proofreading spelling lists, proofreading a story, and writing a paragraph from dictation.

FOR YOUR PLAN BOOK
Proofreading Unit—Grade 4

I. *Informal Diagnostic Test*
- **A.** *Objectives:*
  1. *Introduce the importance of proofreading for increasing spelling accuracy in compositions.*
  2. *Discover a student's proofreading skills.*
- **B.** *Procedures: Give each child a copy of the following letter. Explain that there are spelling errors in the letter, and you want the children to look carefully at the words and draw a line under each incorrect spelling. Explain that the procedure of reading written material and correcting spelling is called proofreading. After the children have underlined the words, have them write the correct spelling above each word, if they know it. Ask the children to write "proofread for spelling" at the end of the letter.*
- **C.** *Materials:*

Deer Tom,	*1*
Have you playd much basbal	*2*
lately? I am know on the Red	*1*
Arows teem. My bruther is two.	*4*
We both bated Fryday nite. I	*3*
gut a hit but he diddn't.	*2*
We are coming to yur town	*1*
next week to viset. I hope we	*1*
cen play basbal together then.	*2*
Yur frend,	*2*
Tony	*19 errors*

- **D.** *Evaluation: Make a proofreading folder for each child. Check each child's proofreading accuracy, noting the types of errors the child is able to detect, and the types of errors he does not correct. Return the papers to the class and discuss the need for developing proofreading skills.*

II. *Dictionary Skills—Arranging Words in Alphabetical Order*
- **A.** *Objective: To review the use of a dictionary in order to improve proofreading skills.*
- **B.** *Procedures: Explain to the children that the dictionary is used to check all guesses they make in spelling. Review the procedures for arranging words in alphabetical order. First alphabetize words according to the first letter of the word: **cattle, bottle, safety, another, timber, doctor, meat, laughter.** Next, alphabetize words according to the second letter of the word: **builder, better, brake, battle, blimp, box, bicycle.** Finally, review the practice that if two words begin with the same letters, you alphabetize according to the first letter that is different. Example: **clever, clatter, clergy, climate, cloudy, cloak, clothes, clump, club, claw, click.***

C. *Materials: Divide the class into teams. Give each team a set of words written on cards to put into alphabetical order. The first team to put their words into correct order scores a point. Players may move about with the cards, or the cards can be placed on the chalk tray. The cards should progress from easier to more complicated alphabetizing tasks. Eventually, have children put lists of words such as these into alphabetical order:* **interesting, interested, interests.**

III. *Dictionary Skills—Using Guide Words*

A. *Objective: To learn that speed of dictionary use can be improved through the use of guide words.*

B. *Procedures: Have the students open their dictionaries to a specific page. Discuss the two words at the top of the page. For example, on page 955 are the words* **rally** *and* **rang.** *Have students look at the page and decide the purpose of the two words. Help them understand that* **rally** *is the first word on the page and* **rang** *is the last word. Discuss the benefits of using guide words.*

C. *Materials: Each team or individual student needs a copy of the same dictionary. Ask students how rapidly they can find the guide words on the page where the word* **horse** *is found. Continue this procedure with various words. For another activity with individual dictionaries, provide the guide words for a page and ask students if certain words would be found on that page. For example, is the word* **know** *on the same page with the guide words* **knight** *and* **knotting?** *Have the students look the words up in the dictionary to verify their answers.*

D. *Evaluation: The teacher evaluates the student's ability to use guide words accurately. Provide a worksheet on which students need to rapidly find the pages on which specific guide words appear. Informally test student's ability to find specific words rapidly using their ability to alphabetize and use guide words.*

IV. *Proofreading Spelling Lists*

A. *Objective: To learn to apply the proofreading skills to a list of spelling words.*

B. *Procedures: Have students write down a list of words the teacher reads to them slowly. Ask the students to pronounce the words carefully to themselves before they write them. Tell them to underline the words they don't know how to spell. After dictating the list, allow the students about 15 minutes to check the spellings in their dictionaries. Have them write the correct spellings above misspelled words. When they have made their corrections, have them write "proofread for spelling" on their papers.*

C. *Materials: Use the following words for dictation. (The words were taken from the fourth-grade text.)*

read	glow	alone	nice
case	twice	fool	hour
afraid	noisy	chair	fear
eager	worst	pearl	harbor
seed	collar	thread	cover

D. *Evaluation: How well did the students check their guesses? Were they able to distinguish between words that were positively known and words that were estimated?*

V. Proofreading a Story

    A. Objective: To improve students' ability to apply proofreading skills to a written composition.

    B. Procedures: Give each child a copy of a poorly written story. Have each child read through the story, then proofread for spelling errors, underlining each incorrect spelling once and writing the corrected word above. Have them use the dictionary to verify any guesses and ask them to write "proofread for spelling" when they are finished.

    C. Materials:

The fat cat ran swiffly across	1
the gras. He was chaseing a mous.	3
Wood he cetch it? The mouse ran bak	3
and fourth by the flower bed.	1
Suddenly the cat jumbd. The mous	2
dashd behind the roze bush, then	2
dizappeared frum site. He wuz safe	4
frum the fat cat.	1
	17 errors

    D. Evaluation: Observe how well students are able to find errors and to use their dictionaries to find correct spellings. Go over the corrections in class so the students can see how they have done.

VI. Writing a Paragraph from Dictation and Proofreading It

    A. Objective: To practice the proofreading skills following a dictated paragraph.

    B. Procedures: Dictate a short paragraph to the class. Tell them to underline any unknown spellings as they write the paragraph. After the dictation, allow time for the class to check their guessed spellings in the dictionary. Have them write the correct spellings above the word. Have them write "proofread for spelling" on the paper when they are finished.

    C. Materials:

        Mary and her brother Tom went to the
beach last Thursday. The sun was hot, the
sea was cold and the sand was clean. Mary
and Tom had fun swimming and playing. When
they came home they were both very tired and
very happy.

    D. Evaluation: Observe how well the students are able to check their guesses using the dictionary. Have the students state the steps they should follow when proofreading a story or a list of spelling words. These steps should be practiced in every subject area.

This unit is used with permission of Virginia Smith, teacher at St. Michael's Academy, Bryan, Texas.

MAINSTREAMING AND SPELLING

Spelling instruction with its demands for visual and auditory perception, short- and long-term memory, attention to details, and fine-motor capabilities causes considerable problems for learning disabled and mentally handicapped children. Consequently, many teachers provide special instruction in spelling and provide many opportunities for children to dictate their stories to scribes. Table 6–2 presents characteristics of learning disabled children that might interfere with spelling, and corresponding teaching techniques for language arts instruction. Table 6–3 presents the same information for mentally handicapped children. The characteristics and techniques for language arts instruction result from a search of the research literature and recommendations of authorities who work with mainstreamed children.

TABLE 6–2

Mainstreaming and spelling for learning disabled children

Characteristics of Learning Disabled Children	Teaching Techniques for Language Arts Instruction
1. Score lower than normal children on spelling both predictable and unpredictable words (Carpenter and Miller, 1982).	1. Provide extra time for spelling. In-school tutoring in addition to regular classroom instruction may result in gains equal to those made by normal spellers (Bessai and Cozac, 1980). Provide practice and meaningful spelling situations by integrating spelling into the broader curriculum (Cohen and Plaskon, 1980). Try mnemonic aids (Otto and Smith, 1980).
2. Demonstrate deficiencies in visual perception and visual motor integration (Harber, 1979).	2. Use informal and formal tests and trial teaching to determine the best method for spelling instruction (Sapir and Wilson, 1978; Harris and Sipay, 1980; Cohen and Plaskon, 1980).
3. Demonstrate poor visual short-term memory and problems with auditory and/or visual memory (Fisher, 1980; Wallace and McLoughlin, 1988; Johnson and Myklebust, 1967).	3. Try a VAKT approach that emphasizes seeing, saying, hearing, and feeling for learning spelling words (Gearheart, Weishahn, and Gearheart, 1988; Campbell, 1976; and Lerner, 1976).
4. Auditory discrimination problems may cause children to have difficulty hearing fine differences of sound within words (Cohen and Plaskon, 1980).	4. Avoid phonics approaches and try a linguistic spelling or whole-word system (Bader, 1980; Sapir and Wilson, 1978). Provide auditory discrimination training (Otto and Smith, 1980). Try the auditory discrimination suggestions on the chart for listening development and learning disabled children.
5. Anxiety may be created by lack of academic success; anxiety, in turn, may interfere with learning (Frey, 1980).	5. Teaching children a self-relaxation technique may decrease anxiety and increase spelling progress (Frey, 1980).

TABLE 6–3
Mainstreaming and spelling for mentally handicapped children

Characteristics of Mentally Handicapped Children	Teaching Techniques for Language Arts Instruction
1. Below average in spelling as well as other academic areas; spelling ability may be even lower than reading ability (Otto and Smith, 1980).	1. Test students' reading vocabulary with the Dolch Basic Word List or the Johnson Word List. Then test students' spelling ability on same list. The initial target words for spelling growth are the words they can read but cannot spell (Otto and Smith, 1980).
2. May lack phonic-analysis ability.	2. Try a multisensory approach for spelling instruction (Campbell, 1976). Try an approach emphasizing spelling patterns or word families with high utility value (Otto and Smith, 1980). Use informal and formal tests and trial teaching to determine the best method for spelling instruction (Sapir and Wilson, 1978; Harris and Sipay, 1980; Cohen and Plaskon, 1980). Try color cues (Frostig, 1984).

SUMMARY

Many current articles deal with the writing crisis in American schools. Poor spelling is often cited as a major problem with children's writing. The school program must help students learn to spell the words they need to know; provide instruction in reliable spelling generalizations; develop an understanding of word meanings and vocabulary; equip the speller with more than one strategy for spelling-word attack; incorporate spelling into all areas of the curriculum; proceed from sound diagnostic evaluation; and provide for the development of motivation, positive student attitudes, and sound habits for studying spelling and proofreading.

Diagnostic approaches to spelling instruction stress the placement of students on appropriate spelling instructional levels. In addition to instructional placement, the diagnostic teacher also analyzes the types of errors a child makes in writing. This information is used to individualize the spelling program for each child. Informal tests for spelling placement, spelling generalizations, and error analysis can be developed by the classroom teacher.

Spelling words are selected according to several different criteria. The frequency-of-use approach stresses that spelling instruction should be based on the words that are the most frequently used. Consequently, it is believed that words most frequently used in writing should be taught first, words commonly used by children in a specific grade should be taught in that grade, and words needed in other content areas should be taught in the appropriate grade. Another quite different selection criterion emphasizes the phonic regularity of words. The words selected for this approach would follow a consistent spelling pattern. Some linguists and psycholinguists also stress the consistency of spelling patterns in words of similar meanings. Thus, spelling instruction should allow students to compare, contrast, and categorize words according to root words, word origins, and similarities in structural patterns.

Several techniques have proven valuable in a developmental approach to spelling instruction. The corrected-test method allows immediate feedback, and liberates students from the systematic study of words already mastered. The self-study method allows students to master their own individualized spelling words. An inductive approach to teaching spelling generalizations allows students to discover and use the reliable spelling generalizations.

Remedial spelling approaches have been developed for the more disabled speller. The Fernald approach has been useful with learning disabled children. In this multisensory method, the child looks at the word, says the word, traces the word, then writes the word without looking at it. Spelling games are useful for reinforcing and motivating the remedial spelling student. Spelling should not, however, be taught as an isolated subject. Students, whether remedial, regular developmental, or gifted, require many opportunities to use spelling in meaningful situations.

ADDITIONAL SPELLING ACTIVITIES

1. *Review several published spelling programs and spelling curriculums for public schools. What are the objectives for each? In what ways do the objectives match the objectives identified by Petty and Jensen listed on pages 192–93?*
2. *Collect writing samples of three- to ten-year-olds. Try to identify the stages in spelling development according to stages I through V identified on pages 193–94.*
3. *Compare an informal spelling level test that you compile from one specific spelling program with an informal spelling level test that you compile from a second spelling program. Evaluate the similarities and the differences between the two tests. What information would each test provide? Is there a difference in word difficulty or word selection?*
4. *Develop a spelling lesson designed to teach an apostrophe generalization.*
5. *Interview several good, average, and poor spellers. Determine their perceptions about their spelling ability, their attitudes toward spelling, their study approaches, and their techniques for approaching the spelling of an unknown word.*
6. *Develop a spelling unit designed to improve proofreading abilities.*

BIBLIOGRAPHY

American Education. United States Department of Health, Education, and Welfare, Office of Education, October 1976.

Bader, Lois A. *Reading Diagnosis and Remediation in Classroom and Clinic.* New York: Macmillan Co., 1980.

Bessai, Frederick, and Cozac, Con. "Gains of Fifth and Sixth Grade Readers from In-School Tutoring." *The Reading Teacher* 33 (February 1980): 567–70.

Bremer, Bishop; and Stone. *Skills in Spelling.* Dallas: American Book Co., 1976.

Burns, Paul C., and Broman, Betty L. *The Language Arts in Childhood Education.* 5th ed. Chicago: Rand McNally & Co., 1983.

Campbell, Dorothy. "Mode of Response to Tactual Stimuli and Learning Disabled and Normal Pupils' Learning New Words with Phoneme-Grapheme Equivalence." *Journal of Research and Development in Education* 9 (Winter 1976): 29.

Carpenter, Dale, and Miller, Lamoine J. "Spelling Ability of Reading Disabled LD Students and Able Readers." *Learning Disability Quarterly* 5 (Winter 1982): 65–70.

Chomsky, Carol. "Reading, Writing, and Phonology." *Harvard Educational Review* 40 (1970): 287–309.

Cohen, Sandra, and Plaskon, Stephen P. *Language Arts for the Mildly Retarded.* Columbus, Oh.: Merrill Publishing Co., 1980.

Dale, Philip S. *Language Development: Structure and Function.* New York: Holt, Rinehart & Winston, 1976.

Fisher, Dennis F. "Compensatory Training for Disabled Readers: Research to Practice." *Journal of Reading Disabilities* 18 (March 1980): 25–31.

Fitzgerald, James A. *A Basic Life Spelling Vocabulary.* Milwaukee, Wis.: Bruce Publishing Co., 1951.

Frey, Herbert. "Improving the Performance of Poor Readers through Autogenic Relaxation Training." *The Reading Teacher* 33 (May 1980): 928–32.

Frostig, Marianne. "Corrective Reading in the Classroom." In *Readings on Reading Instruction,* 3d. ed. Edited by Albert J. Harris and Edward R. Sipay. New York: Longman, 1984.

Gearheart, Bill R., Weishahn, Mel W., and Gearheart, Carol J. *The Exceptional Student in the Regular Classroom.* 4th ed. Columbus, Oh.: Merrill Publishing Co., 1988.

Gentry, J. Richard. "An Analysis of Developmental Spelling in GNYS AT WRK." *The Reading Teacher* 36 (November 1982): 192–200.

_____. "Learning to Spell Developmentally." *The Reading Teacher* 34 (January 1981): 378–81.

Hanna, Paul R., and Hanna, Jean S. "Applications of Linguistics and Psychological Cues to the Spelling Course of Study." *Elementary English* 42 (November 1965): 753–59.

_____, and Moore, James. "Spelling—From Spoken Word to Written Symbol." *Elementary School Journal* 53 (February 1953): 329–37.

Harber, Jean R. "Differentiating LD and Normal Children: The Utility of Selected Perceptual and Perceptual-Motor Tests." *Learning Disability Quarterly* 2 (Spring 1979): 70–75.

Harris, Albert J., and Sipay, Edward R. *How to Increase Reading Ability.* 7th ed. New York: Longman, 1980.

Henderson, Edmund. *Teaching Spelling.* New York: Houghton Mifflin Co., 1985.

_____, and Beers, James W., eds. *Developmental and Cognitive Aspects of Learning to Spell: A Reflection of Word Knowledge.* Newark, Del.: International Reading Association, 1980.

_____, and Templeton, Shane. "A Developmental Perspective of Formal Spelling Instruction Through Alphabet, Pattern, and Meaning." *The Elementary School Journal* 86 (January 1986): 305–16.

Hillerich, Robert L. "Let's Teach Spelling—Not Phonetic Misspelling." *Language Arts* 54 (March 1977): 301–7.

_____. "Spelling: What Can Be Diagnosed?" *Elementary School Journal* 83 (1982): 138–47.

Hodges, R. E., and Rudorf, E. H. "Searching Linguistics for Cues for the Teaching of Spelling." *Elementary English* 42 (May 1965): 527–33.

Horn, Ernest. "Phonetics and Spelling." *Elementary School Journal* 57 (May 1957): 424–32.

_____. "Phonics and Spelling." *Journal of Education* 136 (May 1954): 233–35.

_____. "Principles of Methods in Teaching Spelling as Derived from Scientific Investigation." *18th Yearbook of the National Society for the Study of Education, Part II,* pp. 52–77. Bloomington, Ill.: Public School Publishing Co., 1919.

_____. *Teaching Spelling.* American Educational Research Association, Department of Classroom Teachers Research Pamphlet. Washington, D.C.: National Education Association, 1963.

_____. *Teaching Spelling: What Research Says to the Teacher.* Washington, D.C.: National Education Association Publications, 1971.

Johnson, D., and Myklebust, Helmer. *Learning Disabilities: Educational Principles and Practices*. New York: Grune & Stratton, 1967.

Johnson, P. L. "The Effect of Group Relaxation Exercises on Second and Sixth Grade Children's Spelling Scores." Doctoral Dissertation, Northwestern State University of Louisiana, 1982. *Dissertation Abstracts International*, 43, 773A. (University Microfilms No. DA8217982.)

Lerner, Janet W. *Children with Learning Disabilities*. 2d ed. Boston: Houghton Mifflin Co., 1976.

Linn, S. H. "Spelling Problems: Diagnosis and Remediation." *Academic Therapy Quarterly* (1967): 62–63.

Otto, Wayne, and Smith, Richard. *Corrective and Remedial Teaching*. Boston: Houghton Mifflin Co., 1980.

Personke, Carl. "Spelling as a Language Art." In *Language Arts Instruction and the Beginning Teacher*, edited by Carl Personke and Dale Johnson. Englewood Cliffs, N.J.: Prentice-Hall, 1987.

_____, and Yee, Albert. *Comprehensive Spelling Instruction*. Scranton: Intext Educational Publishers, 1971.

Petty, Walter T., and Jensen, Julie M. *Developing Children's Language*. Boston: Allyn & Bacon, 1980.

Read, Charles. "Pre-School Children's Knowledge of English Phonology." *Harvard Educational Review* 41 (February 1971): 1–34.

Sapir, Selma, and Wilson, Bernice A. *Professional's Guide to Working with the Learning-Disabled Child*. New York: Brunner/Mazel, 1978.

Snowling, Margaret J. "The Development of Grapheme-Phoneme Correspondence in Normal and Dyslexic Readers." *Journal of Experimental Child Psychology* 29 (April 1980): 294–305.

Spell/Write. New York: Noble & Noble Publishing, 1971.

Templeton, Shane. "Young Children Invent Words: Developing Concepts of Word-ness." *The Reading Teacher* 33 (January 1980): 454–59.

Wallace, Gerald, and McLoughlin, James A. *Learning Disabilities: Concepts and Characteristics*, 3d ed. Columbus, Oh.: Merrill Publishing Co., 1988.

Wood, Margo. "Invented Spelling." *Language Arts* 59 (October 1982): 707–17.

Zutell, Jerry. "Some Psycholinguistic Perspectives on Children's Spelling." *Language Arts* 55 (October 1978): 844–50.

Chapter Seven

After completing this chapter on grammar, usage, and mechanics, you will be able to:

1. Develop a definition of grammar that includes the three parts of grammar.
2. Describe several characteristics of traditional grammar, and instructional activities that would be included in a traditional study of grammar.
3. Describe several characteristics of structural grammar, and instructional activities that would be included in a study of structural grammar.
4. Describe several characteristics of transformational grammar, and instructional activities that would be included in a study of transformational grammar.
5. Define the levels of usage and list the goals for a level of usage appropriate for the classroom.
6. Evaluate a child's understanding and application of grammar and usage.
7. Develop a series of instructional activities for improving students' understanding and application of grammar.
8. Develop an inductive lesson for helping children understand a usage concept.
9. List the punctuation skills usually taught in the elementary grades.
10. List the capitalization skills usually taught in the elementary grades.

Grammar, Usage, and the Mechanics of Writing

G rammar has been an important part of the language arts curriculum for many years, although the emphasis it receives in our educational system varies from time to time. There have been revolutionary changes in educators' approaches to grammar and grammar instruction. These innovations, however, have not gone unchallenged. Grammar is frequently mentioned in the back-to-basics movement, and some critics of American education feel that instruction might profit from a return to a more traditional approach to grammar. For teachers to be prepared to answer the critics and to provide better instruction, they will need to understand the different aspects of grammar, as well as the appropriate methods for improving a child's grasp and application of grammar and usage.

WHAT IS GRAMMAR?

Many college students admit, rather sheepishly, that they have forgotten a great deal about grammar, so we start our study with a definition and review of the grammars that might be included in an instructional program. W. Nelson Francis provided a useful, three-part definition of grammar:

Grammar 1:   The first thing we mean by grammar is the set of formal patterns in which the words of a language are arranged in order to convey larger meanings. It is not necessary that we be able to discuss these patterns consciously in order to be able to use them. In fact, all speakers of any language above the age of five or six know how to use its complex forms of organization with considerable skill. In this sense of the word they are thoroughly familiar with its grammar.

Grammar 2:   The second meaning of grammar is that branch of linguistic science which is concerned with the description, analysis, and formalization of language patterns.

Grammar 3:   The third sense in which people use the word grammar is that of linguistic etiquette. The word in this sense is often coupled with a derogatory adjective; the expression "He ain't there" is "bad grammar." What we mean is that such an expression is bad linguistic manners in certain circles. (1964, pp. 69, 70)

The first part of this definition includes the intuitive knowledge of sentence structure that most children demonstrate by the time they enter school, and is the subject of chapter 2, Language and Cognitive Development. This intuitive knowledge

[handwritten margin note: Kids dev. internalize - do not know doing so overgeneralize]

[handwritten margin note: uses of lang. levels / levels of writing more precise]

222

is the basis for further grammar instruction in school. Grammar 1 can be emphasized by providing many opportunities for children to explore their language through oral language activities. John Savage stated: "Through such explorations, the children discover the wonderful flexibility of language and the possibilities it has for communicative purposes. Looking at language in this way involves 'learning grammar' in the true sense of the word. The children learn about language, as well as how to use it" (1977, p. 334).

The second part of the definition refers to the more formal grammar that is usually thought of as a school subject. Various "grammars" have been developed to describe and analyze sentence patterns and classifications of words. Traditional grammar, structural grammar, and transformational grammar have all influenced instruction in the elementary grades. We look at each of these grammars in the next section, along with examples of the instructional approaches used with each philosophy of grammar.

The third part of the definition, grammar 3, refers to the use of language. Usage deals with language standards for different occasions, audiences, and purposes. For example, we use a different level of usage when we have casual conversations with friends than when making a formal presentation to an academic audience. The instructional emphasis in usage should help children make appropriate usage decisions in both oral and written language.

Grammars in Grammar 2   *3 kinds of grammar – currently combine all three*

1) Traditional grammar   Traditional grammar, according to Tiedt and Tiedt (1975), has its roots in the eighteenth-century work of Joseph Priestley, Robert Lowth, George Campbell, and Lindley Murray. These writers were prescriptive, and their rules were based on Latin, which they saw as a perfect language. Traditional grammar is called prescriptive because it attempts to tell people how to speak or write. Traditional grammar classifies words as verbs, adjectives, nouns, adverbs, pronouns, prepositions, conjunctions, or interjections.

Instructional activities in traditional grammar might require children to underline all the nouns in a sentence, to draw an arrow between a pronoun and its antecedent, to mark the verb in a sentence, to define the classification of the verb, to change the verb to passive or active voice, to conjugate verbs through all of their tenses, to identify adverbs in a sentence, to indicate the words that adverbs modify, to find the prepositions in a sentence, or to identify the conjunctions in a sentence.

2) Structural grammar   Structural grammar began with the work of Leonard Bloomfield, and has been further developed by linguists such as Charles Fries and James Sledd. The structuralists suggested a new way to classify language. Structural grammar is not prescriptive, like traditional grammar; instead, it is a "descriptive" grammar.

The structuralists describe the words in a language according to their form classes, the sentence patterns, or the position and function of a word in a sentence. As we look at each of these categories, we also review some of the terms of structural grammar.

Adding a suffix to a word may change the part of speech or the tense of the root word.

The structuralist looks at the behavior of words, and divides them into groups according to the way the words change form, the position of the words in a sentence, and the kind of structure word (determiner, preposition, conjunction) that precedes the word in a sentence. The structuralist identifies four form classes resembling nouns, verbs, adjectives, and adverbs. Let us take the example of a noun, and see how it might be described by structural grammar:

1. A noun changes form by adding *s* or *es* to show plural and *s* or ' to show possession. For example:

    father—fathers (plural)—father's (possessive)

    Other groups of words, such as verbs, adjectives, or adverbs, cannot be made either plural or possessive.

2. Nouns also occur in certain positions in a sentence. For example, in the sentence, "This (determiner) boy (noun) ran (verb) away," the noun falls after the determiner and before the verb. Any word placed into this specific position would be a noun.

    This _____ran away.

Nouns also occur after the verb, as object, as in "Jerry mailed *letters*," and after a preposition, as object of the preposition, as in "We are at *home*."

3. A noun often appears with certain function words called determiners. These determiners frequently precede nouns. For example:

articles: *an* airplane   possessives: *my* sister
            a story   *your* brother
            the city   *their* wagon

demonstratives: *this* house
                  that tree
                  these balls

Structuralists, then, define a noun by describing its changes in form, its position in a sentence, and its relationship to certain structure words.

A verb may also be described in terms of changes in form, its position in a sentence, and its relationship to structure words. For example:

1. A simple verb may change its form by adding *s* for present tense (she walks) and the form *ed* for past tense (she walked). (There are a great many irregular verbs: these verbs are changed to past tense without adding *ed*, as in *ring, rang, rung*.)
2. A verb is usually located after the noun that is the subject of the sentence. In the sentence, "Jimmy ran to school," there is only one form class that would fit into this position:

Jimmy _____ to school

3. The structure words that can come before verbs are auxiliaries, such as *can*, *would*, *were*, and *has been*, as in the sentence,

"Mother *has been* painting all morning."

An adjective may be described by a structuralist in this way:

1. Adjectives containing one or two syllables usually change form to show comparison by adding *er* and *est*. For example:

small   small*er*   small*est*

Adjectives containing more than two syllables usually change form by adding the structure words *more* or *most*.

appealing   *more* appealing   *most* appealing

2. Adjectives often occur before the nouns they modify, as in:

Let's climb the *biggest* tree.

Only one form class may fit into: Let's climb the _____ tree. Adjectives may also appear after a linking verb, as in the sentence "The sky is cloudy." The sky is _____ .

3. Intensifiers such as *very*, *quite*, and so forth, frequently come before adjectives. (This is also true for adverbs).

The final form class is adverbs.

1. Adverbs, like adjectives, may change their form by adding *er* and *est*.
2. Adverbs are often found at the end of the sentence. For example: Call your mother immediately.
3. Adverbs, like adjectives, may also be marked by intensifiers.

Structural grammar also describes grammar according to sentence patterns. Paul Roberts (1962) developed a sequence of ten patterns to describe existing sentences. Examples of these sentence patterns are frequently used in textbooks that use a structural grammar approach. Pattern 1, for example, is shown as follows:

Determiner	*Noun*	*Verb (intransitive)*	*Adverb*
The	woman	walks	rapidly

Instructional activities in a structural grammar text might require the students to change the form of verbs by adding *s, ed,* or *ing.* (These may be taught inductively). Existing sentences might be identified as to their appropriate sentence pattern. The teacher might place on the board a sentence such as, "The tree was tall and _____ ." Children could then discuss the characteristics of words that might fit the empty position and provide examples of appropriate words. Similar sentences with empty positions might be developed for each of the patterns in the grammar. According to Ralph Goodman (1965), patterns are frequently provided as models for the analysis of existing sentences.

Transformational grammar   Transformational grammar is based on the work of Noam Chomsky (1957). It builds on the work of the structuralists, but extends structural grammar into the meaning of language. Whereas structural grammar is concerned primarily with syntax, transformational grammar is more concerned with semantics and the generating of sentences. According to Pose Lamb, "transformational grammar purports to provide rules for producing new sentences as well as patterns for the analysis of existing sentences. Although sentence patterns are suggested in transformational grammars, these serve a different function than they do in structural grammars" (1977, p. 100). Transformational grammar is based on a set of directions for generating sentences. The term *transformational* refers to the division of sentences into basic sentences and the "transforms," or variations, that can be developed from these basic sentences.

The basic sentences, or kernel sentences, have several definite characteristics, and are formed by placing words together in certain patterns. The second part of transformational grammar consists of directions or rules for rearranging or combining the kernel sentences into new, more complicated structures. We look first at the basic characteristics of the kernel sentences, then at their patterns, and, finally, at the rules for transformation. p. 239 & 243 - sentence combining act.

Characteristics of the kernel sentence   The formula for a kernel sentence is:

Sentence = Noun Phrase + Verb Phrase

This kernel sentence formula could also be diagramed as follows:

Noun Phrase	**Verb Phrase**
The car	is red

If we look at the above kernel sentence, "The car is red," we see that it has several definite characteristics:

1. In the kernel sentence, the subject is followed by the predicate.
    Subject (The car)   Predicate (is red)
2. A kernel sentence is a statement, not a question.
    (All questions are transforms.)
3. A kernel sentence is affirmative.
    (All negative sentences are transforms.)
4. A kernel sentence is in the active voice.
    (All passive-voice sentences are transforms.)
5. A kernel sentence contains a single predicate.

Patterns of kernel sentences   Transformationalists also suggest patterns or basic sentence types. John Mellon (1964) named five basic-sentence patterns, while Kean and Personke (1976) identified four types of kernel sentences, according to the positions or slots that words may fill. Any sentence that does not follow one of the four patterns would be considered a transformation:

1. The Be Sentence:

Noun Phrase	*Verb (Be Class)*	*Noun Phrase, Adjective, or Prepositional Phrase*	*Optional Adverb*
The squirrel	is	a rodent	
The squirrel	is	lively	today
The squirrel	is	in the tree	

2. The Intransitive Verb Sentence:

Noun Phrase	*Verb (Intransitive)*	*Open, Adverb or Prepositional Phrase*	*Optional Adverb*
Jerry	ran		rapidly
Jerry	ran	outside	quickly
Jerry	ran	into the field	

3. The Transitive Verb Sentence:

Noun Phrase	*Verb (Transitive)*	*Noun Phrase (Direct object)*	*Optional Adverb*
The girls	mailed	the package	yesterday

4. The Linking Verb Sentence:

Noun Phrase	Verb (Linking)	Noun Phrase, or Adjective	Optional Adverb
Judy	became	the president	
Judy	became	frightened	today

The transformationalists see these patterns as rules or directions for sentences that have not as yet been spoken or written, whereas the structuralists use the patterns for categorizing existing sentences. The transformationalists use the kernel sentences to describe ways that sentences may be altered.

Transforming basic sentences Both single-base transformations, which act upon one kernel sentence, and double-base transformations, which combine two kernel sentences, are possible with the rules of transformational grammar (Kean and Personke, 1976). The following single-base transformations are possible:

1. Question Transformation—The kernel sentence, "The car is red," may be easily transformed into a question by rearranging the words into "Is the car red?" The sentence also becomes a question by changing the intonation, to "The car is red?" The words *do, does,* or *did* may transform the kernel into a question; for example, "Did Judy become the president?" Question transforms may begin with *who, what, when, where,* and *why.* Thus the kernel, "Jerry ran into the field," could become "Why did Jerry run into the field?" or "When did Jerry run into the field?"

2. Passive Transformation—The kernel sentence must be in the active voice, so this change affects sentences that contain transitive verbs. The sentence must be transformed so that the subject receives, rather than performs, the action expressed in the verb. The kernel sentence, "The teacher read the book," would become the transformation, "The book was read by the teacher." As you can see, the direct object of the original kernel sentence is now the subject of the transformation.

3. Negative Transformation—Our original kernel sentence, "The car is red," becomes a transformation merely by inserting *not* or *n't* after the verb. This transformation results in the sentence, "The car is not red." *Do, does,* or *did* may be used along with the *not* to transform the sentence, "Jerry ran outside quickly," to the negative transform, "Jerry did not run outside quickly."

Sentences may also be transformed through double-base transformations, in which the transformation results from combining two kernel sentences into one transformation. More complex sentences result from this transformation. The following double-base transformations are examples of the possibilities:

1. Coordinating Transformation—The coordinating transformations often use coordinating conjunctions, such as *and, but, or* or sentence connectors, such as *however, therefore, moreover,* and so forth, to combine kernel

sentences in equal or coordinating positions. For example, the two kernel sentences, "Jerry has a bicycle" and "Jerry has a kite," combine in the transform to become "Jerry has a bicycle and a kite."

2. Adjective Insert—Two kernel sentences may also be joined to form a more interesting and complex transform by combining a kernel sentence containing an adjective with a kernel sentence containing a noun phrase. For example, the two kernel sentences, "The squirrel is a rodent" and "The squirrel is lively today," may be transformed into "The squirrel is a lively rodent today."

3. Subordinating Transformation—This transformation is more complex than the coordinating transformation. Subordinating connectors, such as *although, whenever, until,* or *because* are used to connect the two kernel sentences. For example, the two kernel sentences, "The boy was quiet" and "He was happy," might combine in the transform "The boy was quiet, although he was happy," or "Although he was happy, the boy was quiet."

4. Relative Clause Transformations—The relative clause begins with a relative adverb or relative pronoun, and is inserted in or at the end of the sentence. For example, the kernel sentences, "The car is red" and "I have a car," may combine to form the transform "I have a car that is red." Similarly the two kernels, "The man is old" and "The man lives on our street," may combine to form the transform "The man who lives on our street is old."

Instructional activities in a transformational grammar would relate closely to the descriptive elements of the grammar. Because the basic sentence is *noun phrase plus verb phrase,* instructional activities would help children develop an understanding of noun and verb phrases. The major emphasis of instruction would be to guide children in experiencing and working with grammar by helping them examine their own sentences, as well as those written by others. They would learn how sentences are formed, changed, combined, and improved. There would be a number of inductive learning strategies, rather than rules of acceptable syntax. We look further into this type of instruction as we proceed with some instructional activities in the area of grammar.

REINFORCEMENT ACTIVITY

With a group of your peers, select several language arts textbooks designed for elementary and middle-school children. Analyze the texts to see whether they are based on a traditional, structural, or transformational philosophy. Find examples of learning activities that you feel represent that particular viewpoint. Compare their similarities and differences. Which do you feel is most teachable? Which do you feel would result in the best understanding of language? Share your findings with the rest of your language arts class, and be prepared to support your conclusions about the grammar with which you would feel most comfortable in the classroom.

Grammar 3, or Usage

Discussions related to grammar usage create considerable controversy. DeHaven stated, "Perhaps no other aspect of language teaching generates more intense discussion than usage. The controversy stems from a lack of agreement in (1) defining an acceptable level of usage and (2) setting instructional goals" (1983, p. 69). Unlike grammar 2, which describes the order and arrangement of words in sentences, usage is concerned with choices of words, phrases, and sentences. For example, the sentence "He ain't there" may correspond with the appropriate syntactic arrangement of words in a sentence, but it would be considered inappropriate in certain speaking or writing circumstances and appropriate in other circumstances. Usage deals with the attitudes and language standards of a group, rather than with the way words are structured to convey meaning. Usage choices are made in response to forces external to the language. Sara Lundsteen (1976) identified three such forces that affect usage: the speaker or writer; the subject; and the audience.

When we think about these three forces, we realize also that a level of usage appropriate for one subject and audience might not be appropriate for another subject and audience. All of us utilize several levels of usage in our daily lives, depending on the formality of the occasion in which we find ourselves. We would, for example, use words differently during a casual conversation or when writing to a friend than we would in presenting a formal speech to a college class or writing a letter of inquiry to a prospective employer. By the time we reach college level, we should have learned when to use one level of usage and when to use another. Most adults are able to switch levels of usage quite easily.

The number of usage levels differs according to which grammarian is speaking. Lodge and Trett (1968) postulated three levels of usage: nonstandard English, general English, and formal English. Language standards change both from one group to another and with time. What was considered almost illiterate a number of years ago may be virtually ignored today; for example, when to use *may* and *can,* or *shall* and *will.* According to Savage, "In the final analysis, the 'correctness' or appropriateness of usage must be judged in the social context in which the language is used" (1977, p. 369). Usage instruction in the elementary school is usually designed to help a child move from one usage situation to another because different levels are appropriate for different audiences and purposes. Instead of teaching that some form is incorrect, modern usage instruction stresses appropriateness for certain purposes and audiences. Modern usage instruction, then, is flexible, and tries to increase the range of levels of usage available to the child.

Tiedt and Tiedt recommended that, to teach this flexibility, "we teach educated forms of standard English without undue stress on picayune points that really achieve little" (1975, p. 30). But what are these educated forms of English? According to Kean and Personke, "Among the many attempts to sort out the more important usage 'violations' for teachers who guide children's language development, perhaps the most noteworthy to date is a list of twenty-five goals for elementary language instruction prepared by English educator Robert Pooley" (1976, p. 290). Although Pooley compiled this list in 1960, and warned at the time that some of the goals would eventually require modification, his list is still valuable. For classroom usage, Pooley recommended these goals:

1. The elimination of all baby talk and "cute" expressions.
2. The correct uses in speech and writing of *I, me, he, him, she, her, they, them.* (Exception: It's me.)
3. The correct uses of *is, are, was, were,* with respect to number and tense.
4. Correct past tense of common irregular verbs such as *saw, took, brought, bought, stuck.*
5. Correct uses of past participles of the same verbs and similar verbs after auxiliaries.
6. Elimination of the double negative: "We don't have no apples," etc.
7. Elimination of analogical forms: *ain't, hisn, hern, ourn, theirselves,* etc.
8. Correct use of possessive pronouns: *my, mine, his, hers, theirs, ours.*
9. Mastery of the distinction between *its,* possessive pronoun, and *it's, it is.*
10. Placement of *have* or its phonetic reduction to *v* between *I* and a past participle.
11. Elimination of *them* as a demonstrative pronoun.
12. Elimination of *this here* and *that there.*
13. Mastery of use of *a* and *an* as articles.
14. Correct use of personal pronouns in compound constructions: as subject (Mark and I), as object (Mary and me), and object of preposition (to Mary and me).
15. The use of *we* before an appositional noun when subject; *us* when object.
16. Correct number agreement with the phrases *there is, there are, there was, there were.*
17. Elimination of *he don't, she don't, it don't.*
18. Elimination of *learn* for *teach, leave* for *let.*
19. Elimination of pleonastic (redundant) subjects: *my brother he, my mother she, that fellow he.*
20. Proper agreement in number with antecedent pronouns *one, anyone, everyone, each, no one.* With *everybody* and *none* some tolerance of number seems acceptable now.
21. The use of *who* and *whom* as references to persons. (But note, "Who did he give it to?" is tolerated in all but very formal situations. In the latter, "To whom did he give it?" is preferable.
22. Accurate use of *said* in reporting the words of a speaker in the past.
23. Correction of "lay down" to "lie down."
24. The distinction between *good* as adjective and *well* as adverb: "He spoke well."
25. Elimination of *can't hardly, all the farther* (for "as far as") and *Where is he (she, it) at?"*

In addition to noting these points of usage to include in instruction, Pooley listed eight items that are no longer considered "incorrect," and that may even be in quite general use:

1. Any distinction between shall and will.
2. Any reference to the split infinitive.
3. Elimination of *like* as a conjunction.
4. Objection to the phrase "different than."
5. Objection to "He is one of those boys who is . . ."
6. Objection to "the reason . . . is because . . ."

7. Objection to *myself* as a polite substitute for *me*, as in "I understand you will meet Mrs. Jones and myself at the station."
8. Insistence upon the possessive case standing before a gerund.[1]

EVALUATION OF GRAMMAR AND USAGE

In the chapters on Language and Cognitive Development, and Oral Language Development in this text we discuss informal methods for evaluating oral language that encourage the gathering of oral language samples and that provide sources for the analysis of how children use word patterns and how they convey meaning with those patterns.

Children's knowledge of and application of grammar may also be analyzed through informal approaches. Many of the activities suggested later in this chapter may provide source material for such informal analysis. Application can be tested only by analyzing actual speech or written samples produced by the individual child. According to Savage:

> Evaluation of grammatical application centers on the sentences that the children themselves produce: whether they can produce individual sentences based on models, the sentence types and sentence variety that are evident in their written work, how ideas are combined into sentences and how sentences are grouped into paragraphs, and how effective the writing is through the use of grammatical constructions. (1977, p. 380)

Topic variance, however, may influence the quality of written samples. Hirsch (1982) maintained that sentences are more varied and coherent, spelling and punctuation are better, and syntax is improved if the topic is easy rather than difficult or unfamiliar. Consequently, he believed that students may need to write on as many as five different topics before valid judgment can be made.

State and local mandates are influencing both the teaching of and the assessment of grammar and usage. Looking at state or local goals and objectives and tests that may be developed to assess them provides insights into what instruction may be required and assessed. For example, Illinois "State Goals for Learning" (Public Act 84–126) amended the school code of Illinois to include a definition of schooling and a requirement that the goals for learning be identified and assessed. Two of the language arts goals deal directly with grammar, usage, and mechanics of language. Goal three states that "as a result of their schooling, students will be able to write English in a grammatical, well-organized and coherent manner for a variety of purposes" (p. 29). Under this goal third-grade students are expected to do such activities as write for various audiences; use correct capitalization and punctuation marks when writing; and review a piece of writing to correct spelling, punctuation, and grammar. By eighth grade students are expected to use appropriate transitions within

---

[1]From Robert Pooley, "Dare Schools Set a Standard in English Usage?" *English Journal* 49 (March 1960): 176. Copyright © 1960 by the National Council of Teachers of English. Reprinted by permission of the publisher.

and between paragraphs; use conventional forms of standard English; correct fragments and run-on sentences; revise written work to correct spelling, punctuation, and grammar; and to meet the needs of audience and purpose.

In a related context, goal six states: "As a result of their schooling, students will be able to understand how and why language functions and evolves" (p. 53). Under this goal third-grade students are expected to recognize cultural differences among people and the unique qualities of individuals as expressed in their communications, and use regular verb forms correctly in writing and speaking. Sixth-grade students are expected to understand how verbal and nonverbal cues affect meaning; to identify cultural differences and similarities among people in their communication behaviors; to enlarge their speaking, reading, and writing vocabulary through the study of roots and affixes; to identify specific purposes of a variety of oral messages; to draw inferences from all forms of communications; to understand that personal values and points of view influence what is said, heard, and read; and to determine the parts of speech of words and phrases by their positions in a sentence.

The State Board of Education Rules for Curriculum mandated by the Texas State Legislature (chapter 75) are more detailed and explicit than the Illinois guidelines. Both teaching and assessment are to be provided in the following areas listed for each grade under "Developing Skill in Using the Grammar of English for Effective Oral and Written Communications":

- Grade One: Use correct singular and plural forms of regular nouns; and use correct forms of regular verbs.
- Grade Two: Use irregular plurals of nouns correctly; use subject-verb agreement in person and number; use modifiers (adjectives and adverbs) correctly; produce basic sentence patterns and variations; and use the fundamentals of grammar, punctuation, and spelling.
- Grade Three: Use irregular plurals of nouns correctly; use subject-verb agreement in person and number; use modifiers (adjectives and adverbs) correctly; produce basic sentence patterns and variations; and use the fundamentals of grammar, punctuation, and spelling.
- Grade Four: Use correct possessive forms of nouns and correct nominative, objective, and possessive forms of pronouns; use correct forms of irregular verbs; use modifiers (adjectives and adverbs) correctly; produce basic sentence patterns and variations; and use the fundamentals of grammar, punctuation, and spelling.
- Grade Five: Use correct agreement between pronouns and antecedents; use correct forms of irregular verbs; use modifiers (adjectives and adverbs) correctly; use all parts of speech correctly; produce a variety of sentence patterns; and use the fundamentals of grammar, punctuation, and spelling.
- Grade Six: Use correct agreement between pronouns and antecedents; use correct subject-verb agreement with personal pronouns, indefinite pronouns, and compound subjects; use modifiers (adjectives and adverbs) correctly; produce, coordinate, and subordinate sentence elements appropriate to meaning; and use the fundamentals of grammar, punctuation, and spelling.

Assessment procedures related to the various goals and objectives for grammar instruction vary according to state and local requirements. Many local school districts are developing methods of assessment, such as informal assessments of students' writings. Other schools rely on state developed and mandated tests. In any case, proceed with caution to assure that whatever type of assessment is used, it matches the goals for instruction and assesses children's application of grammar during realistic writing situations. Lamb noted, "Those responsible for selecting tests should recognize the possible discrepancy between what is taught and what is tested" (1977, p. 114).

One test for evaluating a child's ability to combine sentences is the Syntactic Maturity Test, by Roy O'Donnell and Kellogg Hunt (*Measures for Research and Evaluation in the English Language Arts,* Urbana, Ill.: National Council of Teachers of English, 1975). This test relates to transformational grammar, and may be used before teaching sentence combining, or afterward, to evaluate instructional effectiveness. The Syntactic Maturity Test can be administered to students from beginning fourth-graders to adults. It contains a paragraph of thirty-two very short, three-to-five-word sentences (e.g., "Aluminum is a metal. It is abundant."), and asks the student to rewrite the paragraph by combining the sentences any way he can, without omitting any information. The new paragraph is then analyzed for mean T-unit length (independent clauses and their modifiers).

The evaluation of usage involves a child's ability to use appropriate forms of grammar according to occasion, audience, and purpose. Usage is evaluated by listening to the child's speech in several different situations, and by analyzing the writing he has produced for various purposes. An informal checklist (see table 7–1) may be the most valuable evaluation technique for usage. This checklist can be based on the child's understanding and demonstrated use of several levels of usage, and on Pooley's list of appropriate usage for the classroom.

WHICH GRAMMAR SHOULD WE TEACH?

Research does not show that teaching formal rules of grammar improves either speaking or writing. Educators emphasize the knowledge that children gain from reading literature and from writing. We know that there is a strong correlation between reading and writing. Researchers such as Krashen (1984) and Stotsky (1983), who investigated writing competence, emphasized that reading ability and experience consistently correlate with writing skill and that the abstract knowledge that proficient writers have about text comes "from large amounts of self-motivated reading for interest and/or pleasure" (Krashen, p. 20).

In fact, there is a great deal of argument over the value of teaching grammar and which grammar, if any, should be taught. Research reveals no advantage in teaching traditional grammar, but there has been some indication that instruction in sentence combining yields positive results in writing. John Mellon (1969) found that a technique using transformational sentence combining and grammar produced beneficial results. Frank O'Hare (1973) also used a sentence-combining approach to improve the quality of students' writing. (We look at these approaches later in this section.)

TABLE 7–1
Informal checklist for usage

	Yes	Some-times	No
Name			
Grade			
1. Understands there is more than one level of appropriate usage	———	———	———
2. Is able to change level of spoken usage according to occasion, audience, and purpose	———	———	———
3. Is able to change level of written usage according to occasion, audience, and purpose	———	———	———
4. Has eliminated baby talk	———	———	———
5. Uses correct form of *I, me, he, him, she, they, them*	———	———	———
6. Uses correct form of *is, are, was, were*	———	———	———
7. Uses correct past tense of irregular verbs	———	———	———
8. Uses correct past participle	———	———	———
9. Does not use double negative	———	———	———
10. Does not use *ain't, hisn, hern,* etc.	———	———	———
11. Uses correct possessive pronoun	———	———	———
12. Uses *it's* and *its* correctly	———	———	———
13. Does not use *them* as a demonstrative pronoun	———	———	———
14. Does not use *this here* and *that there*	———	———	———
15. Uses *a* and *an* correctly	———	———	———
16. Uses the correct personal pronoun in compound constructions	———	———	———
17. Uses *we* (subject) and *us* (object) correctly	———	———	———
18. Does not use *he don't, she don't, it don't*	———	———	———
19. Uses number agreement in *there is, there are, there was, there were*	———	———	———
20. Uses *learn* and *teach,* and *leave* and *let* correctly	———	———	———
21. Does not use *my brother he,* etc.	———	———	———
22. Uses proper agreement with antecedent pronouns	———	———	———
23. Uses *said* in past tense	———	———	———
24. Uses *good* as adjective and *well* as adverb	———	———	———

Most critical usage problems are:

_____

_____

_____

Children need many opportunities for meaningful writing.

The National Council of Teachers of English (1983) believes grammar should be taught not only to improve writing, but also to increase understanding of the language. Kean and Personke stated, "The only justifiable reason for teaching grammar is to lead children to want to explore their language and discover how it works." They explain that the process of communication is the objective of all language study, and that "grammar can play a significant role in such a program if children are guided to experience and work with grammar directly rather than studying it as a complex system that must be memorized" (1976, p. 313). Krashen (1984) maintained that "conscious knowledge of rules of grammar and usage helps only at the editing stage and is limited to straightforward, learnable aspects of grammar" (p. 27).

If we believe that grammar can expand a child's understanding of language, which grammar should we teach? Pose Lamb (1977) recommended that teachers compare the various grammars, note their similarities and differences, and judge for themselves which is most teachable and which will result in better understanding and appreciation of our language. She feels each of the newer grammars has merit.

Several authors recommend choosing the best aspects of several grammars. John Savage, for example, recommended an eclectic grammar:

In grammar, eclecticism involves drawing on some of the content and techniques that characterize new grammar in helping children master the grammar they are expected to learn as part of the language arts. It involves, for example, using the signals employed by the structural grammarian as an aid in helping children identify parts of speech, or engaging in some of the sentence-combining activities used by the transformational grammar in helping children generate sentences. It attempts to "plug up" some of the leaks in traditional grammar. (1977, p. 338)

We follow this eclectic approach as we examine some instructional activities for helping students understand and use their language.

INSTRUCTIONAL ACTIVITIES

Early Elementary Grades

Linguistic blocks   Use of linguistic blocks can help develop and reinforce the basic sentence patterns in English, and help children understand the movable and immovable parts of language. A linguistic block has a different word printed on each of its sides; all the words on the block belong to the same form class. The teacher can make these blocks by writing on wooden blocks, or even on milk cartons, words the children have learned by reading or during language experience activities. Color coding the blocks will help children see relationships and patterns within a sentence. Barbara Graves (1972) used color-coded blocks for the U.S. Office of Education First-Grade Study; she used blue for nouns, red for verbs, green for adjectives, white for noun determiners, yellow for conjunctions, orange for adverbs, gray for prepositions, purple for pronouns, and brown for punctuation.

The blocks can be used for a great many experiments with sentence construction and meaning. For example, beginning with the simplest noun-verb pattern of a kernel sentence, students could build sentences such as "Jerry swam." By changing the noun block, they will discover that all the names on the noun blocks fit that same position: e.g., "Jane swam"; "Tag swam"; "Bill swam"; and so forth. They will also discover they cannot reverse the words in this sentence and still make sense (swam Jerry). Next, the children can try other words in the second position and talk about the kind of words that fit that position; for example, "Jerry sang"; "Jerry called"; "Jerry moved"; and so forth. Allow them to suggest other simple combinations.

Next, they can try noun blocks that are not people's or pets' names; a noun block with the words *dog, cat, man,* and *girl* can be substituted for the proper-noun block. Now they must look at the resulting sentence, "dog swam," and decide whether it sounds appropriate. By trying each of the common-noun blocks, they will discover that something must be added to make a complete sentence. They usually provide examples such as, "My dog swam"; "This dog swam"; "The dog swam"; and so forth. The teacher can then introduce another block marked with determiners, such as *the, my, this, your, these,* and lead the students, inductively, to the realization that certain types of words are used before nouns that are not names.

The children can also experiment with other sentence patterns. Can we add more to our simple noun, verb, or determiner noun, verb blocks? (It is not necessary to use these terms with children.) The children can expand orally by telling *how* someone ran, swam, sang, moved, and so forth. Having thus introduced them to the adverb, the teacher may then discuss what this kind of word does and how it adds information to the kernel sentence. The children will thus develop sentences such as "Jerry swam quickly," or "Jane moved quietly."

In order for children to understand the changes in meaning that result when some sentence sequences are reversed, the teacher can introduce, with the linguistic blocks, the transitive-verb sentence. Remember that all the verbs on a given block must belong to the same verb class: they must all be transitive verbs requiring a direct object, or linking verbs, or *be* verbs, and so forth. Sentences such as "The girls mailed the letter" can be formed with transitive-verb blocks. Then the children can experiment by exchanging the noun blocks, "girls" and "letter," so that the sentence reads: "The letter mailed the girls." The teacher would then discuss how the meaning changed when the two words were exchanged. This activity demonstrates how closely meaning is related to a specific sequence of words.

Linguistic blocks can also be used to transform kernel sentences into questions, negatives, and so forth. Linguistic blocks allow students to experiment with building more complex sentences, to experiment with various ways of expanding sentences, and to visualize how expansion changes meaning.

Data shows that such linguistic manipulation activities tend to improve even quite young students' written sentence structures. An unpublished doctoral dissertation (Baele, 1968) evaluated children's writing achievement after a special language program using linguistic blocks in first, second, and third grades. Baele found that children in a reading program that included the language supplement showed a significant increase in language quality in sentence structures, compared to those students who did not use the supplementary program.

Other ways to build sentence patterns   A cloze technique may also be used for building an understanding of sentence patterns. In this approach, the teacher writes a model sentence, omitting a specific type of word, and the students suggest words that might fill that position. Such an approach inductively builds knowledge of form classes.

For example, a second-grade teacher used the following examples to introduce the noun-form class:

The _____ was in the zoo.

The teacher asked the children to think of words that would fit in the blank. They provided the words *zebra, monkey, lion,* and so forth. The class read the sentences and discussed the types of words that fit into the blank. The teacher can also introduce another pattern for nouns, in which the noun is the object, as in the following:

I could see the _____ .

The children can offer and discuss words such as *car, train, house, clouds,* and so forth. They can also work with the pattern in which the noun is the object of a preposition:

The man was in the _____ .

The cloze technique can also be used for other class forms; for example:

Verbs: Susan _____ home.
                (ran, walked, limped, hopped)

Adjectives: The _____ car was on the street.
                (big, small, blue)
                The food tasted _____ .
                (good, hot, spicy, delicious)

Adverbs: The cat ran _____ after the bird.
                (quickly, rapidly, quietly)

The teacher can use the various patterns of form class to devise similar exercises.

Instruction in sentence transformation   Students in lower-elementary grades can do simple transformations as oral activities and, later, use the same principles in written activities. The sentence transformations described earlier with transformational grammar provide the information for this activity. For example, the coordinating transformation might be presented by showing and discussing the following kernel sentences:

Susan has a ball.
Susan has a bat.

Next, the teacher would lead a discussion of how the two sentences could be combined into one better sentence; for example, "Susan has a ball and bat." Numerous combinations such as this one could be suggested orally.

Scrambled sentences   Sentences may be scrambled, then rearranged to make sense. This activity is much like a game, because the scrambled sentences can be quite humorous; in fact, the teacher can present it as a game. For example:

walked the slowly turtle
(The turtle walked slowly.)
hopped rabbit the
(The rabbit hopped.)
her lost Amy game
(Amy lost her game.)

Children can also scramble their own sentences for another student to unscramble; their sentences may become quite complex as the students progress in ability and grade levels.

Changing word order but retaining meaning   Students can practice rearranging a sentence's word order without changing its original meaning:

I ate my breakfast this morning.
= This morning I ate my breakfast.
The third grade girls won the race yesterday.
= Yesterday the third grade girls won the race.

Expanding sentences The teacher starts with a simple sentence and expands it to supply more information. Pantomime or pictures are good introductions to this activity. For example, one second-grade teacher wrote "Mrs. Brown _____" on the chalkboard, and asked the children to watch her to see if they could fill in the blank. First, she walked to the door. The sentence now became, "Mrs. Brown walked," then, "Mrs. Brown walked to the door." She next asked the children, "How did Mrs. Brown walk?" The children watched a second time, and responded, "Mrs. Brown walked slowly to the door." Next she asked, "When did Mrs. Brown walk slowly to the door?" The class now responded with, "Mrs. Brown walked slowly to the door this morning," or "This morning Mrs. Brown walked slowly to the door." The teacher continued the activity by having some of the children pantomime. Each sentence was expanded to include not only *who,* but also *what, where, when, how,* and *why.*

    Another time, the group looked at large pictures and expanded simple sentences that introduced a subject without telling anything about the subject, such as "Snoopy slept," "The bear ran," and "Mother cooked." The children also drew their own pictures and expanded sentences to describe them.

Middle- and Upper-Elementary Grades

Many of the preceding activities are useful, also, in the middle- and upper-elementary grades. Additional sentence-building and combining activities should receive major emphasis. Several research projects in the sixties and seventies led to the conclusion that students' writing could be improved with a sentence-combining curriculum that did not require more formal grammatical instruction. A study by Mellon (1969) suggested that the normal rate of syntactic development could be hastened with sentence-combining activities such as those developed for fourth graders by Kellogg Hunt and Roy O'Donnell (1970). Stoddard (1982) found that sentence-combining activities developed for fifth- and sixth-grade students improved both the overall quality of writing and syntactic maturity. Research by Frank O'Hare (1973) also showed that seventh-grade children improved their writing when sentence combining appeared in the curriculum. In addition to improving the quality of written composition, Mackie (1982) improved the reading comprehension of fourth-grade students who used a sentence-combining approach. In a recent review of sentence-combining research, Hillocks (1986) reported that the majority of studies indicate that sentence combining promotes gains in syntactic fluency. It should always be remembered, however, that the goals of sentence combining are (1) to improve students' syntactic control and (2) to make sentence construction more automatic. Syntactic control and automaticity allow students to focus on writing content rather than on mechanics of sentence construction.

    The series of lessons that Hunt and O'Donnell presented to fourth graders required about fifteen minutes per day, about three times a week, over the school year. Grammatical terms were not used during the classroom activities; the instruction was described only as combining two or more sentences into one. The students practiced oral production of about twelve sentence-embedding transformations, and the free writing of three to five sentences into one sentence. The first lesson required the teacher to show the following two sentences on the chalkboard or a transparency:

> I rode in a boat.
> The boat leaked a little.

The teacher read the two sentences as she presented them, and presented a third sentence that combined the first two:

> I rode in a boat that leaked a little.

Next, the teacher read aloud a similar pair of sentences, and called on the children to combine them like the model. Twelve such combinations were covered in the first lesson.

The transformation lessons progressed in the following sequence:

1. Relative clause modifying object:
   Judy lost a book.
   The book was red.
   > Judy lost a book that (which) was red.
   > Judy lost a red book.

2. Relative clause modifying subject:
   The girl is selling cookies.
   The girl is tall.
   > The girl who is tall is selling cookies.
   > The tall girl is selling cookies.

3. Relative clause reduced to adverb of place:
   The monkey eats bananas.
   The monkey is in the tree.
   > The monkey that is in the tree eats bananas.
   > The monkey in the tree eats bananas.

4. Relative clause reduced to past participle:
   The boy won the race.
   The boy is named Joe.
   > The boy who is named Joe won the race.
   > The boy named Joe won the race.

5. Relative clause reduced to present participle:
   The lion is scary.
   The lion is sitting on the log.
   > The lion that is sitting on the log is scary.
   > The lion sitting on the log is scary.

6. Relative clause reduced to nonrestrictive appositive:
   Mr. Jones directs traffic.
   Mr. Jones is a policeman.
   > Mr. Jones, who is a policeman, directs traffic.
   > Mr. Jones, a policeman, directs traffic.

7. Relative clause with nonsubject relativized:
   The sunset was beautiful.
   We saw the sunset.
   > The sunset that we saw was beautiful.
   > The sunset we saw was beautiful.

8. Free combining of sentences (children may combine them however they wish):
    Children have Christmas stockings.
    The stockings are red.
    The stockings hang by the fireplace.
    Santa fills the stockings.
    Nuts and candy are in the stockings.
    (This activity covers several lessons.)

9. Coordination of noun phrase, verb phrase:
    Mortimer plays football.
    Henry plays football.
       Mortimer and Henry play football.

10. Review of coordination and modifiers of nouns.

11. Free writing, combining three sentences.

12. *That* plus subject as object:
    He said something.
    His name is John.
       He said that his name is John.
       He said his name is John.

13. Extraposition with *it:*
    We lost.
    This disappointed me.
       It disappointed me that we lost.

14. Free writing, combining four sentences.

15. Nonembedded yes-no questions:
    The cat will play the flute.
       Will the cat play the flute?"
    Embedded in direct quotation:
       He asked, "Will the cat play the flute?"
    Embedded as indirect question:
       He asked (if, whether) the cat would play the flute.

16. Change statements to questions by replacing *some* with a word like *who, what, which, where, when:*
    Someone will eat the trainer.
       Who will eat the trainer?

17. Adverb clauses, *as, before, after, until, if, unless, because:*
    The game started.
    My friend came in.
       As the game started my friend came in.
       After the game started my friend came in.
       Before the game started my friend came in.

18. A story of twenty-six short sentences is put on the board and students combine the sentences in free writing.

The whole program from the aforementioned curriculum can be found in Hunt and O'Donnell's *An Elementary School Curriculum to Develop Better Writing Skills.* A sentence-combining program for seventh graders, which includes examples of sentences for each lesson, can be found in O'Hare's *Sentence Combining: Improving Student Writing without Formal Grammar Instruction.*

Guidelines for Sentence-Combining Activities   *Transformational grammar*

Sentence-combining activities are found in numerous textbooks designed for use in elementary through college courses. William Strong (1986) provided guidelines and exemplary activities that use sentence-combining approaches to teach twenty different types of sentence-combining activities. Some of these activities are similar to the structured sentences developed by Hunt and O'Donnell. Other activities emphasize open responses in which students generate a range of grammatical responses. These activities help students explore, discuss, and evaluate various stylistic options. Strong recommended the following guidelines for use when developing any type of sentence-combining activities:

1. Discuss purposes for sentence combining that include creating good sentences, becoming flexible in writing, and exploring ways to transform sentences.
2. Encourage students to try new patterns and to take risks with solutions to sentence-combining problems.
3. Provide a positive environment for risk taking by accepting various solutions; when marginal solutions are offered, refer them to the class for judgment.
4. Use signal exercises, context clues, and oral prompts to help students understand how sentence combining functions.
5. After modeling how sentence-combining exercises function, place students in pairs to work through exercises orally.
6. When students are working in pairs, have one student act as scribe for the other; then reverse roles. Have students discuss problems and work out solutions.
7. Have pairs of students develop options for a sentence-combining activity and then agree on the best solution. Ask them to explain why they prefer particular sentences.
8. In round-robin combining, encourage students to listen closely and then give as many solutions to a sentence-combining cluster as they can. Ask students to vote on and discuss reasons for their choices.
9. Be specific in your praise of good sentences; tell students what you like about these sentences.
10. Welcome mistakes as opportunities for group problem solving and use mistakes as basis for skill development in editing workshops.
11. Use transformations handed in by students for in-class workshops.
12. Assign an exercise for homework and then ask several students to put their work on ditto masters. Compare their versions in class.
13. Brainstorm with the class about how an exercise could be made more specific and detailed by adding additional information. Put these details

between exercise sentences, or elaborate sentence-combining clusters and then compare results.

14. Have students write out solutions to various sentence-combining clusters on note cards. Shuffle the cards and ask students to rearrange them into a clear, coherent paragraph.

15. Use sentence combining as a springboard for journal writing. Have students do an exercise each day and then extend that exercise with sentences of their own.

16. Have students combine sentences in a lean, direct style. Contrast the effects of active and passive voice.

17. Ask students to compare their style with that of professional writers. Rewrite a passage into kernel sentences and ask students to recombine the kernel sentences and then compare their versions with the original.

18. Analyze a sentence-combining exercise for tone, cohesion, method of development, and logical patterns.

19. Create original sentence-combining exercises focused on specific transformations, course content, or discourse patterns.

20. Emphasize transfer learning by drawing sentence-combining activities from student texts and literature being studied. Have students revise their writing with a focus on particular sentence-combining skills. Follow sentence-combining lessons with parallel writing tasks for application.

Open Response Exercises in Sentence Combining

Sentence-combining activities can be structured sentences with definite correct responses or open responses in which students generate a range of acceptable grammatical responses. Open responses help students explore, discuss, and evaluate various stylistic options.

When introducing open response exercises, Strong (1986) first explains to the students that there are many right answers when combining sentences and that making mistakes is expected because we learn from our errors. Next, he asks students to think of different ways that they can combine two sentences such as "Carol was working hard on her test. Sue slipped her a note." Before they do this task, he tells the students that they can (1) add connecting words, (2) take out unnecessary words, (3) move words around, and (4) change word endings. He then shows students how combining can be accomplished by modeling various types of transformations such as "Carol was working hard on her test, and Sue slipped her a note"; "As Carol worked hard on her test, Sue slipped her a note"; and "Carol was working hard on her test when Sue slipped her a note." During this modeling he presents and discusses several awkward or confusing transformations such as "Working hard on her test, Carol was slipped a note by Sue"; "It was Carol, hard at work on her test, who was slipped a note by Sue"; and "A note was slipped from Sue to Carol, who was working hard on her test." Finally, he models and discusses unacceptable sentence transformations such as "Carol working hard on her test, Sue slip her a note"; "On her test Carol was hardly working, and a note Sue slipped to her"; and "Slipping a note to Carol was Sue, who was working hard on her test" (p. 26).

Next, Strong leads an oral discussion during which the students try several other examples for sentence combining such as those found in table 7–2. Strong then pairs students as they search for various sentence combining solutions to additional short sentences from table 7–2. After five minutes, two or three transformations are transcribed on the board. This activity can be expanded as students combine the sentence transformations into extended text such as "Carol was hard at work on her test when Sue slipped a note to her. Not wanting her teacher to see, she unfolded it carefully" or "Carol was working hard on her test when Sue slipped her a note, which she carefully unfolded because she didn't want her teacher to see" (p. 27).

During this sentence combining exercise, the writing teams can be challenged by oral prompts. For example, when working with the second cluster of sentences the teacher can give oral prompts such as: begin with *carefully,* begin with *not wanting,* begin with *the paper,* begin with *to keep,* use *because* as a connector, use a semicolon and *therefore,* use *so that* as a connector, use *in order to* as a connector. These prompts help teach sentence variety. In addition, discussions about the resulting quality of sentences help students realize that some sentences sound better than others and that these judgments about quality frequently depend on the context of the preceding sentences. After the sentence combining activity in table 7–2 is completed it can be handed in for credit, shared in small groups for proofreading or judging

TABLE 7–2
Sentence-combining exercise

<table>
<tr><td colspan="2" align="center">**Value Judgment**</td></tr>
<tr><td>1.1</td><td>Carol was working hard on her test.</td></tr>
<tr><td>1.2</td><td>Sue slipped her a note.</td></tr>
<tr><td>2.1</td><td>Carol unfolded the paper carefully.</td></tr>
<tr><td>2.2</td><td>She didn't want her teacher to see.</td></tr>
<tr><td>3.1</td><td>The note asked for help on a question.</td></tr>
<tr><td>3.2</td><td>The question was important.</td></tr>
<tr><td>4.1</td><td>Carol looked down at her paper.</td></tr>
<tr><td>4.2</td><td>She thought about the class's honor system.</td></tr>
<tr><td>5.1</td><td>Everyone had made a pledge.</td></tr>
<tr><td>5.2</td><td>The pledge was not to cheat.</td></tr>
<tr><td>6.1</td><td>Carol didn't want to go back on her word.</td></tr>
<tr><td>6.2</td><td>Sue was her best friend.</td></tr>
<tr><td>7.1</td><td>Time was running out.</td></tr>
<tr><td>7.2</td><td>She had to make up her mind.</td></tr>
<tr><td>8.1</td><td>Her mouth felt dry.</td></tr>
<tr><td>8.2</td><td>Her mouth felt tight.</td></tr>
</table>

Assignment: Finish the story. Explain the reasons behind Carol's judgment.

SOURCE: From William Strong, *Creative Approaches to Sentence Combining.* Urbana, Ill.: National Council of Teachers of English, 1986. Reprinted by permission.

practice, published and compared with other published examples, enriched with details generated by students during prewriting activities, or used as a springboard for an in-class follow-up assignment.

The sources for other sentence-combining activities may be chosen from content area classes and from literature. This is a way to reinforce the content and to enrich children's writing experiences. For example, the sentences to be combined might include information from social studies or from science, recommendations from writing-style handbooks, and kernel sentences selected by rewriting literature paragraphs into their kernel components. Kernel sentences rewritten from any of these sources can be clustered into groups of two as in table 7–2, into groups of four or five, or can be listed without clustering. Clustering provides more structure for younger students or for students who require more guidance. When kernel sentences are written from literary sources, the students may compare their solutions with those of the original authors and discover the impact of various stylistic techniques.

REINFORCEMENT ACTIVITY

1. *Develop a series of kernel sentences from the following paragraph selected from Jerome Wexler's science information text **From Spore to Spore, Ferns and How They Grow** (1985):*

    Fossils are traces we have found of animals and plants that were once alive. They may be bones or teeth or shells, or imprints left in rocks, or animals and plants preserved in ice, or even insects preserved in amber. A petrified tree trunk is a fossil, and so is a footprint left by a dinosaur. (p. 3)

2. *Describe how you would model the introduction to the sentence-combining activity. Share your modeling example with your class.*
3. *Identify sources of materials from literature and from social studies. What would you emphasize with each sentence-combining activity. Develop the kernel sentences that would accompany the materials. Share your activity with a group.*

Teaching About Cohesive Devices

Words such as *he, she, it, we* (personal pronouns); *this, these, those* (demonstrative pronouns); *who, whom, whose* (relative pronouns); *here, there* (location noun substitutions); and *now, then, former,* and *later* (temporal noun substitutions) are used frequently in writing. These terms reduce redundancy, clarify relationships, and link ideas within the text. These connectors, or cohesive devices, tie together information within sentences as well as information within paragraphs and longer passages. When authors provide clear, logical cohesive devices (e.g., referents such as pronouns, conjunctions that link two ideas, substitutions that replace one word with another word that add to the meaning of the first word, and repetitions that restate information) the comprehension of a text increases. In contrast, when the connectors are vague or inferred, the comprehension for many students decreases. Alden Moe and Judith Irwin (1986) stated that teachers must not only be aware of these cohesive devices within texts, they must also help students understand and use cohesive devices.

James Baumann and Jennifer Stevenson (1986) recommended the following four-step procedure that should be used when teaching cohesive devices: (Notice how the steps are similar to the modeling introduced in chapter 3. In both approaches the teacher moves from full teacher responsibility, to shared student/teacher responsibility, to full student responsibility.)

1. Introduce the lesson by telling the students what skill related to cohesive devices will be taught, by providing an example of the cohesive device, and by explaining why the acquisition of that skill is important to the students.
2. Provide direct instruction in which the teacher models, shows, demonstrates, and leads the instruction in the skill.
3. Provide students with teacher-guided application of the skill previously taught, but use materials not used during previous instruction.
4. Provide independent practice in which the students apply the skill in their comprehension of text. Trade books and other primary sources as well as content area texts provide effective materials for this activity and increase the transfer of the skill to additional contexts. Trade books also allow students to practice their skills within longer passage-length texts rather than using isolated sentences that do not consider prior information.

FOR YOUR PLAN BOOK
Comprehension of Cohesive Devices

Objectives
To teach students to comprehend personal pronouns and to apply this knowledge when reading or listening to picture story books.

Introduction
Teacher: When people write they often use words that stand for other words. These words allow writers to repeat an idea without repeating the same words. Look at the following sentences written on the board and read the sentences with me:

> *Barbara likes to write stories. She writes funny stories about animals.*

*In the first sentence find the word **Barbara**. In the second sentence find the word **she**. This is an example of a sentence in which one word replaces another word. She means the same thing as Barbara. Words like she are called pronouns. We can check to see if we are right in our identification of the meaning of the word by reading both sentences, but replacing she with Barbara in the second sentence. Now read these sentences with me:*

> *Barbara likes to write stories. Barbara writes funny stories about animals.*

Notice that the two sentences still mean the same thing. If we replace she with another girl's name, the second sentence would have a different meaning. These sentences are examples of what we will learn about today. We will learn that words such as she, he, we, her, you, they, their, and it stand for other words. This is important to know because it helps you understand what a writer is trying to tell you. It is also important so that you can use this skill in your own writing.

Teacher-Directed Instruction

(Write sentences on the board that include the pronouns to be emphasized. Choose the sentences according to the reading and interest levels of your students.) Teacher: Look at the sentences that are written on the board:

1. *Cinderella lived with her stepmother and her stepsisters. She sat in the ashes in the kitchen and that is how she got her name.*
2. *The prince wanted to marry one of the girls in the kingdom. He gave a ball to meet the girls.*
3. *The stepsisters and the stepmother dressed in their best clothes. They went to the ball.*
4. *Cinderella was unhappy because she wanted to go to the ball. She was granted a wish by her fairy godmother.*
5. *A magical carriage took her to the ball. It turned back into a pumpkin at midnight.*

Read the sentences in number 1 with me. Look for the words used in place of Cinderella. Notice that the words are underlined. Now listen as I tell you how I figured out the meanings of the words. First, I noticed that her was used twice in the first sentence. I thought about whose stepmother and whose stepsisters the sentence meant. I replaced both of the hers with Cinderella and the sentence made sense. Then I noticed that she was used twice in the next sentence. I thought about the first sentence and I asked myself if the she and the her meant Cinderella, the stepmother, or the stepsisters. I knew that she could not mean the stepsisters because the word would have to be they. I used the meaning of the first sentence to help me identify that she meant Cinderella. I knew from the story about Cinderella that Cinderella was given her name because she sat in the cinders and the ashes in the kitchen fireplace. Now let us read the two sentences together and replace she and her with Cinderella. Do the sentences have the correct meaning? Also notice how much better the sentences sound when she and her replace Cinderella. These sentences also show us that we often need to read the sentences before or after a pronoun to understand what the author means.

Now look at sentence number 2. Read the sentences with me. Now I want you to think through the meaning of the sentences. What is the word that is substituted? (He) Good. Who gave the ball? (the prince) Good. How do you know that he refers to the prince? (students provide evidence and reasoning). (Continue with a teacher-led discussion for sentences 3 through 5. Add additional sentences if students have not mastered this portion of the lesson.)

*Teacher: We are now going to read some examples of these pronouns used by authors of the books in our school library. I have written several paragraphs on transparencies. Look at these paragraphs from **Millions of Cats** (Wanda Gág, 1928, 1956):*

"But we can never feed them all," said the very old woman, "They will eat us out of house and home."

"I never thought of that," said the very old man, "What shall we do?"

The very old woman thought for a while and then she said,"I know! We will let the cats decide which one we shall keep."

"Oh yes," said the very old man, and he called to the cats, "Which one of you is the prettiest?"

"I am!"

"No, I am!" (p. 20 unnumbered)

Teacher: *In the Cinderella example we discovered that we sometimes have to read sentences before the pronoun or sentences after the pronoun to know what the pronoun means. In this example from* **Millions of Cats** *we will need to read into the paragraph before we know what the first pronoun means. Read the first sentence with me. What is the first word that is underlined? (We) Good. Can anyone tell me what we refers to? This is an example in which we need to read beyond the word. Now let us read the whole selection together. Then we will go back and try to identify what each underlined word means. Now who can tell me what we means and how you know what we means? (Student response and reasoning) Now who can write the words that mean we on the right place on the transparency? (Student writes woman and man above we on the transparency or on the board if board work is preferred to a transparency.) Good. Now who can read the sentence by replacing we with the words that mean the same as we? (Student response) Good. (Continue with the remainder of the story, stopping at each underlined word to discuss the meaning and to write the word above the pronoun. Reread the total selection to verify the meaning of the underlined words. Discuss how the use of pronouns improves the beauty of the text.)*

Teacher-Guided Application

Teacher: *We are now going to look at different examples of paragraphs taken from library books. The words that you need to identify are underlined. I want you to read the paragraphs, to think about the meaning for the pronouns, and to write the correct meaning over each underlined word. We will do the first one together to make sure that we all understand what to do. Then I want you to do the rest of the sentences by yourself. Be sure to read the whole selection to make sure that your choice makes sense and is accurate. When you are finished, we will read the paragraphs together and discuss your answers.*

The Mother's Day Mice (Eve Bunting, 1986)

Biggest Little Mouse wakened first. It was early in the morning and still almost dark. He tugged gently on the whiskers of Middle Mouse, who slept next to him. "It's Mother's Day," he whispered. "Time to get up and go for our presents."

Middle Mouse tugged gently on the whiskers of Little Mouse, who slept between him and the wall. "Mother's Day," he whispered.

They crept out of bed and tiptoed past Mother's room. (p. 1)

The Man Who Could Call Down Owls (Eve Bunting, 1984)

By day the man worked in his owl barn. There were wings to be mended and legs to be splinted.

"How do you find the owls that need your help?" the boy asked.

The man smiled. "They find me." He held a screech owl. "When an owl is sick and frightened you must hold it firmly. Then it can't hurt itself or you. Always remember that." (p. 7)

(After the students complete this activity independently, conduct a guided discussion of this phase of the lesson. Encourage discussion that focuses on the reasons for the replacements as well as on the correct answers.)

Independent Practice

Teacher: We have completed several paragraphs in which authors use pronouns to replace other words in the stories. I have a selection of books in which I want you to find examples of pronouns. When you find an example of a pronoun, write the sentence on your paper. Leave space to write the meaning of the word above the pronoun. We will share these examples in class tomorrow. Be ready to read your selection to the class and to tell the class what each pronoun means.

Examples of books that may be used for this activity:

> *Betty Baker's **Rat Is Dead and Ant Is Sad** (1981)*
> *Marc Brown's **Arthur's Baby** (1987)*
> *Denys Caset's **A Fish In His Pocket** (1987)*
> *Valerie Flournoy's **The Patchwork Quilt** (1985)*
> *Mary Rayner's **Mrs. Pig Gets Cross and Other Stories** (1986)*
> *Jean Van Leeuwen's **Oliver, Amanda, and Grandmother Pig** (1987)*

REINFORCEMENT ACTIVITY

*Develop a cohesive comprehension lesson plan appropriate for older students. To identify a needed area for instruction, analyze several books written for middle- and upper-elementary students. What cohesive devices do the authors use? For example, Patricia Lauber's 1987 Newbery Honor book **Volcano: The Eruption and Healing of Mount St. Helens** uses the "it" referent nine times in the first two paragraphs. It, however, may refer to volcano, Mount St. Helens, earthquake, rock, molten rock, or magma. The meanings of the referents are clear as long as the reader reads the text carefully.*

Include in your lesson plan the following four phrases: introduction, direct instruction, teacher-guided application, and independent practice. Whenever appropriate, include modeling in the lesson plan.

Providing Instruction in Usage

According to Gerald Duffy (1969), elementary schools should teach usage better than we have been teaching it. He believed that to accomplish this, we should cease our attempts to indoctrinate pupils with the notion that there is only one right way to say something. Instead, we should show students that what is appropriate on the playground may not be appropriate when addressing the principal or mayor. According to Duffy:

> In using this strategy, we are actually teaching children the use of two dialects—one which is acceptable in the home and the neighborhood and another which admits him to the company of educated society. Approaching usage instruction from such an analytical standpoint makes more sense to the linguist and is to him a better way to teach usage than the traditional endless drill on proper usage. . . . The linguist is saying that the most severe usage errors should be eliminated in the elementary school, with the refinements of the 'prestige dialect' being reserved for secondary school instruction. (p. 37)

Usage errors may conflict with the informal level of English practiced in the classroom (Pooley, 1960). For children to form the habit of a classroom level of usage, they need both an understanding of when to use that form, and practice in doing so.

Role playing   Role-playing activities can help children improve their understanding of various levels of usage. Role playing also allows children opportunities to think about and discuss the roles of speaker or writer, audience, and subject matter, in regard to appropriate levels of usage. Teacher and students can suggest a number of situations requiring different levels of usage. A middle- or upper-elementary class might try the following roles:

1. Jerry's friend Jimmy has joined a neighborhood softball team. Jimmy has invited Jerry to go with him to meet the team, because the team needs a pitcher. Jerry wants them to accept him as the newest member of the team. Pretend Jerry and Jimmy are talking on their way to a vacant lot or park. Then make up the conversation that might occur between Jimmy, Jerry, and the rest of the team.
2. You have just found out that your school will have a science fair. The winner of the science fair will receive a microscope that you have wanted for a long time. In addition, the winner will represent the school in the regional science fair. You have an idea for a project and want to talk about it with the science teacher. You want to convince him that you have thought about your ideas, and you want him to give you some additional out-of-school help.

After role playing, the students should discuss the appropriateness of the communication according to subject matter and audience. What are the consequences of an inappropriate level of usage?

Inductive teaching of usage   Inductive approaches to usage that help children analyze their problems are more helpful than isolated drills, in which they merely underline the correct choices on a worksheet. As an example, let us look at a lesson that might be built around one of the appropriate usages recommended by Pooley—the correct use of *is*, *are*, *was*, and *were* in number agreement (singular and plural) and tense (present tense, past tense).

An Inductive Approach to Was and Were—
Singular and Plural Forms in Past Tense

First, the teacher can write the following sentences on the chalkboard:

1. Carol *was* at the movie last night.
2. Jane and Julie *were* also at the movie last night.
3. They *were* at the movie together.
4. Jim *was* at home last night.
5. I *was* also at home.
6. We *were* at home together.
7. You *were* at the park.

Next, the teacher reads the sentences (use numbers 1–6 first) to the students, asking them to listen carefully. Then the students read the sentences to the teacher. Now, the teacher tells them to look carefully at the underlined words in sentences 1–6. Then she asks, "When did the activity in each sentence take place?" They should conclude that it was last night, which is in the past, and not something that is happening right now, in the present. Next she asks, "How many people were in each of the sentences?" They should conclude that sentences 1, 4, and 5 each concern one person, and sentences 2, 3, and 6 concern more than one person. The teacher writes just this portion of each sentence on the board, in two columns:

One person	More than one person
Carol *was*	Jane and Julie *were*
Jim *was*	They *were*
I *was*	We *were*

Now, the teacher asks the students if they can think of a guiding suggestion to help them remember when to use *was* and when to use *were*. They will usually produce a generalization like this:

When we are talking about something that happened in the past, we use *was* if we are talking about one person and we use *were* if we are talking about more than one person.

Now, the teacher asks them to look at sentence 7. "Does this sentence follow our guide? What is wrong?" They should conclude that the word *you* refers to only one person, but uses *were* in the past tense as if the word *you* referred to more than one person. Ask the students where to place the sentence on the list. They will usually suggest that, because it is an exception, placing it in a third column would help them remember. Now you should ask if there is anything they need to add to their guiding rule. The fact that *were* follows the word *you* can be added as an exception. Finally, the teacher should ask the students to create additional sentences that conform to the same patterns. They will also need frequent opportunities to practice with oral and written activities, rather than with worksheet drills.

THE MECHANICS OF WRITING

Punctuation

Studies show that punctuation—particularly the use of commas and periods—is frequently a problem for elementary-age children (Hazlett, 1972; Porter, 1974). Punctuation is important because it clarifies meaning. During an oral interchange, the

Learning about punctuation.

listener hears signals—pauses, speech stops, rising and falling voice tones—that help him comprehend meaning. The writer replaces these vocal signals with punctuation.

The early activities in which children dictate short stories or sentences to place under pictures provide readiness for and instruction in punctuation. We should emphasize that, like other aspects of composition, punctuation and capitalization are most effectively learned when they are taught in conjunction with meaningful writing activities. Table 7–3 shows the punctuation usually taught in the elementary grades.

REINFORCEMENT ACTIVITY

Look at a scope-and-sequence chart in a language arts textbook. Make a punctuation chart that could be used with a specific elementary-school level. Design your chart to show the signal given by the punctuation mark and where it would be used. For example, you might finish the following chart:

What Do Punctuation Marks Signal?

Name	Example	Signal	Where Used
Period		Come to a full stop	At the end of a sentence that tells something

Capitalization

Capitalization is another signal that clarifies meaning. We expect the writer to signal the beginning of a sentence with a capital letter and the end of a sentence with a period, question mark, or exclamation mark. We also expect the writer to signal important titles and names by capitalization. If the writer capitalizes Northeast, we know he is referring to a specific section of the country; if he does not capitalize northeast, we know the reference is to a direction. When we are skimming a page, looking for information about a particular person or place, the beginning capital letters may be one of our best cues to quick identification.

Most children seem to have fewer problems applying capitalization rules than in applying punctuation rules. The National Assessment of Educational Progress concluded: "Capitalization errors tended to appear less frequently in the 9-year-old papers than most of the other kinds of errors" (Hazlett, 1972, p. 12). The National Assessment found that most high-quality papers were free of capitalization errors; this was also true of middle-quality papers. The lower-quality papers were, however, inconsistent; the first word in a sentence was often left uncapitalized, and many common nouns, such as "deer" or "tree," were capitalized. The judges concluded that children who wrote the lower-quality papers had some idea of when to use capitals and when not to, but their ideas were incomplete. It thus appears that most children will learn about and apply appropriate capitalization in their writing, while a few will require more concentrated assistance.

Punctuation Rule	Example
1. A period is used at the end of a sentence.	1. We went to the zoo.
2. A period is used after numbers on a list.	2. 1. Show and Tell 2. Reading
3. A period is used after abbreviations.	3. Sun. Mon. Tues. Mr. Mrs.
4. A question mark is used after a sentence that asks a question.	4. What was the name of Billy's dog?
5. A comma is placed between the day of the month and the year.	5. Today is October 1, 1980.
6. A comma is used between the names of a city and a state.	6. We live in Ithaca, New York.
7. A comma is used after the salutation in a letter.	7. Dear Sandy,
8. A comma is used after the closing of a letter.	8. Your friend, Jimmy
9. Commas divide items in a list.	9. Susan brought an acorn, a colored maple leaf, and an ear of Indian corn.
10. A comma is used before a coordinating conjunction (upper elementary).	10. He wanted to eat the whole pie, but he knew he would be sick.
11. A comma is used to signal the subject of the sentence (upper elementary).	11. Even though the day was hot, we all decided to go on a picnic.
12. A comma is used when there is a slight interruption which offers additional information (upper elementary).	12. Jim Black, our basketball coach, is a very nice man.
13. A comma is used between two equal adjectives that modify the noun (upper elementary).	13. That swimmer is a strong, healthy athlete.
14. An exclamation mark follows a statement of excitement.	14. Help! Help! Jimmy is chasing me with a snake!
15. An apostrophe shows that letters have been omitted in a contraction.	15. I will = I'll
16. An apostrophe is used to show possession.	16. This is Judy's basketball.

TABLE 7–3
Punctuation taught in elementary grades

Children are introduced to capitalization in their early, dictated stories. When the teacher writes a sentence on the board or chart, she can point out the capital letter at the beginning of each sentence. When writing names of people, cities, states, or holidays, the teacher can again indicate the importance of the capital letter.

Table 7–4 shows the capitalization rules usually taught in the elementary grades.

Instruction in Mechanics

Both punctuation and capitalization clarify written communication for the reader. Children need to understand why mechanical skills are necessary, but they must also be

TABLE 7–4
Capitalization taught in elementary grades

Capitalization Rule	Example
Primary Grades	
1. Capitalize the first word in a sentence.	1. The first grade is planning a puppet show.
2. Capitalize the first and last names of people.	2. Sandy Smith
3. Capitalize the pronoun "I".	3. I do not think I will go.
4. Capitalize the name of a street.	4. Our school is on Southwest Parkway.
5. Capitalize the name of the child's city and state.	5. We live in San Francisco, California.
6. Capitalize the names of months, days, holidays.	6. Today is Monday, June 1. Christmas Hanukkah
7. Capitalize the first word and other important words in titles.	7. We are reading A Snowy Day.
8. Capitalize a title such as Mr., Mrs., Miss or Ms.	8. Our teacher is Mrs. Chandler.
9. Capitalize the greeting in a letter.	9. Dear Martin,
10. Capitalize the first word in the closing of a letter.	10. Your friend,
Middle Grades	
11. Capitalize the titles of persons.	11. Doctor Erickson General Washington
12. Capitalize all cities, states, countries, continents, and oceans.	12. United States Pacific Ocean
13. Capitalize proper adjectives.	13. French
14. Capitalize the names of organizations.	14. Girl Scouts Boy Scouts
15. Capitalize the names of sections of the country but not directions.	15. The Northeast has many large cities. Lake Superior is northeast of Minneapolis.
16. Capitalize words referring to the Bible and the Diety.	16. God the Bible

aware that writing is primarily concerned with the development of content and ideas. The teaching guide, *English Language Arts in Wisconsin,* stated: "Mechanical skills in writing are important, but must not be stressed to the point where they make a student lose his interest in writing or stifle his creative instinct. Nor must they be neglected to the extent that the student becomes careless in his written word" (1967, p. 197).

How can the teacher maintain balance without overemphasizing one aspect of writing to the detriment of the other? Kean and Personke (1976) offered three general guidelines for instruction in mechanics:

1. "Several of the mechanics of written language are learned as children write, read, and observe, making formal classroom presentations unnecessary in many cases." (Individual help may be required for students who did not naturally learn them.)
2. "Individual variation in readiness makes it likely that learning will be most efficient when instruction is conducted among small groups of children who are at similar stages of readiness." (This means small-group instruction of children who demonstrate similar needs, rather than whole-class instruction.)
3. "Instruction in mechanics needs to begin and culminate in a real writing situation; if isolated practice is necessary, it should be in the intermediate learning stage."[2]

If we follow these guidelines, we would call the children's attention to various punctuation and capitalization items as they encounter them in reading or in dictating charts and stories. This provides readiness for a skill they will need in writing, and may enable some children to apply the skill accurately in their own writing. For example, many children are able to begin sentences with capital letters and end them with periods after these mechanics have been pointed out by the teacher during frequent language experience activities. Some children, however, will require additional direct teaching of these skills before they can apply them consistently. When direct teaching is necessary, it should take place in groups, with children who show the same readiness or need for a new or reviewed skill.

Table 7–5 shows one method for teaching the separation of words in series with a comma.

Additional Activities for Writing Mechanics

Some children need visual assistance with punctuation and capitalization. Charts that show various writing conventions are helpful. For example, the colors of traffic signals can be used to mark sentences written on charts. The capital letters at the beginning of sentences, which also signal "go," can be marked in green; commas, signaling a short pause, can be marked in yellow; and periods at the end of sentences, signaling "stop," can be marked in red.

---

[2]From John M. Kean and Carl Personke, *The Language Arts: Teaching and Learning in the Elementary School* (New York: St. Martin's Press, 1976). Reprinted by permission.

TABLE 7–5

Steps in instruction in mechanics of writing for "comma separates words in series"

Guideline	Example
1. Attention called to punctuation or capitalization rule in reading or dictated stories.	1. Find sentences that contain words in series in reading. Point them out in language experience. Example: Oranges, grapefruit, and lemons grow in Florida.
2. An observed mechanical need in an actual writing situation.	2. Students are writing series of words without using the appropriate punctuation.
3. Development of individual or small-group instruction designed to eliminate the problem.	3. Teacher has students read four or five sentences with words in series. Students inductively state rule. Teacher writes new sentences on chalkboard without commas. Students put in commas, using colored chalk. Teacher may reinforce with games, etc.
4. Assign a real writing task that requires application of the skill.	4. "Write a sentence telling us what you had for breakfast"; "Write a sentence listing the members of your family"; etc.
5. Teacher observation to see if mechanical problem has been eliminated.	5. Observe future writing. Is skill being used? If skill not used appropriately, may require further direct teaching.

Children can work with a paragraph written without punctuation, without capitalization, or without both punctuation and capitalization (depending on ability level). As the teacher reads the paragraph, the children mark the punctuation according to the verbal signals of the teacher's voice.

Upper-elementary children can take dictation from kindergarten and first-grade children. To produce a short, written story that expresses what the younger child means, the older child will have to supply appropriate punctuation and capitalization signals. The older child should then read the selection back to make sure the meaning is what the younger child intended.

Proofreading symbols are often effective with upper-elementary students. The teacher can ask someone connected with a newspaper to talk to the children, and show them proofreading symbols. They are usually impressed to learn that adult writers also need to proofread and correct their writing before it is published in a paper. The children can also create their own classroom proofreading symbols, or use the standard symbols, which can be found in a dictionary.

Ideas for Teachers from Teachers: Elementary Language Arts published by the National Council of Teachers of English (1983) describes several activities developed by classroom teachers. For example, punctuation pins can be made from clip-on clothespins. (A punctuation mark is painted on each clothespin.) Sentences without punctuation are printed on tagboard strips. Children then clamp the correct clothespin markers, in the correct location, onto the sentences. Another effective activity is based on dialogue created to accompany pictures students bring to class. Each student brings a picture of a character, teams with another student, creates dialogue between the pictured characters, and transcribes the resulting dialogue. Children are encouraged to use appropriate punctuation.

REINFORCEMENT
ACTIVITY

1. *Identify a mechanics-of-writing need, such as use of quotation marks or capitalization of proper nouns. Design a small-group instructional activity that would result in appropriate use of that skill.*

2. *Design a writing-mechanics game that reinforces understanding of a punctuation or capitalization rule.*

3. *Develop a proofreading lesson for use with older-elementary children. You may wish to refer to the "Proofreading for Spelling" unit in the spelling chapter. A similar approach can be used for introducing the proofreading of punctuation or capitalization.*

MAINSTREAMING AND GRAMMAR

Grammar, usage, and mechanics are not ends in themselves. As was the case with handwriting, the major purpose of grammar is effective communication. All students need many opportunities to use their language in meaningful context. This is especially true for some mainstreamed children who may have difficulty using syntactic information and generalizing limited grammatical knowledge in a variety of settings. If children demonstrate restricted language ability and vocabulary development, they require many meaningful learning opportunities designed to broaden their language experiences.

Table 7–6 presents characteristics of learning disabled children that might interfere with grammar, and suggests teaching techniques for language arts instruction. Table 7–7 presents the same information for mentally handicapped children. The teaching techniques are reported in research or by authorities who work with mainstreamed children.

TABLE 7–6

Mainstreaming and grammar for learning disabled children

Characteristics of Learning Disabled Children	Teaching Techniques for Language Arts Instruction
1. Deficit in use of grammatical form class and use of syntactic information (Fay, 1981; Henderson and Shores, 1982).	1. Have students read aloud materials in which words ending in *ed* and *ing* are underlined. Have pupils chart errors, identify correct forms necessary to complete sentences (Henderson and Shores, 1982). Provide systematic training in the arrangement of inflectional forms of verbs (Cartelli, 1980). Prepare activities developed around generative-transformational grammar and context (Knott, 1979). Generate transformations of kernel sentences and encourage children to experiment with sentence pattern variations. Encourage many meaningful oral language activities such as role playing, show and tell, puppetry, choral speaking, interviewing, and storytelling (Lerner, 1976).
2. Demonstrate disability in comprehending information signaled by grammatical morphemes (McClure, Kalk, and Keenan, 1980).	2. Emphasize comprehension available through function words, word endings, and word stem changes (McClure, Kalk, and Keenan, 1980). Use pictures to help children build morphological generalizations (Lerner, 1976).

SUMMARY

We have discussed a three-part definition of grammar: grammar 1, a child's familiarity with a set of formal patterns; grammar 2, the linguistic science concerned with description, analysis, and formalization of formal language patterns; and grammar 3, the use of language according to language standards for different occasions, audiences, and purposes.

Grammar 2 includes the various grammars developed to describe and analyze sentence patterns and to classify words. This is the grammar taught in most school systems. Traditional grammar is considered prescriptive, because it provides a series of rules for constructing sentences and classifying parts of speech. In contrast, structural grammar is referred to as descriptive grammar, because it describes words according to form classes, and describes sentence patterns or positions of words in sentences. Transformational grammar is built on the work of the structuralists, but extends grammar into the meaning of language. Whereas structural grammar is concerned primarily with syntax, transformational grammar is concerned more with semantics and the generating of sentences. The term *transformation* refers to the division of sentences into kernel sentences and the transforms or variations that can be developed from these basic sentences.

TABLE 7–7
Mainstreaming and grammar for mentally handicapped children

Characteristics of Mentally Handicapped Children	Teaching Techniques for Language Arts Instruction
1. Below average language ability, restricted vocabulary (Gearheart, Weishahn, and Gearheart, 1988).	1. Requires many opportunities to broaden students' language experience. Provide structured, concretely oriented experiences consistent with students' interests and present level of language development (Gearheart, Weishahn, and Gearheart, 1988). Help children develop a sentence sense through easy reading materials, listening activities, and reading their writing orally (Otto and Smith, 1980). Provide opportunities for children to improve language clarity by rearranging grammatical elements, deleting grammatical elements, substituting grammatical elements, and adding grammatical elements (Otto and Smith, 1980).
2. Difficulty generalizing and conceptualizing in a variety of settings (Gearheart, Weishahn, and Gearheart, 1988).	2. Provide many opportunities for functional writing. Diagram letter formats and establish a classroom post office to encourage using writing and punctuation skills (Cohen and Plaskon, 1980).

Grammar 3, or usage, refers to appropriate choices of words, phrases, and sentences. Usage deals with a group's attitudes and language standards, rather than with the way words are structured to convey meaning. Consequently, a level of usage appropriate for one subject and audience might not be appropriate for another. Instead of teaching that some particular form is incorrect, modern usage instruction stresses the appropriateness for certain purposes and audiences, and helps children increase their range of usage.

Grammar instruction should expand a child's understanding of language; consequently, the best aspects of the newer grammars are often recommended for inclusion in the instructional program. Activities using linguistic blocks, sentence patterns, sentence transformations, scrambled sentences, rearranging sentences but maintaining meaning, expanding sentences, sentence combining, and cohesive devices, all help develop an understanding of language and improve spoken and written grammar.

Instruction in usage helps children understand that there is more than one level of usage, although one level may be more appropriate for a specific occasion. Activities such as role playing help develop this understanding. Inductive approaches to usage help children in developing understanding and application of a classroom level of usage. Instead of teaching that there is only one correct usage, modern usage instruction helps children move appropriately from one usage situation to another.

The mechanics of writing, including capitalization and punctuation, relate to the understanding of language and grammar, because both capitalization and punctuation are signals that help the reader understand written communication.

ADDITIONAL GRAMMAR ACTIVITIES

1. *Locate language arts/English textbooks that represent each of the major philosophies of grammar: traditional, structural, and transformational. Identify any sentence-combining activities located in each text. Compare the similarities and differences found in the sentence-combining activities. Why might the activities differ according to the philosophy?*

2. *Locate several language arts/English journals such as* **Language Arts, Research in the Teaching of English,** *and* **College English.** *Read any journal articles pertaining to grammar. What are the issues and concerns? Do these issues and concerns vary according to the audience for the journal?*

3. *Develop a lesson plan that allows children to experiment with sentence pattern variations and understand that various sentence patterns are possible and beneficial when writing.*

4. *Develop a lesson plan that develops children's ability to expand sentences.*

5. *Tape-record several examples of children's oral language. Listen for the levels of usage. Are the children able to vary their levels of usage depending on their audience and purpose? What instructional strategies, if any, would you recommend?*

BIBLIOGRAPHY

Baele, Ernest R. *The Effect of Primary Reading Programs Emphasizing Language Structure as Related to Meaning upon Children's Written Language Achievement at Third Grade Level.* Doctoral dissertation, University of California, Berkeley, 1968.

Bauman, James, and Stevenson, Jennifer. "Teaching Students to Comprehend Anaphoric Relations." In *Understanding and Teaching Cohesion Comprehension,* edited by Judith W. Irwin. Newark, Del.: International Reading Association, 1986.

Cartelli, Lora M. "Reading Comprehension: A Matter of Referents." *Academic Therapy* 15 (March 1980): 421–30.

Chomsky, Noam. *Syntactic Structures.* The Hague: Mouton & Co., 1957.

Cohen, Sandra B., and Plaskon, Stephen P. *Language Arts for the Mildly Handicapped.* Columbus, Oh.: Merrill Publishing Co., 1980.

DeHaven, Edna P. *Teaching and Learning the Language Arts.* Boston: Little Brown & Co., 1983.

Duffy, Gerald. *Teaching Linguistics.* Dansville, N.Y.: Instructor Publications, 1969.

English Language Arts in Wisconsin. Madison, Wis.: Department of Public Instruction, 1967.

Fagan, William T.; Cooper, Charles R.; and Jensen, Julie M. *Measures for Research and Evaluation in the English Language Arts.* Urbana, Ill.: National Conference of Teachers of English, 1975.

Fay, Gayle; Trupin, Eric; and Townes, Brenda D. "The Young Disabled Reader: Acquisition Strategies and Associated Deficits." *Journal of Learning Disabilities* 14 (January 1981): 32–35.

Francis, W. Nelson. "Revolution in Grammar." In *Readings in Applied English Linguistics.* 2d ed., edited by Harold Allen. New York: Appleton-Century-Crofts, 1964.

Gearheart, Bill R.; Weishahn, Mel; and Gearheart, Carol J. *The Exceptional Student in the Regular Classroom.* 4th ed. Columbus, Oh: Merrill Publishing Co., 1988.

Goodman, Ralph. "Transformational Grammar." In *An Introductory English Grammar,* edited by Norman Stageburg. New York: Holt, Rinehart & Winston, 1965.

Graves, Barbara W. "A Kinesthetic Approach to Building Language Power." *Elementary English* 49 (October 1972): 818–22.

Hazlett, James A. *A National Assessment of Educational Progress. Report 8, Writing: National Results.* Washington, D.C.: U.S. Government Printing Office, 1972.

Henderson, Ann J., and Shores, Richard E. "How Learning Disabled Students' Failure to Attend to Suffixes Affects Their Oral Reading Performance." *Journal of Learning Disabilities* 15 (March 1982): 178–82.

Hillocks, George. Jr. *Research on Written Composition: New Directions for Teaching.* Urbana, Ill.: ERIC Clearinghouse on Reading and Communication Skills and the National Conference on Research in English, 1986.

———. "What Works in Composition: A Meta-Analysis of Experimental Treatment Studies." *American Journal of Education* 93 (1984): 133–70.

Hirsch, E. D. "Some Principles of Composition from Grade School to Grad School." In *The English Curriculum under Fire: What Are the Real Basics?,* edited by George Hillocks, pp. 39–52. Urbana, Ill.: National Council of Teachers of English, 1982.

Hunt, Kellogg, and O'Donnell, Roy. *An Elementary School Curriculum to Develop Better Writing Skills.* Washington, D.C.: U.S. Department of Health, Education, and Welfare, Bureau of Research, 1970.

Kean, John M., and Personke, Carl. *The Language Arts, Teaching and Learning in the Elementary School.* New York: St. Martin's Press, 1976.

Klein, Marvin. "Sentence and Paragraph Building in the Language Arts." In *Language Arts and the Beginning Teacher,* edited by Carl Personke and Dale Johnson. Englewood Cliffs, N.J.: Prentice-Hall, 1987.

Knott, Gladys P. "Developing Reading Potential in Black Remedial High School Freshmen." *Reading Improvement* 16 (Winter 1979): 262–69.

Krashen, Stephen D. *Writing: Research, Theory, and Applications.* Oxford: Pergamon Press, 1984.

Lamb, Pose. *Linguistics in Proper Perspective.* 2d ed. Columbus, Oh.: Merrill Publishing Co., 1977.

Lerner, Janet. *Children with Learning Disabilities.* 2d ed. Boston: Houghton Mifflin Co., 1976.

Littlefield, Violet. "Developing a Classroom Dialect." Madison, Wis.: Bulletin Number 2, Wisconsin Department of Public Instruction, 1967.

Lodge, Helen C., and Trett, Gerald L. *New Ways in English.* Englewood Cliffs, N.J.: Prentice-Hall, 1968.

Lundsteen, Sara W. *Children Learn to Communicate.* Englewood Cliffs, N.J.: Prentice-Hall, 1976.

Mackie, B. C. *The Effects of a Sentence-Combining Program on the Reading Comprehension and Written Composition of Fourth-Grade Students.* Doctoral Dissertation, Hofstra University, 1982 (University Microfilms No. DA 8207744).

McClure, Judith; Kalk, Michael; and Keenan, Verne. "Use of Grammatical Morphemes by Beginning Readers." *Journal of Learning Disabilities* 13 (May 1980): 34–49.

Mellon, John. *The Basic Sentence Types and Their Simple Transforms.* Culver, Ind.: Culver Military Academy, 1964.

———. *Transformational Sentence Combining: A Method for Enhancing the Development of Syntactic Fluency in English Composition.* Champaign, Ill.: National Conference of Teachers of English, 1969.

Moe, Alden J., and Irwin, Judith W. "Cohesion, Coherence, and Comprehension," in *Understanding and Teaching Cohesion,* edited by Judith W. Irwin. Newark, Del.: International Reading Association, 1986.

National Council of Teachers of English. *Ideas for Teachers from Teachers: Elementary Language Arts.* Urbana, Ill.: National Council of Teachers of English, 1983.

O'Hare, Frank. *Sentence Combining: Improving Student Writing without Formal Grammatical Instruction.* Urbana, Ill.: National Council of Teachers of English, 1973.

Otto, Wayne, and Smith, Richard J. *Corrective and Remedial Teaching.* Boston: Houghton Mifflin Co., 1980.

Pooley, Robert. "Dare Schools Set a Standard in English Usage?" *English Journal* 49 (March 1960): 176–81.

Porter, Jane. "Research Report." *Elementary English* 51 (January 1974): 144–51.

Roberts, Paul. *English Sentences.* New York: Harcourt Brace Jovanovich, 1962.

Rose, Shirley K. "Down from the Haymow: One Hundred Years of Sentence-Combining." *College English* 45 (September 1983): 483–91.

Savage, John F. *Effective Communication.* Chicago: Science Research Associates, 1977.

Schiller, Andrew. "Grammar in the Grammar Schools?" In *Forum for Focus,* edited by Martha King, Robert Emans, and Patricia Cianciolo. Urbana, Ill.: National Council of Teachers of English, 1973.

Stoddard, E. P. "The Combined Effect of Creative Thinking and Sentence-Combining Activities on the Writing Ability of Above Average Ability Fifth and Sixth Grade Students." Doctoral Dissertation, University of Connecticut, 1982. *Dissertation Abstracts International,* 1982. (University Microfilms No. DA 821 3235)

Stotsky, Sandra L. "Research on Reading/Writing Relationships: A Synthesis and Suggested Directions." *Language Arts* 60 (1983): 627–42.

Strong, William. *Creative Approaches to Sentence Combining.* Urbana, Ill.: National Council of Teachers of English, 1986.

Tiedt, Iris, and Tiedt, Sidney. *Contemporary English in the Elementary School.* Englewood Cliffs, N.J.: Prentice-Hall, 1975.

Weimer, Wayne, and Weimer, Anne. *Reading Readiness Inventory.* Columbus, Oh.: Merrill Publishing Co., 1977.

CHILDREN'S LITERATURE REFERENCES

Baker, Betty. *Rat Is Dead And Ant Is Sad.* Illustrated by Mamoru Funai. New York: Harper & Row, 1981.

Brown, Marc. *Arthur's Baby.* Boston: Little, Brown & Co., 1987.

Bunting, Eve. *The Man Who Could Call Down Owls.* Illustrated by Charles Mikolaycak. New York: Macmillan Co., 1984.

———. *The Mother's Day Mice.* Illustrated by Jan Brett. New York: Clarion, 1986.

Cazet, Denys. *A Fish in His Pocket.* New York: Watts, 1987.

Flournoy, Valerie. *The Patchwork Quilt.* Illustrated by Jerry Pinkney. New York: Dial Press, 1985.

Gág, Wanda. *Millions of Cats.* New York: Coward-McCann, 1928, 1956.

Lauber, Patricia. *Volcano: The Eruption and Healing of Mount St. Helens.* New York: Bradbury, 1986.

Rayner, Mary. *Mrs. Pig Gets Cross and Other Stories.* New York: E. P. Dutton, 1986.

Van Leeuwen, Jean. *Oliver, Amanda, and Grandmother Pig.* Illustrated by Ann Schweninger. New York: Dial Press, 1987.

Wexler, Jerome, *From Spore to Spore: Ferns and How They Grow.* New York: Dodd, Mead & Co., 1985.

Chapter Eight

OBJECTIVES OF WRITTEN COMPOSITION
VIEWPOINTS ON WRITTEN COMPOSITION
THEORETICAL BASIS FOR PROCESSING
Classroom Procedures in Processing
DEVELOPING COMPOSITION SKILLS
THROUGH THE PROCESS OF WRITING
Developing Your Own Understanding of the Writing Process ▪ *Prewriting* ▪ *While Students Write* ▪ *After Students Write*
EVALUATING WRITING
SUMMARY

After completing the chapter on composition, you will be able to:

1. State the objectives of written instruction.
2. Identify several viewpoints on the teaching of writing.
3. Describe the theoretical base for the writing process.
4. Identify classroom procedures related to the processing approach to writing.
5. Describe your own understanding of the writing process.
6. Describe and develop prewriting activities.
7. Describe and develop activities that focus on the writing of the first draft.
8. Describe and develop activities that focus on revision of writing.
9. Describe and develop methods for evaluating students' writing.

Composition: The Writing Process

Written composition is receiving renewed interest in research, and in both technical and nontechnical publications. For example, Hillocks's (1986) text *Research on Written Composition: New Directions for Teaching* allows educators to review past research studies, to select the most effective approaches to teaching writing, and to add strategies for teaching within the writing process that actually increase the effectiveness of process writing. Professional organizations such as the National Council of Teachers of English are publishing texts that focus on the teachers' role in the writing process, the various stages that need to be considered in helping students during the writing process, and activities that help teachers improve students' writing.

Some local writing improvements are so spectacular that newspapers are providing detailed coverage. For example, *The Dallas Morning News* (Kruh, 1988) proudly covered a story showing that elementary students significantly improved their writing through an approach called RADAR writing. Using this approach students are taught specific ways to approach various writing tasks, are taught to use transitions in their writing development, are provided guidance during independent writing, are encouraged to share their writing with an audience of their peers who provide critical feedback, and are encouraged to revise and improve their writing. The article included interviews with teachers and students, examples of students' writing, and test results.

The information provided in this chapter, combined with that in chapter 9, will help you understand the writing process and develop activities that enhance the writing of young children, as well as continue writing improvement into the upper grades. We also refer to effective writing activities such as sentence combining and modeling described in previous chapters, and consider where these activities fit into the teaching of writing. Within these two chapters we consider the development of narrative, expository, descriptive, and persuasive writing. Chapter 8 focuses on the writing process and outlines each of the steps in detail. In addition, the chapter provides information on assessment of composition; research on effective approaches and modes of instruction; on teacher-pupil and peer editing group conferencing; and on publishing students' work. Chapter 9 extends this writing process into specific strategies that provide assistance for teachers in the upper grades and in specific types of writing.

OBJECTIVES OF WRITTEN COMPOSITION

The National Council of Teachers of English (1983) provided guidelines for the enhancement of written composition. According to the NCTE "Essentials of English," students should:

1. Learn to write clearly and honestly.
2. Recognize that writing is a way to learn and develop personally as well as a way to communicate with others.
3. Learn ways to generate ideas for writing, to select and arrange them, to find appropriate modes for expressing them, and to evaluate and revise what they have written.
4. Learn to adapt expression to various audiences.
5. Learn the techniques of writing for appealing to others and persuading them.
6. Develop their talents for creative and imaginative expression.
7. Recognize that precision in punctuation, capitalization, spelling, and other elements of manuscript form is part of the total effectiveness of writing. (p. 246)[1]

Careful analysis of these recommendations suggests both demands for numerous competencies and requirements for many, varied writing experiences. Although everyone agrees that children should have many opportunities to write, there is disagreement about effective methods for developing and improving writing ability. The varied recommendations for teaching written composition and for diagnosing children's needs often seem confusing. Considerable research and application of research during writing institutes, however, identify effective approaches for developing writing capabilities.

VIEWPOINTS ON WRITTEN COMPOSITION

Reviews of research in composition and viewpoints of experts in writing indicate that there is no one favored approach to teaching writing. Some educators emphasize approaches that deal mostly with the process of writing, whereas others emphasize the goals of writing. Applebee (1984) argued for a more eclectic approach to teaching writing that considers both the process of writing and the resulting products of writing. Applebee stated, "We must develop models of writing that more explicitly take account of topic knowledge and of interaction between the writing process and the goals of the writing event. The separation of process and product, though perhaps a needed step in the development of our research, seems now to be a major stumbling block to future progress" (p. 591).

Following a review of the findings gained from the National Writing Project, Myers and Gray (1983) concluded that teachers use writing activities that include processing (writing focuses on a problem-solving cognitive process that includes specific stages such as prewriting, writing, and postwriting), modeling (writing focuses

---

[1]From the National Council of Teachers of English, "Forum: Essentials of English," *Language Arts* 60 (Feb. 1983): 244–48. Copyright © 1983 by the National Council of Teachers of English. Reprinted by permission of the publisher and author.

on the imitation of written samples), and distancing (writing focuses on the relationships between writer and subject, and between writer and audience).

It becomes clear from the discussions in chapters 8 and 9 that effective writing practices frequently include elements from processing, modeling, and distancing. Myers and Gray concluded their review of effective writing practices:

> Many practices also overlap. Writing to an audience has both a processing function (the words often flow more easily in social interaction) and a distancing function. Imitation of texts has a modeling function (the student is modeling the pattern in the text) and a distancing function (the student is adopting a particular point of view). The best classroom techniques probably include all three approaches. Writing groups and a classroom publication program are two ways to combine approaches. Both techniques encourage discussion of the ideas (processing), provide an actual audience that one can study (distancing), and generate a variety of student models that other students can imitate (modeling). (1983, p. 43)

It is interesting to note that the most effective mode of instruction identified by Hillocks's (1986) research does account for topic knowledge and the interaction between the writing process and the goals of the writing event. This effective mode may also include components from processing, modeling, and distancing.

Hillocks's comprehensive review of research on composition is one of the most frequently cited studies related to composition, to effective approaches, and to implications for the types of activities that should be included in the writing process. A brief review of Hillocks's findings illustrates the importance of many of the activities and principles discussed in this chapter.

First, Hillocks found that the least effective mode of instruction was one characterized by teacher lectures and teacher-dominated discussions. The next most effective mode was characterized by considerable free writing about personal interests, by writing for audiences of their peers, and by teachers' responses to whatever students write. The most effective mode was characterized by instruction (1) focusing on such processes as prewriting, composing, and revising; (2) focusing on prewriting activities that help develop the skills that will be applied during the ensuing writing; (3) focusing on specific objectives for learning; (4) emphasizing activities which help students learn the procedures for using those forms during the writing process; and (5) focusing on interaction with peers and feedback during the total writing process rather than primarily at the end of a composing activity.

In summary, the most effective instruction is characterized, according to Hillocks, by specific objectives (e.g., to increase the use of figurative language in writing), by materials and problems selected to engage the students in the processes important for that specified aspect of writing, and by activities such as small-group problem-centered discussions. In effective instruction, teachers frequently give brief introductory lessons about the principles to be studied and applied during small-group activities and then during independent work. Both peer and teacher feedback is a part of this mode of instruction.

It is worthwhile to note Hillocks's assumptions about the most effective composition approaches as we approach the development of teaching strategies for the writing process:

1. Teachers can and should actively seek to develop identifiable writing skills in students.
2. Writing skills are developed by using them orally before using them in writing.
3. One major function of prewriting activities is to develop those skills.
4. The use of skills such as generating criteria to define a concept are complex, and therefore may require collaboration with and feedback from others.

As you read the following definition of the writing process and the instructional activities that develop it, notice how findings by Hillocks and Myers and Gray and concerns by Applebee are included in the discussions and recommendations.

THEORETICAL BASIS FOR PROCESSING

Current practice in written composition is shifting away from a total emphasis on the products of writing to a concern for the processes that writers go through during the act of writing. The information-processing model for written composition is based on the work of cognitive psychologists such as Bruner (1956), Gagné (1970), and Ammon (1977). Researchers in this area ask what is involved in the act of writing, how skilled and unskilled writers approach the different tasks faced in a writing project, and how educators might assist writers as they approach and complete various writing tasks.

Cognitive psychologists view the brain as a system for processing and storing information that is very much like a computer. In this view, the computer's program is considered a map or an internal hierarchy (Simon, 1981). Cognitive research focuses on the writer's mind and researchers identify, analyze, and describe steps or stages a writer goes through during various phases of the writing process. Researchers also try to identify and describe the qualities of mental maps that help writers organize information into meaningful units.

The processes involved in writing have been identified and analyzed by observing writers and asking them to think aloud as they approach various writing tasks. The writing process model defined by Flower and Hayes is one of the most frequently cited models. Flower and Hayes (1978) defined writing as a problem-solving cognitive process, which includes making plans, operating, searching memory, using procedures that get things done, and testing or evaluating the results of the plans and operations. The model shown in figure 8–1 shows the structure of this model as defined by Hayes and Flower (1980). Notice the interactions of the processes as writers approach the problem, and then engage in planning that emphasizes generating ideas, organizing, and goal setting. At the same time, they draw from their memories to obtain knowledge of the topic, the audience, and the writing plans. As the process continues, they compose or translate with interactions still taking place. Finally, they review, evaluate, and revise their writing. Again interactions take place because the reviewing stage requires information considered during other stages in the writing process.

Many researchers who observe the writing process have developed similar definitions and models that differ only slightly from the one envisioned by Hayes and

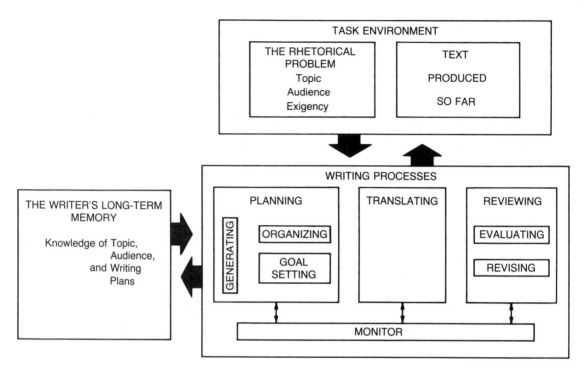

FIGURE 8–1

Stages in the writing process (from "Writing and Problem Solving" by J. Hayes and
L. Flower, 1980, *Visible Language, 14,* pp. 388–390. Reprinted by permission.)

Flower. For example, Graves (1981), following the observation of primary-age
students, defined the writing process as "a series of operations leading to the solution
of a problem. The process begins when the writer consciously or unconsciously starts
a topic and is finished when the written piece is published" (p. 4).

Britton, Burgess, McLeod, and Rosen (1979) identified the stages in the writing
process as conception, which is a long or a brief time during which a specific incident
provokes the decision to write; incubation, which is the preparatory period; and
production, which includes the actual writing. Britton et al. identified memory,
influences of written and printed sources, and revision as other important aspects of
the writing process.

Glatthorn (1982) identified four similar phases of the writing process: exploring,
planning, drafting, and revising. Notice that the terms used may be different, but that
the actual stages or phases are very similar. Later in this chapter, when we develop
instructional sequences related to the writing process, we explore ways that teachers
may help students during each of these stages.

Classroom Procedures in Processing

Instructional activities that focus on processing emphasize the steps in the writing
process. Many of the effective procedures used during this sequence of events were

developed as a result of large writing projects such as the Bay Area Writing Project and The National Writing Project. For example, Proett and Gill's (1986) sequence of events and recommended activities resulted from information compiled during the Bay Area Writing Project and the University of California Project.

Proett and Gill sequenced the steps in the writing process according to activities that teachers and students should do before students write, while students write, and after students write. The before-writing activities encourage students to discover and explore, to choose a focus for writing, to gather and record ideas, to classify information, to structure information, and to apply general truths to specific cases. The during-writing activities encourage students to record ideas into a tentative, first-draft shape; to make choices depending on the audience, the purpose, and the desired form of the writing; and to consider concerns such as choice of words and sentence structure. The after-writing activities encourage students to scrutinize their first drafts, to get peer responses from editing groups, to rethink, to revise, and to polish and proofread. Finally, they share their writings through various activities. Figure 8–2 shows this writing process as envisioned by Proett and Gill and presents the types of activities that would be appropriate for each phase. (Sections later in this chapter illustrate various activities that are part of this writing process.)

In addition to the steps in the writing process, Proett and Gill identified the following principles that guide the selection of activities under this process approach:

1. Research suggests that neither guided composition on an occasional basis nor frequent unguided writing produces marked growth in writing skill. But experience would strongly argue for the combination of frequent, clearly purposeful, guided writing experience.

2. Students need to produce more writing than a teacher of classes of twenty-five to thirty can possibly evaluate carefully. Some compromise with the practice of thoroughly marking every paper is necessary.

3. Students can profitably practice the steps in the writing process (brainstorming, focused timed writing, sentence combining, small-group editing, etc.) one at a time, but they also need to experience the entire process on a single writing project in order to develop a sense of how to work from start to finish on a paper on their own. When the steps in the process are undertaken separately, these steps need to be connected to students' understanding of the whole process so they can see how each step can be independent but also integral in the production of a final written piece.

4. Although writing is a legitimate homework activity, all elements of writing process should be undertaken at some time during class so the teacher can coach and monitor progress. Real teaching must continue after the assignment is given, through all stages of the process.

5. Writing is a tremendously complex activity involving the simultaneous application of a whole series of intellectual and physical activities. It is important in the classroom to break these activities into manageable segments so the learner has a chance to handle them successfully in some consecutive fashion (e.g., clustering for ideas, then seeking coherence through some logic of organization, then identifying controlling elements of

	Content and Idea Building			
	Observing	Brainstorming	Dramatizing	
	Remembering	Clustering	Reading	
Before	Researching	Listing	Mapping	
	Imagining	Detailing	Outlining	
	Experiencing	Logging	Watching films and other media	

Before Students Write

Content and Idea Building
Observing	Brainstorming	Dramatizing
Remembering	Clustering	Reading
Researching	Listing	Mapping
Imagining	Detailing	Outlining
Experiencing	Logging	Watching films and other media

Development and Ordering
Developing with details, reasons, examples, or incidents
Ordering by chronology, space, importance, logic
Classifying
Applying a general truth to a specific case (deductive)
Generalizing from supporting details (inductive)
Structuring by cause and effect

While Students Write

Rhetorical Stance
Voice: Who am I? How do I feel? What do I know? How sure am I?
Audience: Who's listening? What are they ready for? What help do they need? How are they feeling?
Purpose: What do I want to happen? What is likely to result? What effect do I seek? What am I willing to accept?
Form: What form fits this message?

Linguistic Choices
Specific, concrete word choice
Figurative language
Sentence structure: length, opener, verb structure
Sentence type: simple, expanded, complex, compound, cumulative
Syntax: phrase and clause structures, word order; connections and transitions; modification and subordination

After Students Write

Revision
Getting responses: editing groups, read-arounds
Raising questions, expanding, clarifying
Testing against criteria; using a rubric
Proofreading and polishing

Highlighting
Sharing	Posting	Mailing
Publishing	Filing	Reading

FIGURE 8–2

The writing process (Jackie Proett and Kent Gill, *The Writing Process in Action: A Handbook for Teachers.* Copyright © 1986 by the National Council of Teachers of English. Reprinted with permission.)

274

stance, then drafting paragraphs for peer review, and finally revising in light of valid criticism).

6. Teachers and curriculum experts have long sought the ideal linear sequence for the writing program, asking what logically must precede what in a course or a series of grade levels. This search has apparently not been successful, because there is no clearly accepted linear sequence for writing skill development, either in textbooks or curriculum guides. In fact, because writing is a holistic act, one requiring the application of the whole range of skills to every writing project, we could not expect to find a step-by-step schema to define the instructional program. Instead, a recursive model seems more appropriate, one which gives students experience with most of what a writer does early on but which then circles back periodically to give the maturing learner another, perhaps more sophisticated, experience with the skill employed earlier (pp. 2–4).

DEVELOPING COMPOSITION SKILLS THROUGH THE PROCESS OF WRITING

In this section we consider some of the activities that take place during the writing process (presented in outline form on p. 274) and try to develop an understanding of the writing process. We also include both unstructured and structured activities as students and teachers progress through the stages from prewriting, to composing, and to postcomposing activities. Before we begin developing specific aspects of the writing process for instruction, we should consider Suhor's (1984) warning: "The sequence in the Writing Process Model, as flexible as it is, cannot be followed slavishly. The sequence is not a stairway progression. . . . It should be thought of as 'looped,' not tightly linear, since certain aspects—discussion, conferencing, note-taking, revising, proofreading—may take place at several points, not just in the slots assigned on the visual model" (p. 101).

Developing Your Own Understanding of the Writing Process

Before you try to teach writing to students it is worthwhile to develop your own understanding of the writing process. The model developed by Hayes and Flower (figure 8–1) was developed by asking experienced writers to describe what they did as they approached and completed a writing task. The model developed by Proett and Gill (figure 8–2) was developed through various writing workshops and research with students. Daniels and Zemelman (1985), who developed workshops for training teachers in the writing process, stated that asking adults to create their own models of the writing process "is one of the main mechanisms by which participants formalize the discoveries they have made from their own writing experiences" (p. 78).

REINFORCEMENT ACTIVITY

Before you proceed with this chapter, try one of two approaches for developing your own model of the writing process. Remember that models do not need to be identical with those presented in this text. The models may vary somewhat depending on your purpose for writing, your topic, your background knowledge, and your audience.

1. *Assuming you have a writing assignment in one of your classes, keep a log describing all of your activities and actions as you approach the writing assignment, write various drafts, and finish the paper in the form to be handed in to the professor.*
2. *Within your language arts class, complete a writing task that is to be eventually shared with a group of your peers through your college or university newspaper. Choose for your topic an issue of concern on your campus. You may share your writing with your language arts class to gain recommendations before you submit your article to the newspaper.*
3. *After you have completed either one of these tasks, form groups within your language arts class. On large sheets of paper or on the chalkboard, list the steps you went through as you approached and completed your writing tasks. You may find it helpful to first list all of the actions that you accomplished, then try to categorize them into types of activities, and finally put them into a chronological order proceeding from first to last. Like Hayes and Flower, you may find that there are various interactions and repetitive actions performed during different stages in your writing process.*
4. *Share and discuss your models of the writing process in your language arts class. Are there any commonalities among the models? Look especially at what you did before writing, while writing the first draft, and any revising and reviewing you did before you were satisfied with your paper.*
5. *Finally, compare your models of the writing process with the one developed by a group of teachers during Daniels and Zemelman's writing workshop (see figure 8–3). What are the similarities and differences?*

Prewriting

This first stage in the writing process includes activities and experiences that stimulate ideas, that motivate students to write, that generate ideas for writing, and that focus the students' attention on the desired subject or objective of the writing. This phase allows students to become engrossed in a topic, to extend their knowledge, to heighten their awareness, and to develop appropriate thinking skills. Proett and Gill (1986) included in this prewriting phase activities such as semantic mapping or webbing, brainstorming, dramatizing, experiencing and observing, reading, and watching films. These activities are designed to help students build ideas and create content. These activities frequently stimulate students to pull knowledge from their long-term memories and to apply

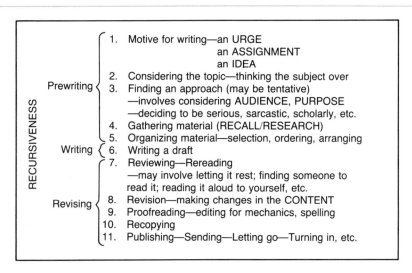

FIGURE 8–3

Writing process model (from Harvey Daniels and Steven Zemelman, *A Writing Project: Training Teachers of Composition From Kindergarten to College.* Copyright 1985 by Heinemann Educational Books. Reprinted with permission.)

    6. After you have compared your model with the one identified by Daniels and Zemelman, compare the Daniels and Zemelman model, the Hayes and Flower model, the Proett and Gill model, and your model. What are the commonalities across models? Why do you believe there are such similarities? Do all of the models have categories that take into account prewriting, writing, and postwriting activities? Are these stages important when you are teaching writing composition to a group of students? Why are they important? How could this knowledge influence your teaching?

that knowledge during the new writing experience (see schema theory, chapter 11). They also include activities such as developing through details and reasons, ordering by chronology or space, classifying, applying, generalizing, and structuring cause-and-effect relationships. These activities are designed to help students in their development and ordering of the written composition.

    Consequently, the type of prewriting activity will differ depending on the age of the students and the objectives for the writing. The activity may vary from going on a class excursion and then talking about the experience, to planning and then drawing a picture, to reading and then discussing a model story or poem written around a specific structure that students will first identify, then try to duplicate and evaluate in their own writing.

The prewriting activities may be an extension of an ongoing class activity such as a science or social studies unit that could motivate writing. The activities may result from common experiences such as birthdays, emotions, and families. Or the activities may be new experiences created by the teacher. Whatever the motivation, Graves (1977) recommended that teachers question and interview students during this precomposing activity so as to clarify their ideas, assist them in thinking about procedures for gathering information, and in making plans for future writing. Extensive oral exchanges of ideas during such activities as brainstorming and semantic mapping help students clarify and extend their ideas.

Examples of prewriting activities developed with students' drawings and prewriting   When analyzing the written production of kindergarten children, Medearis (1985) found that their writing occurred most frequently in the art center. Artwork created by students is one of the earliest motivators used to stimulate young students' writing and to help them plan for written compositions. The prewriting activities include planning for the artwork, and completing and talking about the artwork. As students draw something that interests them they have opportunities to think about the details in the picture, the person or the object in the picture, and something that they might want to disclose about the picture. Stories develop naturally from such experiences. In addition, students enjoy sharing both their artwork and their dictated or individually written stories.

Semantic maps and prewriting   The semantic maps discussed on pages 41, 147, and 148 in this text are equally good for idea building and organizing ideas during the prewriting phase of composition. Heimlich and Pittelman (1986) stated that semantic mapping helped students identify information regarding a topic of interest, identify main ideas and supporting details of the topic, organize prior knowledge onto a semantic map, and write paragraphs from the completed map.

Semantic mapping or webbing may be used prior to writing as a total class, small group, or individual activity. For a total class or smaller group activity the teacher writes the desired topic onto the center of a web on the chalkboard or on a large sheet of paper. Next, the teacher leads a discussion in which main ideas or categories are elicited from the students and drawn in circles that extend from the center of the web. Supporting ideas and details are added as the discussion continues. Then the students review the information that is on the web and write or dictate their stories.

The first processing experience with mapping and writing is usually a group activity. The teacher demonstrates how easily ideas developed during brainstorming may be charted, how easily ideas may be expanded, and how easily writing may be organized. For example, a writing activity in first or second grade might generate from a group trip to a farm. The teacher could ask for a logical title such as "We Had Fun On the Farm" and then begin the map. Questions could help students remember important events during the visit: "What animals did we see at the farm? What buildings did we visit? What machinery did we see?" The web in figure 8–4 was developed following this type of activity.

FIGURE 8–4

A web about a farm trip helps organize and expand ideas

After developing the web, the class wrote a group story following the organization presented on the chalkboard. Through this procedure, the webbing and the group-written story acted as a model for future group and individual activities. The students learned a great deal about organization, grouping ideas into paragraphs, and supporting main ideas with important details. Individual writing could be extended as students write descriptive accounts of each of these main ideas. Students could choose to write about the animals, the buildings, or the machinery. Each of these areas could be further extended using a webbing and discussing process. The semantic mapping technique is equally helpful before an individual writing activity. The web in figure 8–5 was generated by an older student whose assignment was to write a humorous story about her summer vacation.

Brainstorming Brainstorming activities emphasize quick-paced responses from the students without initial evaluation of what is said. They are especially good during prewriting because they act as brain teasers, as avenues for generating numerous ideas, and as ways of retrieving information from long-term memory.

Brainstorming activities may be structured to obtain specific types of responses and to encourage certain types of thinking. For example, Suhor (1984) recommended a two-part brainstorming/language game involving comparison and contrast. The teacher first writes the following sentences on the board: "How is a _____ like a _____ ? How are they different?" In the first step the teacher provides five to eight pairs of nouns such as oak tree, rabbit; tricycle, jet plane; zoo, schoolyard; jogging, studying. The students then provide rapid responses to the initial question. In the second step the students write single nouns on slips of paper that are placed into a grab bag from which a volunteer randomly picks two slips. The questions on the chalkboard are then applied to the randomly selected pairs. Students are then encouraged to use their imaginations for more ingenious comparisons and contrasts.

This warmup activity leads to the discussion of comparison and contrast topics that interest students such as two television shows, two cities, two countries, or two poems. This type of activity logically leads into characteristics of comparison and contrast writing and writing paragraphs that compare and contrast two subjects.

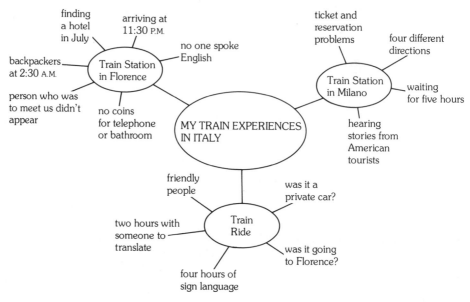

FIGURE 8–5
A web helps an older student develop her story

Brainstorming is especially good before trying to write certain types of poetry. Students might brainstorm rhyming words before writing limericks or brainstorm opposite nouns, describing words, and action words before writing poetry in the diamante formation.

Modeling during prewriting   The previously mentioned limerick and diamante illustrate the necessity for using models during some prewriting activities. To understand this requirement stop now and write a limerick, a cinquain, or a diamante. Can you do it without referring to either a model or a specific definition for each type of poetry form? Unless you have been studying and writing these forms of poetry recently, you probably cannot accomplish the task. This is equally true for elementary students.

The following models for limericks, cinquains, and diamantas were used with elementary students. The teacher first read numerous examples of limericks to students. Edward Lear's *Nonsense Omnibus* (1943, 1846), David McCord's *One At A Time: Collected Poems for the Young* (1977), and N. M. Bodecker's *A Person from Britain Whose Head Was the Shape of a Mitten and Other Limericks* (1980) provide excellent examples for this activity. Next, she and the students discussed the form of the limerick as shown below. Then she led a brainstorming activity during which time the students listed words that rhyme with several key words they might want to use in their own writing. Finally, they wrote their own limericks, shared them with their class, made any revisions after this initial exchange, and then placed them into their own class book of limericks.

Limericks are short, funny poems that allow children to work with rhyme. A limerick is a five-line poem, in which the first, second, and last lines rhyme with each other. The third and fourth lines are shorter, and rhyme only with each other. This form results in a poem that looks approximately like this:

```
Line 1._____a      a lines rhyme
Line 2._____a
Line 3._____b             b lines rhyme
Line 4._____b
Line 5._____a
```

The following limerick was written by a fifth-grade student:

There once was a computer named Zeke
Who turned out to be quite a freak.
When you told him to add,
He would call for his Dad
So his outlook in life was quite bleak.

During other class periods, the teacher followed the same reading, modeling, brainstorming, writing, and sharing formats with the following cinquain and diamante formats for poetry.

The cinquain also provides students with a form to follow for each line. These lines are not difficult to write, and they stress certain types of vocabulary. Unlike the limerick, the cinquain does not use rhyming words.

The cinquain is composed of five lines that meet the following requirements:

1. A word for a title
2. Two words to describe the title
3. Three words to express action
4. Four words to express feeling
5. The title again, or a word like it

A diagram of the cinquain would look like this:

```
                    title
              describe    title
         action    action    action
      feeling    about    the    title
                    title
```

The cinquain offers a chance for inductive teaching of such language concepts as descriptive words and action words. Their cinquains will improve if the students talk about the kinds of words that make-up each line, and brainstorm examples.

The following cinquains were written by fourth-grade students:

Clouds

Fluffy, white
Floating, moving, falling
Clouds are pillowy soft
Clouds

Scrubbrush

Low, brown
Tumbling, scratching, dying
Scrubbrush is painfully ugly
Scrubbrush

The diamante is another form of poetry that has certain requirements for each line. Deborah Elkins (1976) described this poetry:

line 1:   noun
line 2:   two adjectives
line 3:   three participles
line 4:   four nouns or phrase
line 5:   three participles indicating change
line 6:   two adjectives
line 7:   contrasting noun (p. 222)

Elkins believed teachers should help students with this format by drawing it and illustrating the type of words used, rather than merely using the terminology of the parts of speech. Thus, in line one, the teacher will have to present examples of a noun; in line two, examples of adjectives related to the first line noun; in line three, of participles, as verb forms ending in *-ing, -ed, -en.* The nouns in the fourth line contrast with the first noun, so that readers' thoughts move from the subject of the first noun to the subject of the contrasting noun on line seven. The fifth line contains three participles that correspond with the seventh-line noun, and indicate change from the first noun. The sixth line contains two adjectives that correspond with the final noun and, thus, contrast with line two. The final noun contrasts with line one.

A diagram of this poetry would look like this:

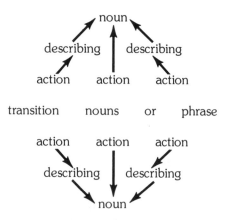

Besides illustrating the parts of speech used in this poetry, the teacher must also make clear the concept of contrasts. Contrasting nouns are necessary, in addition to contrasting adjectives and contrasting verb forms that correspond with the appropriate noun. An instructional activity leading up to this form of poetry begins with identifying the form and suggesting contrasting nouns, such as war–peace, freedom–slavery, giant–dwarf, and so forth. The class should also discuss contrasting descriptive words and contrasting action words. The students will find a thesaurus quite helpful when working with this form of poetry.

The following diamante was written by a fifth-grade student:

fantasy
magical, mysterious
dreaming, fooling, inventing
An Unidentified Flying Object
proving, trying, testing
genuine, true
reality

As you can see, fantasy and reality are opposites. "Magical" and "mysterious" describe "fantasy," and contrast with "genuine" and "true." The participles "dreaming, fooling, inventing" show action related to fantasy, and contrast with "proving, trying, testing." This student chose for the middle line the phrase "An Unidentified Flying Object," which might be either fantasy or reality, and thus provides transition between the two words.

Many of the activities listed under Proett and Gill's "Development and Ordering" phase of prewriting lend themselves to models or to brief instruction in which models of certain kinds of paragraph development, plot structure, or figurative language, for example, are first presented in short introductory lessons similar to the lessons just described for specific poetry forms. Following this, the knowledge gained during this introductory lesson is applied during the writing phase and is used to provide criteria during the revision process as students share their writings with peer groups, with editing groups, and with teachers in conferences, and then revise their writings. This is also the most effective approach identified in Hillocks's research discussed earlier. Numerous short lessons and additional models for various types of written composition are presented in chapter 9.

Listing/categorizing   Activities that encourage students to list and to categorize, to classify, or to order the information provide opportunities for both content and idea building and development and ordering of information. For example, a third-grade teacher used this strategy to help students identify and categorize the incidents that occurred before, during, and after a field trip. The class decided that division by time would be most appropriate, and developed these three categories of events:

1. Things we did to get ready for our train ride.
2. Things we did on the day of our train ride.
3. Things we did for several days after the train ride.

Next, they changed these categories into questions, and wrote each question on a separate section of the chalkboard:

1. What did we do to get ready for our train ride?
2. What did we do on the day of our train ride?
3. What did we do for several days after the train ride?

In the next step, the children answered the three questions by listing information under the appropriate question. They developed the following sentence outline:

A Ride on the Train

I. What did we do to get ready for our train ride?
   A. We studied about railroads.
   B. We called the railroad depot to get train times and ticket cost.
   C. We wrote a note to our parents asking permission to go on the train ride.
   D. We called the school bus company to get a bus to pick us up at school, take us to the depot, and pick us up at the next train depot.
II. What did we do on the day of our train ride?
   A. We rode on the bus to the train depot.
   B. We bought our train tickets.
   C. We each got a seat on the train.
   D. The conductor took our tickets.
   E. The conductor told us about the train.
   F. We saw the dining car, the observation car, and the Pullman cars.
   G. When we got off the train we looked at the engine.
   H. We rode the bus back to school.
III. What did we do for several days after the train ride?
   A. We drew pictures of the train and the depot for a bulletin board.
   B. We wrote thank-you notes to the train conductor.
   C. We wrote a story so we could remember the train ride.
   D. We read more books about trains.
   E. We learned some songs about trains.

After developing this outline, the students wrote individual stories about their trip. Children used the outline, but added their own details to make the story personal. As you can see, this activity also became an introduction to outlining and writing a longer factual report.

Another way to motivate interest, categorize ideas, and help students organize materials and know what references they need from library sources is to have children list some questions they would want answered if they were going to study a general topic. One of my graduate students, who also teaches fourth grade, used this approach before her students wrote reports about Native Americans. The students first listed a number of questions they would like to have answered, then grouped the questions into categories. Part of the list follows (the full list contained over 100 questions):

Native Americans

1. Clothing
   a. What did Native Americans wear?
   b. How did they get their materials?
   c. How did they sew?
   d. Did all Native Americans dress alike?
2. Education
   a. Did Native Americans have to go to school?
   b. What did they need to learn?
   c. Who taught the Native American children?
   d. What did they write on?
3. Houses
   a. What were tepees made from?
   b. How did they cook and wash?
   c. Did all Native Americans live in the same kinds of houses?
4. Medicine
   a. Did Native Americans get sick?
   b. Why did the medicine men dance?
   c. Did Native Americans take medicine?

5. Food
   a. What did the Native Americans eat?
   b. Where did they get their food?
   c. Was their food tasty?
   d. How did they store their food?
   e. What else did they use for hunting besides bows and arrows?
6. Tribes
   a. How did you get to be a chief?
   b. Where did the different Native American tribes live?
   c. How were they alike or different?
7. War and Peace
   a. Why did the Native Americans fight the settlers?
   b. Were any Native Americans friendly with the settlers?
   c. How did the settlers treat the Native Americans?
   d. What would happen if you went on a reservation today?

When students selected a category they wished to investigate, the questions on the list provided both a starting point for the information search and a topic for writing.

In chapter 9 we explore additional ways that may be used to stimulate students during prewriting experiences such as reading and responding to literature, biographical and autobiographical approaches to writing, and writing to learn through the content areas.

While Students Write

The prewriting experiences allow students to explore, to imagine, to consider initial structure, and to think about details and examples that will go into their writing. Writing, however, demands some additional considerations. Proett and Gill (1986) stated that

> the writing stage needs to be viewed in two very different ways. Seen in one way, it is the flowing of words onto the page, easily, naturally, rapidly. But it is also a time of making decisions, of choosing what to tell and what to leave out, or thinking about who is speaking and who is listening, of determining what order, what structure, what word works best. In some ways these functions even seem contradictory; the first needs to be fluid and fast while the other calls for deliberation and reason. The teaching task is to help the writer coordinate these two functions. (p. 11)

In this section we begin with ways of helping students make decisions about audience, purpose, form, organizational patterns, and linguistic style. Next we consider ways of helping students write that first draft rapidly.

Deciding on audience and purpose   If we want to entertain our audience, or if we want to inform the audience about a new scientific development, would we write a composition in the same way? Should we write a composition for a second-grade audience using the same words we would use for an adult audience? The answer to both of these questions is an obvious "no." To help students realize the importance of audience and purpose, they need many opportunities to write for different types of real audiences ranging from themselves, to their peers, to the teacher, to other known audiences, and to unknown audiences. (This concern is important in both processing and distancing approaches to writing.)

Unfortunately, studies in both Britain and the United States show that the audience and the purpose for writing are very narrow. For example, Britton, Burgess, Martin, McLeod, and Rosen (1979), who analyzed 2,122 pieces of writing by British students ages eleven to eighteen, found that 95 percent of the writing was allocated to writing for the teacher. Writing to unknown audiences in which the writer must consider the requirements of an unknown reader accounted for 1.8 percent of the writing; writing to known laymen or peer groups accounted for 0.3 percent; writing for oneself as audience accounted for 0.5 percent. Unfortunately, when writing for the teacher audience much of the writing was for examination purposes rather than for purposes of interaction with the teacher during the writing process.

Similar audiences for student writing were reported in Applebee's (1984) survey of writing in American schools. Applebee found that the teacher-examiner was the primary audience for student writing in all subjects. About one third of those papers resulted in some type of teacher-learner dialogue. The rest were addressed to the teacher as examiner. Only 10 percent of the teachers reported that student writing was read regularly by other students.

These findings suggest that students need many opportunities to write for varied real audiences, to write for purposes other than to be examined on their knowledge, and to write for audiences that will respond to the writing in ways that go beyond the examination purpose.

There are many opportunities in school to encourage students to write for real audiences and for different purposes. For example, they can write letters or stories to and for their families. They can have pen pals in other classrooms and even in other school districts. They can write stories and other types of articles to be displayed in the library, on bulletin boards, and in a school or classroom newspaper. They can write diaries and autobiographies for their own pleasure. They can send reports to and correspond with local historical societies, newspapers and television stations, government officials, and businesses. They can develop their own books, literary magazines, comic books, newspapers, and other types of journals. These products then are shared with larger audiences. Students themselves will think of many additional ways to create real audiences for their writing.

In addition to having many opportunities to write to real audiences, students need to realize the importance of audience characteristics and purposes when they are making decisions about writing. To help them realize the importance of audience and purpose, students can make lists of questions they need to answer about the characteristics of the audience and the purpose of the composition. A group of fifth graders developed the following list:

1. How old is my audience?
2. How much knowledge of the subject does my audience have?
3. Is the subject going to be interesting to my audience?
4. Would the subject add new knowledge to my audience?
5. Does the audience want to be informed or entertained?
6. How do I want the audience to react? Do I want them to be persuaded to my viewpoint? Do I want them to laugh? Do I want them to follow directions? Do I want them to have new ideas? Do I want them to have new information? Do I want them to be sad?

After they developed the questions, they discussed why and how the answers to the questions would influence their writing.

Another processing and distancing activity that helps students understand requirements for audience and purpose is to have the students list several different types of writing, then discuss the purpose of that type of writing, the requirements for it, and the intended audience. One group of fourth-grade students put this information into table form as shown in table 8–1.

Deciding about form   After discussing information similar to that shown in table 8–1, students have clearer ideas about possible content for each type of writing, as well as ideas about purpose and audience. They will also understand that writing can take on numerous forms. You can expand this list to include other forms of writing such as poetry, cartoons, picture story books, nonfictional textbooks, essays, interviews, reviews, songs, and sketches.

Next, bring to class examples of these various forms for writing. Share the forms with students as they consider the differences in form and why certain forms may be more appropriate for specific purposes and audiences.

It is also interesting to try to change forms and then compare content and any changes in purpose and impact on the audience. For example, after reading historical fiction about early British heroes, some of my college students wrote book reports in the form of ballads and then asked fifth- and sixth-grade students to complete a similar task. The students learned that ballads were an appropriate way of sharing heroic character information during that time period, discovered information about the ballad as a form of writing, determined that content might be similar or different depending on the form, and recognized that the influence on the audience may depend on the form chosen to deliver the message.

Other types of interesting activities with changing form include writing telegrams from news articles, writing science fiction from scientific factual information, writing diary entries from novels, writing letters from biographies, and writing cartoons from any type of literature. Whatever the motivation, students conclude the activity by comparing information, characteristics of form, and appropriateness of form for different purposes and audiences.

TABLE 8–1

Writing activity, purpose, requirements, and audience

Writing	Purpose	Requirements	Audience
1. Weather and temperature chart	1. Comparison of conditions over nine-month period.	1. Take temperature and observe clouds at same time of day. Strive for accuracy.	1. Science class.
2. Personal letter	2. Exchange accounts of experiences with friend.	2. Informal writing level. Interesting personal information.	2. A friend who knows me.
3. Invitation	3. Inform and invite audience to a program, party, etc.	3. What, who, why, when where (Proofreading important.)	3. Parents, neighbors, friends.
4. Puppet play	4. Entertain.	4. Original story that entertains; puppets.	4. Class, parents, other invited guests.
5. Diary	5. Record daily experiences and feelings.	5. Short notes on things important to me.	5. Myself.
6. News story	6. To inform about news, events, sports, etc.	6. Accuracy—who, what, when, where. Clear, brief reporting. (Editing and proofreading important.)	6. Whole school. If city paper, large unknown audience that wants information.
7. Filling out forms	7. Obtain social security card, drivers license. Send for information.	7. Accurate information, clear handwriting. (Proofreading important.)	7. People who don't know me, government.
8. Imaginative	8. Entertain, enjoy.	8. Fresh ideas, imagination, feelings.	8. Class; if published, a large, unknown audience.
9. Autobiography	9. Tell others about myself. Compile class book.	9. Facts about myself, feelings, wishes. (Proofread.)	9. Class and parents. If published, a large, unknown audience.
10. Book reports	10. Report author, title information, summary, my reactions.	10. Careful reading and writing for summary. Outlining important information. (Proofreading important.)	10. Share with class. If book review in a paper, people who might want to read the book.
11. Writing reports and projects in social studies, science, etc.	11. Share new facts and information.	11. Research several sources, collect data, develop main ideas, details. Accuracy of facts, best order for ideas. Proofreading and rewriting.	11. Presentation to class, or journal or magazine article. People wanting that information.
12. Biographies	12. Report knowledge about well-known person.	12. Research several sources. Compare information. Accuracy. Proofreading.	12. Class. If published, a large, unknown audience.
13. Writing editorials	13. Persuade someone to my viewpoint.	13. Know ways to persuade, but also use facts. Rewriting until it can clearly persuade.	13. People who read the newspaper. Some will agree with the writer, some will disagree.
14. Minutes for meeting	14. Keep a record of what happened.	14. Accurate listening and reading.	14. Members of the class, group, or organization.

Deciding about organizational patterns   Students frequently pay little attention to organizational patterns used in their own or in others' writing. The organizational patterns in our earlier example of limericks, cinquains, and diamantes were obvious and had to be followed for authenticity.

There are other, less obvious and less structured organizational patterns that help students provide logical order in their writing. Writers frequently use several of these patterns within the same paper or book. As shown by the titles of these major organizational patterns, they also lend themselves to writing in many of the content areas. Teachers may introduce these forms through modeling activities in which they model the identification of various organizational approaches, diagram paragraphs according to the organizational pattern, and lead discussions in which students identify and describe the organizational pattern. Then they can write compositions using organizational patterns that they believe are appropriate. Our example of the train ride used a chronological order of events. Other types of compositions such as those reporting historical events or those telling the autobiographies of students lend themselves to chronological ordering in which the earliest events are reported first.

A second method of organization is through the spatial concepts of direction or physical details. This order might proceed from inside to outside, from right to left, or from top to bottom. A report might discuss everything that happened in a specific area, then move on to another area. A diagram of this order might look like this:

Bottom	The Old Man in the Park
↓	↓ New and shiny black shoes
Top	↓ Bare white ankles
	↓ Cuffed, baggy, and torn pants
	↓ Faded flannel shirt
	↓ Expression of contentment
	and satisfaction

Or this:

Inside	Water distribution
↓	On the inside of the earth.
Outside	↓ On the earth's surface.
	↓ In the atmosphere.

A third method is to start with the simplest, or most familiar, ideas and progress to the most complex, or unfamiliar, ideas. The writer may begin by reviewing information or concepts the audience already knows, then progress to new information that might be harder to understand without background information. Here is a diagram of this order:

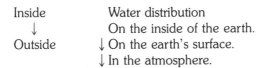

Familiar	Wind energy
↓	Trees swaying
Unfamiliar	↓ Kites flying
	↓ Windmill pumping water
	↓ Wind turbine connected to
	a generator

Social studies writers often use two other patterns of organization that allow for logical development of main ideas and supporting details. The report might first identify a problem, continue by suggesting causes for the problem, and conclude with possible solutions. A diagram would look like this:

Problem	Air Pollution
↓ Cause of problem	↓ Automobiles—Industrial waste
↓ Possible solutions of the problem	↓ Catalytic converters—Filtering smoke in factories

After identifying a problem, the report might discuss its effects rather than its causes. This report form also concludes with possible solutions to the problems; for example:

Problem	Air pollution
↓ Effect of problem	↓ Lung diseases—Plant diseases
↓ Possible solutions to the problem	↓ Catalytic converters on cars— Filtering smoke in factories

A final useful arrangement is the question-and-answer form. In this pattern, the writer asks a question and answers it, then asks another question and answers it, and so forth; for example:

Question	What clothes should you take to
↓ Answer	↓ keep dry when camping?
Question	Discuss characteristics of a
↓ Answer	good raincoat or poncho

If we look at the work of professional authors, we see that most use a combination of several organizational patterns. The patterns help authors develop their ideas and readers follow those ideas. Understanding the advantage of various organizational patterns will help students when they come to actually writing their compositions. Organizational patterns may be considered later during editorial groups and writing conferences.

Deciding about linguistic style   The sentence combining and modeling activities developed in chapter 7 help students explore various linguistic styles, develop fluency in writing, and develop a variety of appropriate solutions to linguistic choices. Teachers may use these approaches during group lessons or introductory activities, or during individual conferences when students need help making linguistic decisions.

Teachers frequently use examples from students' writing to show them how sentence combining and other types of transformations and word choices can improve fluency and clarity. Teachers may write their own compositions reflecting any student problems, place the compositions on transparencies or handouts, and lead discussions during which students identify and use various sentence-combining strategies to improve the writing.

Models from literature are good sources for observation and discussion. Through the models students can discover various techniques used by authors to create interest, fluency, and clarity.

Writing the first draft   Terms on writing process models or instructional sequences such as "composing period" and "while students write" followed by "postwriting experiences" or "after students write" are at first confusing or even misleading. The objective of this phase of the writing process is the fairly rapid writing of ideas generated, developed, and refined during the prewriting phase of the process. The objective is not to have a finished composition; that objective takes place in the postwriting or revision phase of the process. (All writing, however, does not need to lead to a finished composition. Students frequently choose from among their writings selections those they would like to revise.)

Your students may find it helpful if you show them how you write your own first drafts. For example, my own first drafts are written with pencil on legal size pads. I leave enough space between lines to make changes in word choices or to rewrite sentences. I draw arrows if I change my mind about the ordering of sentences in paragraphs or the ordering of paragraphs. I frequently insert ideas between sentences, on the next page, or even on the back of the page. Again arrows show me where to locate and place this material. I erase and cross out words, sentences, and even paragraphs. If I cannot think of a specific word or need to add a specific reference, I draw a line. The word will come later; the reference must be found and added. I draw lines if I need to check spelling or a convention in punctuation or grammar. In this case I draw a line under the questionable word or punctuation, place an abbreviation (sp., gr., or pn.) above the line, and continue with my flow of ideas. The lines and abbreviations are cues for areas that must be checked during revision. From an example such as this, students should see that first drafts are frequently tentative, often rather messy, and require editing and revisions. They should also have an understanding of some of the techniques that they may use in developing their first rough drafts.

I use semantic mapping or webbing to develop my own ideas and to organize my writing, so I frequently go back and forth between the web and the first draft. As I write, I ask myself questions such as: Did I include those ideas and develop those points? What should come next in the development of the writing? Do I have more information on my web than I can develop thoroughly in a paper? Would it be better if I omitted this portion of the web? If I omit this portion, do I need to change my introduction or my conclusions? The web in this example helps jog memory, improves organization, and provides key evaluators for the revision process. The collaboration between the web and the written composition may also suggest that the original ideas are not working, that major modifications must be made, and that it may be necessary to start over. As we all know, writing is not easy.

During this writing period the teacher encourages writers, provides support, and asks and answers questions that help students clarify their ideas.

After Students Write

As suggested by our earlier definition of the writing phase, there are still major portions of the writing process remaining. During this final phase students develop self-evaluative skills as they reread, review, and revise their compositions. They hopefully have opportunities to share their writings with peer editing groups and with teachers during teacher conferences. They proofread their writing and share their finished examples orally or through various publishing and presentation techniques.

Revising Experienced writers use several techniques that are equally effective with less experienced student writers. Experienced writers may reread their compositions aloud to check for fluency and flow of ideas. They frequently put their writing aside and try to approach it later from the viewpoint of the intended audience. They may share it with a critical audience who is asked to respond to clarity of ideas and to ask questions or to give suggestions about areas or points that need to be changed. Through this process experienced writers may add, delete, change, or rearrange ideas. If the written project is very important to the experienced writer, the writer may go through this revision process several times before the paper is ready for the final proofreading and sharing with the intended audience.

Teachers can duplicate these same areas of the revision process in the classroom. Sharing your own writing experiences with students will place you in the role of teacher model. It is advantageous to demonstrate to students how you changed a written project because you read it aloud to yourself and then shared it with another individual or group. This activity helps students understand the power in this phase of writing.

Conferencing with peers and with teachers is considered very important during the revision process. Hoskisson and Tompkins (1987) identified and briefly described various types of conferences that may be used during this stage in the writing process. Table 8–2 presents these types of writing conferences. We consider in greater depth both the teacher-student and the peer editing conference.

Teacher-student conferences Conferences, whether they are with peer editing groups or with teachers, provide opportunities for students to share their writing during different stages of the writing process, to ask and to answer questions, and to receive feedback and suggestions. Three types of teacher-student conferences are mentioned in the literature. There are informal conferences in which teachers provide immediate help for students who require assistance; there are regularly scheduled conferences for individual students; there are scheduled conferences for groups of students in which short lessons pertaining to writing conventions are taught.

Graves (1983) recommended both informal conferencing as the teacher moves around the room, stopping at individual desks, and regularly scheduled conferencing when individual students come prepared to discuss their work, to ask and answer questions, and to receive both motivation and guidance. Graves believed that teachers should have regularly scheduled conferences with students at least once a week.

Sharing writing with peers helps clarify ideas.

Teachers may structure their conferences on general open-ended questions based on what the writing is about, where the author is in the work, and the kind of help that might be needed, or they may structure the conference around questions dealing with specific details, paragraph organization, and other purposes for the writing assignment. The types of questions and the discussion relating to the writing can easily change depending on where the student is in the writing process.

General, open-ended questions include: "What do you like best about this piece of writing? What questions would you like to ask me? What surprised you as you were writing this draft? Are you having any problems? Which part is giving you problems? What would you like to do with this paper in your next draft?"

TABLE 8-2

Types of writing conferences.

1. *On-the-Spot Conferences*
   The teacher visits briefly with a student at his or her desk to monitor some aspect of the writing assignment or to see how the student is progressing. The teacher begins by asking the student to read what he or she has written and then asks a question or two about the writing. Usually the teacher has several questions in mind before having the conference with the student. These conferences are brief; the teacher spends less than a minute at a student's desk before moving away.

2. *Drafting Conferences*
   Students bring their rough drafts and meet with the teacher at a table set up in the classroom specifically for that purpose. Students bring to the conference examples of specific writing problems that they would like to talk to the teacher about. These short, individual conferences often last less than 5 minutes so it is possible for a teacher to meet with 8 to 10 students in 30 minutes. Often students sign up for these conferences in advance.

3. *Revising Conferences*
   A small group of students and the teacher meet together in writing group conferences. Students read what they have written and ask for specific suggestions from classmates and the teacher about how to revise their compositions. These conferences offer student writers an audience to provide feedback on how well they have communicated. These small-group conferences last approximately 30 minutes or as long as necessary for each student to share his or her writing. Many elementary teachers schedule these conferences periodically in place of reading groups. After all, the purpose of reading and writing groups is virtually the same: to read and react to a piece of writing.

4. *Editing Conferences*
   Students meet with the teacher for an editing conference. In these individual or small-group conferences, the teacher reviews students' proofread compositions and helps them to correct spelling, punctuation, capitalization, and other mechanical errors. The teacher takes notes during these conferences about the problems students are having with mechanical skills in order to plan individual, special instruction conferences.

5. *Instructional Conferences*
   Ten- to fifteen-minute conferences are scheduled with individual students to provide special instruction. Teachers prepare for these conferences by reviewing students' writing folders and by planning instruction on one or two skills (e.g., capitalizing proper nouns, using commas in a series) that are particularly troublesome for individual students.

6. *Conferences with Classmates*
   Students meet with one or two classmates to ask for advice, to share a piece of writing, or to proofread a composition in much the same way that they hold a conference with the teacher. In these student conferences, students are expected to maintain a helpful and supportive relationship with their classmates.

7. *Class Conferences*
   Conferences with the entire class are held periodically to write and revise class collaboration compositions, to practice new conferencing strategies, and to discuss concerns relating to all students. Sometimes, these conferences are planned, and at other times they occur spontaneously as the need arises.

SOURCE: Kenneth Hoskisson and Gail E. Tompkins, *Language Arts: Content and Teaching Strategies* (Columbus, Oh.: Merrill Publishing Co., 1987).

Questions and discussions dealing with specific elements might include the following examples around characterization in a story: "I like the way you described the dog in your story. I could really *see* and hear him. I knew he was unhappy because he was lost. How do you think you could make your readers know more about his master, Davey? How do you think you could make your readers see him and understand how he is feeling because his dog is lost?" Similar comments and questions could lead to discussions on ways to work with setting and plot development.

Other topics for discussion might focus on paragraph development, vocabulary choices, transitional phrases or sentences, story conclusions, and literary form. The conference might focus around other objectives of the lesson and relate to specific lessons taught as part of the prewriting phase such as the organizational patterns presented earlier on pages 289–290 for chronological order, familiar to unfamiliar, problem to cause of problem, problem to effect of problem, and question to answer. The conference might use the forms for poetry introduced to help students focus on their writing. The conference might use the questions asked about audience on page 287 or the conference might focus on the collaboration between a semantic map and the various stages of the revised drafts. If the conference is focusing on proofreading, the teacher and student might explore if all of the words that were underlined were checked for spelling or if conventions of writing were checked.

As suggested by the variety of questions and discussion topics, during writing conferences teachers play varied roles depending on the needs of the students. Harris (1986) stated that teachers encourage student writers by playing the role of coach, commentator, counselor, listener, and diagnostician.

As the coach, the teacher uses comments such as "You've done a good job describing the boy in the picture. Can you do the same thing when you describe the girl?" The coach, according to Harris, "Uses comments to help writers identify what they have to watch out for, what they have to work harder on, what has been working well for them, and what to build on" (p. 35).

In the commentator role, the teacher helps students see where they are in the writing process, and where they need to move in the next draft. Comments include statements such as "Good. The first draft helped you focus your subject on _____ . You are now ready to try your idea to add more detail about _____ ."

In the counselor role, the teacher considers the whole person, including the writer's previous experience, motivation, prior learning, attitudes, and interests when interacting with the student. This role is especially important when the teacher is trying to help students counteract writing blocks or other problems that are interfering with writing.

As a listener the teacher plays several changing roles (Murray, 1979). At prewriting conferences the teacher is a friendly listener who is interested in each student as an individual with ideas to express. As students develop drafts, the teacher-listener becomes a fellow writer who shares writing problems as writers try to focus, shape, and form the written composition. As the writing progresses, the teacher listens closely to the language of the paper. Throughout this sequence, the teacher listens carefully to discover what the student needs to know.

As diagnostician, the teacher probes, asks questions, engages in exploratory conversations, and encourages students to make final decisions about their writing. Harris warned, however, that in this role the teacher should not "unwittingly assume total control, wresting from the student all responsibility for what happens and closing off all avenues for student participation. When this happens, chances for students to improve their writing decrease dramatically" (p. 40). This final role may be the most difficult role as students may expect teachers to provide the major input. The purpose of the teacher-student conference is to encourage active writers who participate in the discussion and develop independent writing skills. This role is not met when students respond passively to teacher comments without becoming involved in the process.

Peer editing groups   Peer editing groups are considered effective in the revision process because groups of three to five students provide a known audience for student writers; they encourage writers to sense the impact of their work, to answer questions about their writing, to clarify points, and to revise writing to meet the needs of real audiences. Moreover, they allow student editors to hone their own editing and revising techniques, to ask clarifying questions about content, and to apply their new skills to their own writing.

Before peer writing groups can function, teachers need to develop trust between students so that they will be willing not only to share their work with others, but to provide critical dialogue when it is their turn to respond to writing.

The first editing group should be modeled by the teacher as participant so that students understand that the role of the editing group is to respond to specific things that they like about the paper, to ask questions about parts that are unclear, and to consider what could be added or changed to clarify or increase enjoyment and understanding. The teacher may use sample papers to encourage this interaction and to show students how to respond effectively as both presenter and editor. The complexity of the tasks expected of the peer editing groups and even the length of such activities depends on the ages, writing capabilities, and understandings of the students. Younger elementary students may focus on what they like about a paper, ask questions to clarify details in the paper, and respond to the effectiveness of revisions. Older students may become much more involved in the connections among the objectives of the lesson, form of the writing, and evaluation of the drafts during stages in the revision process. The following recommendations for peer editing groups come from Daniels and Zemelman's work with older students during the National Writing Project.

Daniels and Zemelman (1985) recommended that peer editing groups be formed by tentatively ranking students from 1 (best prepared for writing) to 25 or whatever number is in the total class (least prepared for writing). Students are then listed numerically in three columns. The teacher forms groups by selecting students across columns. In this way each group includes a cross section of above average, average, and poor writers. These groups may be altered to balance students according to sex or to make changes because of inability to work together. The groups may be kept intact for longer periods or they may be re-formed monthly.

Several guidelines improve the function of these groups. First, the peer editing groups are most effective when writers read their papers aloud so that both writers and listeners hear the effect and clarity of the language. In addition, group members as well as writers should sign the drafts presented and discussed in the peer editing groups. These signatures show that both the author and the group members share responsibility for the writing. Furthermore, peer editing groups may benefit from criteria check lists or forms that relate to the specific kind of writing assignment being revised. These forms may be as simple as spaces to show examples of what the editing group liked about the writing and where clarification could take place to improve the writing. Or, the guides may be more detailed to consider elements in a writing project such as developing characterization, creating plot development, developing believable settings, or creating cause and effect relationships. Scales like the Holistic Scoring Scale (page 303) may help editing groups focus on important aspects of the writing task.

Daniels and Zemelman suggested that editing groups should meet for three different purposes. First, they should meet to discuss the assignment. Next, they should meet to focus on editing each other's rough drafts. Finally, they should meet to proofread the final drafts. Daniels and Zemelman provided the following weekly schedule incorporating these three functions of the peer editing groups and also included focus lessons on particular elements in writing:

Mondays
1. Rough drafts are due for assignment given the previous Friday.
2. Groups meet immediately and begin work. Each paper is read aloud by its author. The other group members ask questions about the content of the paper, which the author records on his rough draft, without actually answering any of them at this time. Both positive comments and constructive suggestions are expected at this time. Students have the entire period for this purpose and are instructed to think over the questions that were asked and make any changes that seem appropriate before coming to class on Tuesday.

Tuesdays
1. "Final" drafts are due.
2. Ten-minute free writing.
3. Groups meet again, this time actually exchanging papers and looking over them silently. Their focus today is on content and grammar/mechanics. No marks are made on another person's paper; instead, group members are to point the problems out and allow the author to decide what he/she wants to do about them.
4. Papers are turned in, signed by the author, and cosigned by the readers. Those who wish to do further rewriting simply sign a list on the teacher's desk and are automatically granted an extension.

Wednesdays
1. Ten-minute free writing.
2. People who have done further rewriting may have their group members go over their new "final" drafts instead of doing the ten-minute free writing on this day.
3. Focus lesson subject matter, writing process, or grammar item.

Thursdays
1. Ten-minute free writing.
2. Complete the focus lesson.

Fridays
1. Student-evaluation (alternating between self-evaluation and group evaluation).
2. Vocabulary due.

REINFORCEMENT ACTIVITY

Simpson (1986), a seventh-grade language arts teacher, described the roles she plays while her students are involved in the writing process. Read the following roles and decide during which phase or phases of the writing process she assumes each of these roles and describe her actions as if you could observe her classroom:

1. *Serving as students' most ardent admirer and astute critic.*
2. *Teaching writers to internalize the reading role so that they communicate their intended meaning to others.*
3. *Aiding students so that they write in a tone that readers recognize as genuine.*
4. *Protecting authors as they offer their writings to the audience for comments and helping them screen suggestions and comments as to the appropriateness of the suggestions.*
5. *Modeling appropriate group responses, questions, and techniques.*
6. *Helping students focus their attention on the task that is appropriate for their stage in the writing process.*
7. *Integrating skill instruction within the context of revision.*
8. *Monitoring individual progress.*
9. *Maintaining the human connection.*

3. Students are allowed to work independently, reading their novels, working on the next paragraph, or revising papers. Individual conferences are held at this time. (pp. 167–68).[2]

Sharing writing   If the audience was carefully identified in the earlier part of the writing process, the students now share their writing with this genuine audience. Letters, editorial comments, and reactions to books are mailed. Poetry and short stories are read to peers. Stories are placed in the library or posted on the bulletin board. Class newspapers and magazines are published. Writing is filed in folders for later review.

Students enjoy making permanent collections of their poems and stories. One teacher had each child develop an accordion-pleated poem book. To construct the books, the students folded large sheets of heavy drawing paper in half, connected several sheets with tape, and printed their poems and an accompanying illustration on each page.

Other classes have made their own books by constructing covers in various appropriate shapes, cutting paper to match the shapes, and binding the cover and

---

[2]From Harvey Daniels and Steven Zemelman, *A Writing Project: Training Teachers of Composition from Kindergarten to College.* Copyright 1985 by Heinemann Educational Books. Reprinted with permission.

pages together. A group of second graders placed Halloween stories inside a jack-o-lantern book, fourth graders wrote city poems and stories inside a book resembling a skyscraper, and third graders placed humorous mythical animal stories and poems inside a book resembling a Dr. Seuss beast.

Although it is not necessary to extend writing to any other activity, students often enjoy using their own writing for choral-reading and readers theatre arrangements, art interpretations, or dramatizations.

EVALUATING WRITING

Researchers in the writing process such as Flowers, Hayes, and Glatthorn contended that it is possible and advantageous to identify the problems of weak writers in terms of their procedures during the writing process instead of only focusing on their errors.

Studies that analyze the composing process of skilled and unskilled writers provide important insights for language arts teachers. In addition, the studies identify characteristics of writers who might benefit from activities stressing the writing process. Glatthorn (1982) synthesized the results from several studies and reached several conclusions about the writing processes of skilled and unskilled writers. The following observable characteristics describe skilled and unskilled writers' actions during each of Glatthorn's four stages of the writing process:

1. Exploring: The beginning stage in which writers think about a topic, discover what they know, collect information, consider audience, and reflect about approaches.

Skilled Writers	*Unskilled Writers*
Take time to explore the topic and use many exploring strategies.	Do little exploring and do not consider it important or useful.

2. Planning: The stage in which tentative decisions are made. Writers make preliminary choices about content, organization, and proportion.

Skilled Writers	*Unskilled Writers*
Take time to plan and use a variety of planning techniques such as listing, sketching, and diagramming.	Do little planning either before writing or while writing.

3. Drafting: The psychomotor process by which the words are written and the sentences crafted.

Skilled Writers	*Unskilled Writers*
Write in a way that is less like speech, show more sensitivity to the reader, and usually spend more time drafting.	Write in a way that imitates speech, show less concern for the reader, and seem preoccupied with spelling and punctuation.

4. Revising: The stage in which the writer evaluates what was written and makes changes in the form and content.

Skilled Writers
Either revise very little or revise extensively. When they make extensive revisions, they focus on larger issues of content and reader appeal.

Unskilled Writers
Either revise very little or revise only at the surface and the word levels. They see revising as error hunting and copying over in ink.

Students who exhibit the characteristics of unskilled writers identified by Glatthorn might benefit from a processing approach to composition. In addition, Myers and Gray (1983) stated that writers who lack fluency and who have difficulty generating or expanding ideas should engage in processing activities. One of the most effective ways to assess growth is through the accumulation of writing samples. Each child's examples may be kept in a folder, compiled over long periods of time, so that both teacher and child can look at and evaluate changes in the writing. If changes are not occurring, the teacher should suspect that the instruction is probably not appropriate for that particular child.

Phelps (Gleason and Mano, 1986), an advocate of teaching the writing process, emphasized that writing instructors should analyze drafts in terms of potential for writing development, not in terms of errors. Such an analysis looks for clues to changes going on within the writing and considers ways to help students proceed toward the next step in their writing development.

Effective evaluation should center around the purposes the teacher identifies for each writing activity. For example, if the objective of a poetry-writing lesson is for children to become aware of their own uniqueness through self-expression, the instruction is effective to the degree that children write their own thoughts and emotions in poetic form. Because this uniqueness is individual, poetic originality should be evaluated by comparing a child's poetry with her previous poetry, not some standard for all children. Rupley (1976) suggested that, if the teacher's purposes relate to developing an interest in writing, fostering creative thought, or finding pleasure in writing, evaluation should also relate to these purposes.

In order to analyze the influence of both audience and purpose, Goodman (1985) recommended that teachers keep copies of children's letters written to different audiences for different purposes. These letters can be analyzed according to the following considerations: the appropriateness of the language for the specific purpose, the degree to which children change their language and style to meet the needs of the specific audience, the increase in conventional spellings, the changes in grammatical complexity of the sentences, and the concern for legibility.

In Williams's evaluation of teaching composition in open classes in England, she concluded that merely dating the children's materials allows them to see their own improvement over a period of time. "The teacher could not have developed a more motivating feedback system," she said, "than this chronological record of the student's composition efforts. Handwriting and quantity and quality of work had all improved noticeably" (1978, p. 3).

The writing folder should include samples of many types of writing. The teacher may compile a checklist of some of the areas to observe for written composition. Such a checklist would look something like table 8–3.

TABLE 8–3
Checklist for informal evaluation of writing

	Yes	Some-times	No
1. Sentences express his or her ideas.			
2. Ideas flow from one sentence to another.			
Purpose and Audience:			
3. Understands the purpose for an audience when writing.			
4. Age, knowledge, and interest of audience influenced the composition.			
5. Purpose of the composition influences the writing.			
Organizing Ideas—Paragraphs:			
6. Groups and classifies related ideas.			
7. Understands need for and chooses main idea that unifies a paragraph.			
8. Supports main ideas with necessary facts and details.			
9. Puts main idea and supporting details into logical order.			
Organizing Ideas—Longer Compositions:			
10. Narrows subject to one that can be realistically covered.			
11. Gathers ideas and information before trying to write composition.			
12. Understands and develops important questions that need to be asked on topic.			
13. Organizes important questions and supporting data into logical sequence.			
14. Progressing toward ability to write composition with interesting introduction, main ideas and supporting details, and logical conclusion.			
Rewriting:			
15. Reads orally while developing composition and makes changes to improve clarity of writing.			
16. Responds to teacher questions and makes appropriate writing changes.			
17. Reads composition silently while developing it and makes changes.			
Vocabulary:			
18. Has extensive writing vocabulary.			
19. Uses context clues to develop new vocabulary understanding.			
20. Uses precise vocabulary in factual writing.			
21. Uses few slang expressions in factual writing.			
22. Does not rely on trite expressions.			

Holistic scoring scales are frequently used by teachers for scoring and evaluating compositions. The six-point scale shown in Table 8–4 was developed by Pritchard (1987).

REINFORCEMENT ACTIVITY

Collect several written compositions from the same child and evaluate the writing using the "Checklist for Informal Evaluation of Writing" (table 8–3) and for "The Holistic Scoring Scale." Tabulate your findings.

SUMMARY

Current research in written composition is analyzing both major approaches to teaching written composition and effective approaches for teaching written composition. Studies of the writing process and corresponding models of the writing process focus on tasks writers do as they identify the topic for a writing project, generate ideas, plan and organize their writing, write rough drafts, review and evaluate their writing, and revise their writing.

This chapter focused on developing composition skills through the writing process by emphasizing activities that help students during each phase of the writing process. During the prewriting phase, activities emphasized stimulating ideas, motivating students to write, generating ideas, and focusing attention on desired subjects or objectives. Examples of prewriting activities discussed included drawings, semantic mapping, brainstorming, modeling, and listing and categorizing. Activities designed to improve students' progress during the writing phase included deciding on audience and purpose, deciding about form, deciding about organizational patterns, deciding about linguistic style, and writing the first draft. The after-writing-phase activities focused on revising, teacher-student conferences, peer editing groups, and ways of sharing writing with an audience.

ADDITIONAL WRITING PROCESS ACTIVITIES

1. *Observe an elementary student approach a writing task. Try to identify the activities the student goes through in approaching and completing the task.*
2. *Interview an elementary, a middle-school, and a high-school language arts or English teacher. Ask them how often they believe students should write, what they believe is the purpose for writing, and who they believe is the appropriate audience for writing. Share your findings in your language arts class. How do these results compare with both the American and the British studies discussed in this chapter?*
3. *Develop a file of ideas and resources that you could use during the prewriting stage of the writing process.*

Point Score	Characteristics
6	Has a thesis Concrete details effectively used Fluent in words and ideas Varied sentence structure Satisfactory closing statement Generally clear mechanics
5	Has a central idea Specific facts, details, or reasons Consistent development Less insightful, imaginative, concrete, or developed than a 6 Generally clear mechanics, errors do not interfere with overall effectiveness
4	Has several clear ideas Relevant and specific details Evidence of fluency, but not of unified development May be overly general or trite May have simple sentence structure or vocabulary Mechanical errors do not affect readability
3	Has at least one idea, few, if any supporting details Less fluent, developed, or detailed than a 4 Sentences, vocabulary, and thought may be simplistic Mechanical errors do not affect readability
2	No thesis Has a sense of order, but order may be only that of plot summary Fluency and thought are minimal Has at least one relevant idea May have many mechanical errors but paper is readable
1	No thesis and, or course, no support for thesis No sense of organization Simplistic or vague language May be unreadable due to spelling, handwriting, or other mechanical problems

TABLE 8–4

Holistic scoring scale for writing

SOURCE: "Effects of Student Writing of Teacher Training in the National Writing Project Model by R. J. Pritchard, 1987, *Written Communication, 4,* p. 58. Printed with permission.

4. *Develop a modeling activity that would be appropriate during the prewriting or the writing stage. Share your activity with your language arts class.*

5. *Lead a prewriting activity in which you help a group of children or an individual child develop a semantic map or web that will be used for a writing assignment.*

6. *Collect examples of various forms such as diaries, editorials, and short stories that are appropriate for writing projects. List the characteristics of each form and identify writing assignments in which the forms would be appropriate.*

7. *If possible, visit a classroom during a teacher-student conference or during peer editing groups. What is the focus of the activity? How do teachers or peers respond to papers? How do students respond when they present or discuss their writings? What questions do they ask? Can you identify where students are in the writing process? What are the major writing problems experienced by the students? How are students receiving help with those writing problems?*

8. *Develop a file of different ideas for sharing writing with an audience. Your file might include directions for making books, for developing writing bulletin boards, and for oral presentations emphasizing original student work.*

BIBLIOGRAPHY

Ammon, Paul. "Cognitive Development and Early Childhood Education: Piagetian and Neo-Piagetian Theories." In *Psychological Processes in Early Education,* edited by H. L. Hom and P. A. Robinson. New York: Academic Press, 1977.

Applebee, Arthur N. "Writing and Reasoning." *Review of Educational Research* 54 (Winter 1984): 577–96.

Boyer, Ernest. Conference on Teaching Excellence. College Station, Tex.: Texas A&M University, 1983.

Britton, James; Burgess, Tony; Martin, Nancy; McLeod, Alex; and Rosen, Harold. *The Development of Writing Abilities (11–18).* Schools Council Research Studies, London: Macmillan Co., 1979.

Bruner, Jerome; Goodnow, Jacqueline; and Austin, G. A. *A Study of Thinking.* New York: John Wiley & Sons, 1956.

Daniels, Harvel, and Zemelman, Steven. *A Writing Project: Training Teachers of Composition from Kindergarten to College.* Portsmouth, N.H.: Heinemann, 1985.

Elkins, Deborah. *Teaching Literature: Designs for Cognitive Development.* Columbus, Oh.: Merrill Publishing Co., 1976.

Flower, Linda, and Hayes, John. "The Dynamics of Composing: Making Plans and Juggling Constraints." In *Cognitive Processes in Writing,* edited by Lee W. Gregg and Erwin R. Steinberg. Hillsdale, N.J.: Lawrence Erlbaum Associates, 1978.

Gagné, Robert. *The Conditions of Learning.* 2d ed. New York: Holt, Rinehart & Winston, 1970.

Glatthorn, Allan A. "Demystifying the Teaching of Writing." *Language Arts* 59 (October 1982): 722–25.

Gleason, Barbara, and Mano, Sandra. "Issues in Development: An Interview With Wetherbee Phelps." *The Writing Instructor* 5 (Winter 1986): 45–50.

Goodlad, John. *A Place Called School: Prospects for the Future.* New York: McGraw-Hill Book Co., 1987.

Goodman, Yetta. "Kidwatching: Observing Children in the Classroom." In *Observing the Language Learner, edited by Angela Jaggar and M. Trika Smith-Burke, pp. 9–18. National Council of Teachers of English, 1985.*

Graves, Donald. *A Case Study Observing the Development of Primary Children's Composing, Spelling, and Motor Behaviors During the Writing Process.* Final Report, NIE Grant No. G–78–0174. Durham, N.H.: University of New Hampshire, 1981. ED 218 653.

_____ . "Research Update—Language Arts Textbooks: A Writing Process Evaluation." *Language Arts* 54 (October 1977): 817–23.

_____. *Writing: Teachers and Children at Work.* Exeter, N.H.: Heinemann, 1983.

Harris, Muriel. *Teaching One-To-One: The Writing Conference.* Urbana, Ill.: National Council of Teachers of English, 1986.

Hayes, John, and Flower, Linda. "Writing as Problem Solving." *Visible Language* 14 (1980): 388–90.

Heimlich, Joen E., and Pittelman, Susan D. *Semantic Mapping: Classroom Applications.* Newark, Del.: International Reading Association, 1986.

Hillocks, George. *Research on Written Composition: New Directions for Teaching.* Urbana, Ill.: National Conference on Research in English, 1986.

Hoskisson, Kenneth, and Tompkins, Gail. *Language Arts: Content and Teaching Strategies.* Columbus, Oh.: Merrill Publishing Co., 1987.

Kruh, Nancy. "The Challenge: Teach Two Classes of Kids How to Write." *The Dallas Morning News.* Sunday May 22, 1988, section F, pp. 1, 10. Additional news articles on subject F, pp. 8–9, 8, and 11.

Mancuso, S. "Audience Awareness in the Persuasive Writing of Gifted and Nongifted Fifth Grade Students." Doctoral Dissertation, Louisiana State, 1986. *Dissertation Abstracts International* 47, 03A. (University Microfilms No. 86–10,649)

Medearis, L. "The Written Production of Four Kindergarten Children in a Whole Language Classroom." Doctoral Dissertation, North Texas State University, 1985. *Dissertation Abstracts International* 46, 09A. (University Microfilms No. 85–25,570)

Murray, Donald. "The Listening Eye: Reflections on the Writing Conference" *College English* 41 (1979).

Myers, Miles, and Gray, James. *Theory and Practice in the Teaching of Composition: Processing, Distancing, and Modeling.* Urbana, Ill.: National Council of Teachers of English, 1983.

National Council of Teachers of English. *Language Arts* 60 (February 1983): 244–48.

O'Hare, Frank. *Sentence Combining: Student Writing without Formal Grammar Instruction.* Urbana, Ill.: National Council of Teachers of English, 1971.

Olson, David R. "Children Language and Language Teaching." *Language Arts* 60 (February 1983): 226–33.

Pritchard, Ruie Jane. "Effects on Student Writing of Teacher Training in the National Writing Project Model." *Written Communication* 4 (January 1987): 51–67.

Proett, Jackie, and Gill, Kent. *The Writing Process in Action: A Handbook for Teachers.* Urbana, Ill.: National Council of Teachers of English, 1986.

Rupley, William H. "Teaching and Evaluating Creative Writing in the Elementary Grades." *Language Arts* 53 (May 1976): 586–90.

Simon, Herbert A. *The Sciences of the Artificial.* 2d ed. Cambridge, Mass.: The M.I.T. Press, 1981.

Simpson, Mary K. "What Am I Supposed to Do While They're Writing?" *Language Arts* 63 (November 1986): 680–84.

Strange, R. "An Investigation of the Ability of Sixth Grade Students to Write According to Sense of Audience." 1986 *Dissertation Abstracts International,* 47,05A. (University Microfilms No. 86–17,066)

Suhor, Charles. "Thinking Visually About Writing: Three Models for Teaching Composition, K–12," in *Speaking and Writing, K–12,* edited by Christopher Thaiss and Charles Suhor, pp. 74–103. Urbana, Ill.: National Council of Teachers of English, 1984.

Williams, Laura E. "Methods of Teaching Composition in Open Classes—England, Canada, and the United States." *Innovator* 9 (January 1978): 1–3.

CHILDREN'S LITERATURE REFERENCES

Bodecker, N. M. *A Person from Britain Whose Head Was the Shape of a Mitten and Other Limericks.* London: Dent, 1980.

Lear, Edward. *Nonsense Omnibus.* New York: Warne, 1943 (from earlier editions dating from 1846).

McCord, David. *One at a Time: Collected Poems for the Young.* Boston: Little, Brown & Co., 1977.

Chapter Nine

EXPRESSIVE WRITING
Activities Designed to Encourage Expressive Writing

IMAGINATIVE AND POETIC WRITING
Why Teach Poetic, Imaginative Writing? ▪ Developing the Creative Writer ▪ Improving Specific Aspects of Imaginative, Poetic Writing

EXPOSITORY WRITING
Writing a Report ▪ Writing and the Content Areas ▪ Precise or Summary Writing ▪ Writing Book Reports

DEVELOPING A UNIT BASED ON COMPOSITION

SUMMARY

After completing this chapter on expressive, poetic, and expository writing, you will be able to:

1. Define terms related to expressive, poetic, and expository writing.
2. Identify and develop writing approaches that encourage expressive writing.
3. Identify and develop writing approaches that encourage imaginative and poetic writing.
4. Describe the importance of the environment for developing the creative writer.
5. Describe and develop writing approaches that improve the writer's ability to create plot development, setting, and characterization.
6. Describe and develop writing approaches that develop and improve expository writing.
7. Develop a unit that focuses on a composition approach to the study of biography.

Composition:
Expressive, Poetic,
and Expository Writing

*I*n chapter 8 we considered modes of instruction and the stages in the writing process as students proceed from prewriting, to writing, to revising, to sharing. In this chapter we emphasize specific types of writing that may be developed within those previously discussed stages of the writing process.

Specifically, we focus on expressive writing, poetic or imaginative writing, and expository or transactional writing. Our coverage goes from the expressive function in which young children, especially, write as they speak, to the imaginative and poetic function in which we use language as an art medium and try to concurrently inform and entertain, and to the expository or transactional function in which we inform, record, report, and explain. Within this chapter we consider ways that writing may be used to entice learning across the disciplines.

As in the writing process, there are no absolute divisions among these three types of writing. Expressive writing may be very poetic, just as expository writing may have a strong sense of poetic language. As you approach this chapter, the divisions are not as important as the ideas for stimulating writing.

EXPRESSIVE WRITING

Expressive writing is writing that is very close to speech and, consequently, very close to the writer. It is relaxed, intimate, and comprehensible to ourselves or to others who know us very well. It reveals the writer and verbalizes the consciousness. It is important, according to Britton, Burgess, Martin, McLeod, and Rosen (1975) because "not only is it the mode in which we approach and relate to each other in speech, but it is also the mode in which, generally speaking, we frame the tentative first drafts of new ideas: the mode in which, in times of family and national crisis, we talk with our own people and attempt to work our way towards some kind of resolution" (p. 82).

Expressive writing, with its close relationship to speech, is considered the most accessible form of writing for young children. Children come to school with knowledge of speech and are used to speaking to known audiences on intimate terms. Britton et al. maintained that children's writing in the early stages should be in the form of written-down expressive speech. From this starting point children can proceed into the transactional and poetic forms.

Expressive writing is frequently characterized as thinking aloud on paper. Expressive writing activities encourage writers to explore their reactions, express their feelings, give their opinions, and express their moods. The form for this personal expression may be a diary, journal, or personal letter written to a friend. If the writing is shared beyond oneself, it is addressed to a known audience or at least to an audience that shares the values, opinions, interests, and understandings of the writer. On the adult level, such writing could be in the form of gossip columns, special interest articles, or writing in which the author approaches the readers as if they are personal friends.

Activities Designed to Encourage Expressive Writing

Activities recommended for the stimulation and production of expressive writing encourage students' personal responses and emotional reactions. The activities may be written only for the student audience as in a diary that is not shared with anyone. Or, the activities may be written in journal or log format that encourages student-teacher dialogue, that focuses on listing of ideas that lead to other writings, or that encourages students to react to content area assignments.

Journal writing   Journal writing is recommended by advocates at all levels of education. Even adult writers keep journals in which they write or draw setting descriptions, characterizations, conversations, and historical information that may be used later in their own books. For example, when Maurice Sendak was a boy, he drew pictures of other students playing. Later, he used these pictures to get ideas for his books.

Watson (1987) described a kindergarten journal activity in which teachers first prepare journals for each student by folding eight sheets of 12×18-inch paper in half, adding a cover of heavier paper and stapling the pages so that the pages lie flat when the journal is opened. The journal activity is introduced to students by explaining to them that they will be keeping a monthly journal in which they will write stories (using inventive spellings or dictating entries to the teacher or aid), personal anecdotes, or summaries and reactions to various school activities. They will also be reading the entries to the teacher. Volunteers will be able to publish their entries by reading them aloud to the class. In addition, the students will be writing in a different journal during each month of the year. These journals will remain in school until the end of the year at which time they may be taken home. However, the journals will be available to the students and for parent-teacher conferences.

These journals may be used to extend the writing process by asking students to select short journal entries that they want to expand and refine. The journals are kept on a monthly basis, and consequently provide a means of assessing individual growth.

Journal writing is frequently used at all levels to encourage students to express ideas and to react in personal ways. Some teachers use short daily writing times to have students write about topics of their choosing. Bromley (1985) described a technique called Sustained Spontaneous Writing (SSW) in which students write for five to fifteen minutes on subjects of their own choice. This writing, in the form of a personal journal or diary, is dated but not shared with anyone unless the student volunteers to read or share the writing. Dating is recommended so that students can review their own responses during different time periods.

Journal writing may be more structured and encourage students to brainstorm with themselves, to consider what they already know or feel about a subject, and to list or categorize ideas that may be used later to develop writing projects. Teachers may provide ideas for journal writing such as listing smells or sights you like, listing famous people you would invite to a party and identifying topics or questions you would like to discuss with them, writing about wishes, identifying favorite colors or music and writing why they are favorites, listing favorite characters from books and explaining why you like them, and going for a walk and writing down your impressions. In these examples, journals become incubators for ideas that could formulate writing projects.

Dialogue journals (Gambrell, 1985) are specifically designed to increase the interaction between student and teacher. As the name implies, students write questions and comments about or reactions to something that is happening in their personal lives or something that is happening in school. These comments are written with the knowledge that they will be read by the teacher. The teacher then writes brief responses directly in the journal. Due to time constraints, teachers frequently read and respond to the journals on a rotating basis.

Learning logs   Learning logs focus on specific content and responses to what was learned in a daily lesson, a weekly project, or a unit of study. The logs are completed at the end of the class although they do not need to be completed every day. In many ways learning logs form bridges between expressive writing and expository writing as students write about both their personal responses to a learning experience and the information gained from the lesson. From the expressive viewpoint, students can respond to what they liked best about the lesson, to how the lesson could be improved, to why they think that lesson or information is important, to how they will use that lesson or information in the future, and to what they would like to know more about. From the expository viewpoint, students can respond to the main points of the lesson, to what was learned during the lesson, and to what questions still need to be answered or researched.

Learning logs may be used to help students retain material and to focus on the development of understanding. The logs may also provide ways to generate ideas for writing projects and to help students realize that their viewpoints and responses are worthwhile. We consider the learning log or journal entry format later when we discuss how students can use these formats to respond to literature.

Writer's journals   We conclude the expressive writing section with a description of writer's journals because they form a natural bridge between expressive and poetic or creative writing. The information in the writer's journal provides sources for story ideas, characterizations, rich language choices, and increasing sensory experiences.

In *The Young Writer's Handbook: A Practical Guide for the Beginner Who Is Serious About Writing* (1984) Susan and Stephen Tchudi provided ideas that may be used for students and teachers who are developing writer's journals. The Tchudis recommended that writers, whether age 8 or 80, use their journals to explore their experiences, their world, and their language, and to create an inventory of their experiences. These experiences, descriptions, and reactions frequently become the basis for fictional stories, poetry, and other forms of writing.

Under entries that encourage writers to explore their own experiences, writers should think about their past, probe their memories, write down these early experiences, and then react to those memories by thinking about how the experiences helped to shape their lives or affected them in some way. When trying to capture current experiences, they should write as many details as they can in order to recreate the sights, sounds, smells, settings, and people. When exploring their experiences, they should record their dreams, daydreams, and nightmares as well as their opinions about various topics. Finally, under exploring experiences they should encourage the development of their sensory experiences by sitting in various places and recording everything around them. They should record detailed descriptions of what they see, hear, smell, feel, and taste. When they describe these sensations, they can try to make comparisons with other things in their environment. These comparisons may later become important similes and metaphors in their writings.

Under entries that encourage writers to explore their world, writers move beyond their own experiences and their own day-to-day events, and try to seek new ideas and look at things in new ways. Here they should describe impressions of new experiences and analyze broader ideas such as fear and friendship. They should analyze movies and television. They should extend their experiences through newspapers and write their reactions and responses to news stories and editorials, as well as feature and sports articles.

Under entries that encourage writers to explore language, writers should study the language and use their journals to heighten their awareness of language and to store unusual or interesting expressions they hear or create in their own minds. They can collect dialogues among people that may surface later in their own writings. They can write down dialects and word choices that will later add authenticity to characterizations. They can collect interesting expressions and quotes that will add vitality to stories.

Journal entries are then used to inventory writers' experiences. The Tchudis recommended that writers set aside a few pages of their journals to list areas of expertise (hobbies, reading, interests, locations), firsts and landmarks in their lives (first day of school, trip to dentist, ride on an airplane, meeting grandparent), people who are special or important in their lives, as well as imaginary people who interest them, places that are real or imaginary that they can describe or write about, memories of pleasant and unpleasant experiences, book titles that were enjoyed and why they were enjoyed, favorite media presentations and why they were enjoyed, lists of things the writer would like to change, and lists of questions that fascinate the writer.

Journals that are this rich with detail, descriptions, and ideas are natural bridges between expressive and poetic writing. A few students may create such journals on their own. For the most part, however, teachers will need to create experiences that encourage students to look at their environments, to gather details and descriptions, and to interact with their total senses.

REINFORCEMENT ACTIVITY

1. *Choose an area in which learning logs are appropriate. What questions would you use to heighten expressive writing? Try keeping a learning log for one week in one of your college classes. Include both expressive and expository reactions.*

2. *Reread the description of the writer's journal. How would you modify the journal for younger elementary students? What would you emphasize in upper elementary and middle school? Begin a writing journal for your own observations. Include entries that encourage you to explore your experiences, world, and language. What topics could entice you to write a paper using details and information from your own journal?*

IMAGINATIVE AND POETIC WRITING

Terms such as poetic writing, creative writing, and imaginative writing are used to describe writing that may be fictional prose, poetry, or drama. In Horace's terms, the purpose for such writing is "to inform and delight." A similar belief is expressed by contemporary author and Pulitzer Prize winner Donald Murray who stated that this type of writing "not only communicates information, it makes the reader care about that information, it makes him feel, it makes him experience, it gets under his skin" (1973, p. 523). Creativity, Murray believed, cannot be taught; rather, it must be developed by encouraging students to discover who they are and what they have to say. This creative process involves awareness of all aspects of life, as well as a discovery of meaning in life. The good writer uses these perceptions to create meaning through words, and communicates to the reader both his feelings and the information he is trying to convey.

Poetic and imaginative writing, then, includes original writing that uses both imaginative and experimental thinking. The child writing an original poem, a fairy tale, a puppetry script, or developing a personal experience, is engaging in this creativity. Unfortunately, creativity does not develop without a great deal of nurturing.

Why Teach Poetic, Imaginative Writing?

The process of writing helps children become aware of their own uniqueness because poetic writing is an original product. Through it, they learn more about their own feelings, and about their reactions to life around them. Through experiences that lead to creative writing, a child's creative expression can be stimulated. Vocabularies can be enriched when children describe experiences related to sight, touch, smell, hearing, and taste. Vocabularies can be refined when they search for and discover the most appropriate word to express an idea. Writing poetry and original drama allows experimentation with the sounds and impact of well-chosen words. Just as reading a piece of literature provides vicarious enjoyment, writing an original story that is enjoyed and appreciated by an audience provides children with an enjoyable and rewarding experience. Writing original drama and literature helps students understand the development of plot and characterization. As we discover in this chapter, writing can be improved by teaching students to respond to story structure and to elaborate on descriptive details. Finally, writing improvement can be taught in the context where it actually belongs—in the process of meaningful writing.

Developing the Creative Writer

Creative writing cannot be taught in the same way that many other skills are taught. Instruction in creative development must encourage and stimulate the student in his self-awareness, as well as in his written skills. What do we need to consider in order to encourage creative writing development? In this section we discuss the factors of environment; the teacher's role in creative development; useful types of stimulation; an example of a creative writing activity, including a motivational period, discussion time, writing period, and sharing experience; and the development of the creative poetry writer. We discover that the previously discussed stages related to the writing process apply in the development of the poetic, imaginative writer. The prewriting phases may include stimulation of creative processes, models for literary forms, and lessons designed to teach various conventions that may be applied and evaluated during writing and revision (Hillocks, 1986).

The environment   We have already mentioned that a rich environment is necessary for successful development of oral language skills. Such an environment is equally necessary for the development of creative writers. Oral expression flourishes in a trusting atmosphere that allows many opportunities for purposeful communication; this is true as well with creative writing.

The environment should also offer opportunities to write for many different purposes and many different audiences. You would not, for example, write in the same way when composing an informal note as you would when writing an original story for publication. Thus, development of an audience sense, or distancing, is clearly important for skillful writing.

The proper educational environment for fostering creativity will, according to Murray (1973), allow children to progress through seven stages of development at their own pace. Examination of these stages reveals numerous implications for the instructional environment.

First, children need to develop an awareness of life. This means the development of sensitivity to life through activities that allow them to experience such wonders as visual stimulations, sounds, smells, tastes, movements, and feelings. The teacher might take the children outside to experience a tree or a busy city street. Children look, listen, and feel, then share descriptive words and phrases about their sensations. Many of my students have found that developing descriptive vocabularies requires this type of activity. Fine literature, poetry, and music also expand awareness. Art and building materials also allow children to manipulate and feel before they discuss and write. These activities are part of the exploration and prewriting process discussed in chapter 8.

Second, the environment should encourage the child to care about others. Many awareness experiences help children develop empathy for the feelings and experiences of others. Role-playing activities allow the child to experience what it would be like to be in another's postion. Role playing often leads to exciting creative writing.

Third, the environment allows time for the child to consider his new awareness. Murray called this an "incubation period" that allows the mind to think both consciously and unconsciously. Many of you have had the experience of thinking about an almost impossible task, until all of a sudden, the solution seems very clear. This is applicable to writing. Teachers should know that it often requires longer than half an hour for a child to create a product that he feels is satisfactory. Some children may need a much longer awareness period than others. Some children need many awareness and oral language experiences before they tentatively dictate or write their first piece that demonstrates feeling and originality.

Fourth, the environment allows children to make their own discoveries. They need opportunities to ask questions and find answers. We talked about this need in respect to oral language when we stressed the need for children to search for and develop a background before they are required to develop a concept. This is equally true with all types of composing. The child needs opportunities to be involved in open-ended discussions in which there is not one correct answer, but from which many questions arise and many answers appear.

Fifth, the environment must stimulate the writer to create meaning with words. Many opportunities to try various types of writing in all content areas must be provided. The teacher needs to be involved in the students' writing through interaction during writing conferences and other phases of the composing. (See pages 292–296, chapter 8.) The environment, when necessary, must provide guidelines for writing that can be applied during writing and revision stages. This is especially important when introducing certain types of poetry writing and other characteristics of literary elements.

Sixth, after the child has created meaning by writing a selection, the environment should allow for the development of detachment, so the author can involve himself in the evaluation process through reexamination and rewriting of his creation until it most effectively portrays his feelings and meaning. The teacher's role in this process is essential. (See revision phase, page 292, chapter 8.)

Finally, the environment should allow the writer an opportunity to test the effectiveness of his communication. Does the composition convey both the intended information and feelings when it is presented to an audience? Children need many opportunities to write and evaluate their work with specific audiences in mind, and to present their writings to these audiences. Something wonderful happens when a child's finished product, that expresses his feelings effectively, also meets with awe and pleasure from the audience. I will never forget the class response when a group of third graders had finished a biography of Bucky Badger. After the selection was read to the audience, the rest of the class asked excitedly, "Did you really write that? How did you learn to write so good? May we have a copy of the story? Will you sign my copy of the story?" The writers saw that effective writing could influence others, and all their hard work on the story of Bucky Badger seemed worthwhile.

Classroom topics that stimulate writing   There are many types of classroom topics that provide motivation and stimulation for writing. If students are already involved in a unit or in a study of a topic, they have an in-depth experience and more information on which to base their writings. Caprio (1986) found that kindergarten and first-grade students wrote both longer and higher quality compositions if their writing originated from first-hand experiences rather than textbook assignments. Science and social studies topics and literature units are excellent sources for stimulation. If students are already reading, talking about, and listening to references about the topic, it is logical to have them do all types of writing about it as well. For example, one of my second-grade student teachers developed a unit on the post office and mail service. As the unit progressed, students constructed a classroom mailbox, learned about zip codes, went to the post office, and investigated mail transportation from pony express days to the present. This unit stimulated a great deal of writing. The children wrote and mailed letters in the real and classroom post offices; wrote creative stories about what happened to letters when Mr. Zip was ignored; wrote diaries, from the viewpoint of a pony express rider crossing dangerous mountains and traversing Indian territory; and wrote stories depicting a letter's journey down the Mississippi on a riverboat. Some children wrote about the adventures of pilots and letters in the early airmail service, and other students looked into the future to write fanciful versions of the mail service in the year 3000. Instead of conducting a unit that, at first glance, sounds rather dull and ordinary, this teacher provided many opportunities for creative writing, which stimulated further study and investigation; that study, in turn, motivated still more creative writing. The children in her class wrote from a background rich in knowledge, and the ongoing classroom experiences provided enough diversity so that all the students were motivated by at least a portion of the topic. (Early planes, the pony express, and futuristic travel especially appealed to the boys.)

Using everyday experiences   A second type of stimulation comes from everyday experiences. These are the common things that children know best: their neighborhood; their street; going shopping; their family; their hopes and fears; their likes and dislikes. For example, some very fanciful stories came out of a reading of Dr. Seuss's *And to Think That I Saw It on Mulberry Street.* After hearing the story, the children discussed their own streets. They talked about what they saw every day, and how the street and the things on it might appear to a person with a vivid imagination. Finally, they wrote very imaginative stories from this new perspective.

The theme of personal feelings has been used by a number of my university students to stimulate children's creative writing. A third grader lets us know very clearly what he considers trouble (the spelling is as in the original):

Trouble Is

Trouble is runing in the hall becase
the principal will cach you.
Trouble is I can't wright a poem
becase its hard.

Trouble is getting a tooth pulled
  becase it hurts.
Trouble is when someone yells at you
  becase you get punished.
Trouble is ending a poem
  becase its hard.

Happiness is another emotional theme that is very close to children. Here is a dictated poem, created by a group of five first-grade students:

Happiness Is

Happiness is having pets.
Happiness is having a cat.
Happiness is going skiing.
Happiness is riding a horse.
Happiness is loving a snake.
Happiness is having food to eat.
Happiness is going ice skating.
Happiness is having your own bike.
Happiness is having your own puppy.
Happiness is having your own bedroom.
Happiness is learning how to ride a bike.
Happiness is jumping in a pile of leaves.
Happiness is getting your allowance every week.
Happiness is catching a frog in a swimming pool.
Happiness is being nice to your brother and sister.
Happiness is sharing your jelly beans with no fighting.
Happiness is going snake hunting and finding big snakes.
Happiness is having your own swimming pool in your backyard.
And best of all, happiness is being alive.

Another topic close to a child's everyday experience is recess and the playground equipment they use. The following imaginative writing resulted when a third-grade teacher asked each child to imagine that he was ball and to write a story from the ball's point of view:

The Life of a Ball

Hi, I am a ball. And you know what balls are used for, bouncing. Well, I'm not. I just sit in the closet all day long.

Hey, here comes somebody and they're taking me outside. Ouch! You kicked me. I'm not going to make a basket for you. Oh no, they're going inside and that was the last recess. Now they're getting on the bus and going home.

Burr. It's starting to snow. Ouch! A big hunk of ice hit me. I wish I had a set of ball muffs and a jacket. It's starting to rain and I wish I had some goloshes and a raincoat. Oh. Here comes some wind and I'm blowing under the car. Ouch! I hit the tire. I'm going to try to get into the school.

At last I got inside with my friends the football, and the base ball and my mom the basket ball. It was so warm that I went to sleep.

Something just woke me up. Oh no, it's those kids again. I'm being taken outside but last night was so hard on me that I'm flat and I won't bounce. They say I'm dead and they are putting me back in the closet. I guess I'll just have to stay here until someone blows me up again. But there are worse places than this closet.

The teacher introduces new experiences A third category for stimulation is the teacher's introduction of new experiences. These experiences may be in the form of short lessons such as those described in Hillocks's environmental mode for teaching composition. They may include models of writing and literature that will be applied during writing experiences. Many sense awareness activities fall into this category: going for a walk; listening to, smelling, and touching the environment; discussing one's sensations and writing descriptive phrases; and finally, composing stories or poems about the experience. Teachers may also bring in unique objects, or use music, literature, or pictures and photographs for stimulation.

Many of the research studies that Hillocks (1986) identified as the most effective approaches for teaching writing would fall under teacher-introduced approaches; the teacher first introduces a concept in a short structured lesson and then the students apply that concept during the writing and revision stages of the writing process. Hillocks identified Sager's study in which she increases sixth-grade students' abilities to elaborate details as an excellent example of the environmental mode that he found most effective. Consequently, we look carefully at Sager's strategies. In this example, Sager (1973) asked students working in small groups to read the following story:

The Green Martian Monster

The Green Martian monster descended on the USA. He didn't have a mouth. "Who goes?" they said. There was no answer. So they shot him and he died.

After reading the story, which the students had been told scored 0 on elaboration, the students were asked to do the following tasks:

1. Quickly list all the reasons why a mouthless, green Martian monster might land in the USA.
2. List all the places the Martian could have landed.
3. Who could "they" have been? List all possibilities.
4. List all the thoughts "they" could have been thinking when they saw the Martian.
5. What could have happened between the time the Martian was shot and and the time he died? List all possibilities.
6. Look at your lists. To be interesting and easy to understand, a story needs details such as you have written. Add some of these details to the story and take turns reading the story the way you would have written it. (p. 95)

Hillocks emphasized the importance of this example because students did far more than rate a composition. "They found problems with the writing, generated ideas which would help to correct those problems, and synthesized those problems with the existing frame. Finally, they considered the principles underlying what they had done. Using such materials as the above composition, the experimental groups in the study worked with scales dealing with vocabulary, elaboration, organization, and structure" (p. 123).

The "Sager Writing Scale" used in Sager's study is described in *Measures for Research and Evaluation in the English Language Arts* published by ERIC and the National Council of Teachers of English (1975) and may be ordered through ERIC documents.

As a result of her research Sager (1977) identified the following four factors that contribute to effective writing in the upper elementary grades:

1. A rich and varied vocabulary that allows the writer to express thoughts in such a way that it holds the reader's interest. A rich vocabulary uses exact words, synonyms, words that appeal to the senses, descriptive words and expressions, and word combinations and comparisons.
2. An organizational ability that presents ideas in a logical arrangement, stays with the subject, and presents ideas in effective sequence.
3. The ability to elaborate so that ideas are fully developed and flow smoothly from one to the next. The ability to elaborate reveals itself in an abundance of related ideas and vivid details.
4. The ability to use a variety of sentences and thus state ideas both accurately and fluently.

You will note that many of these factors are discussed as part of the writing process in chapter 8. These areas may also be developed through classroom activities.

William Harpin (1976) offered guidelines for the types of teacher-introduced stimulation effective with seven-to-ten-year-olds. First, Harpin found that photographs are effective if the student can identify with the photograph. The photograph has to be open-ended enough so that responses show variation, and the children can give different interpretations. Photographs must have a certain level of ambiguity; too much, and they are confusing; too little, and the responses are almost identical.

A second effective teacher-introduced stimulus is pictures. Sequences of pictures, photographs, and cartoons are especially reliable for developing narrative writing with young children. I have found, for example, that pictures in Charlie Brown coloring books are highly motivating, especially for the more reluctant writer. We have laminated individual pictures (so they can be used many times), and used them to stimulate discussion and to develop dictated and individually written stories. These pictures have been effective even with seventh and eighth graders in summer remedial programs. Several especially good pictures are one of Charlie Brown and Snoopy in which Charlie says, "Get a hit or there'll be no dessert," a picture of Charlie Brown running out into the waves, shouting "Snoopy! Stop teasing that porpoise," and another of Charlie in which he looks very perplexed and says, "I always pick myself up, dust myself off and fall right down again."

A third teacher-introduced stimulus is music. Harpin warned that too much of the musical selection may have a numbing effect on subsequent writing. Often, a few bars are enough to stimulate a creative response, although the whole selection may be required with narrative records such as *Peter and the Wolf.* One of my university students used "Night on Bald Mountain," by Mussorgski, which stimulated an imaginative sixth grader to write the following description:

Night on Bald Mountain

It is midnight, everyone is asleep, except, bald mountain! As the church bell strikes, the devil of mountain tops opens his wings and smiles evily onto the small grave yards, the evil gleaming of his eyes seem to say, "Thou shalt come out of hiding and appraise the devil of bald mountain!" As if in answer, ghostly shadows and skeletons begin to float up. When seeing this he cackles in delight, and suddenly thrusts his fingers at a huge boulder, and where the boulder had been, there was a fiery pit of flame.

A fourth successful type of stimulus involves objects that provoke a great deal of discussion and speculation. Several that Harpin found especially provocative were a Tibetan prayer wheel, a collection of hats, a stuffed fox with one leg missing, a model stage coach, a boomerang, and a small nineteenth-century writing case with a hidden compartment. Harpin concluded that the key to success in using objects is the development of questions and discussion prior to the writing. Children need opportunities to think about what the object is, who made it, how and why it was used, who owned it, and so forth. As an example of this type of stimulation, a teacher might bring into a class that is studying frontier America, a collection of dolls and other toys characteristic of the period. Some might be handmade, such as a cornhusk doll, whereas others could be replicas of period toys. Having studied it, the children would have background and knowledge of the period, so they would understand what it might be like for a child to grow up on the frontier. They could look at the toys and describe them, discuss the adventures the toys might have shared with their owners, and the feelings the owners might have had about their toys. They could talk about which toy they would choose if they could take only one on a covered-wagon journey. After discussion, they can write their stories, from the viewpoint of either the toy or the child.

A fifth stimulus is literature. Literature is not only an excellent stimulus for creative writing, is also provides a model for writing improvement (this aspect is discussed later in the chapter). One fourth-grade teacher used a study of fables to stimulate creative writing. Because a child cannot write a fable without understanding what it is, the teacher first did a modeling activity in which she read a number of Aesop's Fables to the class. The class developed criteria for fables: they are short tales in which animals talk as humans; the animals represent human nature (e.g., the fox is sly and cunning); and fables have a moral. The class read and discussed many fables. Finally, the teacher had the children write their own fables. The finished fables were typed and reproduced in a booklet so that each student had the accumulated class collection. Here are two examples from this fourth-grade collection.

An unusual picture introduced by the teacher stimulates both oral discussion and creative writing.

The Cat and the Birds

Once there was a cat that picked on all the birds in his neighborhood because they were smaller than he was. One afternoon the cat was chasing a bird when suddenly he felt a striking pain in his back. He turned around and saw a little sparrow on his back. From that day on he never judged anyone by his size.

Love and Smile

Once there was a mother cat and a father cat and their little kitten. The kitten's name was Julie. She got everything she wanted. Then her mom had a baby. It was a girl. Her name was Jamie. She wanted to go everywhere with Julie. Julie hated her sister and her sister knew it.

One morning Julie was going to take a letter to the post office. Her sister was following her. She didn't know she was. Then all of a sudden Jamie ran up and grabbed the letter and said, "Race you to the post office." Julie froze and watched her sister. She was running to the street. Then a car hit her. Julie screamed. They had to take her to the hospital. The doctor said Julie could see her. Jamie said to Julie, "Why are you here?"

> *Julie said, "Because I love you." That day Jamie died with the happiest smile*
> on her face.
> Moral—*LOVE AT FIRST OR YOU WILL LOSE IT!*

For creative writing, the instructional sequence is equally as important as the type of stimulation. These stimulation activities need to be placed within the framework of the writing process discussed in chapter 8.

Instructional approaches for poetry writing   Why is that some children produce quality poetry, while others do not? The answer, according to Duffy (1973), lies primarily with the teacher's dedication to the development of poetic self-expression. Just as in presenting other forms of creative writing, the teacher needs to structure the environment so that children are not punished for novel and creative thinking. The school experience is often criticized because many children apparently learn to conform, and thus learn at an early age not to express their real selves. If children are to develop creative imagination, Khatena (1975) believed they must be taught to break away from the usual and commonplace, to restructure, and to synthesize.

What are some instructional approaches teachers can use to develop creative poetry writers? First, the teacher needs to stimulate an interest in and a love for poetry. None of us can write poetry without first developing a feel for it. The teacher must begin by reading poetry to children so they will learn to appreciate and enjoy this means of expression. Duffy (1973) maintained that there is no stimulation technique as important as reading good poetry to children. He believed the teacher should begin by reading humorous poems, which establish a positive image, and also by reading simple free verse poems, which serve as a model for beginning writing. Popular humorous poets include Jack Prelutsky, Shel Silverstein, William Jay Smith, John Ciardi, and Laura Richards. Sources for poetry selections include textbooks by Norton (1987), Sutherland and Arbuthnot (1986), and Huck (1987).

Awareness and observational powers   In addition to reading poetry to children, the teacher needs to structure activities that nurture both their awareness and their powers of observation. For example, one teacher wanted her students to write a visually descriptive piece about a fall afternoon. She did not just tell them to take out paper and pencil and write a page about a fall day. Instead, she described her creative writing instructional procedures in these words: "We began our meeting outside in a field, which at the time was bursting with being, and generally screaming fall colors, sounds, and scents. We talked about and made lists of sight, sound, and smell words and word clusters (distinguishing between insect, plant, and mechanical life). We made lists of the images we saw in the clouds and watched them change. We investigated our 'buggy' surroundings; finding, catching, and describing daddy longlegs, crickets, and katydids."

The sequential instructional activity in which students went out into a fall environment and discussed its sights, sounds, and smells, also resulted in observations that later took a poetic form. After observing a cricket, one student wrote the following poem:

Cricket

creeping, putting
one leg first,
then another,
and another,
till all six
have moved
and he starts
again.

ing, le
p   a
m   p
u   i
j   n
g
flying through
the air
and then
Bump
he lands
and
jumps
again.

Do you notice how this sixth-grade student has captured the essence of the cricket? Another student used her powers of observation to watch an eagle in the sky. She jotted down her feelings as she watched, and later put her observations and feelings into this poetic form:

Eagle

I like to watch an eagle soar,
How he just glides alone,
I wonder what it's like,
gliding freely through the air,
his wings spread widely,
and, he looks so nice and proud
but, pretty soon he'll be extinct,
and he'll glide and soar no more.

This type of activity required that students observe and describe in a new way things they see every day. For example, one third-grade student teacher planned a lesson around an orange. He began by asking the children where oranges came from, what they were used for, and whether they liked oranges. Then he gave an orange to each child and explained that they were to take their time discovering all they could about an orange through touch, taste, sight, and smell. He told them: "Discover all you can about the outside of your orange by looking at it; look at the pores, the color, the shape, the scars, and the soft spots. Now feel it, roll it in your hands and on the table, squeeze it, rub your fingers over the pores, close your eyes and feel it. Smell it. Slowly begin to peel it; how does it smell? Contrast the feel and sight of the outside of the skin to the inside; taste the inside of the skin. Now look at the inside of the orange. What shape is it? What color is it? Feel the inside, gently squeeze it, taste the inside of the orange. What is inside some of the sections? Put one in your hand. What color is it? What shape is it? Is it hard or soft?''

After the students talked about all the things they had discovered about an orange, they were asked to write a poem, beginning each line with ''An orange.'' These are some observations that resulted from this activity:

An Orange

An orange is round like a ball
An orange is the color of the sun
An orange is bumpy on the outside
An orange is squeezy and yummy
An orange rind is bitter as a lemon
An orange has squishy pulp
An orange has tough, slippery seeds.

This type of observation and poetry development was influenced by Kenneth Koch's *Wishes, Lies, and Dreams,* in which he described his experiences teaching poetry to children in New York City. The book describes such techniques as developing wish poems, noise poems, and color poems, and includes examples of poetry written on these themes by first- through sixth-grade children. Listening to poetry written by other children encourages elementary students and arouses their enthusiasm.

Koch found that themes the children can relate to provide stimulation for writing poetry. The theme of wishes is one that all children are able to relate to. The same format used in ''An Orange'' can help children develop their wishes into poems. The following poem developed after the teacher had read a number of ''Wish'' poems from Koch's book. After hearing the poems, the students talked about various wishes of their own, then wrote their own wish poems.

I Wish . . .

I wish I was a towering giraffe
So I could see everything
I wish I was a swift cheetah
So I could run like the wind
I wish I was a majestic tree
So all the birds could rest on me
—A fourth-grade wish

Developing imagery   Much of the poetry revolves around imagery. The poet often creates word pictures by comparing his thoughts to experiences and likenesses and differences in life. This third-grade author uses various comparisons to describe a giggle:

Giggles

Giggles are good tasting fellows
They smell like brownies
Just coming out of the oven
They feel like a new Bugs Bunny doll
Giggles look like a red clown at the circus
Giggles are shaped like bubbles

Another child developed the image of a gulp:

Gulps

A gulp is very heavy,
 especially on rainy days.
A gulp tastes like chocolate,
 smells like coffee,
 and feels like a magnet.

In order to develop imagery, children will need practice in describing objects in terms of other objects. One teacher introduced color description through a brainstorming activity, in which the class rapidly provided associations with a color. For example,

It is as red as—a stop sign, blood, a valentine, danger, a lobster, a sunset,
 an apple, a strawberry

Similarly, with another concept:

As quick as—drinking a glass of lemonade on a hot day, a shooting
 star, a mongoose

The children can also make comparisons with the following expressions: as tired as; as hungry as; as thin as; as quiet as; as silly as; as sneaky as; as soft as. After this kind of practice with comparative writing, one student described a house as "so empty, you could hear the spiders spin webs."

The experiences of these classes demonstrate the interaction of a stimulation activity and the brainstorming of ideas and concepts prior to writing. The stimulation and discussion activities not only developed creative awareness, they also developed the descriptive vocabularies so necessary for creative writing. Student teachers as well as classroom teachers report that their creative writing activities are not very successful when they ignore this oral exchange of ideas.

Other forms of poetry   Several of the student poems we have presented use a basic form such as starting each line with "Happiness is," or ending each line with a rhyming word. Children enjoy several other forms of poetry, and find them not too difficult to write. These forms may also provide some children with a little more guidance and structure. (Notice the use of modeling and processing in these activities.)

Forms for limericks, cinquains, and diamantes were presented on pages 281–282 in chapter 8 as examples of poetry forms that could be used to provide models during the writing process.

FOR YOUR PLAN BOOK
Developing Creativity Through Poetry

The following five lesson plans are from a series of lessons designed to develop creativity with first graders. The writing samples resulted from the particular lesson.

Lesson Plan 1

I. *Purposes of teaching this lesson:*
    A. *To develop self-image*
    B. *To develop awareness of feelings*
    C. *To develop a group-dictated writing about a feeling*
II. *Materials needed:*
    A. *Filmstrip,* **Circle of Feelings**—focus on self-development
    B. *Poem "Here Is Jack O'Happy"*
    C. *Pictures to illustrate each line of the poem*
III. *Steps in teaching:*
    A. *Show the filmstrip and discuss the feelings it deals with.*
    B. *Present the poem and have the children tell about times they have been happy, sad, mad. Have the children tell about other feelings they have and act out these feelings.*
    C. *Encourage a group-dictated experience related to one of the feelings.*
    D. *Group reading and individual reading of experience story*
IV. *Group-dictated creative writing*
V. *Revision and editing group story*
VI. *Sharing group story orally and placing story in permanent collection.*

Happiness Is

When your mom tells you to play with your baby sister and feed her.
When your friends come over to play with you.
Seeing the flowers grow in the sun.
When the sun is out and you can play in the clubhouse.
Having a new cuddly toy.
Going to a parade.
Getting a new swimming pool and swimming in it.
Our cottage and our boat.
Swimming in the river.
Going camping and going down to the beach finding shells.
Learning to read.
Being glad to grow up.
Having people to like.

Lesson Plan 2

I. *Purposes of teaching this lesson:*
    A. *To heighten awareness of environment*
    B. *To make the child aware of color and shapes in the environment*
    C. *To draw a picture seen through imaginary glasses*
    D. *To write a short story about the picture*

II. *Materials needed:*
   A. *Filmstrip,* **The Magic Glasses**
   B. *A pair of cardboard glasses for each child*
III. *Steps in teaching:*
   A. *Show the filmstrip and discuss.*
   B. *Show the glasses and have the children tell what real things they see through the glasses. Concentrate on one color at a time, e.g., "What things do you see that are blue?"*
   C. *Have the children look through the glasses again, this time to see imaginary things that they would not be able to see without wearing "magic glasses." Discuss what they see.*
   D. *Have the children draw a picture of what they see in their magic glasses.*
   E. *Help the children write short stories about their pictures. Some children need to dictate their stories; others may write their own.*
   F. *Help children revise and edit their stories.*
   G. *Share pictures and stories.*
   H. *Make a "Magic Glasses" bulletin board.*

Lesson Plan 3

I. *Purposes of teaching this lesson:*
   A. *To help children become familiar with rhyming poetry and rhyming words*
   B. *To help children express themselves in poetry*
II. *Materials needed:*
   A. **I Met a Man** *by John Ciardi*
   B. *Rhyming word wheels*
III. *Steps in teaching:*
   A. *Read some poems from* **I Met a Man;** *discuss rhyming words.*
   B. *Present the word wheel and have the children work with it, making new words.*
   C. *Interact with children as they write or dictate their own short rhyming poems.*
   D. *Interact with children as they revise and edit their poems.*
   E. *Have children read poems to the group.*
IV. *Example of an individual rhyming poem written during lesson:*

> *Mr. Mada-doo*
> *Had a cad - a - doo*
> *That ate so much*
> *It got fat - a do*

Lesson Plan 4

I. *Purposes of teaching this lesson:*
   A. *To encourage children to ask questions*
   B. *To write a cumulative poem*
II. *Materials needed:*
   A. *Filmstrip,* **Sometimes I Wonder**

III.   *Steps in teaching:*
    A.   *Introduce the filmstrip; talk about things the children have wondered about.*
    B.   *Show the filmstrip and discuss.*
    C.   *Have the children comment on what they have wondered about. Write their ideas for a cumulative piece.*
    D.   *Orally read their chart story to make certain it says what they mean.*
    E.   *Interact with children as they revise and edit their stories.*
    F.   *Have students, in a group and individually, read the story.*
IV.   *Group-dictated creative writing that resulted from lesson:*

I Wonder

I wonder why my hair turns yellow in the summer, brown in the winter.
I wonder why it's so dark in here.
I wonder why scrambled eggs have yolks in them.
I wonder why Brian and Michael fight all the time.

I Wonder Why

I wonder why I'm so silly.
I wonder why my eyes turn blue—I don't know why they do—
    But my sister said it's true.
I wonder how come the spring is so beautiful—like when the leaves come
    up, and flowers come out, and rainbows too.
I wonder why my dog runs away.

Lesson Plan 5

I.   *Purposes of teaching this lesson:*
    A.   *To develop awareness of the colors, textures, and sounds of clothing*
    B.   *To develop individual poems about personal clothing*
II.   *Materials needed:*
    A.   *Clothing of different textures, colors, purposes*
III.   *Steps in teaching:*
    A.   *Discuss the clothing the children are wearing.*
        1.   *What color shoes are you wearing?*
        2.   *How do they feel?*
        3.   *How do they look?*
        4.   *What can you do in your shoes?*
    B.   *Show clothing of different textures, colors, etc. Have them touch the articles. List all descriptive words the class mentions.*
    C.   *Allow children to dictate or individually write a poem about a piece of favorite clothing.*
    D.   *Interact with the children as they write their poems.*
    E.   *Interact with the children as they revise and edit their poems.*
    F.   *Have the children share their completed poems. Have the children create a bulletin board of the poems and examples of clothing.*
IV.   *Example of an individually written poem that resulted from lesson:*

Shoes

Shoes, shoes, shoes.
Brown, white, black shoes.
Smooth, rough, hard, and soft.
Puffy, dirty, clean, shiny, buckled.
Tied, clear, squeaky shoes.
Shoes, shoes, shoes.

Improving Specific Aspects of Creative Writing

Thus far in this chapter, we have stressed the need for developing a background of awareness, descriptive technique, and information in order to write. In this section we review some techniques for improving creative writing. We look at ways to improve writers' vocabularies, the use of plot and characterization to improve writing, and the use of mythology for achieving an understanding of symbolism and allusion.

Vocabulary development for creative writing   Creative writing has often been termed visual writing. How can we make the reader visualize a character, scene, incident, or poem without using trite and overused expressions? Vocabulary, of course, is the key, and the student needs practice in describing objects, scenes, and people. One activity many teachers use for developing descriptive vocabulary is to have children reach into a bag that holds a number of items—such as a ball, a penny, a pencil, a rubber band, sandpaper, a glass, and so forth. The child describes the item without looking at it, focusing on texture, size, shape, firmness, in such a way that the rest of the group will be able to guess what the item is.

In one class, we listed a number of descriptive phrases that really do not give the reader any information; the class described what they thought the words meant, and suggested better descriptive phrases or words. Some of the more meaningless phrases were as follows: a beautiful morning; a handsome man; a big dog; a lovely girl; a beautiful scene; and an exciting moment. The phrase, "a beautiful scene," conjured up as many different pictures as we had students in the group. It was a good activity to show the class the need for greater clarity and more precise descriptive phrases if the reader is to perceive the same visual picture as the writer.

Overuse of adjectives and adverbs is a common problem in writing. As description, "The great green expanse of meadow was filled with stately, brown, soft-eyed deer eating their sweet smelling breakfast" is so overdone that it is ineffective. To avoid this pitfall, it is important that children develop a variety of descriptive techniques. An important technique is the use of vivid action words, in both story writing and poetry. Duffy (1973) recommended that teachers emphasize instruction in verbs, because the brevity of poetry often precludes the use of a great number of adjectives. In fact, the verb may be the most descriptive word in a line of either prose or poetry.

How can we, as teachers, help children learn to choose more descriptive verbs? If you look at children's writing, especially in the early elementary grades, you will find specific words used over and over. One such word is *said*. We see "Mother said," "Father said," and so forth, used repeatedly. But does "said" describe *how* a person is saying anything? In one class, the children compiled a list of over 100 words to use in place of "said." They decided that *argued, yelled, growled, whined, declared,* and so forth were much more descriptive of how a person talked and felt than the overused "said." This class became aware of visual verbs; the children started to look for them in their reading and to use them in their writing.

The children talked about other verbs that do not communicate as well as they should. They read sentences such as, "The boy *walked* down the street," and discussed *why* and *where* he might be walking. Then they substituted other, more visual, verbs to paint a clearer picture. For example:

The boy *strolled* down the street.	(He was not in a hurry and just wanted to go for a walk.)
The boy *strutted* down the street.	(He had just won a baseball game and was very proud.)
The boy *tramped* down the street.	(He was angry because his father told him he couldn't go to a movie.)

The children worked with other verbs, such as:

The girl *drank* a large glass of lemonade.

The girl *sipped* a large glass of lemonade.	(She was at a party and wanted to be polite.)
The girl *gulped* a large glass of lemonade.	(She had been running and was very hot and thirsty.)

Jim *sat* in the large chair.

Jim *sprawled* in the large chair.	(He was tired and by himself and he wanted to relax.)
Jim *crouched* in the large chair.	(Jim was ready to leave the chair in a hurry.)

This kind of exercise improves the students' use of visual verbs and makes them more observant when they read literature.

REINFORCEMENT ACTIVITY

Design a series of activities to help children develop more descriptive vocabularies. Present the activities to a group of children or share them with a group of your peers.

Improving story writing   Sager's approach designed to increase elaboration of story details emphasized the importance of going beyond minimal information when developing plot, setting, and characterization. There is also a strong connection between reading and listening to literature and story writing. Schema theory (see pp. 424–426, chapter 11) supports the need for reading and telling stories to students, and for filling in knowledge gaps if they are to develop understandings of literary elements required for creating believable and enjoyable stories.

In chapter 11 we stated, "Implications from schema theory emphasize the importance of activating relevant prior knowledge and of helping students fill in gaps in their knowledge before they read. Valuable instructional strategies include prereading discussions that focus on students' prior knowledge, information that helps them use their prior knowledge during the reading experience, and questions that help teachers identify any gaps in students' knowledge. If gaps are identified or if prerequisite knowledge is required, it should be presented before students approach the literature selections. Maps, pictures, films, and historical time lines are useful" (p. 425).

Before you read further in this chapter, go back to this paragraph and replace reading terms with writing terms. Notice how closely these implications now match the prewriting phase of writing that emphasizes remembering, clarifying ideas, and presenting information that will be applicable during the writing phase. In addition, notice the implications for teacher interaction and for filling gaps in knowledge. If we apply these schema implications to imaginative, poetic writing, we understand the need to read literature to students so that they will have knowledge of plot structures, settings, characterization, and literary elements, such as simile and metaphor, that increase understanding and appreciation in their own writing. Many of the activities developed in chapter 10, Literature, and in chapter 11, Reading and Literature, also improve writing. This is especially true for ways to share literature and approaches for developing literary understanding such as webbing, plot structuring, and modeling.

A bulletin board helps students visualize story elements.

Plot development   Some very simple stories may be used to introduce the concept of plot. Stories that have strong, logical sequence are good for this beginning activity. The following sequence of instruction was used with a fifth-grade class. First, the teacher read a simple story, *Three Billy Goats Gruff.* He asked the children to listen carefully to identify the characters and incidents in the story. After listening to the story, they identified its four characters: the littlest billy goat, the middle-sized billy goat, the great big billy goat, and the troll. Next, they listed the incidents that occur in the story:

1. The goats want to eat the grass on the other side of the bridge, but there is a terrible troll under the bridge.
2. After being threatened by the troll, the littlest goat crosses successfully.
3. The middle goat crosses successfully, also after being threatened by the troll.
4. The largest billy goat is confronted by the troll and knocks the troll off the bridge.
5. The three goats eat grass on the other side of the bridge.

Then the students investigated the order of the five incidents. To do so, the teacher asked them to act out the five major incidents. Then he asked them whether the action had to be in that specific order. The students rearranged the order in several ways and tried to act out each new order. The five incidents were written on strips of cardboard and arranged in the order being demonstrated. The students concluded that if the story was to retain logical development, the incidents could not be presented in a different order. Some of their reasons were as follows: (1) the beginning had to start with the goats on one side of the bridge and a need to cross the bridge; (2) the littlest goat had to cross the bridge first so he could tell the troll to wait for his middle-sized brother, who would be fatter and taste better; (3) the middle-sized goat had to come next so there would still be a larger goat remaining; (4) the biggest goat had to cross so the troll could be eliminated; and finally, (5) the troll had to be eliminated so the goats could eat happily on the other side. The group also concluded that if the biggest billy goat had eliminated the troll in the beginning, the story would not have been interesting because there would have been too little conflict.

This activity produced a realization of and definition of plot connecting the characters and action in logical sequence. The students listened to other short stories and identified the sequence of events, from which they began to realize that many stories develop their plots through conflict.

In the next planned activity, the class looked again at their lists of characters and incidents from *Three Billy Goats Gruff.* They identified which characters were in conflict: the three goats against the ugly old troll. They identified the reason for the conflict: the goats wanted grass on the other side of the bridge, but the troll lived under the bridge and wanted to eat the goats. They identified the series of conflicts: first, the little goat against the troll; next the middle-sized goat against the troll; and finally, the biggest goat against the troll.

The teacher then read the story, "Nomusa," from *Drumbeats* (Field Educational Pub., 1970). The students listened for the characters who were in conflict: a girl named Nomusa who lived in a village in Africa, a bird, and a cobra. They listened for the obstacle in the story: a deadly cobra is prepared to attack Nomusa. They identified the series of incidents that develop the plot:

1. Nomusa wakes up and needs water from the stream.
2. Nomusa is attracted by a parrot's feather and leaves the path to catch it.
3. The parrots send a warning that something is wrong.
4. Suddenly a deadly cobra appears in front of Nomusa. He is ready to strike.
5. Nomusa dares not move; she looks for a means of defending herself, but is afraid to strike the cobra because his spirit would haunt her.
6. A little bird suddenly flies down and attracts the cobra's attention.
7. Nomusa dives in the water and is safe.

The teacher introduced the concept that a plot has beginning incidents, middle incidents, and ending incidents. The class, with the teacher's assistance, drew a diagram to show the progress of a plot (see figure 9–1). After they drew the diagram and discussed plot development, they placed the incidents from ''Nomusa'' on the diagram (see figure 9–2).

Finally, students wrote their own stories, using beginning, middle, and ending incidents. The teacher interacted with each student during the composing phases, and the children shared their finished products. The plot development models were used during different phases of the writing process to help students focus on the key elements in plot development. During prewriting students listened to stories, developed the model, and applied the model to other stories such as additional folktales and the example in figure 9–2. During the writing phase they referred to the model to develop stories with beginning, middle, and ending incidents and to include rising action. After writing their first drafts they referred to the models during teacher conferences and peer editing groups. The models encouraged students to focus on areas that were well-developed as well as areas that needed to be revised for clarity. The models encouraged self-evaluation because students were able to consider elements in plot development and to analyze whether their writings included these elements.

Additional plot structures are shown in chapters 10 and 11. The plot structure in person versus self conflicts (problem → struggle → struggle → self-realization → achievement of peace and truth) is diagrammed in figure 10–6, page 406. Three other

FIGURE 9–1
The diagram of a plot

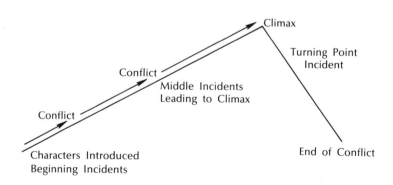

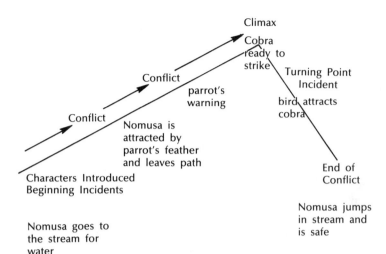

FIGURE 9–2
Plot diagram for "Nomusa"

patterns of action are used in plot structures: (1) the action moves from one incident to another incident; there is a final climax but the author leaves the reader with a sense of uncertainty ⎯⎯⎯⎯ , (2) the action moves with a tension that does not develop a climax; there is minimal suspense ⎯⎯⎯⎯ , and (3) the action moves in a lifeline that does not have one dramatic climax but may have several dramatic episodes ⌁⌁⌁ . These additional plot structures are especially interesting when working with older students. Literature written for younger students usually follows the plot structure shown in figures 9–1 and 9–2.

REINFORCEMENT
ACTIVITY

---

  1. *Develop a writing lesson that follows the plot development shown in figures 9–1 and 9–2. Develop a model using examples other than the two presented in this chapter. Develop a list of children's literature selections that follow that specific plot development. What types of children's books develop this plot structure?*
  2. *Choose one of the other plot structures previously described. Identify several literature selections that follow that specific plot structure and place the incidents from the story on the plot structure. Develop a writing lesson that helps students understand and apply the plot structure.*

Setting   Several of the ideas for entries in the Writer's Journal described earlier in this chapter (pp. 310–311) emphasized sitting in a specific location or describing a new experience by writing vivid details related to sight, sound, smell, and touch. One of the prewriting activities for poetry focused on taking students outside where they could observe nature, talk about what they see, cluster words into categories, and then write poems reflecting these observations. These same activities provide valuable prewriting experiences before students write settings for stories.

Pictures and picture storybooks are other excellent sources for discovering information about settings and for writing vivid descriptions of settings. The RADAR approach to writing (Davis, 1988) described in chapter 8 uses a structured approach to pictures to help students describe both settings and characters. In this approach students are instructed to look at a picture, divide the picture into three parts, and then circle each part. For example, if the picture includes a dog chasing a cat into the woods, the students might circle the dog, the cat, and the woods. Next, they number each part and describe each of these parts in detail. They are instructed to start with the first part and to write their descriptions moving from top to bottom. (This is one of the organizational patterns diagrammed and discussed on p. 289 in chapter 8.) The students are also encouraged to use transitional terms of location such as "beside," "over," "under," and "to the right," when they write their descriptions.

Picture storybooks, with their vivid illustrations, are among the best classroom sources for developing observational and writing skills related to setting. An activity that encourages writing about setting can vary from the young child's simple description of a setting to the older student's creation of a setting that conveys its purposes: creating a mood, developing an antagonist, and providing historical background.

When developing these activities, teachers should introduce the books, read the stories while encouraging students to look carefully at the settings, and then describe these settings in detail. This discussion can also emphasize students' reactions to the settings, as well as descriptive words and phrases that they believe describe the settings. Students may even try to make comparisons through simile and metaphor. Finally, students write their own descriptions of settings that include the points made during the discussion. It is advisable to have the first setting description developed as a group activity in which the teacher models descriptive words and phrases and highlights various things that the students may look for.

Books that have familiar settings are excellent for younger students. If the settings are within their experiential backgrounds they may even compare the settings in the illustrations with similar settings in their own environments. There are so many illustrated books that familiar settings should be easily located. For example, at least some of the following settings should be familiar to many children: Jane Yolen's *Owl Moon* (1987, rural woods in winter), Dayal Khalsa's *I Want A Dog* (1987, city neighborhood and home), Lois Lenski's *Sing A Song of People* (1987, city environment), Merle Peek's *The Balancing Act: A Counting Song* (1987, an amusement park), Marc Brown's *Arthur's Baby* (1987, family home, preparing for a baby) and Valerie Flournoy's *The Patchwork Quilt* (1985, a close, three-generation Black family).

After students have mastered writing simple descriptive settings, they may explore describing and writing settings that have specific purposes. For example, many settings create the mood of a story. Students may use these visual settings to try to match the feeling of the setting through their choice of words. Marcia Brown's illustrations for Cendrars's *Shadow* (1982) are excellent for exploring how setting can create a frightening, spooky mood. The illustrations in sharp collage seem to reach out menacingly. After students try to describe the settings in their own words they should analyze how the author creates the mood as the shadow staggers, grabs, and teems like snakes and worms.

A warm, gentle nostalgic mood is created in Cynthia Rylant's *When I Was Young in the Mountains* (1982). This mood is enhanced by descriptions of grandfather's kiss on a front porch and grandmother's holding a child's hand in the dark. Pictures of country kitchens, country stores, and swimming holes suggest a leisurely, secure, and happy way of life for children.

A humorous mood is created in Catharine O'Neill's *Mrs. Dunphy's Dog* (1987) through illustrations that show lamp posts and fence posts drawn at angles, animals in caricature, and literal interpretations of the news.

Students can explore how both illustrations and text create moods. They can brainstorm lists of words and phrases that suggest different moods and try to match the mood of the illustrations in their own writings. Teachers may also share only the illustrations in books, ask students to describe the mood they think the illustrations suggest, write descriptions reflecting this mood, and then compare their descriptions with those developed by the author.

After completing several writing activities of this sort, a fifth-grade class listed and categorized words that reflected specific moods in books and in their own writings that accompanied the same illustrations. The students included both words that depicted mood in setting and in characterizations. The following chart illustrates part of this activity:

Words that Suggest Moods for Settings		
Frightening Mood	**Warm, Happy Mood**	**Funny Mood**
Cendrar's Shadow: prowler, stolen like a thief, body dragging, teeming like snakes	McPhails's *The Dream Child* moon like warm breath, Tame Bear, Dream Child, drift, sleepy, hugs, licks	Small's *Imogene's Antlers* Imogene had grown antlers antlers decked with doughnuts antlers drying towels
Our Words: squirming black fingers, clawing branches, ghostly white silent shapes	Our Words: cuddly little bear, secure in mother's lap, cradling boat, drifting in sky	Our Words: silly, preposterous unbelievable, stupefied, speechless

Books with illustrations that show setting as the antagonist include the previously mentioned *Shadow*. In Margaret Hodges's *The Wave* (1964), Blair Lent creates a large, dark swirl of a tidal wave. Robert McCloskey's *Time of Wonder* (1957) shows the destructive force as a hurricane bends the lines of the trees. Students can use these illustrations to write settings that reflect the danger in the setting.

Picture books that may be used to stimulate writing as historical setting include Donald Hall's *Ox-Cart Man* (1979, early nineteenth-century New England), John Goodall's *The Story of Main Street* (1987, England from Medieval period through contemporary time), Tomie de Paola's *An Early American Christmas* (1987, New Hampshire, early 1800s), Byrd Baylor's *The Best Town in the World* (1983, nineteenth-century rural Texas) and Diane Stanley's *Peter the Great* (1986, Russia, late 1600s through early 1700s). These books provide excellent sources because students can see how an illustrator has to provide authenticity through illustrations and then they can try to make readers see the same historical settings through words.

Understanding characterization   Another problem emerges for the young author when he tries to write a story by weaving incidents together to form a plot. The characters in the story are intrinsic to the incidents that occur. The characters may be people or animals, but the young author often has problems developing his characters so they seem real. Again, the teacher may use literature to help the student. How do authors of some of the children's favorite books make the characters seem so real? Most authors use the following methods to develop characterization:

1. The author tells about the person through narration.
2. The author records the character's conversation with others in the story (dialogue).
3. The author describes the person's thoughts.
4. The author shows others' thoughts about the character.
5. The author shows the character in action.

Teacher and class can use these five methods to explore characterization and add new dimensions to the student's writing. In preparation, the students can read books about memorable characters with whom they are familiar, and notice how the author develops the character. They will undoubtedly discover basically the same techniques listed previously.

One class, for example, developed large charts, each of which depicted one of the methods of characterization. On the charts, the class described each method in greater detail and recorded examples from various pieces of literature. Finally, they experimented with that type of character writing. The following charts are examples of this activity as it was conducted with a fifth-grade group of accelerated students.

The Author Tells about the Character

Look For:
1. The author tells you what he thinks about the character.
2. The author describes what the character looks like.
3. The author describes the setting to let you know more about the character.

Example: Little Town on the Prairie (Laura Ingalls Wilder)[1]
A description of Mary, after she becomes blind—"Even in the days before scarlet fever had taken the sight from her clear blue eyes, she had never liked to work outdoors in the sun and wind. Now she was happy to be useful indoors." (p. 8)

---

[1]This and other excerpts from LITTLE TOWN ON THE PRAIRIE by Laura Ingalls Wilder. Copyright, 1941, as to text, by Laura Ingalls Wilder. Renewed 1969 by Roger L. MacBride. By permission of Harper & Row, Publishers, Inc.

The Author Records Conversations with Other Characters
(Dialogue)

Look For:
1. Speech lets you know about attitudes, dialect (where the character lives), education, mood, wisdom, *etc.*
2. What does the character actually say?
3. How does the character say it?

Example: Little Town on the Prairie
Ma's attitude toward drinking—" 'It's a pity more men don't say the same,' said Ma. 'I begin to believe that if there isn't a stop put to the liquor traffic, women must bestir themselves and have something to say about it.'

Pa twinkled at her. 'Seems to me you have plenty to say, Caroline. Ma never left me in doubt as to the evil of drink, nor you either.' " (p. 55)

The Author Describes the Person's Thoughts

Look For:
1. The thoughts describe the person's inner feelings.
2. Do they agree with the dialogue, or does the character have a hidden side?

Example: Little Town on the Prairie
Laura is not sure that there is always something to be thankful for—"Laura thought, 'Ma is right, there is always something to be thankful for.' Still, her heart was heavy. The oats and the corn crop were gone. She did not know how Mary could go to college now. The beautiful new dress, the two other new dresses, and the pretty underwear, must be laid away until next year. It was a cruel disappointment to Mary." (p. 106)

The Author Shows Others' Thoughts about the Character

Look For:
1. How do other people react to the character? Do they like him or her? Do they trust him or her? Do all people react the same way?
2. How does your reaction change after you read another's reaction?

Example: Little Town on the Prairie
Laura's various reactions to her sister Mary—"Mary had always been good. Sometimes she had been so good that Laura could hardly bear it." (p. 11)

" 'Oh, Mary,' Laura said. 'You look exactly as if you'd stepped out of a fashion plate. There won't be, there just can't be, one single girl in college who can hold a candle to you.' " (p. 96)

The Author Shows the Character in Action

Look For:
1. Actions suggest what the character is really like, what he or she is doing, and what he or she is planning.
2. Do the actions agree with the dialogue?

Example: Little Town on the Prairie
Laura is very conscientious and trustworthy—"Laura stopped, aghast. Suddenly she had realized what she was doing. Ma must have hidden this book, Laura had no right to read it. Quickly she shut her eyes, and then she shut the book. It was almost more then she could do, not to read just one word more, just to the end of that one line. But she knew that she must not yield one tiny bit of temptation." (p. 141)

These charts and examples provided references throughout the writing process as students wrote character sketches using these various techniques. Students may choose undeveloped characters from literature such as Snow White or Cinderella. It is interesting to write a more in-depth look at Cinderella through her thoughts and the thoughts of her stepsisters, her stepmother, and the prince. Character sketches may come from people students know or people they read about in the newspapers or see on television. It is interesting to speculate about what two sports stars are thinking as they shake hands or what two world leaders are like.

Studying these aspects of characterization obviously has two purposes: (1) it increases the students' understanding of literature and (2) it strengthens their knowledge of writing techniques.

REINFORCEMENT ACTIVITY

Select plot development, setting, or characterization. Develop a series of lessons that will improve understanding of the components of plot, the purposes for setting, or the methods an author uses to show character. Use your lessons with children, and share the results with your language arts class, or present your lessons to a small group of your peers.

Understanding symbolism and allusion through mythology   Both creative writing and literature appreciation profit from an understanding of symbolism and allusion. Research with gifted children shows that the creative writing of fifth-grade children improved when they were taught to symbolize their personal characteristics by using concrete objects. Andressen (1973) believed that if children looked at themselves symbolically, they would be stimulated to a more profound internalization, and, consequently, a higher degree of creativity in writing. In his study, Andressen had children discuss symbolism in places of worship and in advertisements. Next, they discussed and wrote about possible symbols for classmates, then they chose and wrote about symbolisms that would apply to themselves.

Greek and Roman mythology present heroes and gods as symbols for certain human characteristics. Jane Yolen (1977) believed that the study of mythology is important for building an understanding of ancient cultures, and for acquiring the background necessary to understand many of the allusions used in writing. A child cannot fully understand why some mountains are called volcanoes, or what "as strong as Atlas" or "as swift as Mercury" mean if he has no experience with mythology.

Mythological gods are symbols of many characteristics, so I have developed a series of lessons that allow children to study mythology, understand the symbolism in mythology, relate the mythological characters to well-known present-day or historical people, and relate the characteristics portrayed by the mythological characters to their own self-images.

The following activities were developed with sixth-grade students. They had been studying Greek and Roman mythology in Reading-Language Arts, so the lessons were a natural outgrowth of a classroom activity. First, the students reviewed characters from Greek mythology, and discussed their characteristics and what they symbolized:

Poseidon	god of the sea; Greeks prayed to him to protect them when they took a sea voyage.
Ares	god of war, symbolized by the vulture; Hades liked him because he increased the underworld population.
Heracles	extremely strong Greek hero; accomplished such tasks as killing the nine-headed Hydra.
Zeus	ruler of Mount Olympus, Supreme ruler, god of the wealthy.
Apollo	god of the sun, truth, music, medicine, and prophecy
Aphrodite	goddess of love and beauty; presided over girlish babble and tricks.
Hermes	swift, messenger of the gods; god of commerce, orators, and writers.
Demeter	goddess of crops
Artemis	goddess of the moon, guardian of cities, young animals, and of women
Athena	goddess of wisdom
Dionysus	god of wine, symbol of revelry; he gave Greece the gift of wine; driven mad by Hera.
Hephaestus	god of fire and artisans; an inventor, showed great workman-ship.

The students searched for advertisements that used mythological characters and discussed why they were good symbols. For example, "Why is Mercury a good symbol for a car? Why is Atlas used as a symbol for a tire company? Why has a camera company chosen the name Argus?" (Argus was a giant with a hundred eyes.)

Next, the students described what products might use names like Heracles, Hermes, Apollo, Poseidon, and so forth, as advertising symbols, and wrote advertisements for their created products.

In the next part of the sequence, the students discussed well-known personalities from the present and past. Each chose a personality and listed the characteristics they felt described the person. As a final activity in that session, they wrote short descriptions of a person in terms of Greek heroes. One of the students wrote the following symbolic description of Hitler:

> *Characteristics of Hitler: paranoid, insane, greedy, hated the Jewish people, powerful speaker, influential, a killer, an egotist, thought his was the master race.*

Hitler and Mythology

> *Since Hitler was a madman who envisioned conquering the world, I would say that Ares and Dionysus would describe Hitler well. Ares was the god of war; since Hitler started World War II, Ares would be a good mythical symbol. I believe that Hades would have also liked Hitler because the war would have sent so many new people to the underworld. Dionysus also is an excellent symbol for Hitler because Dionysus caused a great deal of revelry with his gift of wine but he later went insane. Hitler also caused revelry when he first influenced the German people, but he went insane and his actions caused a great deal of misery.*

Finally the students thought about themselves and identified characteristics that might be used to describe them. They wrote short symbolic descriptions of themselves in terms of mythology. One boy wrote:

The Way I See Myself and Mythology

> *I personally think that the Greek gods who closely resemble my beliefs are Apollo, Athena, and Hephaestus. Apollo was the god of music and medicine. I hopefully will become a doctor and an accomplished musician. Apollo was also the patron of truth, which is essential for friends to have between one another. Athena was the goddess of wisdom. I always want to know more about everything. Hephaestus was the god of fire and the artisans. He was a great engineer. Hopefully I'll be able to invent patentable items.*

REINFORCEMENT ACTIVITY

1. Read several selections from mythology. (Do not use Greek mythology.) Make a chart showing the characteristics of the heroes and what they might symbolize. Describe yourself in terms of these characteristics.
2. Make a collection of current advertisements that utilize symbols from mythology. Why (or why not) are they appropriate?

EXPOSITORY WRITING

Studies by Britton et al. (1975) and Applebee (1984) reported that expository writing is the most common writing used in classrooms. Consequently, it is essential that we cover ways to develop writing that reports, persuades, and explains.

According to Frye, Baker, and Perkins expository writing is "Explanatory writing, the presentation of facts, ideas, or opinions, as in short forms like the article or essay, or in longer nonfictional forms like the history, scientific treatise, travel book, biography, or autobiography. Expository writing is often treated as a larger category including exposition as a major element, but making use also of argument, description, and narration. It is contrasted with imaginative writing or creative writing" (1985, p. 184).

There is currently renewed interest in the role of expository writing in all areas of the curriculum. Ideas for developing writing to learn are found in current journals as well as in books published by the National Council of Teachers of English. For example NCTE publications such as Gere's *Roots in the Sawdust: Writing to Learn Across the Disciplines* (1985) includes ideas for all content areas, Golub's *Activities to Promote Critical Thinking: Classroom Practices in Teaching English* (1986) includes ideas for speaking and writing across the curriculum, and Fleming and McGinnis's *Portraits: Biography and Autobiography in the Secondary School* (1985) includes both reading and writing activities, many of which may be adapted to upper-elementary and middle-school classrooms.

In this section we consider ways to help students write reports and ideas for writing in the content areas. We conclude with a writing approach for the study of biography. As we progress through different approaches we also see a close relationship to the writing process discussed in the previous chapter.

Writing a Report

The third-grade students who wrote an outline of their class trip, using questions and answers, were actually developing a report (see pp. 283–284, chapter 8). The structure of the report included a central theme: the train ride. Each paragraph had a main idea relating to the central theme. The answers to the main-idea questions provided supporting ideas, which developed the three main ideas under the central theme. This exercise was an excellent readiness activity for the time when the students would have to produce individual, well-planned reports or other expository compositions. As the elementary students progress into longer expository writing they will need to consider many of the topics previously discussed under the writing process, including deciding on the audience or purpose for the writing, deciding on the subject, narrowing the subject to one that can be covered realistically in a short composition, gathering ideas and information, organizing ideas, and writing the report.

Deciding on the audience and purpose   Table 8–1 showed numerous writing activities, purposes, requirements, and audiences that are included under expository writing. Activities that focus on expository writing requirements include weather and temperature charting, writing news stories, filling out forms, writing autobiographies and biographies, developing book reports, composing editorials, writing class minutes, and composing reports for projects in science, social studies, and so forth. The audiences could range from known class members, teachers, parents, and friends to wider unknown audiences. The purposes for most of these ideas are to inform, to compare, to compile, to convince, and to report. The requirements emphasize accuracy, research, and organization.

As pointed out in our earlier discussion of the writing process, students need to make decisions about the audience for their reports, the purpose for their report, and the requirements for that type of writing.

Deciding on a subject   Choosing the subject for a composition that requires collecting information and organizing ideas is usually more challenging than deciding on a subject for a short paragraph that can be written quickly. First, the student should be interested in the subject. Students will be more motivated to write if they consider the subject both interesting and important. General subjects arise from many of the content areas. Often, an area closely related to a student's personal interests provides an excellent subject. For example, a group of fourth graders became very concerned because a stream near their homes was cluttered with bottles, cans, and other offensive garbage. They investigated the dangers of this kind of pollution, what other communities had done in similar circumstances, and recommended ways to improve their stream. Very good reports developed from their study, and they even wrote some "letters to the editor" and to community groups to attract community interest.

Students must not only be interested in the subject, they must also be able to uncover information about it. They should look for information in textbooks, journals, magazines, encyclopedias, newspapers, and so forth. Students should look through class, school, and public libraries to see how much information is available to them on various subjects.

Narrowing the subject   Many subjects are just too broad for the elementary student. Topics such as horses, energy, water, transportation, air travel, and the Revolutionary War are obviously too broad to be approached in a short composition. These topics would have to be narrowed in scope to find some main ideas that might be developed in a composition.

There are several ways to help students narrow a subject. A processing technique in which the students develop a "Web of Interest" is one way to narrow a broad subject into manageable subtopics. Brainstorming helped a group of fifth graders look at the general subject of "Water" and find specific topics and main ideas that could be developed in short compositions. The fifth graders developed a Web of Interest.

They narrowed the broad subject of water to four topics that might be studied: water sports, sea life, energy, and weather. Each specific topic has several main ideas to use in organizing the composition. This technique is successful with elementary students; it can be used with various grade levels by changing the subject matter.

Brainstorming is used to identify interesting water-related topics and narrow the subject to one that can be developed in a composition.

Students can also narrow topics by studying the tables of contents in several books on a general subject. For example, the subject of camping is interesting to many elementary children, because they often go camping with their families or with scout troops. After searching several books on camping, a fourth-grade class developed the following list:

General Topic: Camping

1. Camp cooking
   a. building a campfire
   b. camping stoves
   c. cooking utensils
   d. selecting food for camping
   e. camping recipes
2. Camping equipment for sleeping
   a. tents
   b. sleeping bags
   c. cots and air mattresses
   d. pickup campers
   e. trailers
   f. motor homes
3. Special clothes to take camping
   a. hiking shoes and clothing
   b. raincoats or ponchos
   c. cold weather wear
   d. clothing for horsepacking
4. Skills needed for camping
   a. using an ax
   b. using knives
   c. using a rope—knots, hitches, splices
   d. reading a map and compass
   e. building a fire
   f. packing a backpack
   g. setting up and taking down camp

5. Safety in camp
   a. first aid
   b. pure water
   c. water safety
   d. insects
   e. snakes
   f. wild animals
6. Survival
   a. navigation
   b. getting help
   c. emergency food
   d. emergency shelter
   e. finding water
7. Transportation
   a. backpacking
   b. canoes
   c. bicycles
   d. horsepacking
   e. boats
   f. trailers
8. Where to camp
   a. national parks
   b. state parks
   c. wilderness areas
   d. private campgrounds

This general search for topics yielded many specific, interesting topics that could be turned into meaningful compositions. As another advantage, this approach teaches use of the table of contents for locating information; many students will not have had much practice with the reference skills important for locating information easily.

Another way to narrow a subject is to have children list some questions they would want answered if they were going to study a general topic. Page 283 presents such an approach under prewriting activities that encourage students to list and categorize.

Gathering ideas and information   The three methods for narrowing a topic proceed naturally into the next step: searching for ideas and information to include in the composition. After the students who were studying Native Americans (see p. 285) changed or added questions to the categories they had chosen, the teacher explored the library with them. They looked through card files, tables of contents, indexes, and so forth, to discover sources of information. From these resources, the students compiled information to answer each of their questions.

After narrowing the subject and developing the important questions, students might write each of their questions on a separate index card or sheet of paper. Then, when they find information that pertains to the question, they can write it, in their own words, on the appropriate card or paper. Placing the items on separate cards helps students rearrange their questions and information into the most logical order. They find it easier to deal with one main idea at a time, rather than with a disorganized pile of information. For example, a card asking a main idea question about camping might look like this:

An Example of Facts Found for Survival While Camping

How can you make an emergency shelter if you are lost?
Start looking for your shelter in the day.
Shelter should protect from cold, wind, and rain.
Build a lean-to from about 1-inch thick poles tied together with vines or fishline.
  Draw a picture to show this in my report.
Put evergreens or leaf branches on poles to protect yourself from rain.
  Draw a picture to show this in my report.
Make the lean-to very steep.
A cave or overhanging rock makes a shelter.
Build a fire in front of the shelter.
A fire will make you warm and be a signal.
If you are in snow dig a cave in the side of a drift.
Put pine boughs on the floor of the snow cave.
A snow cave is a good insulator.
Leave air holes in the cave.
Mark the outside of the cave so searchers will know you're there.
  Draw a picture to show this in my report.

Organizing ideas and deciding about organizational patterns   After students finish gathering information and ideas on cards or sheets, they are ready to start analyzing, selecting, and ordering their information. (If they have not put their information under main idea questions, they need to classify the information under several main ideas before proceeding. Webbing, discussed previously, is another good way to accomplish this task.) They may find they have too little or too much information about one main idea. Some information may be important, whereas other information may seem trivial. Some information may require verification from sources other than those they have already used.

Students now need to make decisions about sequencing information and organizational patterns that may improve the presentation of their information. As students arrange and rearrange their information cards they frequently discover the best organizational patterns to use. The various organizational patterns presented on pages 289–290, chapter 8, should be reviewed or introduced. These patterns include ordering by chronological order; spatial concepts; familiar to unfamiliar; problem to cause of problem, to possible solutions of the problem; problem to effect of problem, to possible solutions to the problem; and question to answer formats. These formats provide logical organizations for reports in many content areas.

Turning the outline into a composition   Now the writer must develop the main ideas and important details into a closely related unit. In literature the plot has a beginning, middle, and end. The same progression appears in factual writing, but we do not think of it as building the action to a climax; instead, the beginning introduces the reader to what the writing will be about, and catches her attention; the middle section develops the main ideas and supporting information; finally, a summary, or concluding, paragraph helps the reader pull together the ideas in the composition and gives the reader the feeling the composition is finished.

The introductory paragraph is important; it both introduces the theme of the composition and attracts the audience's attention. When we pick up a magazine, how do we choose the articles we are going to read? The author may ask a question that arouses our interest, or he may make a startling statement or present an interesting, little-known fact. The author may show us how important something is to us personally. Questions such as "Would you like to save $100 a month on your grocery bill?" or "Do you know that the new _____ cars are considered unsafe and are being recalled?" usually attract our attention because the implied subject matter of the article affects each of us in a highly personal way. If these two questions were used, we would expect the introductory paragraph of the first example to introduce us to a piece about various ways to save money in the grocery store, and of the second, to introduce us to a critical report on the dangers of a specific new car. The writer has presented the issue to the reader, and the reader expects the remainder of the article to develop facts and information that will help solve the problem. No matter how attention-getting an opener may be, it is a poor introduction if it raises false expectations about the content of the composition that follows.

The body of the work develops the main ideas related to the introductory paragraph. If students grasp the concept of a paragraph, writing the body is not too difficult. If students understand the need for rational sequencing of ideas, writer as well as reader will be able to move logically from one point to the next.

Some vocabulary terms will help students move from one idea to another. If students are writing in chronological order, for example, they may use words such as *first, second, third, before, during, after, yesterday, today, tomorrow, next, last,* or dates, such as *in 1980, in 1982, in 1989* and so forth. These words signal the developing order of the ideas. If students organize their thoughts according to problem, causes, solution, or problem, effect, solution, they will want to show the relationships among the ideas. Words such as *because, therefore, since, too,* and *when* signal the relationships in this kind of organization.

Sometimes, certain words signal that the composition is drawing to a conclusion. Terms such as *in conclusion* or *to summarize* let us know that a report's main ideas are going to be summarized, or that the beginning idea is going to be restated (the restatement is not in the identical words used previously in the report). Writers are thus completing the contract they made with the reader by introducing the article the way they did, and are letting the reader know the contract has been fulfilled.

Let us review the three parts of the composition by putting them into diagram form (see figure 9–3). This diagram is quite different from the diagram for a novel. In the written report, we want to present an introduction that quickly interests our reader, then develop the ideas that move toward a conclusion or summary.

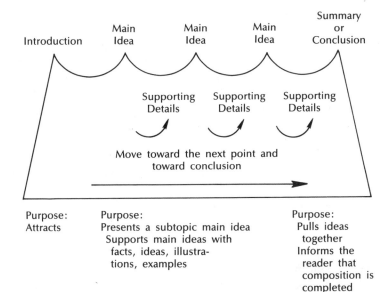

FIGURE 9–3
A diagram may be used to identify the major parts of a composition

Revising   In teaching composition skills, the teacher is concerned with the student's ability to develop clear ideas and to organize and elaborate on them. Consequently, teacher feedback, student evaluation of the writing, and any rewriting should focus on the clear development of those ideas. Schwartz defined the rewriting connected with composition in this way:

> Rewriting is not just recopying neatly, minus a few punctuation errors. It is not just fixing what is wrong. Rewriting is finding the best way to give your newly discovered ideas to others; it's a finishing, a polishing up, and it should be creative and satisfying as any job well done. (1977, p. 757)

If teachers interact with students during the entire writing process, students will have both positive feedback and an opportunity to make improvements during each step of the process, instead of having to wait for teacher reaction to the finished product. The writing conference provides opportunities for the teacher to interact with students when they select a topic, narrow the topic, gather information, organize information, develop a working outline, and when they write the final composition. The teacher can react as students discuss and write during each part of the process. When students write and the teacher reacts, the students can decide on changes before the final writing, so that only minor changes may be necessary at that point.

Writing and the Content Areas

Two different types of writing focus on the needs of students in various content areas and the specific content materials. Some activities use content area sources to improve students' understanding of writing and the writing process, whereas other activities emphasize using writing to learn in the content areas. Many content writing activities reflect both goals, although one of the goals may be more important for a specific teacher and a specific content area. As you read the various activities described in this section, consider how you might include both goals in your classes.

To show the differences in the two goals, Gere (1985) stated that writing to learn approaches are designed to do more than develop writing fluency; the writing to learn strategies are designed to foster abstract thinking, to encourage learning, and to encourage teachers to guide students in their quest for knowledge. There are numerous ways learning through writing may be accomplished. Gere stated, "there is no 'right way' to use writing to learn. Although the general approach is solidly grounded in theory, there are no quick fixes or rigid systems for implementing writing to learn. Rather, as is true in all good teaching, it is the responsibility of the individual instructor to select from a wide range of approaches those which seem best suited to accomplish course goals" (p. 5).

In this section we consider some of the ways teachers can both expand writing opportunities and increase learning. Hopefully, most of these activities will increase both writing and learning.

Stimulating writing and learning through nonfictional texts   Activities developed around nonfictional literature help students consider the author's point of view, realize the importance of using facts to support nonfictional writing, and identify strategies that could help students in their own expository writing. Hairston (1986) used simulation activities with older students to place them in the role of a specific professional writer and to have them consider how that author may have responded during the writing process. For example, Hairston selects a good example of nonfictional literature and then helps students construct imaginary scenarios in which they try to reconstruct the writer's actions, both mental and physical, from the time of the initial idea to the time of the completed product. Within this scenario, students consider what might have motivated the author to write the text; how and why the author chose a certain approach to attract the readers' attention and gain the readers' sympathy or understanding; what techniques the author used to substantiate the specific point of view including examples, experiences, comparisons, details, and images; why the author chose those techniques; and how the author shaped the text to develop that specific point of view.

During this analysis and simulation, students should discover that good professional authors of nonfiction support their writing with facts and specific details rather than with generalizations and unsupported opinions. In addition, students discover how professional authors deal with some of the same problems that may cause them difficulty. To extend understanding of specific writing techniques used by authors, teachers can develop files of literature, articles, texts, and xerox pages that illustrate solutions to various writing problems such as gaining attention, developing support for positions, shaping articles to develop positions, using figurative language and images to persuade, and developing conclusions. These techniques can then be applied in the students' own writing.

Children's literature authors of nonfictional literature that are excellent for this activity include Franklyn Branley, Joanna Cole, Russell Freedman, Patricia Lauber, Hershell and Joan Nixon, Dorothy Patent, Laurence Pringle, Jack Denton Scott, Millicent Selsam, Seymour Simon, and Jerome Wexler.

You might try this activity with the 1988 Newbery Award winner, Russell Freedman's *Lincoln: A Photobiography* (1987). This book was selected by the American Library Association as the most distinguished contribution to children's literature for that year; students will gain an understanding of literary quality as well as characteristics of excellent nonfictional literature.

To teach the reading of nonfictional literature to remedial reading classes, Carnes (1988) developed units in which students write nonfictional books modeled after content area textbooks. (I would strongly recommend that you do this type of activity with the literature works of the nonfictional authors previously listed.) The units incorporate the writing process as students proceed from prewriting activities through printing and book binding. At the prewriting stage, they study nonfictional works, list characteristics of nonfictional literature, develop standards for their own books, choose their topics, determine structure for their books, brainstorm ideas and questions related to their topics, arrange categories into topics, subtopics, and important details, and read materials to gain more knowledge about their topics. During the research and notetaking phase, Carnes models skills such as finding key words and concepts and summarizing material before students collect their own notes. During the writing phase, students receive guidance in integrating notes, determining headings, sequencing information, and inclusion of specific facts. During this phase they also place their rough drafts into the word processor. During the revision phase, the students refer to the original standards to help them approach revision and to focus attention during teacher and peer conferences. After completing several rough drafts, the writings are ready for final printing and for binding in the form of books. The books are then shared orally, read by students, compared with previous classes, and second copies placed in the class library.

Precise or Summary Writing

Finding a main idea in written texts and writing a main idea for one's own writings are related skills and are often difficult tasks for elementary students to perform. Taylor (1986) found that fourth- and fifth-grade students had difficulty finding and expressing the main idea when they were asked to write summaries. Teaching children to write summaries integrates reading, writing, listening, and speaking with the various content areas. Hidi and Anderson (1986) found that effective summary writing teaches students how to discover information important to the author and how to analyze, paraphrase, and summarize this information. Key elements in this approach require knowledge of paragraph organization and identification of topic sentences, main ideas, and important details; require identification of key words such as ''in summary''; necessitate identification of text clues such as underlining, italics, and bold face print; and require ability to supply synonyms and paraphrase text.

Hidi and Anderson recommended that teachers begin with short passages and gradually increase passage length as students improve their summarizing skills. In addition, early experiences should use simple narratives and texts that are well-organized so that important text elements are obvious to the students.

Teaching summary writing is enhanced by modeling and by approaches that proceed from introduction of the skill, to direct instruction by the teacher, to teacher-guided application, and finally to independent practice. Notice how this sequence is incorporated into the following effective strategies identified by Bromley and McKeveny (1986) for teaching summary writing: (1) establish purposes and provide rationales for using summary writing; (2) use demonstrations and simulations to develop understandings and to provide opportunities to study topics such as paragraph organization and location of main ideas; (3) use verbal practice to help students to identify key words, supply synonyms, and paraphrase text; (4) try group composition of summaries before expecting students to write individual summaries; (5) use models of acceptable summaries to help students compare, evaluate, and revise their own summaries; (6) use folders in which students keep their summaries for review and for study; and (7) provide study time during which students can use their summaries to review important information from various content areas. Bromley and McKeveny emphasized the importance of long-term instruction in summary writing. Their research showed that "using precise writing twice weekly over an entire semester was more effective than daily 45-minute lessons for three or four weeks" (p. 393).

Writing that requires paraphrasing and summarizing may take place within any content area and may emphasize various types of activities. Shugarman and Hurst (1986) provided the following additional suggestions for paraphrasing and summarizing:

1. Have students elaborate on paraphrasing and summarizing by creating captioned cartoons, graphs, or maps that reflect the information in the text.
2. Ask students to paraphrase and summarize compositions written by other students. Then have them share and discuss their summaries with the original authors.
3. Divide the class into teams and choose a panel of experts. Assign text passages to be read and paraphrased by each team. Have the panel select the best paraphrased text.
4. Have students bring in relevant text from library materials, newspapers, and magazines. Ask them to write summaries of the materials and compare and discuss the various solutions.

Writing Book Reports

Book reports, whether oral or written, are used by almost every elementary teacher. Through the creative use of book reporting, teachers can provide opportunities for many types of writing as well as strengthen the connections among writing, reading, and literature. Literature includes books from all content areas, thus book reporting also heightens learning in the content areas.

As a review of the writing discussed in this chapter, I have provided some suggestions for writing book reports that are in the form of expressive, poetic, and expository writing:

Expressive Book Reports

1. Write a personal response to a book in your journal. Include in your reaction to the book what you liked or did not like about a character, the plot of the story, or the language. How would you change any of these elements if you could rewrite the book?
2. Choose one of the characters in the book and keep a dialogue journal with that character. Write questions you would like to ask that character about the character's feelings, reactions, beliefs, motives, or actions. Exchange your dialogue journal with someone else in the class who has read the same book. Ask the other person to write the answers to your questions as if he or she were that person.
3. Keep a learning log about a piece of nonfictional literature. Write reactions to the author's ability to interest you in the subject and to present information so that you believed it was true. What were the points the author was trying to make? Did he or she accomplish this task?
4. Keep a diary as if you were one of the characters in a book. Include personal reactions to things that happened to you in the story.
5. Pretend that you are the author of the book you just finished and that you were motivated to write the book because of entries in your writer's journal. Write the entries that might have motivated you to develop the setting, the characters, the plot, or the theme of the book.
6. Pretend that you are the author of a book you have read. Write the inventory of writer's experiences from your writer's journal that you would need to have before you wrote the book.

Imaginative and Poetic Writing

1. You are an author like Lloyd Alexander, or Cynthia Voigt, or Lois Lowry who likes to write sequels to stories or likes to tell stories about other members of the family. Choose a book that could have a next installment. Choose one of the characters from the book and tell us what happens after the first story ends.
2. Choose a character from your book and write a letter that character might write to another character in the story. Be sure that you write the letter from the viewpoint of that character. You might choose a problem that your character must solve and have the character write to a friend or relative in the book asking for advice or explaining why your character responded in a certain way.
3. Write a limerick about one of the characters in your book. (This activity should follow writing limericks in class.)
4. Identify the main theme in your book. Write a cinquain about that theme. For the title select one word that expresses that theme. On the next line write two words that describe the theme. On the third line write three action words that express the theme. On the fourth line write four words that express your feelings about the theme. Complete your cinquain with

the title or a word that means the same as the title. (This activity should follow writing cinquains in class.)

5. Identify the major conflicting forces in your book (for example, good versus evil). Write a poem in the diamante format that expresses the conflict in your book. Line 1 should be one of the conflicting forces. Line 2 should have two words that describe that force. Line 3 should have three action words related to the force. Line 4 should have four transitional nouns or a phrase about the conflict. Line 5 should have three action words related to the opposite force. Line 6 should have two words that describe the opposite force. Line 7 should be the opposite force. (This activity should follow the writing of diamante formats in class.)

6. Draw and label the plot development from your book.

7. Pretend you are the set designer who will construct the set when your book becomes a stage play. Divide the book into three acts. Write detailed descriptions of the settings for each of the acts. Make your readers "see" the stage settings.

8. Choose a picture book and describe the setting as if you could not use illustrations in the book. Make your readers "see" and "feel" the settings through your choice of words.

9. Create a dictionary of words and phrases that suggest the mood of the story. Add words and phrases that you believe would have been appropriate in addition to those chosen by the author.

10. Choose a character from your book and write a character sketch showing how the author feels about the character, how the character feels about him- or herself, and how other characters feel about that character.

11. Choose an incident in the book in which you would like to interact with the character. Pretend that you are one of the new characters introduced into the story. Write the incident in which you interact with the character. You may use your incident to clarify information or to try to persuade the character to do something. You may not, however, change the character from the characterization developed by the author.

12. Create a postcard that one of the characters might have sent in the story. Select one of the scenes from the book to draw on the front. Write the postcard as if you are that character.

Expository Writing

1. Write an appropriate weather report for several scenes in the book or for each chapter in the book.

2. Pretend you are the author of an informational book or a biography you have just read. Write a paragraph describing the audience for your book. How old is your audience? How much information on your subject does your audience have? How did you obtain information about your audience? How did this information influence the way you wrote your book?

3. Pretend you are the author of an informational book or a biography you have just read. Write a paragraph describing your purpose for writing your

book. What was your purpose? Why did you think that purpose was important? What decisions did you have to make when you wrote for that purpose? After you read your completed book did you still think you were successful in presenting your purpose? How would you know if you were successful?

4. Pretend you are the author of an informational book or a biography. Write three paragraphs about your book: a paragraph describing how you decided on your subject, a paragraph describing how you narrowed your subject to one that could be covered in your book, and a paragraph describing how you conducted research on your book.

5. Pretend you are the author of an informational book. Draw a Web of Interest that you might have developed to help you organize your book.

6. Choose an exciting incident in the story. Write the incident as a news article. Include accurate information on who, what, when, where, and why.

7. Choose an incident in the book that could be controversial. Pretend you are one of the characters in the book and write an editorial to the newspaper in which you try to persuade readers that your point of view is correct.

8. Write a summary of your book in the form of a cartoon, a graph, or a map. Write a caption explaining your content.

9. Diagram your book report according to the major organizational patterns used by the author. These organizational patterns include chronological order; spatial order; familiar to unfamiliar; problem, to cause of problem, to possible solutions of the problem; problem to effect of problem, to possible solutions to the problem; and question to answer formats.

REINFORCEMENT ACTIVITY

1. Select a content area and develop a lesson plan for teaching beginning summarizing skills within that content area. Include in your lesson a rationale for teaching summarizing, a modeling in which you explain how to write a summary, a teacher-guided application, and an independent practice that encourages students to apply the summary skills within the content area. (Note: This lesson may require several days to complete.)

2. Develop a file of examples of expository writing that you may share with students who are having difficulties with writing introductions, substantiating facts, organizing information, presenting arguments, or drawing conclusions.

DEVELOPING A UNIT BASED ON COMPOSITION

This unit is developed from evaluative criteria and teaching suggestions developed by Norton (1987, chapter 12, "Nonfiction: Biographies and Informational Books"), and from recommendations by Johnson (1986) and Fleming and McGinnis (1985). You may combine this unit with "The Literary Approach to the Study of Biography" unit developed in chapter 11 for an extended study of biography.

FOR YOUR PLAN BOOK
A Composition Approach to the Study of Biography

This composition approach to biography first analyzes short biographies to discover characteristics of biographical writing and then encourages students to conduct interviews and to complete research in order to write their own biographies. Discussion and other activities related to the approach encourage students to read reflectively, to analyze the impact of biographical writing, to consider the types of information that make biographical writing interesting, to develop interviewing skills that will obtain that information, to complete research that will obtain that information, and to write factual and interesting biographies. This approach emphasizes the interaction of reading, writing, oral language, and listening through the focus on biographical literature. As you develop the unit with students you may place it in the framework of the writing process.

Unit Objectives

1. *To develop an appreciation for biographical literature.*
2. *To identify characteristics of the biography genre.*
3. *To differentiate between biography and fictional literature.*
4. *To analyze techniques used by successful biographical writers that make their characters seem real.*
5. *To develop interviewing and research skills needed to complete biographical writing.*
6. *To develop and to reinforce composition skills and to apply information gained from biographical literature to writing.*

References Discussed in This Unit

Gish, Lillian. **An Actor's Life for Me!** *As told to Selma G. Lanes. New York: Viking Press, 1987.*

Rothman, John. "Taking It Like Little Troupers," **The New York Times Book Review.** *Sunday, Nov. 8, 1987.*

Fritz, Jean. **Homesick: My Own Story.** *New York: G. P. Putnam's Sons, 1982.*

Numerous biographical sources (found in Norton, 1987; Sutherland and Arbuthnot, 1986; and Huck, Hepler, and Hickman, 1987).

Procedures

1. *Read and analyze the content of several shorter biographies such as Lillian Gish's **An Actor's Life for Me!** Discuss the answers to the following questions:*
    a. *What characteristics make this literature biographical? Are all the characters real people who actually lived? How do you know? Are the dialogue, thoughts,*

and actions authentic and did the author support them with research, facts, and interviews? Is the setting authentic for the time of the actual biographical character? Does the plot follow the life of the real character? (As you lead this discussion, ask students to support their answers with evidence.)

b. *What differences are there between a fictional development of a story and this nonfictional work? Read Jean Fritz's foreword to her fictionalized autobiography,* **Homesick: My Own Story.** *How does Fritz differentiate between fictionalized and nonfictionalized work in this following quote: "Since my childhood feels like a story, I decided to tell it that way, letting the events fall as they would into the shape of a story, lacing them together with fictional bits, adding a piece here and there when memory didn't give me all I needed. I would use conversation freely, for I cannot think of my childhood without hearing voices. So although this book takes place within two years from October 1925 to September 1927, the events are drawn from the entire period of my childhood, but they are all, except in minor details, basically true" (unnumbered foreword).*

c. *Who is the biographical character? What period in the character's life did the biographer emphasize? What techniques did the author use to make you believe that the character was real? Was the biographer successful? Why or why not?*

d. *What types of information did the author include? Were any types of information left out of the biography? What is the impact of this information? Why do you believe the author included or left out specific information? Do you believe that the age of the intended audience influenced the author? What other purposes for writing might influence the author?*

e. *Who are the supporting characters? Who are the people who most influenced the life of the subject? How important are these people in the development of the subject's character? Are they positive or negative influences? What techniques did the author use to make you believe that the supporting characters are real? Was the author successful? Why or why not?*

f. *How did the author authenticate the biography? What documentation is included? What research information is included or inferred in the text? How did the author get "to know" the biographical character? What other ways could an author authenticate this character and this time period? (The Lillian Gish biography provides interesting discussion material because the biographical character told her own story to the author. Consequently, interviewing skills would be necessary. In addition, the biographical character is made to seem very real through numerous personal anecdotes that reveal her emotions, her dreams, her times of happiness and sadness, and her interactions with her family and theatrical friends. The biography includes enough references to famous actors and directors so that students could do additional research about this time period and the people who influenced Lillian Gish.)*

2. *Choose a biographical character or a supporting character from a biography. Give a short oral presentation to the class. Make the class believe that the biographical character is real. For example, choose one of the anecdotes developed in the Gish*

biography. Tell this anecdote as if you were either Lillian Gish or one of the supporting characters involved in the incident.

3. Choose a scene in the biography in which the biographical character interacts with one or two of the supporting characters. With a classmate, prepare the interaction as a presentation for the class. Make the class believe that the interaction is real.

4. Choose an incident in the life of the biographical character or an emotion expressed by the biographical character. Pretend that you have the opportunity to interact with the biographical character. What additional questions would you ask? What information do you think you would learn? Write this incident as if it were an anecdote in the life of the biographical character. For example, in the life of Lillian Gish you might be interested in having her expand on the following belief:

> Our greatest fear was of being separated from Mother. She gave us security. Father brought us insecurity. As I grew older, I often wondered which was the most valuable gift. Insecurity taught me to work as if everything depended on me, and to pray as if everything depended on God. Somehow, given enough insecurity, I learned to do for myself, never counting on others to do things for me. Wherever Mother was, we had love, peace, and sympathy. Yet, without the insecurity Father brought, the blessings Mother provided might have left us weak, dependent, and helpless. (p. 10)

5. Choose one of the supporting characters developed in the biography. Try to find out more information about the character and write a short biography from the viewpoint of that character. Include the original biographical character in your biography, but make the biographical character a supporting character. For example, in the life of Lillian Gish you might try to find information about her sister Dorothy Gish, actress Mary Pickford, film director D. W. Griffith or any of the people involved in the early melodramas that played in small theaters across the country.

6. Read a review of your biography such as "Taking It Like Little Troupers" in the **New York Times Book Review.** What techniques and incidents did the reviewer emphasize? Do you agree with this evaluation of the book? Why or why not? If you were going to write a review of the biography what would you emphasize? Write a paragraph that develops your own evaluation of the biography.

7. Make a list of all of the ways that authors and class members made their biographical characters seem real and also allowed readers or listeners to learn the facts about the person and the time period. Include an example of each technique.

8. Choose a person you know such as a family member, a neighbor, or a classmate. Consider how you would gather information to write a biography about that person. How do you think the biographer of Lillian Gish encouraged her subject to reveal the interesting anecdotes and other factual information? Brainstorm with the class to identify types of questions and questioning techniques that you might use to gather information about someone that you know. After the questions and techniques are identified, role play your interviewing experiences. Consider other sources of information such as school yearbooks, newspaper articles, photograph albums, journals, courthouse documents, and people who know the person. Select an interesting anecdote, a personal characteristic, or an experience that would make an interesting short biography. Compose a rough draft of your biographical charac-

ter. Share the rough draft with a group of your peers. How did you make the character seem real? Are there any other techniques that you could use to improve understanding of the character? Organize your introduction, body, and conclusion. Does each part reinforce the qualities or the experiences you are emphasizing? Continue the rewriting process until the biographies seem like biographies of real and interesting people.

9. *Write a biography about a person you do not know. Choose a person whose biography requires library research and other documentation. Consider possible sources of information and how you will gather that information. Choose a character about whom there is sufficient source material. Choose an interesting period in that person's life. Accumulate as many reference materials as you can find. For example, speeches and writings by the person, reproductions of the front page of a newspaper that might have been available at that time, biographies and autobiographies, newspaper and magazine articles about the person, letters, diaries, and so forth. Write a biography that uses the facts to support a person's possible thoughts, actions, associations, concerns, and so on. Make your readers believe that the person is real. As you are writing your biography, share the biography with your editing group and with your teacher during conferences.*

SUMMARY

The chapter focused on three specific types of writing and methodologies for developing expressive, poetic, and expository writing.

Expressive writing is writing that is very close to speech and to the writer. It is relaxed, intimate, and comprehensible to the writer and to others who know the writer well. Writing activities that stimulate the production of expressive writing encourage student's personal responses and emotional reactions. These activities include journal and diary writing, learning logs, and writer's journals.

Imaginative and poetic writing includes original writing that uses both imaginative and experimental thinking. This type of writing includes fictional prose, poetry, and drama. The chapter focused on developing the creative writer through stimulating fictional story writing and poetry. Methodologies emphasized the enhancement of creative writing through studies of plot development, setting, characterization, and mythology as well as stimulating environments.

Expository is explanatory writing that presents facts, ideas, or opinions. It is the kind of writing that researchers show is dominant in the schools. Methodologies developed in this chapter focused on writing a report, writing and the content areas, and writing book reports. The book reports demonstrated how writing such reports could take on an expressive, a poetic, or an expository function.

The chapter concluded with a unit of study that combined a study of biography and the writing of biographies. The unit included the integration of the language arts through reading, literature, writing, and oral language.

ADDITIONAL ACTIVITIES TO ENHANCE UNDERSTANDING OF EXPRESSIVE, POETIC, AND EXPOSITORY WRITING

1. Develop a lesson in which you introduce journal writing, learning logs, or writer's journals to students. Explain to them why these writing sources are important and how they might be able to use them to expand and enhance other types of writing.

2. Interview a teacher about how he or she teaches poetry writing. How often does the teacher ask students to write poetry? How does the teacher stimulate students' interests? What instructional techniques does the teacher consider effective? How does the teacher encourage students to share poetry?

3. Interview a student about imaginative, poetic writing. What subjects does the student like to write about? How does the student approach a creative writing task? Ask to see examples of writing that the student would like to share with you.

4. Develop a file of poetry sources that you could use to stimulate poetry writing.

5. Write a lesson plan for poetry or story writing. Account for the various stages in the writing process.

6. Diagram several books according to plot development.

7. Develop a characterization reference file in which you include examples of how authors tell about the character through narration, record the dialogue of the character, describe the thoughts of the character, show the thoughts of others about the character, and show the character in action. Choose references that you can use to model characterization with students in the grade levels in which you teach or would like to teach.

8. Read an article from **Gere's Roots in the Sawdust: Writing to Learn Across the Disciplines** (1985). Use the article to develop or adapt a lesson for a writing to learn activity designed for elementary or middle-school students.

9. Develop a book report file of ideas that encourage expressive, poetic, and expository writing.

BIBLIOGRAPHY

Andressen, Oliver. "Creativity in Children's Writing." In *A Forum for Focus,* edited by Martha L. King, Robert Emans, and Patricia J. Cianciolo. Urbana, Ill.: National Council of Teachers of English, 1973.

Applebee, Arthur N. "Writing and Reasoning." *Review of Educational Research* 54 (Winter 1984): 577–96.

Britton, James; Burgess, Tony; Martin, Nancy; McLeod, Alex; and Rosen, Harold. *The Development of Writing Abilities (11–18).* Schools Council Research Studies. London: Macmillan Co., 1975.

Bromley, Karen, "SSW: Sustained Spontaneous Writing." *Childhood Education* 62 (1985): 23–29.

———, and McKeveny, Laurie. "Precis Writing: Suggestions for Instruction in Summarizing" *Journal of Reading* 29 (February 1986): 392–95.

Caprio, J. "The Influences of First Hand Experiences on Children's Writing." Doctoral Dissertation, Univ. of New Jersey at New Brunswick, 1986. *Dissertation Abstracts International* 47,02A. (University Microfilms No. 86-09, 228).

Carnes, E. Jane. "Teaching Content Area Reading Through Nonfiction Book Writing." *Journal of Reading* 31 (January 1988): 354–60.

Davis, Stephanie. "How Radar Writing Works.' *The Dallas Morning News.* Sunday, May 22, 1988, p. 8F.

Duffy, Gerald G. "Crucial Elements in the Teaching of Poetry Writing." In *A Forum for Focus,* edited by Martha L. King, Robert Emans, and Patricia J. Cianciolo. Urbana, Ill.: National Council of Teachers of English, 1973.

Fleming, Margaret, and McGinnis, Jo, editors. *Portraits: Biography and Autobiography in the Secondary School.* Urbana, Ill.: National Council of Teachers of English, 1985.

Frye, Northrop; Baker, Sheridan; and Perkins, George. *The Harper Handbook to Literature.* New York: Harper & Row, 1985.

Gambrell, L. B. "Dialogue Journals: Reading-Writing Interaction." *The Reading Teacher* 38 (1985): 512–15.

Gere, Anne Ruggles, editor. *Roots in the Sawdust: Writing to Learn Across the Disciplines.* Urbana, Ill.: National Council of Teachers of English, 1985.

Golub, Jeff, Chair. *Activities to Promote Critical Thinking: Classroom Practices in Teaching English, 1986.* Urbana, Ill.: National Council of Teachers of English, 1986.

Hairston, Maxine. "Using Nonfiction Literature in the Composition Classroom." In *Convergencies: Transactions in Reading and Writing,* edited by Bruce T. Petersen, pp. 179–88. Urbana, Ill.: National Council of Teachers of English, 1986.

Harpin, William. *The Second 'R,' Writing Development in the Junior School.* London: George Allen & Unwin Ltd., 1976.

Hidi, Suzanne, and Anderson, Valerie. "Producing Written Summaries: Task Demands, Cognitive Operations, and Implications for Instruction." *Review of Educational Research* 56 (Winter 1986): 473–93.

Hillocks, George. *Research on Written Composition: New Directions for Teaching.* Urbana, Ill.: National Conference on Research in English, 1986.

Huck, Charlotte; Hepler, Susan; and Hickman, Janet. *Children's Literature.* New York: Holt, Rinehart & Winston, 1987.

Johnson, Scott. "The Biography: Teach It From the Inside Out." *English Journal* (October 1986): 27–29.

Khatena, Joe. *Creative Imagination and What We Can Do To Stimulate It.* Paper presented at the National Association of Gifted Children. Chicago, 1975.

Koch, Kenneth, *Wishes, Lies, and Dreams.* New York: Vintage Book/Chelsea House Publishers, 1971.

Murray, Donald M. "Why Creative Writing Isn't—or Is." *Elementary English* 50 (April 1973): 523–25, 556.

Norton, Donna E. *Through the Eyes of a Child: An Introduction to Children's Literature.* Columbus, Oh.: Merrill Publishing Co., 1987.

Rothman, John. "Taking It Like Little Troupers." *The New York Times Book Review.* Sunday, November 8, 1987.

Sager, Carol. "Improving the Quality of Written Composition in the Middle Grades." *Language Arts* 54 (October 1977): 760–62.

_____. "Improving the Quality of Written Composition Through Pupil Use of Rating Scale." Doctoral Dissertation, Boston University, 1973. *Dissertation Abstracts International* 34, 04A.

_____. "Sager Writing Scale." In *Measures for Research and Evaluation in the English Language Arts.* Urbana, Ill.: National Council of Teachers of English, 1975. (ERIC Document Reproduction Service No. ED 091 723).

Schwartz, Mimi. "Rewriting or Recopying: What Are We Teaching?" *Language Arts* 54 (October 1977): 756–59.

Shugarman, Sherrie L., and Hurst, Joe B. "Purposeful Paraphrasing: Promoting A Nontrivial Pursuit for Meaning." *Journal of Reading* 29 (February 1986): 396–99.

Sutherland, Zena, and Arbuthnot, May Hill. *Children and Books.* Glenview, Ill.: Scott, Foresman & Co., 1986.

Taylor, K. K. "Summary Writing for Young Children." *Reading Research Quarterly* 21 (1986): 193–208.

Tchudi, Susan, and Tchudi, Stephen. *The Young Writer's Handbook.* New York: Charles Scribner's Sons, 1984.

Watson, Dorothy, ed. *Ideas and Insights: Language Arts in the Elementary School.* Urbana, Ill.: National Council of Teachers of English, 1987.

Yolen, Jane. "How Basic is Shazam?" *Language Arts* 54 (September 1977): 645–51.

CHILDREN'S LITERATURE REFERENCES

Baylor, Byrd. *The Best Town in the World.* Illustrated by Ronald Himler. New York: Charles Scribner's Sons, 1983.

Brown, Marc. *Arthur's Baby.* Boston: Little, Brown & Co., 1987.

Cendrars, Blaise. *Shadow.* Illustrated by Marcia Brown. New York: Charles Scribner's Sons, 1982.

De Paola, Tomie. *An Early American Christmas.* New York: Holiday, 1987.

Flournoy, Valerie. *The Patchwork Quilt.* Illustrated by Jerry Pinkney. New York: Dial Press, 1985.

Freedman, Russell. *Lincoln: A Photobiography.* New York: Clarion, 1987.

Fritz, Jean. *Homesick: My Own Story.* New York: G. P. Putnam's Sons, 1982.

Gish, Lillian. *An Actor's Life for Me!* As told to Selma G. Lanes. New York: Viking Press, 1987.

Goodall, John. *The Story of a Main Street.* New York: Macmillan Co., 1987.

Hall, Donald. *Ox-Cart Man.* Illustrated by Barbara Cooney. New York: Viking Press, 1979.

Hodges, Margaret. *The Wave.* Illustrated by Blair Lent. Boston: Houghton Mifflin Co., 1964.

Khalsa, Dayal. *I Want A Dog.* New York: Potter, 1987.

Lenski, Lois. *Sing a Song of People.* Illustrated by Giles Laroche. Boston: Little, Brown & Co, 1987.

McCloskey, Robert. *Time of Wonder.* New York: Viking Press, 1957.

McPhail, David. *The Dream Child.* New York: E. P. Dutton, 1985.

O'Neill, Catharine. *Mrs. Dunphy's Dog.* New York: Viking Press, 1987.

Peek, Merle. *The Balancing Act: A Counting Song.* New York: Clarion, 1987.

Rylant, Cynthia. *When I Was Young in the Mountains.* Illustrated by Diane Goode. New York: E. P. Dutton, 1982.

Seuss, Dr. *And to Think That I Saw It on Mulberry Street.* New York: Vanguard, 1937.

Small, David. *Imogene's Antlers.* New York: Crown Pubs., 1985.

Stanley, Diane. *Peter the Great.* New York: Four Winds, 1986.

Wilder, Laura Ingalls. *Little Town on the Prairie.* New York: Harper & Row, 1941.

Yolen, Jane. *Owl Moon.* Illustrated by John Schoenherr. New York: Philomel, 1987.

Chapter Ten

OBJECTIVES OF THE LITERATURE PROGRAM
THE LITERATURE ENVIRONMENT
THE LITERATURE IN THE ENVIRONMENT

Appropriateness ▪ Curriculum Needs ▪ Balancing
Selections ▪ Sources that Help Teachers Choose
Literature ▪ Children's Interests and Literature

USING LITERATURE WITH CHILDREN

Developing Enjoyment Through Reading Aloud ▪
Developing Enjoyment Through Storytelling ▪
Developing Enjoyment Through Art ▪ Developing
Student Critics for the Literature They Read ▪
Developing an Understanding of Criteria for Fine
Literature ▪ Developing Understanding and
Appreciation of Literary Genres ▪ Developing
Understanding of Literary Elements ▪ Mapping
Instructional Possibilities ▪ Teaching World
Understanding Through Folktales

SUMMARY

After completing this chapter on literature, you
will be able to:

1. State some of the objectives for a literature
   program.
2. Describe an environment that stimulates interest in literature.
3. Select literature for the literature environment that considers appropriateness, balance
   of selections, and children's interests.
4. List benefits of reading aloud to children,
   discuss selection of books for reading aloud,
   and read a selection aloud to children.
5. Prepare and tell an appropriate story.
6. Describe how art adds to appreciation of
   literature.
7. Prepare students to be critics for the literature they read.
8. Describe ways for developing students' understanding of and appreciation for literary
   genres.
9. Describe methods and develop lessons that
   increase students' understanding of literary
   elements through webbing, modeling, and
   plot structures.
10. Map the instructional possibilities for a child-
    ren's book.
11. Describe how world understanding can be
    taught through folktales.

Literature

*T*he many books written especially for children open up new and wonderful worlds to our students. Through books, they encounter the enchantment of Cinderella, the adventures of Huckleberry Finn, the loneliness and challenges of living on the Island of the Blue Dolphins. Most of us have favorite books from childhood that remain vivid in our memories; often, Mary Poppins, Pooh Bear, or Caddie Woodlawn seem like personal friends. Not all children, however, think of literature as a source of pleasure and a way of meeting new friends. Our goal in this chapter is to show the teacher how to help children develop a love of reading and an appreciation for a wide variety of literature. In addition, this chapter stresses using literature in the language arts curriculum, and suggests ways literature and language arts objectives can be integrated into other subject areas. Chapter 11 continues this emphasis on literature in the curriculum with approaches that bring literature into the reading curriculum.

OBJECTIVES OF THE LITERATURE PROGRAM

Using literature in the curriculum and teaching about literature in the classroom are two exciting yet complex areas. To choose literature that motivates students to read and that stimulates students to appreciate literature requires knowledge about literature, awareness of students' interests, and knowledge about exciting instructional approaches that motivate and stimulate interest. Likewise, to teach about literature requires knowledge about literature and knowledge about exciting instructional approaches that encourage understanding of various genres, story structures, and literary elements such as author's style, characterization, and theme.

As we approach literature and methodologies related to literature, we will be aware of Fishel's (1984) warning for English and language arts teachers: "Of the content areas, English is one of the most demanding in terms of the reading skills required to understand the various genres. In addition, appreciation of the genres is a teaching goal" (p. 9). Within this chapter we consider the selection of literature from various genres, the identification of students' interests that can relate to literature, the stimulation of interest and appreciation of literature, and the methodologies that help students understand particular genres and literary elements within books. In chapter 11, we emphasize additional methodologies that may be used in the reading program. At all times, however, we should remember that the dual role of pleasure and understanding cannot be separated. The methodologies should never turn children off to the excitement of reading literature.

This dual role of enhancing appreciation and developing understanding of literature is emphasized in endorsements by educational groups and by various state goals and objectives. For example, the National Council of Teachers of English (1983) stated that students should gain the following objectives from literature:

1. Realize the importance of literature as a mirror of human experience, reflecting human motives, conflicts, and values.
2. Be able to identify with fictional characters in human situations as a means of relating to others; gain insights from involvement with literature.
3. Become aware of important writers representing diverse backgrounds and traditions in literature.
4. Become familiar with masterpieces of literature, both past and present.
5. Develop effective ways of talking and writing about varied forms of literature.
6. Experience literature as a way to appreciate the rhythms and beauty of the language.
7. Develop habits of reading that carry over into adult life (p. 246)[1]

State goals and objectives may also include literature objectives within the language arts curriculum. For example, the literature goal for language arts in the state of Illinois *State Goals for Learning and Sample Learning Objectives: Language Arts* (1985) stated: "As a result of their schooling, students will be able to understand the various forms of significant literature representative of different cultures, eras, and ideas" (p. 43). The sample objectives identified to encourage students to meet these goals may be divided into objectives that emphasize appreciation, objectives that relate to understanding different genres of literature, and objectives that relate to understanding literary elements. For example, at the appreciation level the third-grade objective states that students will read and enjoy appropriate literary works. At the same grade level, objectives related to genre state that children will recognize the nature of poetry, prose, and biography, and will compare versions of folktales. By the sixth grade, students are encouraged to recognize examples of historical fiction, fantasy, science fiction, realistic fiction, and folk literature. At the literary elements level, the third-grade objectives include recognizing plot sequences and actions, identifying setting, identifying important character traits and explaining how and why characters change throughout a story, recognizing the main idea of a selection, and identifying similes, personification, and onomatopoeia. By the sixth grade, this list of literary elements expands to understanding author's tone, point of view, symbolism, and other types of figurative language. Within this chapter, we

[1]From the National Council of Teachers of English, "Forum: Essentials of English," *Language Arts* 60 (February 1983): 244–48. Copyright © 1983 by the National Council of Teachers of English. Reprinted by permission of the publisher and author.

consider methodologies for developing appreciation, understanding of genre, and literary elements.

In addition to objectives related to developing appreciation, understanding genre characteristics, and understanding literary elements, the literature program has broader educational and personal goals. Folklore collections from around the world, writings of Mark Twain and Robert Louis Stevenson, and poets from Langston Hughes to Edward Lear help us understand and value our cultural and literary heritage. Historical fiction and nonfictional books that chronicle the early explorations and the frontier expansion allow us to vicariously live through world history. Books on outer space, scientific breakthroughs, and information in every field open doors to new knowledge and expand our interests. Fantasy, science fiction, and poetry nurture and expand our imaginations and allow us to visualize worlds that have not as yet materialized or to see common occurrences through the insights of the poet. Realistic fiction, biography, and autobiography allow us to explore human possibilities and to promote our personal and social development. Interactions with many types of literature allow us to expand our language and to enhance our cognitive development. The time spent with literature is among our most rewarding experience. Within this chapter, we include literature from a wide range of genres that has the potential for this broader educational and personal development.

THE LITERATURE ENVIRONMENT

The foundations for a strong literature program include a rich environment in which books are easily located and read, a teacher who loves books and shares them daily with students, and an attitude toward literature that encourages interaction with a variety of books.

Research that analyzes children's responses to literature provides support for this type of environment. Galda (1988) reported that students' responses to literature are limited when teachers insist on only one type of response. In contrast, responses are extended and heightened when students have many opportunities to read and respond to a variety of genres, styles, and authors; when students are in a secure environment that encourages individual differences and provides opportunities to explore and to compare responses; and when students are in an environment that provides time and encouragement for them to respond to literature through various approaches such as writing, discussing, acting out, and drawing.

The value of literature is demonstrated to students if time is provided in the environment for them to read, to listen to, and to discuss books. Mendoza's (1985) results from a survey of 520 elementary students ranging in age from five to thirteen show the importance of an environment that encourages reading to and by students:

1. Children throughout elementary grades enjoy having books read to them. Consequently, teachers and parents should read to children frequently.
2. Role models are important; both teachers and parents should read to children.

3. Teachers and parents should provide opportunities for children to read to other children.
4. Children should have opportunities to select books read to them or read by them to others.
5. Children like information about a book before it is read to them; they should be told who the author is and be given a brief summary of plot, characters, and setting.
6. Children like and should be given an opportunity to discuss books and to read books after they are read aloud.

Educators who use or recommend literature-based programs emphasize the dynamic nature of the environment and the desirability of emphasizing both appreciation and understanding of literature. Taxel (1988) described the literature-based classroom "as fluid and dynamic, . . . a place where educators see literature as central to the curriculum, not as an occasional bit of enrichment undertaken when the 'real' work is completed" (p. 74).

Understanding and enjoying literature are the major goals in a first- through sixth-grade program described by May (1987). May's program for exploring different literature genres begins in kindergarten as students and teachers explore books containing personification, and compare these books to other fantasy stories and to folktales. At this level students consider how the characters and their actions seem real. First graders explore literary patterns within folktales, look at variances among folktales found in different cultures, and examine different picture book versions of the same folktale. They also explore realistic fiction and relate the literature to a unit on the family. Second graders discuss repeated patterns, magical numbers, and settings in folktales. They compare additional versions of folktales and explore picture book versions of folktale adaptations. Third graders read E. B. White's *Charlotte's Web* (1952) and discuss animal characters, personification, fantasy, and point of view. They compare fantasy and realistic stories about animals. Fourth graders explore literature around a social studies unit on India. Fifth graders explore literature as part of a social studies unit on medieval Europe and discuss historical fiction as literature and as interpretation of history. They also consider the importance of setting, symbolism, and stereotypic character portrayals in legends and work with myths and fables. Sixth graders explore differences between autobiography and biography, conduct oral interviews, write short biographies, and share them with the class.

May is enthusiastic about this literature program and shares the following responses from the teachers: "For instance, by the end of the first year the children were reading more library books than before. They were also talking about literature and were 'borrowing' an author's style or symbolism when they admired it. They were more appreciative of an author's writing style. All of the students, even the slow learners, were writing and reading for enjoyment" (p. 136).

Equal enthusiasm is shown by the teacher of a fifth-grade literature program. Five (1988) described her literature-based program in which she begins each session "with a mini-lesson on an element of fiction—character development, setting, titles, flashbacks, and other techniques writers and readers need to know. Often I use the

Books provide enjoyment and increase knowledge.

books that I read aloud to them each day as the basis for my mini-lessons, allowing time for discussions, for making predictions, and interpretations, and for discovering characteristics of a particular author's style that students may wish to apply to their own writing. Different genres are also introduced during these lessons" (p. 105). Some of these mini-lessons involve sharing and discussing, some involve direct teaching, and some include doing specific activities together. The mini-lessons may focus on such topics as book selection, copyright, graphing plot structures, and developing time lines of characters. After the mini-lessons students read related books of their own choosing, discuss the books, and write about their choices. Five enthusiastically evaluated her program: "Today my reading program is dramatically different and so are the results. Today my students read between 25 and 144 books a year. Children listen to each other and seek recommendations for their next book selections. They wonder about authors and look for feelings, for believable characters, for interesting words, and they are delighted with effective dialogue. And my students and I always talk books before school, during school, at lunch and after school, something we never did in my 'workbook' days" (p. 104).

Both of these programs emphasize an environment that is rich with literature, that encourages children to explore various literary devices in a number of ways, and that includes both teacher- and student-directed activities with literature. The programs also show that teachers may use literature to combine reading, writing, and other content areas.

THE LITERATURE IN THE ENVIRONMENT

The previously described literature-based language arts programs and the objectives for the literature program illustrate the importance of including a variety of genres of literature in the curriculum. In addition, this selection of literature should include books that can be read to and by children, and books that meet the instructional goals for developing literary appreciation and understanding. To meet these objectives the language arts curriculum should include picture books, folklore, modern fantasy, poetry, contemporary realistic fiction, historical fiction, biography and autobiography, and informational literature selected to held students develop an understanding of and an appreciation for specific genres. The literature within these genres should reflect our varied cultural and literary heritage, as well as provide insights into personal and social development.

Questions related to appropriateness, to curricular needs, to balancing selections, to sources that help teachers choose literature, and to children's interests, are all basic to the literature in the school environment.

Appropriateness

In the last few years, numerous controversial issues have appeared in discussions of and reactions to children's literature. Much of the concern surrounds contemporary realistic fiction that mirrors the society in which the stories take place. During times of changing societal values, and altering attitudes toward children and their books, controversy is not surprising. Current controversy in realistic fiction centers around issues related to literary merit and appropriate content.

Sheila Egoff (1980) differentiated between the distinguished realistic novels that have strong literary qualities—including logical flow of narrative; delicate complexity of characterization; style; insights that convey the conduct of life as the characters move from childhood to adolescence; and a quality that touches both the imagination and the emotions—and the fiction she categorizes as the problem novel. Egoff criticized problem novels in which the authors focus on problems rather than plots or characters, use a style that is flat and emotionally numb, and write with a self-centered, confessional tone.

The degree to which realistic fiction should reflect the reality of times leads to controversy as writers create heroes and heroines who face problems related to sexism, violence, drugs, sexual activities, racism, and family disturbances such as divorce, death, separation, and cruelty. The last ten years have produced an increase in the number of organized group efforts to censure children's literature and school textbooks. Newsletters from organized groups frequently suggest the harmful effects of sharing certain books with children (Woods, 1978). In contrast, educators and librarians frequently caution against the dangers of censorship. Rather than avoiding all books that contain volatile subjects, Day McClenathan (1979) suggested that wholesale avoidance, in addition to encouraging overt censorship, is inappropriate because (1) a book about a relevant sociological or psychological problem can give young people opportunities to grow in their thinking process and to extend experiences; (2) problems in books can provide children who need them opportunities for identification and allow others an opportunity to sympathize with their peers; and (3) problems in books invite decisions, elicit opinions, and afford opportunities to take positions on issues. The current controversy in children's books is exemplified by the various reactions to Margot Zemach's *Jake and Honeybunch Go to Heaven*. Literary merit is suggested by the *New York Times Book Review*. Negative criticism is suggested by reactions of the book selection committees in Chicago, San Francisco, and Milwaukee, which found the book weak and/or racially stereotyping. The March 1983 issue of *American Libraries* focuses on both sides of the controversy (Brandehoff, 1983). To counter increased efforts for censorship, professional educational journals are responding with articles focusing on freedom to read and advice to teachers (Palmer, 1982).

Another problem in selecting literature is choosing books that represent ethnic and minority groups fairly. Guides and recommendations for selecting literature about minorities are included in the chapter entitled "The Linguistically Different Child and Multicultural Education." Teachers should use these guidelines to choose appropriate books about minorities for all their students to read.

Critics also denounce sexism in children's literature. Frasher (1982) found that literature published after 1970 reflected a heightened sensitivity toward feminist concerns; in contrast, literature published prior to 1970 had more male main characters. In addition, negative comments about females and stereotyping were common.

The school library should certainly offer books that portray women and girls in nonstereotyped roles. There should be biographies of famous women to provide models for girls just as there are biographies of well-known men. Girls should not be taught that it is inappropriate for them to become doctors, lawyers, or members of other professions. The library should also offer stories about little girls who are active and inventive and who do not always follow the initiative of boys.

Now more books are being written that deal with the real problems many children face, such as divorce, death, special needs, and various difficulties associated with growing up. Examples of books based on specific problems include the following:

- Divorce—Beverly Cleary's *Dear Mr. Henshaw* (1983). Corresponding with his favorite author helps a sixth-grade boy overcome problems associated with his parents' divorce. Gary Paulsen's *Hatchet* (1987). Surviving in the Canadian wilderness helps a boy face problems associated with his parents' divorce.
- Death—Eleanor Cameron's *That Julia Redfern* (1982). A girl develops a close relationship with her father and then accepts his death. Marion Dane Bauer's *On My Honor* (1986). A boy faces both his own disobedience and the death of his best friend. Norma Fox Mazer's *After the Rain* (1987). A girl develops a close relationship with her grandfather and then mourns his death.
- Special Needs—Ellen Howard's *Edith Herself* (1987). A girl faces and solves her problems associated with epilepsy in a story set in the 1890s. Jane Madsen and Diane Bockoras's *Please Don't Tease Me* (1983). A physically disabled girl asks for understanding.
- Growing Up—Jean Little's *Different Dragons* (1986). A boy and a girl discover that they both have fears that they must overcome.

Literature selections should take into account the needs of children of different ages and abilities. Although many books are geared to the varied interests of children with average or above-average reading ability, there are fewer books designed for children with beginning skills, or children whose reading ability is several years below grade level. Students who are below grade level in reading ability present a special problem, because their interests are similar to those of children who read at grade level. The teacher must select easier-to-read books for these children. Sources such as *High Interest, Easy Reading* (1984) published by the National Council of Teachers of English provide guidance in selection of materials for older readers who experience reading difficulties.

Curriculum Needs

Literature collections should also include books on topics that will be studied in science, art, social studies, and music. If certain subjects will be suggested for individual study, the library needs books on those topics as well. Materials on specific topics should be available in a wide range of reading levels, even when they are studied at only one grade level. The reason for this is obvious from the test results of six sixth-grade classes' reading ability: reading levels in the six classrooms ranged from second through twelfth grade. In addition, because specific musical selections can enhance the study of a period of history, and because filmstrips or sets of pictures can add visual understanding to a curriculum subject, the library materials should reflect the needs of the overall curriculum.

Balancing Selections

Literature selection for the library, as well as for the individual child, should include books from various classifications: picture books, traditional literature, modern fantasy, poetry, contemporary realistic fiction, historical fiction, biography, and nonfiction.

Picture books such as Mother Goose and alphabet books, as well as storybooks that develop plot and characters, afford children their earliest encounters with literature. Teachers and parents usually read these books to younger children, because picture storybook reading levels are usually at least for third grade. Some excellent alphabet books include Chris Van Allsburg's *The Z Was Zapped* (1987) and Suse MacDonald's *Alphabatics* (1986). Both books expand young children's language development and encourage interaction with the text. Picture storybooks should include old favorites such as Dr. Seuss's *The 500 Hats of Bartholomew Cubbins* (1938), Robert McCloskey's *Make Way for Ducklings* (1941), and Maurice Sendak's *Where the Wild Things Are* (1963). There should also be new stories such as Chris Van Allsburg's *The Polar Express* (1985), Audrey Wood's *King Bidgood's in the Bathtub* (1985), and Arthur Yorinks's *Hey, Al* (1986). These books encourage children to interact with both the illustrations and the text. Picture storybooks such as Denys Cazet's *A Fish in His Pocket* (1987) show children that even young characters can solve their own problems in very satisfying and creative ways. A well-balanced collection of picture books also includes books that use different media for their illustrations: photographs, woodcuts, collages, painted pictures, ink drawings, and crayon illustrations.

The collection should also include materials from well-loved traditional literature—folktales handed down from earlier generations, such as "Little Red Riding Hood," "Three Billy Goats Gruff," and "Cinderella," along with mythology and stories of epic heroes such as King Arthur and Robin Hood. Traditional literature should include folklore from numerous cultures such as John Bierhorst's *Doctor Coyote: A Native American Aesop's Fables* (1987), Van Dyke Parks's adaptation of Joel Chandler Harris's *Jump Again! More Adventures of Brer Rabbit* (1987), and Momoko Ishii's *The Tongue-Cut Sparrow* (1987).

In addition, modern fantasy should be included in the collection. Children should be introduced to classics in this genre such as E. B. White's *Charlotte's Web* (1952), C. S. Lewis's *The Lion, the Witch and the Wardrobe* (1950), A. A. Milne's *Winnie-The-Pooh* (1926), and Beatrix Potter's *The Tale of Peter Rabbit* (1902). Authors such as Lloyd Alexander, Hans Christian Andersen, Michael Bond, Lucy Boston, Lewis Carroll, Susan Cooper, Kenneth Grahame, Rudyard Kipling, Madeleine L'Engle, Ursula LeGuin, and Margery Williams bring excitement to this genre of literature. In addition to these well-known authors, the collection should include newer fantasies such as Bill Brittain's *The Wish Giver* (1983), Robin McKinley's *The Hero and the Crown* (1984), and Brian Jacques's *Redwall* (1986).

The study of our past might be very dull without the addition of historical fiction to our reading. These stories allow us to live vicariously in the past and to understand important themes and values that shaped the time period. We can learn about the Revolutionary War by reading Esther Forbes's *Johnny Tremain* (1943), visit the Pilgrims by reading Patricia Clapp's *Constance: A Story of Early Plymouth* (1968), be persecuted as a witch in Elizabeth Speare's *The Witch of Blackbird Pond* (1958), live in the pioneer wilderness in Laura Ingalls Wilder's *Little House in the Big Woods* (1932), discover Victorian London in Philip Pullman's *The Ruby in the Smoke* (1987) and in Leon Garfield's *The December Rose* (1986), or feel what it might be like to be a slave in Belinda Hurmence's *A Girl Called Boy* (1982).

Biographies add another dimension to the literature program by allowing children to learn more about past and present heroes and heroines. Children can be introduced to biographies in the early grades with picture storybook biographies such as Alice Dalgliesh's *The Columbus Story* (1955) and David Adler's *Martin Luther King, Jr.: Free at Last* (1986). Middle elementary students can read autobiographies such as Lillian Gish's *An Actor's Life for Me!* (1987) and biographies such as Jean Fritz's *Make Way for Sam Houston* (1986). Older children should be encouraged to read biographies such as Russell Freedman's *Lincoln: A Photobiography* (1987) and Polly Brooks's *Queen Eleanor: Independent Spirit of the Medieval World* (1983).

Informational books are available in all areas of knowledge. There are books about the history and culture of the ancient and the modern worlds. There are books about the laws of nature that encourage children to understand their own bodies, observe nature, explore the life cycles of animals, consider the impact of endangered species, experiment with plants, understand the balance of the smallest ecosystem, and explore the earth's geology. There are informational books that answer children's questions about discoveries of the past and the present or provide explanations of how many kinds of machines work. There are also informational books that allow children to learn more about their hobbies and interests. These books often include directions, provide guidelines for choosing equipment or other materials, or give interesting background information.

Sources That Help Teachers Choose Literature

Thankfully, there are numerous sources that help teachers select literature. Children's literature textbooks are the best source for genre-specific literature. Norton's *Through the Eyes of a Child: An Introduction to Children's Literature* (1987), Sutherland and Arbuthnot's *Children and Books* (1986), and Huck, Hepler, and Hickman's *Children's Literature* (1987) discuss literature according to genre and recommend hundreds of titles that will be useful for the literature program. Caldecott and Newbery award-winning books are listed within these texts.

Literature journals are excellent sources for new books and for specialized collections and topics. For example, *School Library Journal, Booklist,* and *Hornbook* contain book reviews and include starred reviews for excellent books. *Booklist* publishes yearly lists of "Notable Children's Books" and "Children's Editors' Choice." In addition, *Booklist* publishes specialty lists such as "Popular Reading-Chapter Books" (Bennett, 1987), "Popular Reading-After Henry Huggins" (Cooper, 1986), "Storytelling Sources" (Corcoran, 1986), "Picture Books for Older Children" (Kiefer, 1986), "Poetry for Young Children" (Phelan, 1988), and "Contemporary Issues-Intergenerational Relationships" (Wilms, 1986). *The School Library Journal* publishes the "Best Books" list for each year, as well as specific recommendations for literature such as "Modern Classics" (Breckenridge, 1988). *The Children's Literature Association Quarterly* publishes articles on specific genres of literature, authors, literary criticism, and issues. The association has also identified a list of "Touchstone" books; these are books that are considered so good by the association that they should provide the criteria by which all other books are evaluated.

Children's literature is also reviewed in *Language Arts, The Reading Teacher*, and *The New Advocate*. Each year the October issue of *The Reading Teacher* publishes an annotated bibliography of "Children's Choices." This list contains books that children from across the United States select as their favorites.

Within this chapter and the next chapter, you will find many literature selections that meet the needs for developing appreciation and for developing understanding of literature.

REINFORCEMENT ACTIVITY

1. *Visit an elementary classroom in which literature is an important part of the curriculum. Describe the environment, the teacher's attitude toward literature, and the children's interactions with books.*
2. *Choose a source for selecting literature. Review the information about books that you can gain from that source.*

Children's Interests and Literature

Understanding why and what children read is necessary if we are to help them select books that stimulate their interests and enjoyment. Teachers can learn about children's interests by reviewing studies of children's interests and preferences, and by talking to children and evaluating their responses on interest inventories.

Each year a joint project of the International Reading Association and the Children's Book Council allows approximately 10,000 children from around the United States to evaluate children's books published during a given year. Each year their reactions are recorded, and a research team uses this information to compile a list of "Children's Choices."

A look at these lists of children's favorites gives teachers a better understanding of the characteristics of books that appeal to children. Sebesta (1979) evaluated books listed and identified characteristics of these books. His evaluation produced the following conclusions:

1. Plots of the Children's Choices are faster paced than those found in books not chosen as favorites.
2. Young children enjoy reading about nearly any topic if the information is presented in specific detail. The topic itself may be less important than interest studies have indicated; specifics rather than topics seem to underlie children's preferences.
3. Children like detailed descriptions of settings; they want to know exactly how the place looks and feels before the main action occurs.
4. One type of plot structure does not dominate Children's Choices. Some stories have a central focus with a carefully arranged cause-and-effect plot; others have plots that meander, with unconnected episodes.
5. Children do not like sad books.
6. Children seem to like some books that explicitly teach a lesson, even though critics usually frown on didactic books.
7. Warmth was the most outstanding quality of books children preferred. Children enjoy books where the characters like each other, express their feelings in things they say and do, and sometimes act selflessly.

Sebesta believed this information should be used to help children select books and to stimulate reading and discussions. For example, children's attention can be drawn to the warmth, pace, or descriptions in a story in order to encourage their involvement with the story.

The various Children's Choices lists also suggest particular types of stories that appeal to young readers. For example, analyzing a recent list showed that the "beginning independent reading" category contains comical stories about more or less realistic family situations, humorous animal stories, stories that develop emotional experiences, action-filled fantasies, traditional stories, counting books, rhymes, and riddles. The "younger reader" category includes realistic stories about families, friends, school, and personal problems; animal stories; fantasies; fast-paced adventures; folktales; and humorous stories. Stories chosen by children in the middle grades include realistic stories about sibling rivalry, peer acceptance, fears, and not conforming to stereotypes; fantasies; suspense; and humorous stories. Popular information books include factual and nonsensical advice about human health, factual information about animals, and biographical information about sports stars. Popular poetry includes collections by Judith Viorst, Shel Silverstein, and William Cole. The Children's Choices includes books from a wide variety of genres.

Research also indicates that children's reading interests are influenced by their reading ability. Swanton's (1984) survey comparing gifted students with students of average ability reports that gifted children prefer mysteries (43 percent), fiction (41 percent), science fiction (29 percent), and fantasy (18 percent). In contrast, the top four choices for students of average ability were mysteries (47 percent), comedy/humor (27 percent), realist fiction (23 percent), and adventure (18 percent). Gifted students indicated that they liked "science fiction and fantasy because of the challenge it presented, as well as its relationship to Dungeons and Dragons'' (p. 100). Gifted students listed Judy Blume, Lloyd Alexander, J. R. R. Tolkien, and C. S. Lewis as favorite authors. Average students listed Judy Blume, Beverly Cleary, and Jack London. As you can see, there are similarities and differences within these favorites. Many additional types of literature may become favorites if an understanding and knowledgeable teacher provides opportunities for students to listen to, to read, and to discuss literature.

Although information from research and Children's Choices provides general ideas about what subjects and authors children of certain ages and reading abilities prefer, teachers should not develop stereotyped views about children's preferences. Without asking questions about interests there would be no way to know about children's specific and unusual interests. Informal conversation is one of the simplest ways to uncover children's interests. Ask a child to describe what she likes to do and read about, note the information on a card, and find several books that might meet the child's interests. Some way of recording the information is usually needed when teachers work with a number of children. Teachers can develop interest inventories in which students answer questions about their favorite hobbies, books, sports, television shows, and other interests. Teachers can write down young children's answers. Older children can read the questionnaire themselves and write their own responses. Such an inventory might include some of the questions asked in table 10–1. (Changes should be made according to children's age levels, and additional information could be discovered if children told why they like certain books.) After the interest inventory is complete, the findings can serve as the basis for helping children select books and extend their enjoyment of literature.

REINFORCEMENT ACTIVITY

Administer an interest inventory to several elementary children. Tabulate their interests and select several literature selections that you feel match their interests.

USING LITERATURE WITH CHILDREN

If we want children to love and to appreciate literature, we must not only provide them with a selection of fine and varied literature, we must also give them many opportunities to read, to listen to, to share, and to discuss literature. In this section we proceed from teaching techniques and activities that emphasize enjoyment and appreciation of literature, to activities that develop student critics of what they read, to teaching techniques that emphasize understanding and appreciation of various literary genres, to teaching techniques that emphasize understanding and appreciation of literary elements, and finally to the development of units that incorporate literature across the curriculum.

TABLE 10–1

An informal interest inventory

1. What do you like to do when you get home from school?
   _____

2. What do you like to do on Saturday?
   _____

3. Do you like to watch television?_____If you do, what are the names of your
   favorite programs?
   _____

4. Do you have a hobby?_____If you do, what is your hobby?_____

5. Do you like to make or collect things?_____If you do, what have you made or
   collected?_____
   _____

6. What is your favorite sport?_____
7. What games do you like best?_____
8. Do you like to go to the movies?_____
   If you do, what was your favorite movie?_____
   _____

9. Do you have a pet?_____If you do, what is your pet?_____
10. Where have you spent your summer vacations?_____
    _____

11. Have you ever made a special study of rocks?_____space?_____
    plants?_____animals?_____travel?_____
    dinosaurs?_____other?_____

12. What are your favorite subjects in school?
    art?_____handwriting?_____social studies?_____physical
    education?_____science?_____music?_____creative
    writing?_____spelling?_____arithmetic?_____other?_____
13. What subjects is the hardest for you?_____
14. What kinds of books do you like to have someone read to your?
    animal stories?_____fairy tales?_____true stories?_____science
    fiction?_____adventure?_____mystery stories?_____
    sport stories?_____poems?_____humorous stories?_____
    other kinds of stories?_____
15. What is your favorite book that someone read to you?_____
16. What kinds of books do you like to read by yourself?
    animals?_____picture books?_____fairy tales?_____true
    stories?_____science fiction?_____adventures?_____mystery
    stories?_____sport stories?_____poems?_____funny stories?
    _____other kinds of stories?_____
17. What is your favorite book that you read by yourself?_____
18. Would you rather read a book by yourself or have someone read to you?
    _____

19. Name a book that you read this week.
    _____

20. What books or magazines do you have at home?
    _____

21. Do you ever go to the library?_____
    How often do you go to the library?_____
    Do you have a library card?_____

Developing Enjoyment Through Reading Aloud

There is probably no better way to interest children in the world of books than to read to them. This is one way for children to learn that literature is a source of pleasure. For children who are just struggling to learn to read, a book may not yet be a source of happiness. In fact, books may actually arouse negative feelings in many children. Reading aloud develops an appreciation for literature that children could not manage with their own reading ability. Because of educational concern with the affective domain, instructional techniques that improve attitudes are highly desirable. Is there a better way for children to see the pleasure they can derive from books than by sharing them in a relaxed environment with a teacher who obviously enjoys good literature?

In addition to pleasure, students gain understandings about story structures from the books they hear. They hear language patterns and words that may be unfamiliar to them. These listening experiences increase their appreciation for different story and sentence structures and prepare them for reading such structures when they encounter them in their own reading.

Another important value of reading aloud is its motivational aspect. After children are excited by hearing a selection, they usually want to read it to themselves. The interaction between an enthusiastic teacher and interested peers is especially important while reading aloud.

This teacher knows the many benefits of reading aloud to children.

Selecting the books Choosing books to read aloud depends on such concerns as the age of the children, their interests, their experiences with literature, the quality of the literature, and the desire to balance the type of literature presented. A book selected for reading aloud should be worthy of the time spent by both reader and listeners. It should not be something picked up hurriedly to fill in the time.

Style and illustrations are both considerations when choosing books to read aloud. The language in A. A. Milne's *Winnie-the-Pooh* (1926, 1954) and in Dr. Seuss's *The 500 Hats of Bartholomew Cubbins* (1938) appeals to young listeners. Likewise, young children enjoy illustrations that are an integral part of the story such as those found in Maurice Sendak's *Where the Wild Things Are* (1963) and in Arthur Yorinks' *Hey, Al* (1986).

Children's ages, attention spans, and levels of reading ability are also important considerations when selecting stories to be read aloud. The books chosen should challenge children to improve their reading skills and increase their appreciation of literature. The numerous easy-to-read books should usually be left for children to read independently. Younger children respond to short stories that can be finished in one reading. Books such as Amy Schwartz's *Oma and Bobo* (1987), Ann Grifalconi's *Darkness and the Butterfly* (1987), and Amy Hest's *The Purple Coat* (1986) are favorites with first and second graders. Books such as Graham Oakley's *The Church Mice in Action* (1982) have enough plot to appeal to second graders. By the time children reach third grade, teachers can read continued stories, although they should complete a chapter during each story time, rather than leave a segment unfinished. Third graders usually enjoy E. B. White's *Charlotte's Web* (1952) and Laura Ingalls Wilder's *Little House* series. Fourth and fifth graders often respond to books like Madeleine L'Engle's *A Wrinkle in Time* (1962) and C. S. Lewis's *The Lion, the Witch and the Wardrobe* (1950). Armstrong Sperry's *Call It Courage* (1940) and Esther Forbes's *Johnny Tremain* (1946) often appeal to sixth and seventh graders.

Teachers should keep records of the books they read to children, and note children's reactions to the stories. These records help teachers both balance the types of books selected and develop an understanding of children's interests during the story time.

Preparing to read aloud Enthusiasm is a vital ingredient during the story hour; if the teacher is uninterested and unprepared, the story time will be spoiled for the children. Preparation requires reading the story silently so that the reader understands the story, the sequence of events, the mood, the subject, and the vocabulary or other concepts.

Next, teachers should read the story aloud to practice pronunciation, pacing, voice characterizations, and so forth. Beginning oral readers should listen to themselves on a tape recorder before they read to an audience; enunciation, pacing, and volume are especially important for young children's understanding.

Finally, teachers should consider how to introduce the story (different introductions are listed under "storytelling" in this section), and what type of discussion or other activity might follow the reading.

Reading the stories   An appropriate environment for story time is also essential. Children in the early elementary grades need to sit close to the teachers, especially if the story is a picture book that the teachers will show them as they read. Teachers also prepare the children for the listening experience by getting their attention and providing appropriate goals for the experience (refer to chapter 4 for suggestions in developing listening skills and the listening environment). They may have to clarify a concept or a vocabulary word before the children will understand the story. They might introduce the story with a question, a discussion about the title, a prediction about the story, or background information about the author.

Just as teachers should draw on a variety of categories of literature, they should also use various methods for sharing stories with the group—reading aloud, and using recordings, films, storytelling, and visual techniques such as puppets, felt boards, and chalk talks.

Developing Enjoyment Through Storytelling

Have you ever sat around a campfire and listened to a storyteller take you to all kinds of magical places? Have you watched as children become entranced listening to a story? What makes storytelling more effective than merely reading the story aloud? A great deal of magic results from the close eye contact with the storyteller, and from the amount of preparation that goes into the activity.

My college classes are introduced to the magic of storytelling by listening to an experienced adult storyteller, who has told my classes such well-loved stories as Rapunzel and Thumbelina. The students sit in wonder, amazed at how these simple stories come to life. Although we cannot expect everyone to have this level of expertise, storytelling is a skill you can master with some practice.

The value of storytelling   To convince teachers to take the time to prepare a story for telling rather than reading, we must also convince them that the value to children is worth the effort. John Stewig (1978) offered three significant reasons for including storytelling in the curriculum:

1. An understanding of the oral tradition in literature. Young children in many societies have been initiated into their rich heritage through storytelling; today, few children encounter such experiences.
2. The opportunity for the teacher to actively involve the children in the storytelling. When the teacher has learned the story, he is free from dependence on the book, and can use gestures and action to involve the children in the story.
3. The stimulus it provides for children's storytelling. Seeing the teacher engage in storytelling helps children understand that storytelling is a worthy activity, and motivates them to tell their own stories.

This last point was made clear to one of my college classes when we invited a group of kindergarten and first-grade children to our class to take part in a story time. One child was particularly delighted with the stories, but was apparently too shy to become involved in any activity or discussion. Her mother informed us a few days later that the child was so stimulated by the experience that when she went home, she told her family all the stories she had heard. She also made flannel-board and puppet figures to accompany some of the stories. Then, she started to practice new stories, and wanted to tell them to anyone who would take a few minutes to listen. Any activity with the power to motivate a child this way is certainly worth the effort.

Selecting the story   When you select a story for telling, you want it to have a strong beginning, to bring your listeners into the story quickly. Your story should also have lots of action. Children enjoy the development of an active plot. In addition, the story's dialogue should appear natural. You should also consider the story's characters; usually three or four characters are enough, because both you and the children may have difficulty distinguishing any more than that. As with all stories for young children, the story should have a definite climax that both you and the children can recognize. Finally, the story should have a satisfactory conclusion.

Try to choose something suspenseful, such as Rapunzel, or a similar kind of folktale, and choose one that has a few alive, vital characters. Rapunzel, for example, has a wicked witch, a beautiful maiden, and a rescuing prince. Finally, choose a story appropriate to the age of the audience and to the time allotted.

When selecting stories, you must consider both the children's age and experience. Robert Whitehead (1968) suggested that preschool through kindergarten children need stories that are short and to the point. The stories should include familiar things—animals, children, homes, machines, people, and so forth. Humorous and nonsense stories and the accumulative tales are good choices—"Three Little Pigs"; "Three Billy Goats Gruff"; "Henny Penny." Ancient and modern fairy tales usually appeal to children ages six through ten—"The Elves and the Shoemaker"; "Rumpelstiltskin"; "The Bremen Town Musicians." Animal tales and stories of children in other lands are also appropriate. Older-elementary children usually like true stories, hero tales, and stories that teach something about personal ideals. These children also enjoy adventure, so myths, legends, and epics are popular—"Aladdin"; "How Thor Found His Hammer"; "Robin Hood"; "Pecos Bill"; "Paul Bunyan."

Some storytellers enjoy telling several stories around a specific subject or theme. A "Hans Christian Andersen Storytelling Festival," for example, could include "The Steadfast Tin Soldier"; "The Nightingale"; "The Tinderbox"; "The Swineherd." A theme of "Forgetfulness" might include "Soap, Soap, Soap," in Richard Chase's *Grandfather Tales* (1948), the story of a boy who can't remember what he is shopping for; and "Icarus and Daedalus" a Greek legend, in which Icarus forgets that his wings are made of wax. If you are interested in choosing several stories about a certain subject, Caroline Bauer's *Handbook for Storytellers* includes an annotated bibliography of stories by subject, as well as recommendations for single stories.

Preparing your story   Now that you have selected a story you would like to tell, the actual process of preparing the story begins. An experienced storyteller, Patti Hubert, recommended the following sequence, which has proven successful for her.[2]

---

[2]This sequence is used with permission of Patti Hubert, drama teacher, San Antonio, Texas.

1. Read the story completely through about three times.
2. Try to list mentally the sequence of events. You are giving yourself a mental outline of the important happenings.
3. Go back and reread the story, taking note of the events you didn't remember.
4. Now go over the main events again and add the details you remember. Think about the meaning of the events and how to express that meaning, rather than trying to memorize the words in the story.
5. When you feel you know the story, tell the story to a mirror. (You will be surprised at how horrible the story sounds the first time.)
6. After you have practiced two or three more times, the wording will improve, and you can try changing vocal pitch to differentiate characters.
7. Try changing your posture or hand gestures to represent different characters.
8. Don't be afraid to use pauses to separate scenes.

Introducing the story   Many storytellers set the mood for story time with a story hour symbol. One librarian used a small lamp; when the lamp was lit, it was also time to listen. If you have a certain place in your classroom for storytelling, just the movement to that storytelling corner may prepare the children and set the mood. Your storytelling corner might include an easel, on which to place a drawing from the story or a motivating question written on cardboard. Some storytellers have discovered they can create a mood with a record or a piano. One of my student teachers successfully used a guitar for this purpose. As soon as he went into the storytelling corner and quietly played a specific song on the guitar, the students came over, anticipating an enjoyable activity.

There are numerous ways to introduce your story. You might ask a question, or tell the students why you enjoy a particular story. You might offer something interesting about the author or the background of the story. Background information about a country or a period of history adds interest to telling a folktale. It is often advantageous to display prominently the book from which your story is taken, along with other stories by the same author and other stories on the same subject. One of your reasons for storytelling is to motivate children to tell stories themselves, so the display adds visual interest.

Many of my students effectively introduce a story with objects. One student used a stuffed toy rabbit to introduce *The Velveteen Rabbit* by Margery Williams (1922). A toy or figure of a cat could be used with *Puss in Boots* by Charles Perrault (1952). Another student used a lariat to introduce a tall tale about Pecos Bill. Artifacts from a country are good introductions to folktales. One of my favorite objects is a small, painted jewelry case from Japan. I often bring out the case when I am telling stories to young children, and we talk about the magic of stories that is contained in the case because it has been to so many story hours. Then I open the box slowly while the children catch the magic in their hands. This magic is wonderful, so they hold it carefully while they listen to the story. When the story ends, each child carefully returns the magic to the box until it is time to use it again for another story.

Telling your story   Now that you have prepared your story and thought about ways to introduce it or to set a mood, you are ready to face your expectant students. Patti Hubert offered the following suggestions for telling your story:

1. Find a place in the room where all the children can see and hear you.
2. Either stand in front of the group, or sit with them.
3. Select an appropriate introduction: use a prop, tell something about the author, discuss a related event, or ask a question.
4. Maintain eye contact with the children. This engages them more fully in the story.
5. Use your voice rate and volume for effect.
6. While telling your story, take a short step or shift your weight to indicate a change in scene or character, or to heighten the suspense. If you are sitting, lean toward or away from the children.
7. After telling your story, pause to give the audience a chance to soak in everything you have said.

Adding pictures and objects to storytelling   Effective storytelling does not require additional props during the story, but you may want to add variety to the story time with felt boards, flip charts, roll stories, objects, or chalk talks.

Children need many opportunities to listen to and to tell stories.

Felt boards   For exciting visual interest, cut out representations of the main characters from felt, flannel, or pellon, and put them on a board covered with felt or flannel. The figures provide cues to the story, giving the beginning storyteller added security. Placing the figures in proper order before starting the story shows the sequence of events. It is better to tell a felt board story rather than read it, because it is awkward to manipulate both figures and book at the same time.

It is easy to make a flannel or felt board by covering an artist's cardboard portfolio with felt or flannel. You can make a firmer felt board by covering a rectangle of light-weight wood, such as fiber board or plywood, with a large piece of felt or flannel. For greater mobility, you can hinge two smaller squares of board together, if you will need to transport the felt board. You can make backdrops for different scenes and locations by chalking in scenery on a loose piece of flannel large enough to cover the board. These scenic flannel pieces are easy to store, and add interest without cluttering the appearance of the felt story.

Certain types of stories are more suitable for the felt board than others. Stories with a great deal of physical action or detailed settings are inappropriate for felt boards; simple stories with only a few major figures and definite scenes are easy to handle. Cumulative tales, which add elements throughout the story, are extremely good. Some stories that meet these requirements are "Goldilocks and the Three Bears," "Three Little Pigs," Three Billy Goats Gruff," "Jack and the Beanstalk," and "The Emperor's New Clothes." All the traditional favorites can be shown with a few simple characters.

When you tell the felt board story, place the board so the whole group can see it. An easel is a good place to position the board for a larger group. Place the figures in proper order before you begin the story, and keep the figures out of the children's sight until you place them on the board. You will need to practice several times so you will be talking to the children, not to the felt board.

Felt board stories allow the students to become involved. They can retell the felt stories, reinforcing oral language and comprehension skills. After they have seen one or two felt board stories, many children want to make their own. These stories are an excellent means of giving an oral book report or illustrating a creative writing story. If you leave the felt board in the room so the children have easy access to it, you will find they use it frequently. (A felt board is also an excellent visual aid for developing concepts in math and science.)

Flip charts   Another technique that adds visual interest to storytelling is illustrating the story on several large sheets of poster board. Like the felt-board stories, the flip chart should be used with stories whose main ideas can be developed with a series of pictures. The charts can be drawn by hand or with the aid of an opaque projector. The most successful flip charts used three-dimensional materials. Yarn, fabric, sandpaper, small stones, bark, and straws add a tactile dimension to the flip chart, and invite the children to touch the pictures.

One of my students drew a flip-chart story entitled "Freddie's Private Cloud." She used rice for shingles on a house, cloth for clothing, real feathers on birds, nylon net for clouds, sandpaper for a lawn-mower engine, spaghetti for a ladder, bark for tree trunks, rolls of real paper for newspapers, beans for stones, sponge for waves, and felt for animals.

Roll stories The roll story is also sequentially illustrated. The pictures are drawn on a large roll of paper, then unrolled as the story is told. The completed roll story is placed inside a box with an opening cut out like a television screen. The roll should have dowels at either end so that it is easy to roll and unroll. Roll stories are also excellent culmination activities after you have read or told a story. The children can form groups to draw pictures, and retell the story with their own pictures.

Roll stories reinforce reading skills as well as storytelling skills. For example, after one first-grade teacher told "Peter and the Wolf" to her students, the students listed the scenes from the story, and each of them chose two sequential scenes to illustrate. The teacher cut a long roll of paper into sections that fit the length of the classroom's work tables, and the children, with the teacher's assistance, measured the distance

needed for each picture. Then they drew their illustrations, in correct sequence, onto the roll. When each child had completed her illustrations, the child dictated to the teacher the part of the story that corresponded to her illustrations. These individually dictated stories were then used for reading instruction. The children practiced reading their own contributions so they could present a "movie" of "Peter and the Wolf." Next, the table-length rolls were taped together and dowels attached to each end. The roll was placed inside a box with a rectangular viewing area, and the children practiced reading their parts as the story unrolled. They added background music from "Peter and the Wolf," and invited another grade to come in and view their "movie" presentation. This is an excellent motivating activity for remedial reading instruction.

Object stories   Certain stories lend themselves to showing objects from the story. For example, while telling Marcia Brown's *Stone Soup* (1947), the teacher or students can place the ingredients in a soup kettle.

Peter, Peter pumpkin eater
Had a wife and couldn't keep her
He put her in a pumpkin shell
And there he kept her very well.

Humpty Dumpty sat on a wall
Humpty Dumpty had a great fall
All the king's horses
And all the king's men
Couldn't put Humpty Dumpty
   together again.

Chalk talks   It is fascinating to watch someone illustrate a story while he tells it. One student teacher invited a cartoonist to visit her fourth-grade class. The cartoonist quickly sketched cartoon figures while telling a corresponding story. This was a highly motivating activity that led many students to try the technique with simple-figure stories. Although you may not have this cartoonist's rapid drawing ability, you might enjoy trying the activity with stick figure characters or simple shapes. The examples above show simple sketches drawn to accompany familiar nursery rhymes. One teacher made up a story about the adventures of an Easter egg, with simple shapes that could be sketched quickly on the chalkboard or on a flip chart placed on an easel.

Choose a story to prepare for storytelling according to the suggestions in this section of the chapter. You can use pictures or objects, if you wish. Tell your story to a group of children or to a group of your peers.

Developing Enjoyment Through Art

Strange and curious worlds, imaginary kingdoms, animal fantasies, space explorations, historical settings, and informational books all lend themselves to artistic interpretations. Coody (1979) stated that "creative art-literature experiences occur in the classroom when boys and girls are moved by a good story well told or read, when art materials are made available, and when time and space are allowed for experimenting with the materials" (p. 92).

Illustrated books showing various artistic media, descriptive passages, and stories that encourage hypothesizing stimulate artistic expression, encourage enjoyment of literature, enhance aesthetic development, and stimulate understanding of various literary elements such as setting and plot development. (A word of caution: art should allow children to expand their enjoyment of a story through self-expression; it should not be used as a forced activity following the reading of every story.)

Illustrated books may be used to show children that artists see their subjects in different ways and use different media to interpret mood and setting. These books help children discover that artistic interpretations are very individual qualities. For example, you can share and discuss Trina Shart Hyman's illustrations for Grimm's *Little Red Riding Hood* (1983), James Marshall's illustrations for *Red Riding Hood* (1987), and Sarah Moon's photographs for Perrault's *Little Red Riding Hood* (1983). In these three examples students discover that illustrations can reinforce a "once upon a time" traditional setting, develop a humorous mood, or create a stark, terrifying mood.

Illustrated books using a particular artistic media may be used to stimulate students to try that media in their own illustrations. For example, collage techniques are used in numerous books. In collage, any object or substance that can be attached to a surface can be used to develop a design. Artists and student artists may use paper, cloth, cardboard, cloth, leather, wood, leaves, flowers, or even butterflies. They may cut up and arrange their own paintings or use paint and other media to add background. When photographically produced in a book, collages still communicate a feeling of texture.

Children's literature selections that may be used to show children how adult authors use collage include Ezra Jack Keats's *Peter's Chair* (1967), *The Snowy Day* (1962), and *The Trip* (1978); Eric Carle's *The Very Hungry Caterpillar* (1971), *Twelve Tales from Aesop* (1980), and *The Honeybee and the Robber* (1981); Marcia Brown's illustrations for Blaise Cendrars's *Shadow* (1982); and Jeannie Baker's *Grandmother* (1978). Giles Laroche's illustrations for Lois Lenski's *Sing a Song of People* (1987) shows the drama of three-dimensional paper constructions.

Books are filled with descriptive passages and vivid characters that entice children to try their own artistic explorations and interpretations. For example, vivid settings from Beatrix Potter's *The Tale of Squirrel Nutkin* (1903, 1986), Kenneth Grahame's *The Wind in the Willows* (1908, 1940), and E. B. White's *Charlotte's Web* (1952) encourage individual drawings or group-developed murals. Monica Hughes's *The Dream Catcher* (1987) could motivate the depiction of a science fiction setting in a world as yet to materialize. Likewise, Janet Lunn's *Shadow in Hawthorn Bay* (1986) could motivate pictorial comparisons between the hills of Scotland around Loch Ness and Hawthorn Bay, Ontario in the early 1800s.

Many stories encourage students to hypothesize about what could happen next and to create a visual interpretation of these actions, settings, or characters. Younger students could expand on one of the nursery rhymes in Tomie DePaola's *Mother Goose* (1985). For example, how do Peter Peter, pumpkin eater and his wife live in the pumpkin shell? Students could extend Tejima's *Fox's Dream* (1987) by illustrating pictures of the same forest in the spring rather than in the icy covering of winter. Older students could extend Robert O'Brien's *Mrs. Frisby and the Rats of NIMH* (1971) by drawing Thorn Valley, the intelligent rat colony that the rats want to develop. Then, they could compare their drawings with the written descriptions in Jane Leslie Conly's sequel, *Rasco and the Rats of NIMH* (1986).

Mythology can inspire students, like illustrators of picture books, to depict the settings for such mythological places as the sacred lake where a Native American boy waits for the buffalo to appear (Olaf Baker's *Where the Buffaloes Begin,* 1981) or the foaming sea that produces the goddess Aphrodite (Doris Gates's *Two Queens of Heaven: Aphrodite and Demeter,* 1974).

Reading stories aloud, storytelling, and artistic interpretations of literature are stimulating ways to develop appreciation for and understanding of literature. These activities also motivate students to read literature.

Developing Student Critics for the Literature They Read

At one time in the history of education, selecting children's literature was not really a problem; in the late 1800s, the problem was finding anything at all for children. In 1800, only 270 books were published for the juvenile market. Thankfully this has changed. Currently, almost 3,000 books for children are published annually. This increase now causes us to be concerned with the quality of literature that students read. In this section we discuss ways to help students become critical of the books they read.

Literature is usually evaluated in terms of plot, characterization, setting, theme, style, and format. Consequently, teachers need to recognize literary elements in the books they share with students. We merely review them quickly because a children's literature course usually covers these subjects in detail.

Plot   The plot of a story develops the action. Plot development should contain a generous amount of action, excitement, suspense, and conflict. The development of the events usually follows a chronological order, although flashbacks may answer questions about a character's background or reveal information about a previous time or experience. Excitement in a story occurs when the main character experiences a struggle or overcomes a conflict. When conflict is added to the sequence of events, the result is called plot. Authors develop plots through four kinds of conflict: (1) person against person, (2) person against society, (3) person against nature, and (4) person against self. The author must describe and develop the conflict so that it is believable. A good plot lets children share the action, feel the conflict, recognize the climax, and respond to a satisfactory ending.

Characterization The characters should seem believable and should develop throughout the course of the story. The credibility of characters such as Laura in the *Little House* books depends on the author's ability to reveal their natures, strengths, and weaknesses. Characterization is developed when the author tells about the characters, records their conversations, describes their thoughts, shows others their thoughts, and shows the characters in action. The most memorable characters usually have several sides to their characters—like real people, they are not all good or all bad.

Setting The setting is the geographic location and the time—either past, present, or future—during which the story takes place. The events should be consistent with what actually occurred during that period, and if the location is identifiable as a real place, it should be presented accurately. Authors of historical fiction, for example, must make the background for their stories as authentic as possible. In addition to depicting a complete historical background, setting may provide an instantly recognizable background such as found in folktales, create a mood, develop conflict, or suggest symbolism.

Theme The theme is the author's purpose for writing the story. Students in literature classes often spend a great deal of time discussing the theme of a story, and analyzing the theme according to several levels or dimensions. Although elementary students do not need to study theme in such depth, the theme should be worthwhile. Children's literature often uses themes such as overcoming fear, searching for love or acceptance, or the process of growing up.

Style Style refers to the way the author uses words and sentences to develop the story. The style, of course, should be appropriate for the characters and the plot. Does the language of the story sound like what the characters would really say? If there is figurative language, is it within the understanding of the children? Reading aloud is a good way to see if the author has a pleasing and appropriate style.

Format Format refers to the physical aspects of the book—cover, printing, illustrations, and size. Illustrations should help the story come alive and add a necessary element to it. The printing should be clear and easy to read. Some bindings are more appropriate for school libraries than others. Although paperbacks are more readily affordable, hardcover bindings last longer, and are usually purchased for more permanent collections.

Developing an Understanding of Criteria for Fine Literature

A sixth-grade teacher introduced her students to an evaluation-of-literature unit by having the children interview their parents and other adults to find out what favorite books adults remembered reading when they were in elementary school. The children listed the books and the adults who chose them on a large chart. Then the students checked book awards, such as the Newbery and Caldecott, to see how many of the books appeared on these lists. Next, each student read a book their parents had mentioned, and discussed the book with their parents in terms of what made it special. Now the teacher introduced the concepts of plot, characterization, setting, theme, style, and format, and the students examined the books they had read for examples of each of these elements. Finally, the sixth graders listed questions to ask in evaluating a book:

Questions to Ask Myself When I Judge a Book

1. Is this a good story?
2. Is the story about something I think could really happen?
3. Did the main character overcome the problem, but not too easily?
4. Did the climax seem natural?
5. Did the characters seem real?
6. Did the characters grow in the story?
7. Did I find out about more than one side of the characters?
8. Did the setting present what is actually known about that time or place?
9. Did the characters fit into the setting?
10. Did I feel that I was really in that time or place?
11. What did the author want to tell me in the story?
12. Was the theme worthwhile?
13. When I read the book aloud, did the people sound like real people actually talking?
14. Did the rest of the language sound natural?

As an additional follow-up activity, students wrote stories and asked these questions about their own writing. This type of activity may be used to introduce a more extensive study of literature; the following activities are designed to teach an understanding of literary genres and literary elements.

Developing Understanding and Appreciation of Literary Genres

Books are divided into categories or types. Children's literature genres are identified and discussed in major textbooks about children's literature according to such categories as picture books, traditional literature, modern fantasy, poetry, contemporary realistic fiction, historical fiction, biography and autobiography, and informational books. Each of these genres has special characteristics related to the characters who are in the stories, the settings, the plot development, and the author's techniques used to create believable literature. Sharing, discussing, and comparing various genres helps students gain appreciation for the author's ability to create that genre and increase their understanding of specific characteristics of that genre. These activities are appropriate for a range of grades and reading levels because books are available in picture book format, in easy to read versions, and in longer novel lengths. The grade level will depend on the literature chosen and the depth of analysis lead by the teacher.

Picture books versus illustrated books   Picture books are excellent choices for initial analysis because, according to Lacy (1986), they "provide enjoyable opportunities for visual exploration, interpretation, and reflection" (p. 2). Although most children's books are illustrated, not all illustrated books are classified as picture books. Activities such as looking at illustrations, discussing the importance of text and illustrations within a book, comparing picture books and illustrated books, and classifying books as either picture books or illustrated books encourage an appreciation for picture books. They also stimulate higher cognitive development as students analyze, compare, and evaluate.

A logical progression for this activity begins with definitions for picture books, extends to analyzing examples of picture books and to discussing how the books do or do not meet these definitions or criteria, and finishes with comparisons between picture books and illustrated books in which students justify why they place books within either category. The emphasis on this activity should be on the discussion, the analysis, and the support for various decisions and not on the decision that one answer is right or wrong. Some books lend themselves to very lively discussions because students may justifiably argue and classify them in more than one way.

Introduce picture books by asking students to give you their definitions for picture books. Next, share several definitions developed by either children's literature authorities or illustrators/authors of picture books and discuss their implications. For example, Sutherland and Hearne (1984) defined a picture book as a book in which the illustrations are either as important as the text or more important than the text. Illustrator Uri Shulevitz (1985) stated that a picture book is closely related to theater and film because the picture book is a dramatic experience that includes "actors" and "stages." In a picture book the characters, settings, and actions are shown through the pictures.

Next, share and discuss examples of picture books. Begin your discussion with wordless books such as Pat Hutchins's *Changes, Changes* (1971), Emily Arnold McCully's *Picnic* (1984), and John Goodall's *The Story of a Main Street* (1987). Ask the students to tell you the story from the illustrations, to describe the setting, to describe the characters, and to describe the conflict, if any, that they find in the stories. They should conclude from this discussion that all of these subjects are revealed through the illustrations. Consequently, these books meet our earlier definitions for picture books.

Proceed from wordless books to books that have minimal texts such as Maurice Sendak's *Where the Wild Things Are* (1963) and Audrey Wood's *King Bidgood's in the Bathtub* (1985). Ask students to look at the illustrations and to notice the information presented in the illustrations that is not found in the written text. For example, the illustrations in Sendak's book show the magnitude and type of mischief and depict the appearance of the wild things. In addition, the illustrations provide drama as they increase in size until the land of the wild things is shown in double-page spreads without text. This book clearly matches Shulevitz's definition for a picture book that includes both actors and stages. Likewise, the illustrations in Wood's book provide maximum detail for minimal text. For example, the text states, "Today we lunch in the tub!" (p. 11, unnumbered). The double-page illustrations show an elegant luncheon with a cake centerpiece that is a model of the castle. You may ask the students to describe the contents of the illustrations either orally or in writing. They will discover that without the illustrations the author would need to write pages of descriptions, as well as to create a humorous and exaggerated mood through the choice of words rather than the creation of illustrations. Students should conclude from this discussion that picture books with minimal text contain numerous important details within the pictures.

Next, proceed to picture books in which the illustrations and the text play equal roles such as Wanda Gag's *Millions of Cats* (1928) and Arthur Yorink's *Hey, Al* (1986). Gag's book is an excellent choice because the written text compliments the illustrations and the repetitive style by creating a feeling of movement. Even though there is more written text in *Hey, Al* than in *Where the Wild Things Are*, the illustrator uses some of the same techniques: the illustrations become larger as the conflict develops, the illustrations cover a double-page spread at the height of the interest, and the illustrations include considerable information about setting and characters. After reading the texts and looking at the illustrations, students should discover that the illustrations and the written text are equally important.

Next, proceed to highly illustrated books that are not classified as picture books. For example, Mildred Taylor's *The Gold Cadillac* (1987) contains an illustration on about every third page and Russell Freedman's *Lincoln: A Photobiography* (1987) contains a photograph or other type of illustration on about every second page. After students listen to or read the texts and analyze the importance of the illustrations and the text, they should discover that although the illustrations add both information and interest, the illustrations cannot stand alone or be classified as equally important to the written text.

Finally, ask students to develop a list of activities and questions that would help them decide if a book is a picture book or an illustrated book. For example:

1. Try to retell the story without looking at the words. Does the sequence of illustrations allow me to understand the story? (picture book)
2. Read the text. Is the story self-sufficient without the illustrations? (illustrated book)
3. Compare the illustrations and the text. Are the illustrations so important that they add information about the setting, the actions, and the characters that is not provided in the written text? (picture book)
4. Think about the characteristics of film and theater. Do the illustrations have a dramatic flow that makes me think of "stages" and "actors"? (picture book)

Additional examples of picture books and illustrated books that may be used for this final comparison include:

- Picture Books: Chris Van Allsburg's *The Z Was Zapped* (1987), Nancy Tafuri's *Have You Seen My Duckling?* (1984), Margaret Mahy's *17 Kings and 42 Elephants* (1987), and Marc Brown's *Arthur's Baby* (1987).
- Illustrated Books: Cynthia Rylant's *Children of Christmas* (1987), Jean Fritz's *Shh! We're Writing the Constitution* (1987), and Michael Foreman's illustrated versions of Rudyard Kipling's *Just So Stories* (1987) and *The Jungle Book* (1987).

Books such as Mavis Jukes's *Like Jake and Me* (1984) and Joel Chandler Harris's *Jump Again! More Adventures of Brer Rabbit* (1987) show how important it is to emphasis support for categories rather than right and wrong answers. Lively discussions accompany these books because students could support the books as either picture books or illustrated texts.

Identifying and comparing other genres of literature   Exciting discussions and analysis occur when students study genres of literature such as modern fantasy, contemporary realistic fiction, historical fiction, and biography; when they identify characteristics of the specific genre of literature; and when they compare similarities and differences among genres. As seen in table 10–2, there are major differences as well as similarities among characters, settings, plot development, and creating believable stories.

To develop an understanding of and an appreciation for these differences and similarities the teacher should introduce and help students identify characteristics of one genre. After they have thoroughly investigated one genre by listening to literature, reading literature independently, and identifying and discussing the characteristics of that literature, they may continue the same type of listening, reading, and discussing activities with the next genre. After the first genre is concluded, however, a new element is introduced as students compare similarities and differences between genres. This activity may be easily adapted to different grade levels and to students with diverse reading abilities by choosing books from picture book formats, easy to read books, and novel type literature.

There are numerous, clearly defined differences between modern fantasy and contemporary realistic fiction; consequently, this is a logical place to begin this portion of the study of literary genres. First, share and discuss several modern fantasy selections and encourage students to identify specific characteristics of modern fantasy that are similar to those shown in table 10–2. Examples of books that show these characteristics and evidence of these characteristics may be placed on the table. The first time this activity is completed, within any genre, it should be lead by the teacher who reads texts or portions of texts to students and then leads a discussion during which students talk about and identify various characteristics and examples of those characteristics from the literature. After this initial activity, additional books from the genre may be read by the teacher or by the students and then added to the charts.

It will help our study if we first define modern fantasy. Modern fantasy is fiction in which the author takes the reader into a time and a setting where the impossible

TABLE 10–2

Differences among genres of literature

	Modern Fantasy	**Contemporary Realistic Fiction**	**Historical Fiction**	**Biography**
Characters: **Literature Selections:** **Literary Evidence:**	Personified toys. Little people. Supernatural beings. People who have imaginary experiences. Animals who behave like people.	Fictional human characters must behave like real people. Fictional animal characters must behave like real animals.	Fictional humans must behave like real people who express understanding of historical period.	All people must be real people.
Setting: **Literature Selections:** **Literary Evidence:**	Past, present, or future. Imaginary world created by the author. May travel through time and space.	Contemporary world as we know it. May have fictional location but must be possible in the contemporary world.	Authentic for historic time period. Setting supported by facts.	True for time the person actually lived. Setting supported by facts.

becomes convincingly possible. Authors of modern fantasy create their settings by altering one or more of the literary elements from what is expected in the real world. Authors of modern fantasy frequently choose to have their characters depart from what we know to be possible in the real world; the characters themselves may be contrary to reality or experience preposterous situations (Norton, 1987).

Modern fantasy selections should include stories that reflect the various types of characters in the modern fantasy genre such as personified toys (Margery Williams's *The Velveteen Rabbit: Or How Toys Become Real,* 1922, 1958, and A. A. Milne's

TABLE 10–2
continued

	Modern Fantasy	**Contemporary Realistic Fiction**	**Historical Fiction**	**Biography**
Plot Development:	Conflict may be against super-natural powers. Problems may be solved through magical powers.	Author creates plot as fictional characters face contemporary problems such as growing up, survival, family problems.	Author creates plot as fic-tional charac-ters cope with problems that are authentic for the his-torical time.	Plot follows life of real person. Conflict follows real problems and dates.
Literature Selections:				
Literary Evidence:				
Creating Believable Stories:	Authors must encourage readers to suspend disbelief.	Authors may rely on relevant subjects, everyday occurrences, or realism.	Authors must use research to create historically authentic settings and problems.	Authors must use research and documen-tation. Objectivity.
Literature Selections:				
Literary Evidence:				

Winnie-the-Pooh, 1926, 1954); little people (Mary Norton's *The Borrowers,* 1952, and Carol Kendall's *The Gammage Cup,* 1959); supernatural beings (Lucy Boston's *The Children of Green Knowe,* 1955); real people who have imaginary experiences (C. S. Lewis's *The Lion, the Witch and the Wardrobe,* 1950, Lewis Carroll's *Alice's Adventures in Wonderland,* 1866, 1984, and Madeleine L'Engle's *A Wrinkle in Time,* 1962); and animals who behave like people (Beatrix Potter's *The Tale of Peter Rabbit,* 1902, Robert Lawson's *Rabbit Hill,* 1944, and E. B. White's *Charlotte's Web,* 1952). (Notice how many of these recommendations are considered classics in children's literature.)

Modern fantasy selections should show students that settings may be in the past, present, or future. Settings may also be in worlds that are totally created by the author such as Narnia in *The Lion, the Witch, and the Wardrobe,* Prydain in Lloyd Alexander's *The Book of Three* (1964), and Demar in Robin McKinley's *The Hero and the Crown* (1984). Characters may also go through time warps as the settings shift from present to past (Janet Lunn's *The Root Cellar,* 1983, Ruth Park's *Playing Beatie Bow,* 1982, and David Wiseman's *Jeremy Visick,* 1981) or from present to future (Margaret Anderson's *The Mists of Time,* 1984).

Likewise, the literature should show students that the conflict may be against supernatural powers such as in *The Lion, the Witch and the Wardrobe* or in *The Hero and the Crown.* The problems in modern fantasy may be solved through magical or supernatural powers as are found in both of these books. Supernatural powers may also help in gentler plots such as *The Velveteen Rabbit: Or How Toys Become Real.*

The books read and discussed should help students understand and appreciate that authors of modern fantasy must make readers believe in worlds, characters, and actions that are not possible in the world as we know it. In literary terms, the author must encourage readers to suspend disbelief. The key questions for students to consider are: "Did the author make me believe that the characters, settings, and conflicts were possible?" and "How did the author create the world, the people, and the conflict so that I believed they were possible?"

After students have developed a clear understanding of modern fantasy, extend the study of genres into contemporary realistic fiction. The definition for contemporary realistic fiction is that everything in the story, including characters, setting, and plot, could happen to real people or animals living in our contemporary world. Contemporary realistic fiction does not mean that the story is true; it means only that it could have happened (Norton, 1987).

Use a similar approach for the study of contemporary realistic fiction as that described for modern fantasy. Select books such as the following examples to help students identify and develop characteristics of contemporary realistic fiction identified in table 10–2: Beverly Cleary's *Dear Mr. Henshaw* (1983) and *Ramona Quimbly, Age 8* (1981); Judy Blume's *The One in the Middle is the Green Kangaroo* (1981) and *Tales of a Fourth Grade Nothing* (1972); Betsy Byars's *Cracker Jackson* (1985); Lois Lowry's *Anastasia's Chosen Career* (1987); Katherine Paterson's *Bridge to Terabithia* (1977); Gary Paulsen's *Hatchet* (1987); Virginia Hamilton's *Zeely* (1967); and Dennis Haseley's *The Scared One* (1983).

After identifying and developing the characteristics of contemporary realistic fiction, encourage students to compare modern fantasy and contemporary realistic fiction.

Continue this study with an analysis of historical fiction. The definition for historical fiction is a fictitious story written about an earlier time in which the setting is authentic in every respect; the actions, beliefs, and values of the characters are realistic for the time period; and the conflicts reflect the historical time period. Although the characters did not actually live, they could have lived during that time period (Norton, 1987).

Follow the sequence of events described under modern fantasy. Select examples of books that help students develop an appreciation and understanding for the characteristics shown in table 10–1: Barbara Brenner's *Wagon Wheels* (1978), Patricia

MacLachlan's *Sarah, Plain and Tall* (1985), F. N. Monjo's *The Drinking Gourd* (1970), Uri Orlev's *The Island on Bird Street* (1984), Philip Pullman's *The Ruby in the Smoke* (1987), Laura Ingalls Wilder's *Little House in the Big Woods* (1932), and Mildred Taylor's *Roll of Thunder, Hear My Cry* (1976).

After analyzing the literature and identifying and discussing characteristics, compare similarities and differences among historical fiction, contemporary realistic fiction, and historical fiction.

Finally, extend the study of genre into biography. The definition for a biography is a book written about a real person, living or dead. The biography must be a true story about the person and the other people and the incidents that influenced that person's life. This is the first time in our study of literary genres that the characters, their actions, and the settings must be true reflections of what actually happened and not reflections of what could have happened. Examples of biographies that may be used to help students identify and develop characteristics of the biography genre listed on table 10–2 include: Russell Freedman's *Lincoln: A Photobiography* (1987), David Adler's *Martin Luther King, Jr: Free At Last* (1986), Lillian Gish's *An Actor's Life for Me!* (1987), Jean Fritz's *Where Do You Think You're Going, Christopher Columbus?* (1980), Joe Lasker's *The Great Alexander the Great* (1983), Betsy Lee's *Charles Eastman, The Story of an American Indian* (1979), Polly Brooks's *Queen Eleanor: Independent Spirit of the Medieval World* (1983), and Tobi Tobias's *Maria Tallchief* (1970).

After examples are found from biographies that support these characteristics and discussions about biographies are concluded, characteristics of biographies as a genre should be compared with characteristics of other genres.

REINFORCEMENT ACTIVITY

Choose another genre of children's literature such as poetry, traditional literature, or informational literature and develop an activity that encourages students to develop appreciation and understanding of the characteristics of that genre.

Developing Understanding of Literary Elements

Developing understanding of literary elements such as setting, characterization, conflict, plot development, theme, point of view, and author's style is both a challenge to the teacher and an exciting approach to literature. My own research with fifth- through eighth-grade students (Norton, 1987), my work with teachers across the country, and my work with graduate and undergraduate language arts, reading, and children's literature students shows that exciting instructional strategies can be used to develop both an understanding of and an appreciation for these literary elements. For example, work with fifth- through eighth-grade students shows that webbing is one of the most effective ways to help students understand the importance of literary elements in a story. Modeling of literary elements is another exciting approach that helps students understand and appreciate these elements. Drawing plot structures, tracing symbolism, writing setting descriptions as if they were travel brochures or encyclopedia descriptions, rewriting stories from a different point of view, and plotting possible literary activities associated with a book are all effective instructional strategies that add both excitement and understanding to the classroom.

Using webbing in guided discussions   Semantic mapping or webbing has been discussed in previous chapters. In chapter 2 the technique was used for vocabulary development; in chapter 8 the technique was used in the writing process. Later in this chapter, webbing will be used to develop units around literature.

When webbing is used during guided book discussions, the webbing helps students understand important characteristics of a story, increases their appreciation of literature, and improves their reading and writing competencies. Prior to the webbing experience, the teacher introduces the literary elements of setting, characterization, plot development, conflict, and theme by reading and discussing folktales with students. Then the teacher draws simple webs that include each of these components, while leading discussions that help students identify the important characteristics that are being placed on the web. For example, "Three Billy Goats" might be placed in the center of the web, and then each of the literary elements drawn from the web. As the teacher reads the story, students listen for the various categories. After the story is completed, the teacher leads a discussion in which the students fill in the various categories on the web. If necessary, the teacher rereads sections to help students identify specific elements and to word them for the web. Teachers may also ask students to focus on only one element in the web. This technique is effective when literary elements and webbing are first introduced. A short story may require rereading several times to complete all of the literary elements on the web. Another approach is to ask groups of students to be responsible for specific categories on the web. Students then listen for specific information and web their specific category before the oral discussion takes place. This type of discussion frequently leads to in-depth analysis as students listen attentively for their own areas of concern.

After this activity has been completed with several simple books and webs, the teacher and students may progress to the webbing of more complex books. For example, John Steptoe's *Mufaro's Beautiful Daughters: An African Tale* (1987) is a logical next choice because the story is short enough to read several times within one class and the basic literary elements are easily identified. First, the teacher introduces the story. Next the teacher draws a web on the board with *Mufaro's Beautiful Daughters* in the center of the web and with the terms setting, characters, conflicts, plot development, and theme extended from the web as shown in figure 10–1. Now the teacher reads the story orally to the students while they listen both for pleasure and for information that may be placed on the web. Following this reading, the teacher may either reread the story, stopping at appropriate places to consider information that should go on the web, or lead a discussion in which the students fill in the various categories on the web. If the total book is not reread before filling in the web, the teacher may still need to reread parts of the story as the students consider what information should be placed on the web. The choice of approach will depend on the

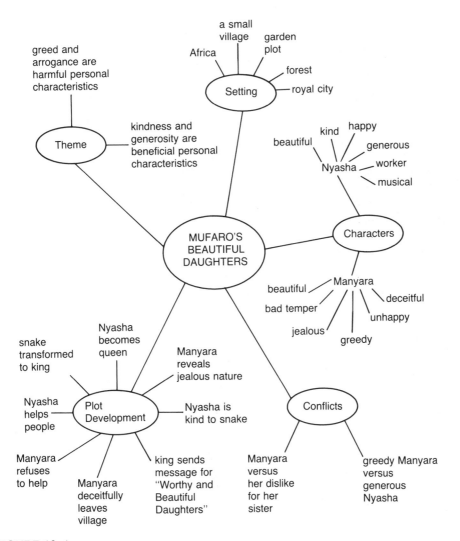

FIGURE 10–1
A literary discussion web for a folktale

students' listening capabilities and their experiences with literary elements. Whichever approach is used, the students develop a web on their own papers as the teacher completes the web on the board using the information identified by the students. Some of these webs become quite large and may cover several chalkboards. In another class period following the development of the web, students may write their own stories about the book. They use the information on the web to help them construct their stories. This is a good place to reinforce point of view. Students may use the basic information on the web to write a story from the point of view of one of the other characters in the story.

Figure 10–1 provides an example of a web developed around *Mufaro's Beautiful Daughters*. Only the first level information is shown on this web. When the web was completed with third- and fourth-grade students, the web extended to several chalkboards as students identified additional characteristics for each of the settings. This type of activity stimulates considerable discussion about literary elements as students support their choices with evidence from the text, discuss relationships within the identified categories, and discuss inferences that led them to make certain decisions. For example, the author does not state directly the themes shown on the web. The text, however, does provide plenty of clues to these themes by describing characteristics of people who are rewarded and characteristics of people who are punished. Likewise, some of the terms describing the characters are words that students chose to describe the characters' actions.

Webs may also be developed by the teacher to provide structure for a literary discussion or to identify specific books or poems that exemplify certain literary elements. For example, figure 10–2 presents a semantic web developed by one of my

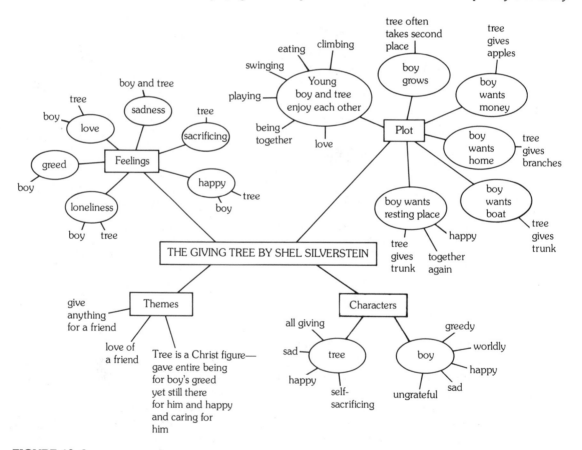

FIGURE 10–2
A web summarizing plot, characterization, themes, and feelings

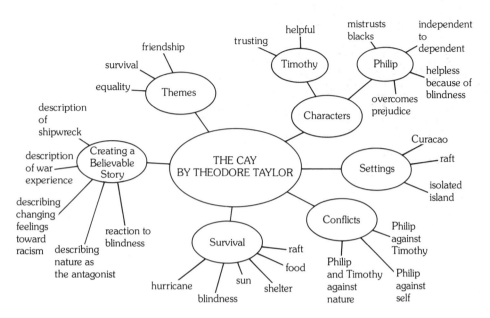

FIGURE 10–3
A web summarizing literary qualities in *The Cay*

students, Pam Buster. It summarizes the plot, the characters, the themes, and the feelings found in Shel Silverstein's *The Giving Tree*. The web in figure 10–3 was developed by another student, Karen Wisher. It summarizes the various literary qualities of Theodore Taylor's *The Cay*. The web in figure 10–4 illustrates the various literary styles found in Judith Viorst's book on poetry, *If I Were in Charge of the World*. This type of web helps the teacher select a poem that exemplifies specific characteristics.

REINFORCEMENT ACTIVITY

Choose a book that would be good for a literary discussion. Make a web according to the literary elements found in the book.

Using modeling of literary elements   Modeling was introduced in chapter 3 as a way of teaching students how to become actively involved in and aware of their thought processing. The example used in chapter 3 described the steps in modeling inferencing of characterization with Patricia MacLachlan's *Sarah, Plain and Tall* (1985). Review the in-depth lesson plan described for that book on pages 73–76. Notice how the teacher in that example is guiding the students through understanding of characterization. The same approach may be used to model the understanding of

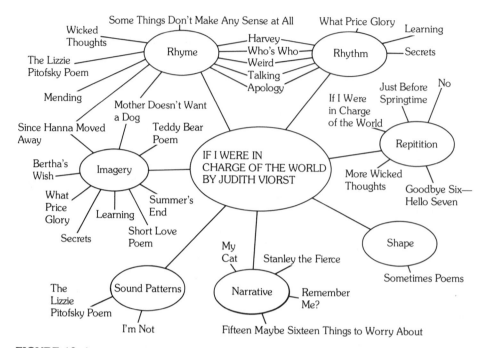

FIGURE 10–4

A web identifying appropriate poems in a book of poetry

any of the literary elements. In this chapter we consider how we could use the same modeling technique to help students understand author's style. A lesson plan format similar to that introduced in chapter 3 is used in the following example.

FOR YOUR PLAN BOOK

Modeling for the Understanding of Similes

Level: Lower Elementary   Book: Swimmy (1963), Leo Lionni.

Requirements for Effective Reasoning

1. *Going beyond the information the author provides in the text.*
2. *Knowing that similes are comparisons that authors make between something in the text and something that is not in the text.*
3. *Knowing that similes are introduced with the clue words "like" and "as".*
4. *Knowing that authors use similes to develop vivid descriptions and characterizations.*
5. *Thinking about how one object is like another object.*

Teacher's Introduction

1. *Introduce similes through a dialogue like this: "Sometimes when authors write, they use a special kind of writing called figurative language. This kind of lan-*

guage is a way for authors to use fewer words but to write more vividly. Similes are a form of figurative language in which authors compare one thing with another thing. Today we will be listening for ways that authors use similes to encourage us to see pictures in our minds.

2. Provide examples and nonexamples of similes: "Similes are stated comparisons in which the author tells us how two things are alike. We say a simile is a stated comparison because the author uses the clue words "like" or "as" to signal the comparison. Many times the comparison will be between one object you know, and one object that you do not know. For example, in 'Sandy is as quiet as a mouse.' I know that the author is using "as" to clue the comparison. I know that a mouse is shy around people so I can visualize that Sandy is a very shy person who does not talk very much. In another example, 'Giraffes have ears like giant leaves.' I know the author is using "like" to clue the comparison. I know even without seeing a giraffe that the ears are shaped like big leaves. I know that big leaves hang down so I can 'see' the giraffe with big, floppy ears that hang down alongside his head. We must listen carefully, however, because an "as" or a "like" in a sentence does not always signal a comparison. For example, in 'We will watch as the moon rises.' are there any comparisons? Likewise, in 'I like hamburgers.' is there a comparison?"

3. Provide a rational for why similes are important to students: "We need to know about similes and what they can tell us as we read or listen to stories so that we can understand what we read. Similes help us appreciate the author's story. Similes are a way that authors paint a vivid picture with words just like an artist paints a vivid picture with paints. We can also use similes to make our own writing more vivid and more meaningful. Can anyone give me an example of a simile that you might use in your own writing? Today I am going to read a story about a fish. We will listen for the similes, tell what we think the comparison means, and explain how we reached our answer. Listen carefully as I do the first one for you. By listening to how I thought through the meaning of the simile you will know what I expect of you when it is your turn."

First Example from Literature: *Read page 1 but do not show the illustration, Ask the question: "What does Swimmy look like?" Answer the question: "Swimmy is shiny and deep black." Provide the evidence: "The author uses a simile clue 'as.' The author compares one of them to the black of a mussel shell." State the reasoning used to reach the answer: "I know that the author is telling me to think about more information because I say a simile. I know the comparison is between Swimmy's color and a black mussel shell. I know from my own experience that there are many shades of black. I also know from seeing mussel shells in the water at the beach that a mussel shell is dark black and shiny. If I close my eyes I can see a small, black, shiny fish swimming among many other fish who are not black." (After modeling the simile ask the students to listen to the page again with their eyes closed. Can they see Swimmy's color? Then share the illustration and compare their visualized picture with that in the illustration.)*

Second Example from Literature: *Read pages 4 through 10, stopping with the sentence, "One day . . . a lobster, who walked like a water-moving machine." (Show illustrations* *after modeling or the students will use the illustrations to gain visual images rather than use the words and their imaginations.) Ask the question: "How does the lobster move? Answer the question: "The lobster walks mechanically through the water, pushing the water ahead of it." Provide the evidence: "The author used simile clue 'like.' The author compares the lobster and a water-moving machine." Provide the reasoning for reaching the answer: "I can 'see' a water-moving machine in my mind that would move mechanically through the water rather than darting and gliding like a fish. This machine would have to push forward against the water. I can almost imagine it moving like a robot with its claws moving the water and its feet plodding along. I am also impressed with how many words the author saved and how he depicted a vivid picture by using just the right comparison."*

At this point make sure that all of the students understand the process involved in answering questions, giving evidence, and providing reasoning used to reach the answer. Follow the same procedures but change the involvement as the teacher asks the questions but the students enter into the discussion with answers, evidence, and reasoning.

Example Three: *Read pages 11–17, stopping with "sea anemones . . . pink palm trees swaying in the wind." Ask the question, "What do the sea anemones look like?" Answer the question: "The sea anemones are pink and have a thin body (trunk). The top of the anemone has thin stringy arms that wave in the water." Continue the discussion as the students provide the evidence for the answer and discuss the reasoning used to reach the answer. Continue this procedure, reading and stopping at page 18 after the simile comparing the new school of fish to Swimmy's school and reading to the end of the text where the comparison is with the biggest fish in the sea. Encourage looking for evidence and thinking through the reasoning process. Students' visualizations of the similes may be different from the artist's illustrations. This provides a marvelous opportunity for students to draw their own interpretations of the similes. They frequently decide that their own interpretations are more exciting than the artist's.*

Modeling may be used to improve the understanding of any of the literary elements. Modeling is effective at any grade level because of the wide selection of books available. My own modeling with students has ranged from first through eighth grades. The following books and literary elements suggest some of the modeling activities that may be developed:

- Developing Understanding of Setting: Armstong Sperry's *Call It Courage* (1940), setting as antagonist; Jean Craighead George's *Julie of the Wolves* (1972), setting as antagonist; Joan Blos's *A Gathering of Days* (1979), setting as historical background; Jean Fritz's *Make Way for Sam Houston* (1986), setting as historical background; Kate Seredy's *The White Stag* (1937), setting as mood; Marcia Brown's illustrations and Blaise Cendrars's poetry in *Shadow* (1982), setting as mood.
- Developing Understanding of Conflict: Byrd Baylor's *Hawk, I'm Your Brother* (1976), person-versus-person and person-versus-nature conflicts; Barbara Cohen's *Molly's Pilgrim* (1983), person-versus-society conflict;

Katherine Paterson's *Bridge to Terabithia* (1977), person-versus-society and person-versus-self conflicts.

■ Developing Understanding of Literary Style: Virginia Lee Burton's *The Little House* (1942), personification of objects; Kenneth Grahame's *The Wind in the Willows* (1908, 1940), personification of nature and animals; Sharon Bell Mathis's *The Hundred Penny Box* (1975), symbolism through music and pennies; Cynthia Voigt's *Dicey's Song* (1982), symbolism through music and boat; Kathryn Lasky's *Sugaring Time* (1983), similes and figurative language; Hugh Lewin's *Jafta* and *Jafta's Mother* (1983), similes.

■ Developing Understanding of Themes: Margery Williams's *The Velveteen Rabbit: Or How Toys Become Real* (1922), love is powerful and can make the impossible possible; Byrd Baylor's *Hawk, I'm Your Brother* (1976), it is important to have a dream and to have freedom; Valerie Flournoy's *The Patchwork Quilt* (1985), families can share happy memories while establishing bonds across the generations; Ann Grifalconi's *Darkness and the Butterfly* (1987), we all have fears that cause us problems, but we must overcome our own fears; Denys Cazet's *A Fish in His Pocket* (1987), personal problems can be solved if we work on their solutions.

Using plot structures to understand conflict and plot development Most plots in stories for younger readers follow a structure in which the characters and the problems are introduced at the beginning of the story, the increasing conflict rises until a climax is reached, the turning point is identified, and the conflict ends. This type of structure is usual in stories in which the main character faces problems caused by external forces. We can help students understand this plot development by having them place key incidents from the story on a plot structure such as that shown in figure 10–5. Stories ranging from "Three Billy Goats Gruff" to the high fantasy of Lloyd Alexander lend themselves to this type of structure.

When characters try to overcome problems caused by inner conflicts we can use the same type of structure but the terminology changes. Cohen (1985) identified four major components in the development of person-against-self conflicts: problem, struggle, realization, and achievement of peace or truth. "The point at which the struggle wanes and the inner strength emerges seems to be the point of self-realization," says Cohen (p. 28). "The point leads immediately to the final sense of peace or truth that is the resolution of the quest. The best books are those which move

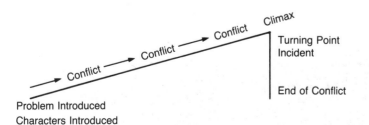

FIGURE 10–5
Diagram of a plot structure

FIGURE 10–6
A person-versus-self plot structure

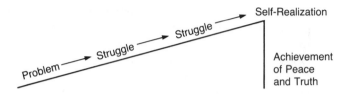

readers and cause them to identify with the character's struggle.'' Cohen's components may be placed on a plot structure such as that in figure 10–6, and students may
place major incidents on these forms. Books such as Paula Fox's *One-Eyed Cat*
(1984) and Marion Dane Bauer's *On My Honor* (1986) lend themselves to this type
of person-versus-self struggle.

Mapping Instructional Possibilities

Mapping or diagramming a book is an effective way to explore various ways that it
may be used for instructional purposes. Some of the categories that may be used for
this purpose include art interpretation, drama, values clarification, personal response,
developing characterization, developing awareness of plot and symbolism, and related
literature or themes. A group of my college students used this technique to map the
instructional potential of several children's books, and they chose creative and
discussion activities from their lists to use in the classroom. The following example
shows how many ideas and activities might arise from a single book.

Instructional Possibilities for O'Dell's Island of the Blue Dolphins

Art Interpretation
Make a model of Karana's weapons.
Make a shadow box of Karana's home on the island.
Make a model of the Aleut's ship.
Make a salt map of the Island of the Blue Dolphins and mark the places Karana went.

Drama
Make masks, and dramatize the opening scene between Karana's father and the Aleuts.
Pantomime Karana's attempts to make weapons.
Pantomime the meeting between Karana and Tutok.

Values Clarification
What takes more courage—to stay on the boat, or to jump off the boat and stay with her
    brother?
Should Karana have broken the tribal taboos about weapons?
Was Karana's judgment about the Aleutian otter killing a fair one?

Personal Response
Have you ever felt as alone as Karana did?
Compare the degrees of loneliness felt after the death of Karana's brother and after Rontu's
    death.
What factors would have persuaded you to go or to stay on the island?

Developing Characterization
How did Tutok see Karana?
How did the missionaries see Karana?
How did Rontu see Karana?
How do you see Karana?

Literary Awareness: Plot Development
Decision to leave or to stay at beginning of the story.
Decision to make weapons.
Decision to trap Rontu.
Decision to talk to Tutok.
Decision to leave or to stay at the end of the story.

Literary Awareness: Use of Symbols
To reveal a secret name causes it to lose its magic.
Island is the shape of a dolphin.
Blue clay symbolic of eligibility for marriage.
Dolphins are protectors.

Related Literature
  Tests of Courage Today:
    Profiles in Courage (Kennedy)
    Rosa Parks (Greenfield)
    My Side of the Mountain (George)
  Incredible Journeys:
    A Wrinkle in Time (L'Engle)
    A Wind in the Door
      (L'Engle)

Kinds of Courage:
  The Cay (Taylor)
  Where the Red Fern Grows (Rawls)
  The Yearling (Rawlings)
  Old Yeller (Gipson)

REINFORCEMENT ACTIVITY

1. *Develop a modeling activity in which you identify a literary element and provide instruction in the understanding of that literary element.*
2. *With a group of your peers, select a book appropriate for children and map its possibilities for instructional activities. Share your suggestions with the entire class.*

Teaching World Understanding Through Folktales

Folklore is that wonderful part of literature handed down to us by the storytellers of many cultures. Folktales enable the teacher to instill an understanding of other cultures and an appreciation for cultures that differ from our own. Folklore is the mirror of a culture, and lets us see that culture from the inside instead of from the outside.

Ruth Kearney Carlson (1972) identified nine ways the study of folklore contributes to world understanding: (1) an understanding of the cultural traditions of the nonscientific mind, through folktales that deal with the mysteries of creation; (2) an understanding of the relatedness of story types and motifs among peoples of the world, through universal tales such as Cinderella, which has variations in many cultures; (3) an aesthetic appreciation for the music, art, literature, and dance of other cultures; (4) an understanding about cultural diffusion, through observation of how different versions of the same tale or hero appear in different world settings; (5) an increase in knowledge about unfamiliar places around the world; (6) a better understanding of the dialects and languages of other countries; (7) the reader's imaginative identification with people in another time and place; (8) an understanding of the conditions of a time and culture, through study of heroes shaped by those conditions; and (9) an understanding of the inherent qualities of goodness, mercy, courage, and industry that folktale heroes possess and, consequently, an intuitive grasp of the better qualities of the human spirit.

These insights are certainly worth imparting to children. We have already examined the study of mythology as a way to improve creative writing skills and to enhance self-understanding. Now, we consider a folktale unit as a way to develop understanding of various world cultures. The following unit was designed for fourth-grade students.

FOR YOUR PLAN BOOK

A Folktale Unit

General Objectives

1. *To encourage children to appreciate a literary form, the folktale, and to encourage them to realize they share common interests, attitudes, and values with their counterparts in other lands.*

2. *To have the children verbalize generalizations—such as the strongly contrasted values of good and evil—regarding the nature of folktales.*

3. *To share the teacher's enthusiasm for folktales by reading certain selections aloud.*

4. *To develop the children's listening skills by asking them to listen for specifics in each story.*

5. *To motivate the children to search for more folktales to read and enjoy.*

6. *To provide opportunity for the children to enjoy silent reading of the folktales of their choice.*

7. *To encourage the children, once they are acquainted with a country's folk literature, to learn more about the people and customs of that country.*

8. *To encourage the children to share what they learn with one another.*

9. *To have the children, individually and as a group, demonstrate their knowledge of and appreciation for folktales to their classmates by creating a bulletin board display and presenting a puppet play, thereby inviting the others to read and share in the enjoyment of such stories.*

Presentation of Folktales from India

To motivate interest in Indian folktales, the teacher greeted the children with a Hindi word, "namaste," which means both "help" and "good-bye." The teacher also displayed

Indian objects—jewelry, metal plates, metal pitchers, a sari, a statue of an elephant—and pictures of musical instruments, such as the sitar (a long-necked guitar), and the banshri (a bamboo flute), as well as pictures of people and buildings. A record of music from India was also part of the motivational preparation. The students talked about the display and listed questions about some of the objects and the people. The students located India on a globe, and on a map of Asia. Next, the teacher read numerous folktales to the children. These included:

> Asian Cultural Centre for UNESCO's **Folk Tales from Asia for Children Everywhere,** Books Four and Five (1976, 1977)
> David Conger's **Many Lands, Many Stories: Asian Folktales for Children** (1987)
> Nancy DeRoin's **Jataka Tales** (1975)
> Virginia Haviland's **Favorite Fairy Tales Told in India** (1973)

After reading several stories, the teacher asked the children if they thought American children could understand and enjoy stories intended for Indian children, and whether or not Indian children would appreciate American stories. As soon as they had a number of stories to draw from, they looked for such things as similarities in plot, style, and motifs. Then the teacher told them there were many other interesting folktales in the books they were using. To motivate independent reading, she read the titles and showed pictures.

While the students read other tales, they were asked to pretend that the folktales were the only way that they could learn about the country. They looked for information about the culture's animals, food, climate, occupations, music, beliefs, and values.

Next they discussed ways to determine whether their information about India and its cultural heritage was accurate. They decided they could compare their information with reference materials, **National Geographic,** and other books and pictures about India, and that they could also invite a visitor from India to come to the classroom. Each child chose some aspect of Indian life on which she prepared a short research report. Next, a graduate student from India, who was attending a nearby university, visited the class. He answered their questions and played the sitar for them.

Because the class would be reading folktales from countries besides India, they began a bulletin-board display entitled, "Read Around the World." At the base of the bulletin board, they put a large world map over a dark-blue background. Each time they read a folktale, they attached to the appropriate country on the map a symbol representing the story. An annotation of the story was positioned against the background and connected to the symbol with a length of yarn. They placed unusual objects from that culture on a table below the bulletin board.

Presentation of Folktales from Japan and Russia

The teacher also included in this unit folktales from Japan and Russia. Each country was introduced with examples of its language, objects, music, and art. The teacher drew on the following sources for folktales:

Japan

Molly Bang's **The Paper Crane** (1985)
Claude Clement's **The Painter and the Wild Swans** (1986)

Virginia Haviland's **Favorite Fairy Tales Told in Japan** *(1973)*
Jane Hori Ike and Baruch Zimmerman's **A Japanese Fairy Tale** *(1982)*
Momoko Ishii's **The Tongue-Cut Sparrow** *(1987)*
Anne Laurin's **The Perfect Crane** *(1981)*
Nancy Luenn's **The Dragon Kite** *(1982)*
Arlene Mosel's **The Funny Little Woman** *(1972)*
Patricia Newton's **The Five Sparrows: A Japanese Folktale** *(1982)*
Sumiko Yagawa's **Momotaro, The Peach Boy** *(1986)*
Sumiko Yagawa's **The Crane Wife** *(1981)*

Russia

Aleksandr Nikolaevich Afanasev's **Russian Folk Tales** *(1980)*
R. Nisbet Bain's **Cossack Fairy Tales and Folktales** *(1976)*
Elizabeth Isele's **The Frog Princess** *(1984)*
Michael McCurdy's **The Devils Who Learned to Be Good** *(1987)*
Charles Mikolaycak's **Babushka** *(1984)*
Alexander Pushkin's **The Tale of Czar Saltan or the Prince and the Swan Princess**
    (1975)
Arthur Ransome's **The Fool of the World and the Flying Ship** *(1968)*
Arthur Ransome's **Old Peter's Russian Tales** *(1916, 1984)*
Maida Silverman's **Anna and the Seven Swans** *(1984)*
Leo Tolstoy's **The Fool** *(1981)*
Boris Zvorykin's **The Firebird and Other Russian Fairy Tales** *(1978)*

Students and teacher read the folktales of each country and recorded the information they gained from them about the people, their beliefs, and values. The children used the information they learned from the folktales to make travel posters of each country, which formed another very colorful bulletin board. They also constructed mobiles depicting various aspects of Japanese life. The students then used other resources to find out more about the two countries' cultures. They especially enjoyed studying Japanese music, dance, and theater. A visitor from Japan shared experiences with the children, and introduced them to samples of Japanese cooking and the practice of eating with chopsticks.

Culmination Activity

The group chose a favorite folktale from each of the countries: India, Russia, and Japan. They divided into three groups, according to which story and country they wanted to represent. Each group, with the teacher's assistance, decided on a method for sharing their story (e.g., a puppet show, a feltboard story, a play). They prepared and practiced the stories, and invited another class to see their presentations. Each group also shared information they had learned about their particular country.

The children continued to maintain the "Read Around the World" bulletin board so they could add to their folktale collection whenever they read a story from another country.

Additional ways of involving children in traditional literature are found in Norton's *Through the Eyes of A Child: An Introduction to Children's Literature* (1987). The text includes ways of comparing different versions of the same folktale and investigating folktales from a single country. A method for using webbing for developing units is also described.

1. Select a group of folktales from a particular country. Read the folktales, and note the information you find about values, beliefs, occupations, transportation, animals, and agriculture.

2. Select a story type or motif from folktales, such as the various Cinderella stories. Find as many folktales as you can from different cultures that tell basically the same story. Compare the different versions. How are they alike? How are they different? How do you account for the similarities?

SUMMARY

Using literature in the curriculum and teaching about literature in the classroom requires knowledge about literature, awareness of students' interests, knowledge about exciting instructional approaches that motivate and stimulate interest, and knowledge about exciting instructional approaches that encourage understanding of story structures and literary elements. Consequently, literature objectives must fill the dual role of developing both pleasure and understanding.

The foundations for a strong literature program include a rich environment in which books are easily located and read, a teacher who loves books and shares them daily with students, and an attitude toward literature that encourages interactions with a variety of books. This chapter included suggestions for balancing the literature selections, sources for choosing literature, and ways to identify students' interests.

The chapter described numerous ways to use literature with children. Under reading to students, the chapter included selecting books to read aloud and guidelines for reading the story. Under developing enjoyment through storytelling, the chapter included storytelling values, preparing the story, introducing the story, telling the story, and adding pictures and objects to storytelling. Under developing enjoyment through art, the chapter emphasized creating visual interpretations of text and using picture books to stimulate using and understanding specific media.

The chapter included techniques to increase students' understanding of literature through such activities as developing student critics for the literature they read, developing understanding and appreciation of literary genres, and developing understanding of literary elements through guided discussions and modeling. Techniques for mapping instructional possibilities and ways to use folktales in a unit that teaches world understanding were discussed.

ADDITIONAL LITERATURE
ACTIVITIES

1. *Interview a librarian about the literature that children enjoy during story hour and the literature that they check out of the library. Are there any differences or similarities between these two groups? Speculate about reasons for either differences or similarities.*

2. *If possible, visit a library or a classroom during a story hour. Note any effective techniques the storyteller uses to catch students' interest, to introduce the story, and to interact with listeners during the story presentation.*

3. *Read a journal article about the use of literature in the classroom or in the library such as those found in **The New Advocate, Children's Literature in Education,** and **The School Library Journal.** What is the main emphasis of the article? Share the article and your review of the article with your class.*

4. *Referring to the journals listed in activity 3, review the articles published during one year in the journal. What are the major topics covered during that time period? Are there any recurring issues or themes? What are the recommendations for using literature in the curriculum?*

5. *Choose a grade level. Develop a list of books that you believe would be appropriate for that grade level. Be sure to balance the selections so that the major genres of literature are included.*

6. *Refer to a journal such as **Booklist, Hornbook,** or **School Library Journal** that includes critical reviews of literature and recommends children's books. Locate three books that have received starred reviews within the journal. Read the books and consider why the books were recommended by the journal board of reviewers. Do you agree with the reviews? Share the books, your reviews, and the journal reviews with your language arts class.*

7. *Interview a student about his or her interests. Then develop a list of books that you would recommend to that student.*

8. *Patricia Wilson (1985) studied the preferences of fifth- and sixth-grade students for literature that is both highly recommended and considered classics in children's literature. She found that the students' favorites included **Charlotte's Web; The Little House in the Big Woods; The Hobbit; The Secret Garden; The Lion, the Witch and the Wardrobe; Heidi;** and **The Borrowers.** Either lead a discussion with a student about one of these books or read one of these books to a group of students. What is the reaction of the students to the book? Why do you believe Wilson's subjects chose these books as their favorites? If one of these books is also your favorite children's book, why is it your favorite book?*

9. *Develop a file of books that are appropriate for stimulating students' enjoyment through art. Include ways that you would use the books to stimulate students' enjoyment.*

10. *Using table 10–2, identify books that may be used for understanding each genre of literature and identify the literary evidence that places the books within that genre. (Choose books that are not discussed in this chapter.)*

II. *Develop a modeling lesson designed to help students understand and appreciate one of the literary elements.*

I2. *Using either the plot structure in figures I0–5 or I0–6, choose a book and place the key events from the book on the plot structure.*

BIBLIOGRAPHY

Agee, Hugh. *High Interest, Easy Reading*. Urbana, Ill.: National Council of Teachers of English, 1984.

Bauer, Caroline. *Handbook for Storytellers*. Chicago: American Library Association, 1977.

Bennet, Kathleen. "Popular Reading—Chapter Books." *Booklist*. November 1, 1987, 488–90.

Brandehoff, Susan E. "Jake and Honeybunch Go to Heaven: Children's Book Fans Smoldering Debate." *American Libraries* 14 (March 1983): 130–32.

Breckenridge, Kit. "Modern Classics." *School Library Journal* 34 (April 1988): 42–43.

Carlson, Ruth Kearney. *Folklore and Folktales Around the World*. Newark, Del.: International Reading Association, 1972.

Cohen, Caron Lee. "The Quest in Children's Literature." *School Library Journal* 31 (August 1985): 28–29.

Coody, Betty. *Using Literature With Young Children*. Dubuque, Iowa: William C. Brown Co., 1979.

Cooper, Ilene. "Popular Reading—After Henry Huggins." *Booklist*. November 1, 1986, 415–16.

Corcoran, Frances. "Storytelling Sources." *Booklist*. April 15, 1986, 1234–35.

Egoff, Sheila. "The Problem Novel." In *Only Connect: Readings on Children's Literature*, edited by Sheila Egoff, G. T. Stubbs, and L. F. Ashley. Toronto: Oxford University, 1980.

Fishel, Carol T. "Reading in the Content Area of English." In *Reading in the Content Areas: Research for Teachers*," edited by Mary M. Dupuis. Newark, Del.: International Reading Association, 1984.

Five, Cora Lee. "From Workbook to Workshop: Increasing Children's Involvement in the Reading Process." *The New Advocate* 1 (Spring 1988): 103–13.

Frasher, Ramona. "A Feminist Look at Literature for Children: Ten Years Later." In *Sex Stereotypes and Reading: Research and Strategies*, edited by E. Marcia Sheridan. Newark: International Reading Association, 1982.

Galda, Lee. "Readers, Texts and Contexts: A Response-Based View of Literature in the Classroom." *The New Advocate* 1 (Spring 1988): 92–102.

Huck, Charlotte; Hepler, Susan; and Hickman, Janet. *Children's Literature*. New York: Holt, Rinehart & Winston, 1987.

Kiefer, Barbara. "Picture Books for Older Children." *Booklist*. September 15, 1986, 138–39.

Lacy, Lyn. *Art and Design in Children's Picture Books*. Chicago: American Library Association, 1986.

May, Jill P. "Creating a School Wide Literature Program: A Case Study." *Children's Literature Association Quarterly* 12 (Fall 1987): 135–37.

McClenathan, Day Ann. "Realism in Books for Young People: Some Thoughts on Management of Controversy." In *Developing Active Readers*, edited by Dianne Monson and Day Ann McClenathan. Newark, Del.: International Reading Association, 1979.

Mendoza, Alicia. "Reading to Children: Their Preferences." *The Reading Teacher* 38 (February 1985): 522–27.

National Council of Teachers of English. "Forum: Essentials of English." *Language Arts* 60 (February 1983): 244–48.

Norton, Donna E. *Through the Eyes of a Child: An Introduction to Children's Literature*, 2d ed. Columbus, Oh.: Merrill Publishing Co., 1987.

Palmer, William S. "What Reading Teachers Can Do Before the Censors Come." *Journal of Reading* 25 (January 1982): 310–14.

Phelan, Carolyn. "Poetry for Young Children." *Booklist.* January 1, 1988, 790–92.

Sebesta, Sam. "What Do Young People Think About the Literature They Read?" *Reading Newsletter,* No. 8. Rockleigh, N.J.: Allyn & Bacon, 1979.

Shulevitz, Uri. *Writing with Pictures: How to Write and Illustrate Children's Books.* New York: Watson-Guptill, 1985.

State of Illinois. *State Goals for Learning and Sample Learning Objectives: Language Arts.* Illinois, 1985.

Stewig, John. "Storyteller: Endangered Species?" *Language Arts* 55 (March 1978): 339–45.

Sutherland, Zena, and Arbuthnot, May Hill. *Children and Books.* Glenview, Ill.: Scott, Foresman & Co., 1986.

———, and Hearne, Betty. "In Search of the Perfect Picture Book Definition." In *Jump Over the*

Moon, edited by Pamela Barron and Jennifer Burley. New York: Holt, Rinehart and Winston, 1984.

Swanton, Susan. "Minds Alive: What and Why Gifted Students Read for Pleasure." *School Library Journal* 30 (March 1984): 99–102.

Taxel, Joel. "Notes from the Editor." *The New Advocate* 1 (Spring 1988): 73–74.

Whitehead, Robert. *Children's Literature: Strategies of Teaching.* Englewood Cliffs, N.J.: Prentice-Hall, 1968.

Wilms, Denise. "Contemporary Issues— Intergenerational Relationships. *Booklist.* May 1, 1986, 1318–20.

Wilson, Patricia Jane. "Children's Classics: A Reading Preference Study of Fifth and Sixth Graders." Doctoral Dissertation, University of Houston, 1985. *Dissertation Abstracts International* 47:454A.

Woods, L. B. "For Sex: See Librarian." *Library Journal* 1 (September 1978): 1561–67.

CHILDREN'S LITERATURE REFERENCES

Adler, David. *Martin Luther King, Jr.: Free At Last.* Illustrated by Robert Casilla. New York: Holiday, 1986.

Afanasev, Aleksandr Nikolaevich. *Russian Folktales.* New York: Random House, 1980.

Alexander, Lloyd. *The Book of Three.* New York: Holt, Rinehart & Winston, 1964.

Anderson, Margaret. *The Mists of Time.* New York: Alfred A. Knopf, 1984.

Asian Cultural Center for UNESCO. *Folktales from Asia for Children Everywhere.* Books Four and Five, Weatherhill, 1976, 1977.

Bain, R. Nisbet. *Cossack Fairy Tales and Folktales.* Mitchell, 1976.

Baker, Jeannie. *Grandmother.* New York: Deutsch, 1978.

Baker, Olaf. *Where the Buffaloes Begin.* Illustrated by Stephen Gammell. New York: Warne, 1981.

Bang, Molly. *The Paper Crane.* New York: Greenwillow, 1985.

Bauer, Marion Dane. *On My Honor.* New York: Clarion, 1986.

Baylor, Byrd. *Hawk, I'm Your Brother.* Illustrated by Peter Parnall. New York: Charles Scribner's Sons, 1976.

Bierhorst, John. *Doctor Coyote: A Native American Aesop's Fables.* Illustrated by Wendy Watson. New York: Macmillan Co., 1987.

Blos, Joan. *A Gathering of Days.* New York: Charles Scribner's Sons, 1979.

Blume, Judy. *The One in the Middle is the Green Kangaroo.* New York: Bradbury, 1981.

———. *Tales of a Fourth Grade Nothing.* New York: E. P. Dutton, 1972.

Boston, Lucy. *The Children of Green Knowe.* San Diego: Harcourt Brace Jovanovich, 1955.

Brenner, Barbara. *Wagon Wheels.* Illustrated by Don Bolognese. New York: Harper & Row, 1978.

Brittain, Bill. *The Wish Giver.* Illustrated by Andrew Glass. New York: Harper & Row, 1983.

Brooks, Polly. *Queen Eleanor: Independent Spirit of the Medieval World.* Philadelphia: J. B. Lippincott Co., 1983.

Brown, Marc. *Arthur's Baby.* Boston: Little, Brown & Co., 1987.

Brown, Marcia. *Stone Soup.* New York: Charles Scribner's Sons, 1947.

Burton, Virginia Lee. *The Little House.* Boston: Houghton Mifflin Co., 1942.

Byars, Betsy. *Cracker Jackson.* New York: Viking Press, 1985.

Cameron, Eleanor. *That Julia Redfern.* Illustrated by Gail Owens. New York: E. P. Dutton, 1982.

Carl, Eric. *The Honeybee and the Robber.* New York: Philomel, 1981.

_____. *Twelve Tales from Aesop.* New York: Philomel, 1980.

_____. *The Very Hungry Caterpillar.* New York: Thomas Y. Crowell Co., 1971.

Carroll, Lewis. *Alice's Adventures in Wonderland.* Illustrated by John Tenniel. Macmillan, 1866, New York: Alfred A. Knopf, 1984.

Cazet, Denys. *A Fish in His Pocket.* New York: Watts, 1987.

Cendrars, Blaise. *Shadow.* Illustrated by Marcia Brown. New York: Charles Scribner's Sons, 1982.

Chase, Richard. *Grandfather Tales.* Boston: Houghton Mifflin Co., 1948.

Clapp, Patricia. *Constance: A Story of Early Plymouth.* New York: Lothrop, Lee & Shepard, 1968.

Cleary, Beverly. *Dear Mr. Henshaw.* Illustrated by Paul O. Zelinsky. New York: William Morrow & Co., 1983.

_____. *Ramona Quimby, Age 8.* New York: William Morrow & Co., 1981.

Clement, Claude. *The Painter and the Wild Swans.* Illustrated by Frederic Clement. New York: Dial Press, 1986.

Cohen, Barbara. *Molly's Pilgrim.* New York: Lothrop, Lee & Shepard, 1983.

Conger, David. *Many Lands, Many Stories: Asian Folktales for Children.* Ruthland, Vermont: Tuttle, 1987.

Conly, Jane Leslie. *Rasco and the Rats of NIMH.* New York: Harper & Row, 1986.

Dalgliesh, Alice. *The Columbus Story.* New York: Charles Scribner's Sons, 1955.

DePaola, Tomie. *Mother Goose.* New York: G. P. Putnam's Sons, 1985.

DeRoin, Nancy. *Jataka Tales.* Boston: Houghton Mifflin Co., 1975.

Flournoy, Valerie. *The Patchwork Quilt.* Illustrated by Jerry Pinkney. New York: Dial Press, 1985.

Forbes, Esther. *Johnny Tremain.* Illustrated by Lynd Ward. Boston: Houghton Mifflin Co., 1943.

Fox, Paula. *One-Eyed Cat.* New York: Bradbury, 1984.

Freedman, Russell. *Lincoln: A Photobiography.* New York: Clarion, 1987.

Fritz, Jean. *Make Way for Sam Houston.* Illustrated by Elise Primavera. New York: G. P. Putnam's Sons, 1986.

_____. *Shh! We're Writing the Constitution.* New York: G. P. Putnam's Sons, 1987.

_____. *Where Do You Think You're Going, Christopher Columbus?* Illustrated by Margot Tomes. New York: G. P. Putnam's Sons, 1980.

Gág, Wanda. *Millions of Cats.* New York: Coward-McCann, 1928.

Garfield, Leon. *The December Rose.* New York: Viking Press, 1986.

Gates, Doris. *Two Queens of Heaven: Aphrodite and Demeter.* Illustrated by Trina Schart Hyman. New York: Viking Press, 1974.

George, Jean Craighead. *Julie of the Wolves.* New York: Harper & Row, 1972.

Gish, Lillian, Told to Selma Lanes. *An Actor's Life for Me!* New York: Viking Press, 1987.

Goodall, John. *The Story of a Main Street.* New York: Macmillan Co., 1987.

Grahame, Kenneth. *The Wind in the Willows.* Illustrated by E. H. Shepard. New York: Charles Scribner's Sons, 1908, 1940.

Grifalconi, Ann. *Darkness and the Butterfly.* Boston: Little, Brown & Co., 1987.

Grimm, Brothers. *Little Red Riding Hood.* Retold and Illustrated by Trina Schart Hyman. New York: Holiday, 1983.

_____. *Red Riding Hood.* Retold and Illustrated by James Marshall. New York: Dial Press, 1987.

Hamilton, Virginia. *Zeely.* New York: Macmillan Co., 1969.

Harris, Joel Chandler. *Jump Again! More Adventures of Brer Rabbit.* Adapted by Van Dyke Parks. Illustrated by Barry Moser. San Diego: Harcourt Brace Jovanovich, 1987.

Haseley, Dennis. *The Scared One.* New York: Warne, 1983.

Haviland, Virginia. *Favorite Fairy Tales Told in India.* Boston: Little, Brown & Co., 1973.

_____. *Favorite Fairy Tales Told in Japan.* Boston: Little, Brown & Co., 1967.

Hest, Amy. *The Purple Coat.* Illustrated by Amy Schwartz. New York: Four Winds, 1986.

Howard, Ellen. *Edith Herself.* New York: Atheneum Pubs., 1987.

Hughes, Monica. *The Dream Catcher.* New York: Atheneum Pubs., 1987.

Hurmence, Belinda. *A Girl Called Boy.* Boston: Houghton Mifflin Co., 1982.

Hutchins, Pat. *Changes, Changes.* New York: Macmillan Co., 1971.

Ike, Jane, and Zimmerman, Baruch. *A Japanese Fairy Tale.* New York: Warne, 1982.

Isele, Elizabeth. *The Frog Princess.* Illustrated by Michael Hague. New York: Crowell, 1984.

Ishii, Momoko. *The Tongue-Cut Sparrow.* Illustrated by Suekichi Akaba. New York: E. P. Dutton, 1987.

Jacques, Brian. *Redwall.* New York: Philomel, 1986.

Jukes, Mavis. *Like Jake and Me.* Illustrated by Lloyd Bloom. New York: Alfred A. Knopf, 1984.

Keats, Ezra Jack. *Peter's Chair.* New York: Harper & Row, 1967.

_____. *The Snowy Day.* New York: Viking Press, 1962.

_____. *The Trip.* New York: Greenwillow, 1978.

Kendall, Carol. *The Gammage Cup.* San Diego: Harcourt Brace Jovanovich, 1959.

Kipling, Rudyard. *The Jungle Book.* Illustrated by Michael Foreman, New York: Viking Press, 1987.

_____. *Just So Stories.* Illustrated by Michael Foreman. New York: Viking Press, 1987.

Lasker, Joe. *The Great Alexander the Great.* New York: Viking Press, 1983.

Lasky, Kathryn. *Sugaring Time.* Photographs by Christopher Knight. New York: Macmillan Co., 1983.

Laurin, Anne. *The Perfect Crane.* New York: Harper & Row, 1981.

Lawson, Robert. *Rabbit Hill.* New York: Viking Press, 1944.

Lee, Betsy. *Charles Eastman, the Story of an American Indian.* Minneapolis: Dillon, 1979.

L'Engle, Madeleine. *A Wrinkle in Time.* New York: Farrar, Straus & Giroux, 1962.

Lenski, Lois. *Sing A Song of People.* Illustrated by Giles Laroche. Boston: Little, Brown & Co., 1987.

Lewin, Hugh. *Jafta.* Illustrated by Lisa Kopper. Minneapolis: Carolrhoda, 1983.

_____. *Jafta's Mother.* Illustrated by Lisa Kopper. Minneapolis: Carolrhoda, 1983.

Lewis, C. S. *The Lion, the Witch and the Wardrobe.* New York: Macmillan Co., 1950.

Lionni, Leo. *Swimmy.* New York: Pantheon, 1963.

Little, Jean. *Different Dragons.* New York: Viking Press, 1986.

Lowry, Lois. *Anastasia's Chosen Career.* Boston: Houghton Mifflin Co., 1987.

Luenn, Nancy. *The Dragon Kite.* Illustrated by Michael Hague. San Diego: Harcourt Brace Jovanovich, 1982.

Lunn, Janet. *The Root Cellar.* New York: Charles Scribner's Sons, 1983.

_____. *Shadow in Hawthorn Bay.* New York: Charles Scribner's Sons, 1986.

MacDonald, Suse. *Alphabatics.* New York: Bradbury, 1986.

MacLachlan, Patricia. *Sarah, Plain and Tall.* New York: Harper & Row, 1985.

Madsen, Jane and Bockoras, Diane. *Please Don't Tease Me. . .* Illustrated by Kathleen Brinko. Judson, 1983.

Mahy, Margaret. *17 Kings and 42 Elephants.* New York: Dial Press, 1987.

Mathis, Sharon Bell. *The Hundred Penny Box.* Illustrated by Leo and Diane Dillon. New York: Viking Press, 1975.

Mazer, Norma Fox. *After the Rain.* New York: William Morrow & Co., 1987.

McCloskey, Robert. *Make Way for Ducklings.* New York: Viking Press, 1941.

McCully, Emily Arnold. *Picnic.* New York: Harper & Row, 1984.

McCurdy, Michael. *The Devils Who Learned to be Good.* Boston: Little, Brown, 1987.

McKinley, Robin. *The Hero and the Crown.* New York: Greenwillow, 1984.

Mikolaycak, Charles. *Babushka.* New York: Holiday House, 1984.

Milne, A. A. *Winnie-The-Pooh.* Illustrated by Ernest Shepard. New York: E. P. Dutton, 1926, 1954.

Monjo, F. N. *The Drinking Gourd.* Illustrated by Fred Brenner. New York: Harper & Row, 1970.

Mosel, Arlene. *The Funny Little Woman.* Illustrated by Blair Lent. New York: E. P. Dutton, 1972.

Newton, Patricia. *The Five Sparrows: A Japanese Folktale.* New York: Atheneum Pubs., 1982.

Norton, Mary. *The Borrowers.* San Diego: Harcourt Brace Jovanovich, 1952.

Oakley, Graham. *The Church Mice in Action.* New York: Atheneum Pubs., 1982.

O'Brien, Robert. *Mrs. Frisby and the Rats of NIMH.* New York: Atheneum Pubs., 1971.

O'Dell, Scott. *Island of the Blue Dolphins.* Boston: Houghton Mifflin, 1960.

Orlev, Uri. *The Island on Bird Street.* Boston: Houghton Mifflin Co., 1984.

Park, Ruth. *Playing Beatie Bow.* New York: Atheneum Pubs., 1982.

Paterson, Katherine. *Bridge to Terabithia.* New York: Thomas Y. Crowell Co., 1977.

Paulsen, Gary. *Hatchet.* New York: Bradbury, 1987.

Perrault, Charles. *Little Red Riding Hood.* Illustrated by Sarah Moon. Mankato, Minn.: Creative Education, 1983.

_____. *Puss in Boots.* Illustrated by Marcia Brown. Charles Scribner's Sons, 1952.

Potter, Beatrix. *The Tale of Peter Rabbit.* New York: Warne, 1902.

_____. *The Tale of Squirrel Nutkin.* New York: Warne, 1903, 1986.

Pullman, Philip. *The Ruby in the Smoke.* New York: Alfred A. Knopf, 1987.

Pushkin, Alexander. *The Tale of Czar Saltan or the Prince and the Swan Princess.* New York: Crowell, 1975.

Ransome, Arthur. *The Fool of the World and the Flying Ship.* New York: Farrar, Straus & Giroux, 1968.

_____. *Old Peter's Russian Tales.* London: Jonathan Cape, 1916, 1984.

Rylant, Cynthia. *Children of Christmas.* New York: Watts, 1987.

Schwartz, Amy. *Oma and Bobo.* New York: Bradbury Press, 1987.

Sendak, Maurice. *Where the Wild Things Are.* New York: Harper & Row, 1963.

Seredy, Kate. *The White Stag.* New York: Viking Press, 1937, Puffin, 1979.

Seuss, Dr. *The 500 Hats of Bartholomew Cubbins.* New York: Vanguard, 1938.

Silverman, Maida. *Anna and the Seven Swans.* Illustrated by David Small. New York: William Morrow & Co., 1984.

Silverstein, Shel. *The Giving Tree.* New York: Harper & Row, 1964.

Speare, Elizabeth. *The Witch of Blackbird Pond.* Boston: Houghton Mifflin Co., 1958.

Sperry, Armstrong. *Call It Courage.* New York: Macmillan Co., 1940.

Steptoe, John. *Mufaro's Beautiful Daughters: An African Tale.* New York: Lothrop, Lee & Shepard, 1987.

Tafuri, Nancy. *Have You Seen My Duckling?* New York: Greenwillow, 1984.

Taylor, Mildred. *The Gold Cadillac.* Illustrated by Michael Hays. New York: Dial Press, 1987.

_____. *Roll of Thunder, Hear My Cry.* New York: Dial Press, 1976.

Taylor, Theodore. *The Cay*. New York: Doubleday, 1969.

Tejima, Keizaburo. *Fox's Dream*. New York: Philomel, 1987.

Tobias, Tobi. *Maria Tallchief*. New York: Crowell, 1970.

Tolstoy, Leo. *The Fool*. New York: Schocken, 1981.

Van Allsburg, Chris. *The Polar Express*. Boston: Houghton Mifflin Co., 1985.

_____. *The Z Was Zapped*. Boston: Houghton Mifflin Co., 1987.

Viorst, Judith. *If I Were in Charge of the World*. Illustrated by Lynne Cherry. New York: Atheneum, 1981.

Voigt, Cynthia. *Dicey's Song*. New York: Atheneum Pubs., 1982.

White, E. B. *Charlotte's Web*. Illustrated by Garth Williams, New York: Harper & Row, 1952.

Wilder, Laura Ingalls. *Little House in the Big Woods*. Illustrated by Garth Williams. New York: Harper & Row, 1932, 1953.

Williams, Margery. *The Velveteen Rabbit: Or How Toys Become Real*. New York: Doubleday & Co., 1922, 1958.

Wiseman, David. *Jeremy Visick*. Boston: Houghton Mifflin Co., 1981.

Wood, Audrey. *King Bidgood's in the Bathtub*. Illustrated by Don Wood. San Diego: Harcourt Brace Jovanovich, 1985.

Yagawa, Sumiko. *The Crane Wife*. Illustrated by Suekichi Akaba. New York: William Morrow & Co., 1981.

_____. *Momotaro, The Peach Boy*. Illustrated by Linda Shute. New York: Lothrop, Lee & Shepard, 1986.

Yorinks, Arthur. *Hey, Al*. Illustrated by Richard Egielski. New York: Farrar, Straus & Giroux, 1986.

Zemach, Margot. *Jake and Honeybunch Go to Heaven*. New York: Farrar Straus & Giroux, 1982.

Zvorykin, Boris. *The Firebird and Other Russian Fairy Tales*. New York: Viking Press, 1978.

Chapter Eleven

THE READING AND LITERATURE
CONNECTION

USING LITERATURE IN THE READING
CURRICULUM

Approaches Supported by the Schema Theory
■ *Approaches that Increase Opportunities to
Learn to Read by Reading* ■ *Approaches that
Encourage Readers to Use the Whole Text to
Create Meaning* ■ *Approaches that Help Students
Improve Their Comprehension* ■ *Approaches
that Encourage Students to Make Predictions*

DEVELOPING A UNIT AROUND LITERATURE
AND READING

SUMMARY

*After completing this chapter on reading and
literature, you will be able to:*

1. *Identify and describe approaches to reading
    and literature that are supported by
    research.*
2. *Describe and develop reading and literature
    approaches that are supported by the
    schema theory.*
3. *Describe and develop reading and literature
    approaches that increase students' opportu-
    nities to read literature.*
4. *Describe and develop reading and literature
    approaches that encourage readers to use
    the whole text to create meaning.*
5. *Describe and develop reading and literature
    approaches that help students improve their
    comprehension.*
6. *Develop a series of lessons that proceed
    from a web, to a guided discussion, and fi-
    nally to a plot structure.*
7. *Develop a semantic map or web that ac-
    companies a book.*
8. *Develop questioning strategies that focus on
    literary elements.*
9. *Describe and develop reading and literature
    approaches that encourage students to make
    predictions about what they read.*
10. *Develop a unit focusing on literature and
    reading.*

Reading and Literature

E verywhere you look, there seems to be renewed interest in the use of children's literature in the reading program. This trend is evident in increased coverage of the topic in conference presentations, journal articles, and books" (Miller and Luskay, 1988, p. 1). This quote from the lead article in *Reading Today* emphasizes the national thrust to include more literature in the reading program.

In addition to professional writings, daily newspapers and popular journals emphasize the need to make literature a dominant force in the reading curriculum. For example, an article released by the New York Times Service (Hechinger, 1988) reported recent studies in reading achievement that conclude, "Children's literature must be the core of every reading program because real literature touches the lives of children. It makes children want to read and to think about and comprehend what they have read" (p. E16).

This renewed interest in literature and reading is attributed to both community and scholarly concern. Educators Miller and Luskay (1988) attributed the interest in literature to a community concern that children are not learning to read and are not reading; to a realization that literature is important to children; to a support for activities that promote the fun and enjoyment of reading; and to an understanding that basal readers have limitations because they lack classic children's literature, contain few good contemporary literature selections, and limit children's interactions with books. They emphasized that children's literature can be used to introduce children to a wide variety of enjoyable literature as well as to support the reading curriculum.

This interest in literature in the reading curriculum is not merely an American phenomenon. Canadian educator Stott's curriculum structure was used in one of the literature-reading programs described in chapter 10 (May, 1987). Sawyer (1987), an Australian educator, reviews both Australian and British studies that support a strong literature-based reading curriculum. Sawyer argued that we can no longer separate learning to read and reading to learn because the two are interwoven. Sawyer defended this position because "researchers have been unable to study how and why children learn to read through literature without at the same time addressing the question of how they acquire competence in dealing with literary structures" (p. 33). Sawyer contended that the story structures chosen to teach reading are important because the structures themselves teach the rules of narrative organization. Within this argument, the materials chosen to be read or to be listened to are just as important as the processes being used.

Sawyer also emphasized the need to develop a closer tie between literature and reading because, "Reading ought not be a school activity while reading stories remains a home one—reading cannot be 'real work' while reading stories is 'just fun.' Another lesson to learn is to bring into our classrooms those very activities that the best practice in literature teaching have encouraged and 'new model comprehension' activities such as text sequencing, cloze procedures, and text prediction. . . . All of these activities are at base a way to teach children about their reading processes" (p. 37).

Meeks (1983), a British educator, added strong support for using literature in the reading curriculum because she contended that students who fail to learn to read have not learned "how to tune the voice on the page, how to follow the fortunes of the hero, how to tolerate the unexpected, to link episodes" (p. 214). Interaction with literature is viewed as one of the most important ways to develop these capabilities in students.

THE READING AND LITERATURE CONNECTION

Studies show that using literature in the classroom does improve reading comprehension. In addition, studies provide implications about approaches that increase reading comprehension, understanding of story structures, and appreciation of reading and literature. For example, Feitelson, Kita, and Goldstein (1986) found that first graders who were read to for 20 minutes each day outscored comparable groups in decoding, reading comprehension, and active use of language. This research, however, shows that an adult needs to help students interpret the literature by elaborating beyond the text. Feitelson, Kita, and Goldstein attributed the success of their research to enriching student's information base, to introducing students to language that may be unfamiliar to them, to increasing their knowledge about various story structures, to exposing them to literary devices such as metaphor, and to extending their attention spans.

My own research with fifth- through eighth-grade students (Norton, 1987) found that a literature-based reading program combining children's literature and teaching strategies that emphasize cognitive processes, story structures, and modeling significantly increases reading comprehension and attitudes toward reading.

Early and Ericson (1988) identified nine findings from reading research that should influence how and why we teach literature in the reading curriculum. A brief review of these findings shows us the importance of many of the strategies described in chapter 10 as well as presents some additional ways to bring literature into the reading curriculum. First, readers use their knowledge of texts and the cues supplied by the text to create meaning during reading (schema theory). Second, readers learn to read by reading. Third, to increase meaning readers need to experience whole texts. Fourth, good readers understand when their reading makes sense, are aware when their reading processes break down, and use a variety of corrective strategies. Fifth, readers improve their comprehension if teachers use techniques such as modeling, direct explanations, and questioning. Sixth, good readers use cues in the text and their own prior knowledge to make predictions. Seventh, analysis of types of inferences supports teaching strategies that focus on how to read literature. Eighth, comprehension of inferences shows the importance of focusing on details. Ninth, the range of students' reading achievement grows wider at each successive grade level. These research findings have strong implications for why literature needs to be added to the reading curriculum and for how teachers can develop effective instructional strategies.

USING LITERATURE IN THE READING CURRICULUM

As we discuss the approaches and methodologies for bringing literature into the reading curriculum, we consider how techniques described in chapter 10 and in this chapter support these research findings.

Approaches Supported by Schema Theory

Implications from schema theory provide valuable information for the teacher of literature within the reading curriculum. The basis for comprehension, learning, and remembering the ideas in stories and in other types of texts is, according to schema theory, the reader's schema or organized knowledge of the world. In this theory readers use their prior knowledge of various kinds of texts, their knowledge of the world, and the cues supplied by the text to create meaning.

Anderson (1985) clarified these connections between prior knowledge and comprehension of new texts: "In schema-theoretic terms, a reader comprehends a message when he is able to bring to mind a schema that gives a good account of the objects and events described in the message" (p. 372).

Anderson expanded on this definition of schema theory by describing six functions of schemata. First, our schema, with its sets of knowledge and expectations, provides the scaffolding that allows us to assimilate new information. When new information fits into our slots of prior knowledge, information is readily learned. When the new information does not fit into our sets of knowledge, information is neither easily understood nor readily learned. Second, our schema provides information for determining what is important in a text and for deciding where within the text we should pay close attention. Third, our schema, with its sets of prior knowledge, provides the basis for making inferences that go beyond the information literally stated in the text. Fourth, our schema allows us to orderly search our memories by providing guides for the type of information we need to recall. Fifth, our schema allows us to produce summaries and to edit information so that trivial information is omitted and important information is retained. Sixth, when gaps occur in our memories, our schema helps us generate hypotheses about the missing information. In addition to these functions, our schema also depends on our age, sex, religion, nationality, and occupation. In other words, our schema depends on the total accumulation of past experiences.

Prior knowledge and past experiences are extremely important for comprehending and appreciating stories whether the stories are in basal readers or in longer literature selections. Stories that are read to students help them develop schema for story structures, for vocabulary, for language patterns, and for other literary elements. Reading stories aloud is important at all grade levels because many of the desired story structures exist in literature that may be too difficult for the students to read independently. These sets of knowledge and expectations about literature are extremely valuable when students face similar structures, vocabularies, and literary elements in their independent reading. Consequently, teachers and parents should read, tell stories, and talk about a wide variety of literature. To provide the maximum knowledge base, this literature should come from various genres and include varied language patterns. Students should also be encouraged to add to their knowledge base and to their expectations by reading a wide variety of literature independently.

Implications from schema theory emphasize the importance of activating relevant prior knowledge and of helping students fill in gaps in their knowledge before they read. Valuable instructional strategies include prereading discussions that focus on students' prior knowledge, information that helps them use their prior knowledge during the reading experience, and questions that help teachers identify any gaps in students' knowledge. If gaps are identified or if prerequisite knowledge is required, it should be presented before students approach the literature selections. Maps, pictures, films, and historical time lines are all useful.

For example, prior to reading Patricia MacLachlan's historical novel *Sarah, Plain and Tall* (1985), prereading discussions could focus on prairie life in the 1800s (see pp. 73-76 for a modeling activity with this book). One technique that encourages both activation of previous knowledge and the identification of any knowledge gaps involves asking students to close their eyes and pretend that they are living in the 1800s. They are sitting on the front steps of a pioneer house located in one of the prairie states. Ask them to use their imaginations and to describe what they see when they look toward the prairie. (They should see prairie grass on gently rolling plains, wheat fields, dirt roads, and expanses of space in which the prairie meets the sky. If there are vehicles on the road, they should be horse drawn. Likewise, any machinery is drawn by horses, mules, or oxen.) Next, ask the students to turn around and to describe the house. They can see the inside of the building because the door is open. The author in *Sarah, Plain and Tall* frequently contrasts the prairie setting and Sarah's familiar Maine coast. Ask the students to close their eyes and to pretend they are sitting on the Maine coast. When they look toward the ocean, what do they see? When they look toward the land, what do they see? This knowledge and understanding is extremely important for comprehension of settings, conflicts, characterizations, and comparisons within the literature. If students do not have the prior knowledge of the historical period or the settings, pictures, time lines, and films will help them fill in the gaps and will lead to greater comprehension and enjoyment.

Picture books and easier stories may also be used to activate students' prior knowledge or to prepare them for story structures or for elements found in more difficult texts. For example, my students have used Selina Hastings's *Sir Gawain and the Loathly Lady* (1985) and Margaret Hodges's *Saint George and the Dragon* (1984) to activate prior knowledge, or to introduce new information about legends, quests, and chivalry before their students read longer legends about King Arthur or other heroic legends.

Approaches that Increase Opportunities to Learn to Read By Reading

Students who demonstrate mastery of higher level reading skills do considerable reading, choose books from a variety of genres including both fiction and nonfiction, and select books beyond required reading (National Assessment of Educational Progress, 1981). As we discovered under schema theory, reading widely and frequently allows students to develop sets of knowledge about the world and about literature, and to develop expectations that in turn help them comprehend new materials. Students who read widely and frequently also develop appreciation for literature and increase the likelihood that they will become life-long readers.

Students need opportunities to read literature during teacher-directed activities in which students choose their own books as well as during teacher-directed reading that teaches specific reading skills. Reading activities that encourage students to choose the literature help them synthesize information from several sources, clarify their own developing literary skills, and share revelations about their own discoveries. In this section we consider approaches to reading and literature such as Uninterrupted Sustained Silent Reading, Recreational Reading Groups, and Thematic or Focus Units that encourage varied and individual reading as well as student and teacher response.

Uninterrupted sustained silent reading (USSR)   Classes or whole schools that motivate students to read through the USSR approach provide about a 30-minute period in the school day or several periods during the week in which students silently read self-selected books. If the total school is involved, the principal, the secretaries, and the custodians stop their normal duties and also read self-selected literature. Within the classroom, everyone reads including the teacher.

For this approach to succeed, students must have access to a wide selection of literature and teachers need information about students' interests and reading levels. In addition, teachers need knowledge about books that they can recommend to their students.

This approach is applicable to all reading levels because even kindergarten students can read the pictures in picture storybooks and in wordless books. Easy to read books allow first- and second-grade students to practice their reading skills. High interest–easy reading books encourage lower ability older readers to extend their knowledge and increase their appreciation of many genres of literature. The book *High Interest-Easy Reading* (Matthews, 1988) suggests sources for older students. The literature available for this activity should include all of the literature genres discussed in chapter 10.

Students and teachers may self-select books from the school, classroom, public libraries, or bring selections from home. Students in one school read books selected

Recreational reading time allows a child to read a book for pleasure and personal satisfaction.

from the local Reading Is Fundamental program (RIF). This activity was especially popular because the books became the students' private property and formed the beginnings for many private collections. Some teachers introduce USSR prior to a visit to the public library. Students then get library cards and their first books from the library at the same time. These books become the sources for USSR. The relationship between the library cards and USSR can be very beneficial as many students are introduced to the library and to the various library services such as story hours, summer reading programs, and knowledgeable sources for information.

The classroom structure for this approach is extremely simple. Everyone, including the teacher, takes out a book during this specified time and reads. Enjoyment is emphasized as students are not expected or even permitted to do other types of reading assignments during this time. Teachers model reading for enjoyment by joining in with their own self-selected reading.

Recreational reading groups   Recreational reading groups have some of the same characteristics of USSR except they usually contain more structure, may encourage students to self-select literature around a specific theme or topic, and include opportunities for students and teachers to verbally respond to their reading experience. Consequently, books may be totally self-selected from a wide range of literature or they may be self-selected from a narrower range of topics, genres, or characteristics of literature. If topics form the focus, groupings are formed according to interest in those topics rather than on students' reading achievement levels. Consequently, it is important to include a range of literature that meets reading levels of all of the students in the group.

If the materials are totally self-selected, the teacher divides the class into about three groups with a leader designated for each group. The students bring their books to one of these three reading circles, read for about thirty minutes (the time varies according to attention spans and reading interests), and then students tell something interesting about the books they have been reading. The first time recreational reading groups are formed it is advantageous to have one large group with the teacher as leader. Then the teacher models independent reading behavior for the total group and begins the discussion by telling something interesting about the book she is reading. After students understand the procedures, however, more interaction develops in smaller groups. By placing competent readers next to less efficient readers, students may quietly get help with unknown words. These groups may contain different students each time they are formed or groups may stay together until their books are finished. The teacher usually changes groups after each session.

My own favorite recreational reading groups are structured around specific topics. This grouping encourages reading of a wider variety of materials and genres than students might otherwise select. This approach also encourages literary appreciation as students and teachers consider why they enjoy or do not like the literature they select.

For example, the topic humor appeals to students, contains enough books to allow self-selection, and encourages them to develop appreciation for various ways that authors develop humor in literature. Students could begin by selecting, reading, and sharing incidents of humor in picture books. Within this category they can share their enjoyment of both the humor developed by the author and by the illustrator. For example, the various books by Theodore Geisel (Dr. Seuss) develop humor through nonsensical characters and word play. Surprising and unexpected situations provide humor in Audrey Wood's *King Bidgood's in the Bathtub* (1985), Catharine O'Neill's *Mrs. Dunphy's Dog* (1987), and Patricia Lee Gauch's *Christina Katerina and the Time She Quit the Family* (1987). Exaggeration provides humor in Patricia Polacco's *Meteor!* (1987) and in James Stevenson's *Could Be Worse* (1977). Caricature in illustrations and ridiculous situations provide humor in James Marshall's illustrations for *Red Riding Hood* (1987) and in the language and situations in Mem Fox's *Hattie and the Fox* (1987) and Paul Galdone's *The Three Wishes* (1967).

Recreational reading groups may explore the preposterous characters and situations in modern fantasy selections such as Carol Sandburg's *Rootabaga Stories* (1922), Astrid Lindgren's *Pippi Longstocking* (1950), and Richard and Florence Atwater's *Mr. Popper's Penguins* (1938).

Other recreational reading groups may experience and explore humor in contemporary realistic fiction such as in the humorous person-versus-person conflict in Beverly Cleary's *Ramona and Her Father* (1977), in the preposterous situations in Beverly Keller's *No Beasts! No Children!* (1983), in the exaggerations found in Helen Cresswell's various "The Bagthorpe Saga" books, and in the humor of role reversals in Lois Lowry's *Anastasia on Her Own* (1985).

Numerous other topics such as animals we like and people we admire may focus attention in recreational reading groups. Authors who interest students may result in recreational reading groups formed around specific authors. Authors such as Maurice Sendak, Beverly Cleary, Virginia Hamilton, Lloyd Alexander, and Tomie de Paola have numerous books in print. Likewise, poets such as Myra Cohn Livingston, Byrd Baylor, Edward Lear, and Jack Prelutsky have sufficient poetry collections in print to provide choices for recreational reading groups.

When students select their materials, read for pleasure, and share what attracts them about the literature or the author, we are encouraging them to learn to read by reading. There is also a strong motivational factor in this approach because students frequently choose books that are highly recommended by their peers.

Focus or thematic units   Units such as "Developing World Understanding Through Folktales" described in chapter 10 help teachers and students focus on specific reading skills and literary attributes. Units also encourage considerable individual reading as students choose related literature for either unit-related projects or for enjoyment. If one of the objectives of the unit is to provide opportunities for students to expand their reading, then units need to include independent reading activities and expanded reading lists as well as teacher-directed activities that help students focus on specific reading and literature skills.

Moss's (1984) publication *Focus Units in Literature: A Handbook for Elementary School Teachers* provides guidelines for developing units that include objectives, teacher-directed group sessions, individualized and independent activities, culminating projects, annotated bibliographies of literature for teacher-directed sessions, and extensive bibliographies of literature designed for independent reading. The examples in Moss's text include detailed focus units for first through sixth grades.

Teachers may develop focus topics or themes that include one genre of literature or several genres. The conclusion of this chapter contains a unit for teaching reading skills and literary appreciation associated with biography. My college classes are currently developing units designed to encourage students to use information from our literary, scientific, and historical past and present and to use that information to help students speculate about the "what ifs" of the future. They are taking topics such as "The Pioneer Spirit," "The Planet Earth," "Our Literary and Cultural Heritage: Past, Present, and Future," and "Exploring the Universe"; searching for literature from all genres that allows them to help students understand our past as well as the present; and developing scenarios that help students consider the critical issues related to their topics that will make the future possible. For example, the group working on "The Pioneer Spirit" is considering the personal, psychological, physical, spiritual, and social qualities that made early exploration possible, that sent colonists from Europe to America, that encouraged people to cross inhospitable terrains, and that placed astronauts into the space frontier. Historical fiction, biographies, autobiographies, informational materials, and even poetry are proving to be exciting sources of literature. College students are developing teacher-directed lessons that help students read the literature, gain insights on the subject, and draw possible conclusions from the past and the present. They are developing scenarios and creating activities that will help students consider what kind of a pioneer spirit and what type of characteristics will be required in a future in which the pioneer spirit may take us to universes not yet imagined or cause us to restructure environments on planet earth. These teacher-directed lessons usually focus on specific literature. There are other activities, however, that require independent reading. Consequently, each group is identifying literature from as many genres as possible related to the topic. They are building into their units opportunities for students to use this literature for independent research, for correlating with other subjects, and for recreational purposes.

Approaches that Encourage Readers to Use the Whole Text to Create Meaning

Literature is certainly one of the best sources for encouraging students to consider longer segments in their judgments of meaning. Characterizations change over the course of a story; themes are often drawn from the whole book; and conflicts gradually develop until problems are overcome or personal conflicts are resolved. Good readers use a number of effective strategies to make these connections. Poor readers have fewer effective strategies from which to choose.

Shorter picture storybooks are ideal sources for introducing the desirability of reading a whole text to develop understanding because the stories can be read, and reread if necessary, to see how conflict and theme are developed throughout a total text.

Let us consider an example in which we are helping students search for important themes in a story and then supporting those themes through specific instances in the book. In this activity we are first using the whole text to create meaning, and second using details from the text to support the broader generalizations gained from the total text. In this initial activity we will also read the text to the students so that they may see the importance of considering the whole book. We will then reread the book to help them search for important support for the theme. After the activity is understood by students and they have several ways to approach evidence of theme, they will be ready to do this activity through independent reading. The activity may also be used at any grade level. Younger students may use simplified terms, whereas older students may search for additional symbolism or higher-level relationships among theme, characterization, and conflict.

Before introducing the specific book tell students that they will be searching for important themes in a story by answering the question: "What is the author trying to tell us that would make a difference in our lives? We may find that the author has more than one important message." After we have answered the first question, we will ask: "How do we know that the author is telling us _____?" To answer the second question we must search for clues and think about all the ways that the author made us believe that something is important. Our evidence from the book may be the illustrations, the character's actions, the character's thoughts, how the story ends, or the author may tell us. The more evidence we can find, the more certain we are about our themes."

Now introduce Ann Grifalconi's *Darkness and the Butterfly* (1987). Tell the students that this is a story about Osa, a young African girl who has a problem that needs to be solved. Ask them to listen carefully to the story to discover what the author is telling us that will make a difference to Osa and that may even make a difference in our lives. Now read the whole book showing the pictures as you proceed. After you finish the book lead a discussion in which the students identify the following themes:

1. It is all right to have fears.. We all may have fears that cause us problems.
2. But we can and must overcome our own fears.

Write each of these themes on the chalkboard, allowing room to place evidence under each theme. Read the story a second time. Encourage students to identify as much evidence for each theme as possible. Your themes and evidence will include lists similar to the following:

1. It is all right to have fears. We all may have fears that cause us problems:
   a. The illustrations show contrasts between the beauty of the world in the daytime without fear and the monsters that haunt Osa's mind at night.
   b. The actions of the mother show that she understands Osa's fear. She gives Osa beads to help Osa feel less afraid.
   c. The actions of Osa show her fear when she contrasts day and night.
   d. The wise woman tells Osa that she too was once afraid, "specially at night!"

2. But we can and must overcome our own fears:
   a. The story of the yellow butterfly, the smallest of the small, that flies into the darkness suggests that even small beings must and do overcome fear (Based on African proverb "Darkness Pursues the Butterfly").
   b. The wise woman tells Osa, "You will find your own way" (Comparisons to butterfly).
   c. The dream sequence allows Osa to face the night and discover new characteristics of the night.
   d. The actions of the butterfly show need to overcome fear.
   e. Osa expresses her own self-realization: "I can go by myself. I'm not afraid anymore."
   f. The author states that Osa, the smallest of the small, "found the way to carry her own light through the darkness."
   g. The butterfly symbolizes the author's belief that the smallest, most fragile being in nature can light up the darkness, trust the night, and not be afraid.

Notice that the support for the theme may be discussed at several levels. Younger readers can discuss concrete examples found in illustrations and characters' actions. Older readers may extend this discussion into the symbolism and the proverb used by the author to develop the theme.

This technique of searching for themes and supporting themes with evidence from the total book is successful with almost any piece of literature. The example used to introduce the skill was chosen because the themes are reinforced in multiple ways and students can discover the desirable interrelationships among theme, setting, characterization, conflict, and illustrations. Multiple approaches to theme detection show students that there are several ways to approach the problem. Poorer readers may benefit especially from pointing out the reinforcement found in illustrations.

Following this teacher-directed activity, provide opportunities for students to read additional short stories, identify the themes found in the whole text, and then support these themes with evidence.

Other approaches that encourage students to acquire meaning from the whole text include using plot structures to develop understanding of conflict and plot development described in chapter 10. When developing plot structures, students must use the total story to develop understanding.

Approaches that Help Students Improve Their Comprehension

Comprehension of text is usually identified as one of the major goals for reading instruction. Yet critics of comprehension instruction frequently claim that comprehension is tested and not taught, that students are not taught the processes required for comprehension, and that comprehension questions are at the lower end of a taxonomy of comprehension and do not encourage students to develop higher thought processes. In this section we consider three major approaches to reading comprehension that are especially appropriate for teaching both reading comprehension and for teaching reading through a literature approach: semantic mapping or webbing, modeling, and questioning strategies.

Semantic mapping or webbing   Many of the approaches for using semantic maps or webs that are appropriate for the reading class were introduced elsewhere in this text. For example, chapter 2, pages 41–45, described ways to increase vocabulary comprehension and development with semantic maps. Similar semantic maps may be developed around the relevant vocabulary in a piece of literature. For this literature text activity, place the title of the book in the center of the web drawn on the chalkboard. On the arms extending from the center, place the important words from the story. During a brainstorming session students fill in words such as synonyms and definitions that expand their comprehension of these key words. This activity may be completed as a prereading vocabulary introduction. The same web may also be used to extend understandings as students consider exact meanings for the words while reading the story, from the prior meanings select terms that seem closest to the meaning developed in the story, add new words or even phrases gained from the story, and use the vocabulary words identified in the web for extension of story comprehension.

In the following example this approach was used with the vocabulary from Tomie dePaola's *The Legend of the Bluebonnet* (1983). Notice how the procedure progresses from a vocabulary web, to a discussion of context clues that relate to the text and the vocabulary words on the web, and finally to a plot development structure in which the vocabulary words are reinforced and placed in the context of the total text. This series of lessons was developed and taught for my research with lower achieving fifth- and sixth-grade students (Norton, 1987). The procedure, however, has been used with students reading at all ability levels and in other grade levels. In addition, the procedure has been used with many other books.

Before reading *The Legend of the Bluebonnet,* a web similar to that shown in figure 11–1 was drawn on the chalkboard. The title of the book was placed in the center of the web and the vocabulary words *drought, famine, selfish, healing, plentiful, restored, valued,* and *sacrifice* were extended from the center. During the prereading activity these vocabulary words were introduced and students identified definitions or synonyms that could be related to these words. (The extended vocabulary shown in figure 11–1 was not completed during this first experience. Some of the meanings were added after reading and discussing the book. The actual web was also larger as more words were added during and following the reading, and the vocabulary words *Comanche, Shaman,* and *miraculous* were added because the group decided that they were extremely important for understanding.)

Following the introductory brainstorming with the vocabulary words, the story was read and the vocabulary words were discussed within the context of the story. In this story cause-and-effect relationships as well as opposites were emphasized. For

FIGURE 11–1
Vocabulary web for a book

example, the first paragraph was written on the chalkboard with the vocabulary word drought underlined:

> 'Great Spirits, the land is dying your People are dying, too' the line of dancers sang. 'Tell us what we have done to anger you. End this drought. Save your people. Tell us what we must do so you will send the rain that will bring back life.'(p. 1, unnumbered)

After reading the paragraph the students identified, circled, and discussed the words that meant drought: "land is dying." Next, they identified, circled, and discussed the words that indicated why the drought must end and showed the dangerous consequences of the drought: "Save your people," "bring back life." Now they discussed the consequences of drought if something that would save the people and bring back life was requested by the most powerful person in the tribe. Picture clues were discussed that showed the hot, yellow sun, and the brown, dry earth. As the reading of the story proceeded, the meaning of drought was extended to include cause-and-effect relationships:

drought→ famine, starving
rain→ grass, plenty

and opposite meanings for drought:

drought→ rain; earth will be green and alive.

This procedure continued until all of the vocabulary words were discussed in context. Now the web was viewed again, terms were added if necessary, and words previously identified were considered for their meanings with *The Legend of the Bluebonnet.*

Finally, a plot diagram was drawn using as many of the vocabulary words as were appropriate. Figure 11–2 shows the plot diagram developed with sixth-grade students. The underlined words were identified by the students during this teacher-led activity. The words in parenthesis are vocabulary words that had to be drawn out by the teacher during the plotting experience. For example, when the group did not identify *drought* in the beginning problem, the teacher said: "What words did you use for the land and the people are dying?"

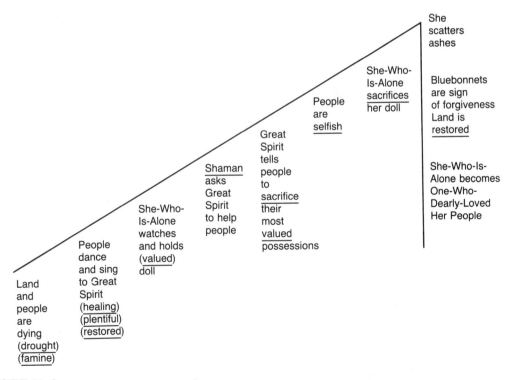

FIGURE 11–2
Plot development highlighting vocabulary

This approach develops comprehension for vocabulary, emphasizes the importance of context and picture clues in developing meaning from text, and relates the meaningful vocabulary back to the plot structure of the story. These interrelationships are especially important for poor readers who may not make these connections.

Another type of semantic map or web that is especially beneficial for developing comprehension diagrams story structure through literary characteristics. This type of web, with examples, was introduced in chapter 4, pages 146–147. Chapter 10, pages 399–401 presented approaches for developing understanding of the literary elements of setting, characters, conflict, plot development, and theme through semantic mapping. The webs shown in those chapters are equally beneficial for developing comprehension of text in the reading class.

Finally, webs may be drawn that show the various types of activities that may accompany a book. This type of web is very beneficial for the teacher who is exploring various ways that a book might be used in the curriculum. Figure 11–3 shows this type of web drawn around a picture book version of a folktale. Notice especially the application of the book in developing reading, cognitive, and language development. Also notice that the book is especially good for using stimulating approaches to literature such as oral storytelling and creative drama. A file of webs showing book possibilities is extremely useful for the language arts teacher.

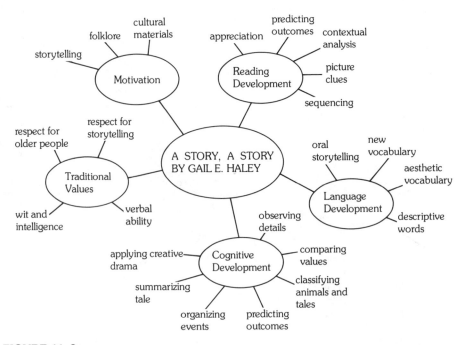

FIGURE 11–3
A web illustrating various activities that might accompany a book

1. *Develop a series of lessons around a literature selection that proceed from a vocabulary web, to a guided discussion of the vocabulary through contextual analysis, and finally to a plot structure that reinforces the vocabulary.*

2. *Develop a semantic map or web that could be used to lead a literary discussion of the book.*

3. *Develop a semantic map or web that shows the various activities that might accompany the book.*

Modeling   Pearson and Camperell (1985) identified three important implications gained from comprehension research that relate to modeling. All of these modeling techniques have been presented previously in this text, consequently we merely review them and identify where they may be found in this text. First, Pearson and Camperell emphasized the influence of connectives or cohesive devices on comprehension. Studies show that cohesive devices can actually improve students' comprehension because they help students make inferences. Consequently, complex sentences containing cohesive devices may be easier to comprehend than shorter, choppier sentences. Well-written literature contains numerous examples of more complex sentences and cohesive devices. Chapter 7, pages 247–250, includes a modeling lesson developed for teaching comprehension of cohesive devices through literature selections.

In addition, Pearson and Camperell identified the important relationship between instruction in sentence combining and comprehension. Chapter 7, pages 243–244, includes instructional strategies in sentence combining that emphasize modeling of sentence-combining strategies.

Furthermore, Pearson and Camperell identified advantages of modeling related to the author's message. They stated, "If teachers want students to get the author's message, they are well advised to model for students how to figure out the author's general structure" (p. 338). Early and Ericson (1988) reinforced this statement when they stated that research shows that teachers neglect modeling the process and explaining or describing what students should do during a comprehension lesson. These modeling strategies are especially important in teaching students how to make inferences because failure to make inferences is "frequently the cause for faulty comprehension" (p. 37).

This modeling approach, based on research conducted by Roehler and Duffy (1984) and Gordon (1985) was presented in detail elsewhere in this text. Chapter 3, pages 73–76, presented modeling of inferences related to characterization. Chapter 10, pages 401–405, described a modeling activity designed to develop understanding of similes. Both of these examples were based on children's literature.

REINFORCEMENT
ACTIVITY

*Review the modeling lessons described. Identify a comprehension skill that would be im-
proved through modeling, select a book that would be appropriate for modeling the skill,
develop the modeling lesson, and share it with your class. Lists of suggested books and
their modeling potentials are included in Chapter 10, pages 404–405.*

Questioning strategies   Critics of comprehension instruction emphasize that questions too frequently encourage students to respond at the lowest levels of the thought
process. Ambrulevich (1986) found that questions in literature anthologies focused on
the two lowest types of thinking on Bloom's Taxonomy. Reading textbooks emphasizes the need for developing questioning strategies that require higher-level thought
processes, and basal readers usually identify the taxonomy level of the comprehension
questions.

Not all literature selections should be accompanied by questioning. Neverthe-
less, teachers responsible for encouraging students to think about and react to
literature in a variety of ways find it helpful to have a framework for designing
questions that assist students in focusing on certain aspects of a story and questions
that require higher-level thought processes. If children's literature is used to reinforce
and strengthen a basal reading program, teachers frequently want to lead discussions
of literature that focus either on a taxonomy of comprehension or on literary elements.

My own students recently developed questioning strategies around books in
which they either used Barrett's Taxonomy of Comprehension (1972) or focused on
developing appreciation of the literary elements of setting, plot development, characterization, theme, and author's style. The following examples show how both of these
discussion frameworks might be used with Beverly Cleary's *Dear Mr. Henshaw*
(1983). After the students developed their questioning strategies they shared them
with the class and compared their questions. They discovered that both approaches to
questioning helped them discover vital information about developing their own
questioning techniques. As you read the following questions, compare the questions
asked and consider any advantages or disadvantages of either approach. How would
the development of both series of questions help students improve their questioning
strategies and their ability to focus on important elements in literature?

The first series of questions illustrates how questioning strategies can be
developed around Barrett's four levels of reading comprehension: literal recognition
or recall, inference, evaluation, and appreciation. The questions listed exemplify each
level on the taxonomy. Additional questions or fewer questions might be advisable
depending on the book and the instructional purpose.

Literal recognition requires students to identify information provided in the
literature. Teachers may require students to recall the information from memory after
reading or listening to a story or to locate information while reading a literature
selection. Literal-level questions such as the following often use words such as who,
what, where, and when:

Effective questioning strategies increase the interaction within a group.

1. *Recall of details:* How old was Leigh Botts when he first wrote to Mr. Henshaw? What is the name of Leigh's dog? What is Leigh's father's occupation? Where does the story *Dear Mr. Henshaw* take place?
2. *Recall of sequence of events:* What was the sequence of events that caused Leigh to place an alarm on his lunch box? What was the sequence of events that caused Leigh to write to Mr. Henshaw and to write in his diary?
3. *Recall of comparisons:* Compare the way that Leigh thought of his mother and the way that he thought of his father?
4. *Recall of character traits:* Describe Leigh's response to Mr. Henshaw when Mr. Henshaw asks Leigh to answer ten questions about himself.

When children infer an answer to a question, they go beyond the information the author provides and hypothesize about such things as details, main ideas, sequence of events that might have led to an occurrence, and cause-and-effect relationships. Inference is usually considered a higher-level thought process; the answers are not specifically stated within the text. Examples of inferential questions include the following:

1. *Inferring supporting details:* At the end of the book Leigh says that he "felt sad and a whole lot better at the same time." What do you think he meant by this statement?

2. *Inferring main idea:* What do you believe is the theme of the book? What message do you think the author is trying to express to the reader?

3. *Inferring comparisons:* Think about the two most important characters in Leigh's life, his mother and his father. How do you believe that they are alike and how do you believe that they are different? Compare Leigh at the beginning of the book when he is writing to Mr. Henshaw with Leigh at the end of the book when he is writing in his diary and writing true stories for school.

4. *Inferring cause-and-effect relationships:* If you identify any changes in Leigh, what do you believe might have caused those changes?

5. *Inferring character traits:* On page 73 Leigh decides that he cannot hate his father anymore. What does this tell us about Leigh and about his father? Why do you think Leigh's father sent him $20.00?

6. *Inferring outcomes:* At one point in the story, Leigh wishes that his mother and father would get back together. What do you think would have happened to the story if that happened? How would Leigh's life have changed? How would his mother's life have changed? What might have been the outcome of the writing contest at school if Leigh had not had the advantages of his advice from Mr. Henshaw?

Evaluation questions require children to make judgments about the content of the literature by comparing it with external criteria such as what authorities on a subject say or internal criteria such as the reader's own experience or knowledge. The following are examples of evaluative questions:

1. *Judgment of adequacy or validity:* Do you agree that an author would take time to write to a child and to take such an interest in him? Why or why not?

2. *Judgment of appropriateness:* Do you believe that Leigh's story, "A Day on Dad's Rig" was a good story for the Young Writers' Yearbook? Why or why not?

3. *Judgment of worth, desirability, or acceptability:* Do you believe that Leigh had the right to feel the way he did toward his father? Why or why not? Do you believe that Leigh was right in his judgment that his father did not spend enough time with him? Why or why not? Was Leigh's mother's judgment right at the end of the book? What would you have done?

Appreciation of literature requires a heightening of sensitivity to the techniques authors use in order to create an emotional impact on their readers. Questions can encourage students to respond emotionally to the plot, identify with the characters, react to an author's use of language, and react to an author's ability to create a visual image through the choice of words in the text. The following are examples of questions that stimulate appreciation:

1. *Emotional response to plot or theme:* How did you respond to the plot of *Dear Mr. Henshaw?* Did the author hold your attention? If so, how? Do you believe the theme of the story was worthwhile? Why or why not? Pretend

you are either recommending this book for someone else to read or recommending that this book not be read; what would you tell that person?

2. *Identification with characters and incidents:* Have you ever felt, or known anyone who felt like Leigh? What caused you or the person to feel that way? How would you have reacted to the theft of an excellent lunch?

3. *Imagery:* How did the author encourage you to see Leigh's home, his neighborhood, and his school? Close your eyes and try to describe your neighborhood or town. How would you describe it so that someone else could "see" it?

These examples are not organized according to any sequence for presentation to students, but they do exemplify the range of questions that are considered when the teacher's focus is on strengthening students' reading comprehension abilities.

The next series of questions was written about the same book, *Dear Mr. Henshaw,* but focuses on specific literary elements within the text.

1. *Setting:* How have Leigh's living conditions changed as a result of his parents' divorce? What are some of the positive, and some of the negative aspects of where he is now living? Leigh's mother describes their new home by saying, "At least it keeps the rain off, and it can't be hauled away on a flatbed truck." How does this statement reflect her opinion of their past living conditions? What is the importance of having a stable environment in which to live?

2. *Plot development:* What is the problem that is causing conflict for Leigh and his family? Why does Leigh write his first letter to Mr. Henshaw? Why does he continue to write letters even after the assigned letter is completed? What does Mr. Henshaw suggest that Leigh do? How do Mr. Henshaw's suggestions help Leigh accept the divorce? What causes Leigh to stop writing "Dear Mr. Pretend Henshaw"? What was the significance of this title when he first began his journal? Why does Leigh have such a problem with his lunch box? What plan does he devise to overcome the lunch box thief? Does the plan work? What happens as a result of his lunch box plan? What does Mrs. Badger say to Leigh that made him feel good? What did Leigh learn from Mrs. Badger? What helps Leigh understand his mother's point of view about the divorce? How do we know he now understands his mother's point of view? What happens to help Leigh understand his father's point of view? What does Leigh do to show that he understands his father's point of view? How does Leigh feel about himself and his family at the end of the story?

3. *Characterization:* How does Leigh view himself in comparison to the other children in his school? What three events help improve Leigh's self image? How does Leigh's self-image change as a result of these events? How does the following passage reflect changes in Leigh's character and reflect his growth? "I don't have to pretend to write to Mr. Henshaw anymore. I have learned to say what I think on a piece of paper. And I don't hate my father

either. I can't hate him. Maybe things would be easier if I could" (p. 73). How does Leigh's real father differ from the father of his dreams? In what ways does Leigh's father disappoint him? How do we know that Leigh's father still cares about Leigh and Leigh's mother? How does the character of Leigh's mother differ from that of his father? What actions show that Leigh's mother wants to care for and to provide for her son?

4. *Theme:* Why do you believe that Beverly Cleary wrote *Dear Mr. Henshaw* in a diary and letter format? Why was this an effective way to present this book? What important message did the letters and the diary reveal about Leigh and the problems that he was facing within himself and his family? How did either the real or the imaginary exchange of letters and ideas with Mr. Henshaw allow the author to develop this message? How did Leigh change because of these exchanges? At what point do you think Leigh started to accept his parents' divorce? At what point in the story does Leigh begin to understand that his parents are not going to get back together? How do we know that divorce is difficult for everyone involved? How do we know that divorce may not be blamed on one person in the family? Based on the theme of the book, what do you think would be another good title for this book?

5. *Style:* From whose point of view is this story told? How do you know that the story is written from the point of view of a young boy? How does the format of *Dear Mr. Henshaw* differ from most other fictional books? How did this diary and letter format allow the author to develop the theme, the conflict, and the characterization? How did the author's style make you believe that you were or were not reading the writing of a real boy who was facing believable problems? This book deals with some painful and serious emotions. At the same time, the author is able to make the reader laugh. How does she accomplish this?

REINFORCEMENT ACTIVITY

1. *Analyze the two series of questions. Are there any overlapping topics? What are the strengths and the weaknesses of each questioning framework? Do you agree with my students that they gain more literary understanding by focusing on the literary elements of setting, plot development, characterization, theme, and style, but they gain insights about balancing questions so that they include all levels of thought processes from the taxonomy of comprehension?*

2. *Choose a book and develop questioning strategies that could accompany that book. Try to include the strengths from both a focus on literary elements and a balance from questions that include the levels on the taxonomy of comprehension.*

Developing card files of questioning strategies Card files of questioning strategies, book summaries, difficult words or concepts, and other activities that can accompany a book are very beneficial. The teacher will need to know a great deal about specific children's books if he hopes to use them effectively in recreational reading, individualized instruction, storytelling, or social studies. College students should begin to read a number of children's books from the various categories of literature. It is easier to file pertinent information about a book when you read it, rather than try to remember it months or even years later. The following format has proven valuable to many of my students:

A Literature Card Format

1. Title
2. Author, Publisher, Date
3. Reading Level
4. Interest Level
5. Category
6. Difficult words or concepts
7. Short summary
8. Questions to ask during a conference. Include factual, inferential, evaluative, and appreciative, with possible answers to each question.
9. Several follow-up activities to use with the book
10. Other books on same subject or by same author

You should put the information on a large filing card, or on a separate page in a loose-leaf notebook, and group the cards by grade level, category, and so forth. Some students also color code their cards, so they can easily choose an appropriate card from their file box. The following card files are examples developed by preservice teachers.

Example of a Card for a Book for Early Elementary Grades

Horton Hears A Who	Second Grade
Dr. Seuss. New York: Random House	Lower Elementary
	Fantasy: Animals

Difficult words: murmured, nonsensical, vigor, vim, hullabaloo

Summary: Horton the elephant hears a cry from a small dust speck flying by him. He decides there must be a small creature on the dust speck, and proceeds to save the small "Who." He puts the small "Who" on a clover and carries it with him everywhere. The other characters make fun of him and think he has gone crazy. At the end of the story, the other characters hear the small Who and Whoville is saved by the smallest of all.

Questions (F = Factual, I = Inferential, E = Evaluative, A = Appreciative):

(F) **1.** Where did the eagle drop Horton's clover? (In a field of clover.)
(I) **2.** What do you think this story is trying to tell you about small people? (Small people are important even though they are small.)

(I) **3.** What do you think made Horton decide to take care of the small Who? (Answers will vary: he wanted to have some new friends; he did not want the Whos to get hurt.)

(E) **4.** Was it important for the other creatures to hear the Whos? Why or why not? (Yes, because every little Who was important and they would not be saved if they could not be heard.)

(A) **5.** How would you have felt if you lived in Whoville? (Answers will vary: I would have been glad that Horton took care of us and made us feel important.)

Activities:
1. Have the students role play an experience with a younger brother or sister to see how it would feel to be small. Have them discuss how they would want people to treat them.
2. Have the students write a creative story telling what they would do to protect the small Whos.

Other Books by Dr. Seuss:
Horton Hatches the Egg, How the Grinch Stole Christmas, And to Think That I Saw It on Mulberry Street, The Sneetches and Other Stories, Scrambled Eggs Supper

Example of a Card for a Book
Appropriate for Third or Fourth Grade

Bill Cosby: *Look Back in Laughter*   Third-Fourth Grade
James T. Olsen   Biography

Summary: Bill Cosby lived his childhood years in a black ghetto. He got his first job shining shoes at age six. He later worked in a grocery store for eight dollars a day; he gave this money to his mother. His brother James died and his father, heavily in debt, left home. Bill did not do well in school, and dropped out to join the Navy. He got his high school diploma in the Navy, and later enrolled at Temple University. He earned money as an entertainer by telling jokes about his childhood. He first became popular during his television series *I Spy*. Since then, he has acquired millions of fans, but he tries to help everyone he comes in contact with. He gives heavily to charities and is also concerned with education. He wants to work with children because "the one child who could have gone the other way but who changed his mind is worth everything."

Factual Questions:
1. How old was Bill when he got his first job? What was it? (6; shining shoes)
2. Did Bill have a happy home life when he was growing up? (no; his father left his family and his brother died)
3. What was Bill's first television show? (*I Spy*)
4. Who starred with Bill on his first show? (Robert Culp)
5. What are some charities Bill contributes to? (Boy Scouts, American Cancer Society, United Fund)

Inferential Questions:
1. Do you think Bill uses his money in a good way? Why? (yes, he helps others)
2. Why did Bill become a medic in the Navy? (they did not have to fight at the battlefront)
3. Why did Bill use his childhood as a subject for jokes? (everyone has some pleasant childhood memories and can relate to children)

Evaluative Questions:
1. Is this a true story? How do you know?
2. Do you think Bill should have given all the eight dollars he made at the grocery store to his mother? Why or why not?
3. Do you think entertaining is a job that is well-suited to Bill Cosby? Explain.
4. Do you think Bill's wife's parents were right in not wanting her to marry him when he was working in a bar? Why or why not?

Appreciative Questions:
1. Have you ever seen or heard Bill Cosby? Did his life story sound like you thought it would? Explain.
2. Did you feel sorry for Bill as a child? Why or why not?
3. What does this story tell you about people who work hard for what they want?

Activities:
1. Write a script with jokes that Bill Cosby might use in a television show.
2. Write about the things a charity would do with the money Bill gave them.
3. Make a time line of important events in Bill Cosby's life.
4. Write a review of this book. Illustrate it, and show it to the class.
5. Compare this biography with others you have read. How are they alike? How are they different?

Other Books About the Lives of People:
Louis Braille: The Boy Who Invented Books for the Blind by Margaret Davidson
The Golda Meir Story by Maragaret Davidson
Franklin D. Roosevelt, Gallant President by Barbara Silberdick
Arthur Mitchell by Tobi Tobias
Lincoln: A Photobiography by Russell Freedman

Example of a Card for a Book Appropriate for Upper Elementary Grades

Across Five Aprils   Sixth Grade
Irene Hunt   Sixth Grade and Up
   Historical Fiction: Civil War

Summary: This is the story of the Creighton family and the Civil War as seen through the eyes of Jethro, the youngest member of the family. Jethro never fought in the war, but saw all his brothers, cousins, and his only surviving sister leave home to fight for either North or South. The book opens in April 1861 and gradually, all Jethro's brothers go to war. Tom, Eb, and John go to fight for the North, while Bill, Jethro's favorite brother, goes to fight for the South. Jethro is left with his parents to work on the farm. Two families in the community blame the Creightons for Bill's view on the war, and one night their barn is burned and their well filled with kerosene. Jethro's sister leaves to marry her fiance, who was wounded while fighting for the North. Jethro keeps up with the war through newspaper reports provided him by the newspaper editor, Ross Milton. Jethro becomes lonely for the past and comes to realize how horrible war is. He has to make many decisions, and grows up in many ways, during the five Aprils in this book.

Factual Questions:
1. In what part of the country does this book take place? (Illinois)
2. Why is the book called *Across Five Aprils?* Is the title appropriate? (the book covers five Aprils of the Civil War)

3. Why was Jethro the only one who did not go to war? (he was too young)
4. Why was the Creighton's barn burned and their well contaminated? (one of their sons was fighting for the South)
5. Why did Jethro want to keep up with his studies? (he idolized Shad, his schoolmaster, and knew study would help him in life)

Inferential Questions:
1. Why was Ross Milton so special to Jethro? (he was an understanding adult; he understood Jethro's need for education)
2. If Jethro had spare time, what do you think he would spend it doing? (educational projects, reading, etc.)
3. How did Jethro feel about the men of the community helping them so much to rebuild their barn? (very grateful; he felt how good it was to have understanding friends.)

Evaluative Questions:
1. Do you think the author portrays the feelings and problems of the Civil War accurately, even though it is through the eyes of a young boy? Explain.
2. Do you think Jethro should have helped the man who burned down the barn? Why or why not?
3. Do you think Jethro had a right to feel overburdened? Did he?

Appreciative Questions
1. How did the author present historical facts in an interesting and real way?
2. Did the Civil War seem more interesting to you? In what way?
3. Did you learn more about the Civil War than just the names and dates of the battles? Explain.

Activities:
1. Make a time line of the five years covered by the book. Include on the time line the important things that happened in Jethro's life.
2. Draw a map of the United States and have students color-code the Union and Confederate States.
3. Write a letter that Jethro might have written to Shad in Washington, D.C., after Shad had been injured fighting for the North.
4. Write a letter to a history teacher and recommend the book for teaching about the Civil War. List several reasons why the book would be a good instruction book about the war.

Other Historical Fiction Books or Books About the Civil War:
The Drinking Gourd by F. N. Monjo
The Slave Dancer, by Paula Fox
Zoar Blue by Janet Hickman
Rifles for Watie by Harold Keith

You can use these literature cards with summaries, questions, and activities in numerous ways. The questions provide a framework for discussion for extending understanding of a book, either with individual children, or during small-group discussions with several children who have read the same book. The activities can be used for enrichment experiences after a child or a group of children have read the book. At times, the teacher may use the cards only to find books to recommend for recreational reading. Other books are listed so the teacher can recommend additional books to a child who has a specific interest, or wants to explore a topic further.

Approaches that Encourage Students to Make Predictions

Good readers use cues in the text and their own prior knowledge to make predictions. These predictive skills may be stimulated and enhanced at any level. For example, chapter 4, page 145, described a procedure in which predictable books were used to teach reading to kindergarten and first-grade students. Predictable books such as John Ivimey's *The Complete Story of Three Blind Mice* (1987) and Scott Cook's *The Gingerbread Boy* (1987) encourage students to join in during repetitive phrases. Predictable story structures such as "The Three Billy Goats Gruff" and "The Three Little Pigs" encourage students to hypothesize about plot developments.

Teaching students to identify foreshadowing, a device that allows the author to suggest plot development before it occurs, is another method for helping students use cues in the text to predict outcomes. A detail in an illustration, the mood of the story, the author's choice of words, a minor episode, or a symbol presented early in the story may foreshadow later action.

Picture storybooks are good introductions to foreshadowing because the illustrations frequently include details that suggest later actions. For example, careful scrutiny of Don Woods's illustrations for Audrey Wood's *Heckedy Peg* (1987) shows the young girl feeding a friendly blackbird through the window; the same blackbird later helps the children by leading their mother to the witch who has transformed them into food. Another illustration in the same book shows the boisterous children playing within their cottage while through the window the observant viewer sees the witch approaching; later, the witch arrives at the door, the children disobey their mother, and the children allow the witch to enter the cottage. Both of these earlier illustrations foreshadow later plot developments and conflict. Students may try to search for these clues while they read the book, or they may reread the book searching for clues that foreshadow later developments. Many other picture storybooks may be used in the same way. For example, Richard Egielski's illustrations of a bird with hands in Arthur Yorinks's *Hey, Al* (1986) foreshadow the fate of Al and his dog Eddie.

Illustrations in picture storybooks frequently foreshadow the ending, the type of story, and the plot development by creating appropriate moods. For example, Marcia Brown's delicate lines and soft colors suggest a magical setting and foreshadow a fairy tale ending in Charles Perrault's *Cinderella* (1954). In contrast, Charles Keeping's stark, black lines create ghostly, terrifying subjects and foreshadow the tale's sinister and disastrous consequences in Alfred Noyes's *The Highwayman* (1981). Other contrasting books such as John Schoenherr's illustrations for Jane Yolen's *Owl Moon* (1987) and Marcia Brown's illustrations for Blaise Cendrars's *Shadow* (1982) allow students to look at the illustrations, to analyze the mood created by the illustrations, and to predict the type of story, the type of conflict, and the possible ending.

Terms used to describe characters may provide clues to plot changes and even foreshadow themes and plot endings. For example, students may predict plot development, changing conflict, changing characterization, and possible themes as they trace the significance of name changes in Sid Fleischman's historical fiction, *The Whipping Boy* (1986). For example, the following names are used in reference to the Prince and to his whipping boy, Jemmy; the names in parentheses indicate who is using the specific name:

	Prince		*Jemmy*
p. 1	Prince Brat (King, royalty)	p. 2	Common boy (narrator)
p. 4	Your Royal Awfulness (Jemmy)	p. 6	Jemmy-From-The-Streets (Prince)
p. 5	You fiddle-faddle scholar (tutor)	p. 6	Contrary rascal (Prince)
p. 8	No friends (Jemmy)		
p. 18	Empty headed prince (Jemmy)	p. 20	My whipping boy (Prince)
p. 21	His Royal Highness (Prince) Changing Roles	p. 21	That ratty street orphan (Prince)
p. 21	Witless servant boy (Jemmy)		
p. 28	Whipping boy (Hold-Your-Nose Billy)		
p. 31	Jemmy-From-The-Streets (Jemmy)		
p. 32	Contrary as a mule (Jemmy)		
p. 33	Prince Blockhead (Jemmy)	p. 33	Imposter (Prince)
		p. 42	Unfaithful servant (Prince)
p. 53	A wounded bird (Jemmy)		
p. 53	Friend (Jemmy)		
p. 61	A brave one (Jemmy)	p. 61	Jemmy (Prince)
p. 61	Friend-O-Jemmy's (Prince)		
p. 70	Heir to the throne (Jemmy)	p. 74	My friend (Prince)
p. 88	Friend (Jemmy)		

In the previous example, notice how the changes in names and references to the main characters follow plot development, suggest changes in relationships, and even develop one of the themes of the book. Students and teachers both enjoy doing this type of activity as they search for clues provided by the author that help them predict the story line.

Books for older students may use more complex symbolism to foreshadow characterization and plot development. In such books students may identify symbols and hypothesize how the symbol may foreshadow characterization and plot development. For example, in Cynthia Voigt's *A Solitary Blue* (1983) students may trace Jeff's changing attitudes toward the heron and predict how these gradual changes in attitude will eventually reveal Jeff's growing self-esteem. Likewise, in Voigt's *Dicey's Song* (1982) students may trace, discuss, and predict the importance of the author's reference to musical selections, to a dilapidated boat, and to a tree.

REINFORCEMENT ACTIVITY

1. Select a grade level and develop a list of books that can be used to encourage students to make predictions.
2. Develop a lesson in which you lead students in a search for clues that will help them predict plot development, characterization, theme, or story ending.

In addition to developing predictive skills, these types of activities enhance students' abilities to use inference skills and to understand authors' structure.

DEVELOPING A UNIT AROUND LITERATURE AND READING

We conclude this chapter with an example of a unit that incorporates many of the goals and approaches described in this chapter and in chapter 10. "A Literary Approach to the Study of Biography" includes activities that enhance appreciation of biography as literature, as well as activities that encourage students to read carefully and to critically evaluate what they read. The unit includes objectives from all of the language arts including reading, oral language, listening, and writing as well as literature. The books included in the bibliography include a range of ability levels including picture biographies to biographies appropriate for upper-elementary and middle school. Teachers may choose the literature and the activities that are appropriate for their specific grade level. The unit includes both teacher-directed activities to be used with the class and a specific book and additional activities that encourage students to read widely and independently. Units based on biographies may include content and concepts from social studies, history, and geography. Additional ideas for developing various approaches to biographies are found in Fleming and McGinnis's *Portraits: Biography and Autobiography in the Secondary School* (1985). Although this text is designed for secondary students, there are many excellent ideas that may be developed for younger children.

FOR YOUR PLAN BOOK
A Literary Approach to the Study of Biography

The literary approach to biography emphasizes an understanding of and an appreciation for the techniques an author uses to create biographies about real, believable people. The literary approach is concerned primarily with the development of characterization, plot, setting, theme, and style. Discussion and other activities related to the approach encourage students to read reflectively and to analyze thoughtfully. Activities encourage the interaction of all the language arts and English skills as students read, analyze, discuss, dramatize, listen, and write.

Unit Objectives

1. *To develop an appreciation for biography.*
2. *To analyze and to critically evaluate plot in biography.*
3. *To analyze and to critically evaluate characterization in biography.*
4. *To analyze and to critically evaluate setting in biography.*
5. *To analyze and to critically evaluate theme in biography.*
6. *To analyze and to critically evaluate style in biography.*
7. *To integrate the language arts and English skills of reading, oral language, listening, and writing within a literature focus.*

Biographies Discussed in This Approach

> Adler, David. **Martin Luther King, Jr.: Free At Last** (1987)
> Brooks, Polly. **Queen Eleanor: Independent Spirit of the Medieval World** (1983)
> Coerr, Eleanor. **Sadako and the Thousand Paper Cranes** (1977)
> D'Aulaire, Ingri and D'Aulaire, Edgar. **Columbus** (1955)
> Dalgliesh, Alice. **The Columbus Story** (1955)
> Fido, Martin. **Oscar Wilde: An Illustrated Biography** (1987)
> Fido, Martin. **Rudyard Kipling: An Illustrated Biography** (1987)
> Freedman, Russell. **Lincoln: A Photobiography** (1987)
> Fritz, Jean. **Where Do You Think You're Going, Christopher Columbus?** (1980)
> Goodnough, David. **Christopher Columbus** (1979)
> Goodsell, Jane. **Eleanor Roosevelt** (1970)
> Grimble, Ian. **Robert Burns: An Illustrated Biography** (1987)
> Hamilton, Virginia. **W.E.B. Du Bois** (1972)
> Hamilton, Virginia (ed.). **Writings of W.E.B. Du Bois** (1975)
> Roosevelt, Elliot. **Eleanor Roosevelt, With Love** (1984)
> Ventura, Piero. **Christopher Columbus** (1978)
> Whitney, Sharon. **Eleanor Roosevelt** (1982)
> Yates, Elizabeth. **Amos Fortune, Free Man** (1950)
> Yates, Elizabeth. **My Diary, My World** (1981)
> Yates, Elizabeth. **My Widening World** (1983)

Procedures

Read, analyze, and discuss a biography according to the following literary elements:

1. *Plot Development*
    a. *What is the order for the plot development? Does author use chronological order? flashbacks? How does the author present important information from the past that influenced the biographical character?*
    b. *What time frame does the author emphasize? Why do you believe the author chose that time period in the character's life?*
    c. *What type of conflict does the biographer develop? Find examples of these conflict types. Why do you believe the biographer chose to emphasize these types of conflict?*
        (1) *Person versus Self*
        (2) *Person versus Person*
        (3) *Person versus Society*
        (4) *Person versus Nature*
    d. *What pattern of action does the author develop?*
        (1) ⌐‾‾‾‾‾‾‾‾‾‾‾‾⌐ *Rising action with identifiable climax and end of conflict.*
        (2) ――――――― *Action moves from one incident to another incident. There is a final climax but author leaves reader with a sense of uncertainty.*

(3) ——————— *Action moves with a tension that does not develop a cli-max. There is minimal suspense.*

(4) 〰〰〰 *Action moves in a life-line that does not have one dra-matic climax but may have several dramatic episodes.*

e. *How does the author develop reader's interest in plot development?*

    (1) *Find examples of interest-creating introductions. For example, how does Brooks attract your attention in this introductory paragraph to* **Queen Eleanor: Independent Spirit of the Medieval World:**

> *The French King, Louis the Fat, lay dying in his hunting lodge where he had been taken to escape the summer heat, the flies, the stench of Paris. His hands shook with palsy, and his bleary eyes could hardly see. Recently he had grown so fat that he could no longer mount a horse or bend over to tie his shoes. (p. 7)*

    (2) *Find examples of author's use of direct quotes from the biographical character or photographs that stimulate interest. For example, how does Freedman attract your attention in the introduction to chapter 1 in* **Lincoln: A Photobiography:**

> *If any personal description of me is thought desirable, it may be said, I am, in height, six feet, four inches, nearly; lean in flesh, weighing, on average, one hun-dred and eighty pounds; dark complexion, with coarse black hair and grey eyes—no other marks or brands recollected.*
> *(p. 1)*

    (3) *Find examples of any other techniques such as foreshadowing of events to come.*

f. *Additional activities that help students and evaluate plot development in biog-raphy:*

    (1) *Draw and label the life-line of the characters such as Yates'* **Amos For-tune, Free Man.** *Draw and label the pattern of action developed by the author. Compare the two drawings. How are they alike and how are they different. Try to account for any differences.*

    (2) *Compare the pattern of action for two biographies written about the same person, but written by two different authors. For example, choose two of the biographies listed about Christopher Columbus or about Eleanor Roosevelt. Try to account for any differences.*

    (3) *Compare examples of introductory paragraphs in several biographies. Choose the example that you think is the most effective. Prepare the paragraph as an oral reading, share it with your class, and tell why you believe the introduction is effective.*

    (4) *Develop a class chart of types of conflicts found in biographies. Catego-rize the biographies according to their predominant conflict. Are there any similarities among biographies that are in a specific category? For ex-ample, are person-versus-society conflicts found more often in biogra-phies about Civil-Rights leaders?*

Cover photograph of Russell Freedman's book *Lincoln: A Photobiography* (from *Lincoln: A Photobiography* by Russell Freedman, copyright © 1987 by Russell Freedman. Reprinted by permission of the Chicago Historical Society and Clarion Book/Ticknor & Fields, a Houghton Mifflin Company)

2. *Characterization*
   a. *How does the author reveal the characterization of the biographical character? Find and discuss examples of any of the following ways that authors can reveal character:*
      (1) Actions of the character.
      (2) Dialogue that reveals the speech of the character.
      (3) Thoughts of the character.
      (4) Thoughts of others toward the character.
      (5) Narrative—Comments about the character made by the author.
   b. *Using each of the aforementioned examples, what did you learn about the character? Was each technique appropriate for the biography? Could the author substantiate that each way of revealing character was authentic? (Look for references in the biography and for sources used by the author.)*
   c. *How did the character change over the course of the biography? What caused any changes? How did the author show any character changes?*
   d. *Who are the characters who influenced the biographical subject? How did the author develop these supporting characterizations?*
   e. *Who is telling the biographical story? What is the relationship of the author to the biographical character? How does this relationship influence the tone of the biography and the attitude toward the biographical subject? For example, answer these questions for Elliot Roosevelt's* **Eleanor Roosevelt, With Love.**
   f. *Additional activities that help students understand and evaluate characterization in biography:*
      (1) Identify scenes in the biography that include dialogue between characters. Dramatize the scenes using the dialogue presented in the literature. Discuss the impact of the dialogue with the class.
      (2) Compare the characterization developed in two biographies written about the same person but by different authors.
      (3) If possible, read something written by the biographical character. How does the viewpoint in the writing compare with the characterization developed in the biography? For example, compare Hamilton's **W.E.B. Du Bois: A Biography** *and* **Writings of W.E.B. Du Bois** *written by Du Bois but edited by Hamilton.*
      (4) Rewrite a scene from the biography but choose to write the scene in a different point of view. Discuss how this different point of view might change the way the character is presented. How would the tone and the attitude change?

3. *Setting*
   a. *How important is the setting in the biography? Why is the setting important or not important?*
   b. *What role does the setting have in the biography? Locate examples, if used, and discuss the importance of each of these purposes for the biography:*
      (1) Providing historical background.
      (2) Setting the mood.
      (3) Providing an antagonist.

        (4)   *Providing symbolic meanings.*

        (5)   *Providing illumination for characterization and clarifying conflict.*

  c. *What are the various settings developed in the biography? How much detail is included?*

  d. *Is the setting in the past or present? How does the author inform the reader of the time period?*

  e. *How much influence does the setting have on characterization?*

  f. *Additional activities that help students understand and evaluate setting in biography:*

      (1) *Locate the specific locations mentioned in the biography. Find these locations on a map and locate descriptions of these places in geography texts or other nonfictional sources. How do the descriptions compare with the settings in the biography?*

      (2) *Locate the specific dates and historical happenings identified in the biography. Check the accuracy of these dates and happenings in other nonfictional sources.*

      (3) *Choose a specific setting described in the biography. Draw the setting as if the setting was a backdrop for a stage production. Is there enough detail to complete the drawing? If not, what information would improve your drawing?*

4. *Theme*

  a. *There may be one primary or main theme in a biography and several secondary themes. Search for and identify any primary and secondary themes in the biography. How are these themes integrated into the biography? Does the title of the biography reflect the theme? For example, what could Adler's title* **Martin Luther King, Jr.: Free At Last** *infer?*

  b. *An author may literally state or infer the theme. Search for evidence of explicit themes in which the author states the theme and implicit themes in which the author implies the theme. Why do you believe the author used either approach? What was the impact to you as the reader?*

  c. *Themes in biographies frequently depict the biographical character's ability to triumph over obstacles or to struggle for identity. Search for evidence of either of these themes.*

  d. *Additional activities that help students understand and evaluate theme in biography:*

      (1) *The themes in biographies frequently include didacticism or instruction. The lives of great people are often used to provide models for readers and to suggest proper values, attitudes, and actions. Analyze several biographies to identify any instructional purposes. Why do you believe the author chose that specific character's life to teach a lesson?*

      (2) *Compare biographies written for younger children with biographies written for young adults. Analyze any differences in themes. Are there any themes that are more consistently developed at either level?*

      (3) *Identify and compare the themes developed in several biographies written about the same character. If the themes are not the same, how do* you account for any differences? Could the differences relate to the author's purpose for writing the biography or to the age level of the reader?

5. *Author's Style*

   a. *Author's style refers to how an author says something rather than what the author says. Authors may vary sentence length, repeat words, or choose descriptive words to suggest mood and to create interest. For example, in her biography* **The Columbus Story** *written for younger readers Alice Dalgliesh repeats words to create interest and excitement. The waves beckon Columbus, "Come, Come, Come!" White sails chant, "Adventure, adventure, adventure!" "day after day after day." Excitement is enhanced by "Land! Land! Land" and "Climbing, climbing, -higher-higher-higher." By repeated words such as "Nothing but sea—and—sea—and sea" Dalgliesh lengthens out the days and creates an image of the vast open sea. Identify examples in biographies in which the author's choice of words enhances the mood of the biography.*

   b. *Symbolism is a more complex literary technique that allows authors to suggest deeper meanings. For example, a dove may symbolize peace and a flag may symbolize patriotism. Authors of biographies, especially those written for older readers, may choose symbols that relate to characterization, conflict, and theme. For example, Eleanor Coerr uses considerable symbolism in her biography,* **Sadako and the Thousand Paper Cranes.** *The cranes are an ancient Japanese symbol for honor, humility, respect, hope, and protection. There is an ancient Japanese myth that reveals that the gods will grant a long and happy life to a sick person who can make one thousand paper cranes. Coerr uses both the symbolic meaning for crane and the belief in the myth to develop Sadako's character and to create plot development as the biographical character tries to overcome the personal ravages caused by the atomic bomb. The symbolism of the crane also suggests and reinforces themes developed by the author.*
   Search for examples of symbolism in a biography. How does the symbolism relate to characterization, plot development, and theme? Do you believe the symbolism is appropriate for the biographical character? Why or why not?

   c. *Figurative language such as similes and metaphors may be used by authors to clarify and enhance descriptions of settings and characters. For example, in* **Sadako and the Thousand Paper Cranes** *Coerr uses a simile when she states, "There was a flash of a million suns. Then the heat pricked my eyes like needles" (p. 18). She uses a metaphor to imply a comparison between the atomic bomb and a thunderbolt. In* **The Columbus Story** *Dalgliesh uses numerous similes to describe the setting in which the sea is smooth and "quiet as a river," the colors are as beautiful "as feathers of birds," and the trees are "as green and beautiful as the trees of Spain in the month of May." Search for examples of similes and metaphors. How do the comparisons relate to the setting or to the characterization? Are they appropriate for the biography? Why or why not?*

d. *Additional activities that help students understand style in biography:*

    (1) *Many biographies are about individuals who are authors. Read a biography about an author. Then, read several works written by that author. Think of the work as revealing information about and emotional responses of the author. Analyze whether or not the biography develops the style, the characterization, the emotions, and the beliefs of the real person. Examples for this activity include Grimble's* **Robert Burns: An Illustrated Biography** *and various collections of Robert Burns's poetry; Fido's* **Rudyard Kipling: An Illustrated Biography** *and Kipling's* **Just So Stories** *and* **The Jungle Books;** *and Fido's* **Oscar Wilde: An Illustrated Biography** *and Wilde's "The Happy Prince" and "The Selfish Giant."*

    (2) *Authors of biographies are frequently prolific writers. Identify a biographer who has written numerous books including, if possible, nonfictional materials that reveal information about the author. Analyze the works according to any similarities in style, themes, and characterization. Elizabeth Yates is an excellent example. Her 1950 biography* **Amos Fortune, Free Man** *won the Newbery award for literature, her autobiographies* **My Diary, My World** *and* **My Widening World** *received critical acclaim, her many other books are available in libraries, and her biographical profile and acceptance speech for the Newbery award are available in the* **Horn Book** *magazine.*

    (3) *Reviews of biographies usually emphasize the author's literary style and ability to develop accurate characterization. Search in sources such as* **Horn Book, Booklist, School Library Journal,** *and* **New York Times Book Review** *to find reviews of biographies that you have read. What do the critics who wrote the reviews emphasize? Do you agree or disagree with the book critic? Why or why not?*

    (4) *Attitudes toward some biographers and biographies differ. Attitudes toward a biography and the author's writing style may also change over time. Yates's* **Amos Fortune, Free Man,** *which won the Newbery award for excellence in literature, exemplifies such diversity and changing attitudes. The book is still highly acclaimed by numerous critics. For example, Zena Sutherland (1986) described the book as "written with warmth and compassion . . . . The details are grim, but Amos Fortune carried suffering lightly because his eyes were on the freedom of the future. It is this characteristic of Fortune's, so clearly depicted by Yates, that causes some modern critics to disparage the book. They disapprove of the quiet way Amos Fortune bore his enslavement with courage and dignity, forgetting the circumstances under which he achieved his own personal integrity and offered the chance to live free to other blacks"(p. 459). In contrast, Donnarae MacCann (1985) stated that the book is written with a "white supremacist attitude" (p. 169) and that "the author's descriptions are condescending, but beyond that they make the African appear almost subhuman" (p. 170).*

*Read a book such as **Amos Fortune, Free Man** and decide for yourself which viewpoint you endorse. Defend your reasons for this viewpoint.*

SUMMARY

There is currently a renewed interest in using children's literature in the reading program. Educators and community leaders recognize that the content of what is read is as important as the reading approach.

This chapter reviewed the literature and reading connection and described and developed approaches and methodologies supported by reading research.

First, the chapter reviewed schema theory and discussed the importance of literature in providing this valuable basis for comprehension, learning, and remembering ideas in stories. Approaches that help teachers use literature to develop understanding of story structures, to activate relevant prior knowledge, and to help students fill in gaps in their knowledge before they read were described.

Second, the chapter described approaches that increase students' opportunities to learn to read by reading. Approaches such as Uninterrupted Sustained Silent Reading (USSR), recreational reading groups, and focus or thematic units were described because these approaches motivate students to read widely and independently.

Third, the chapter described approaches that encourage readers to use the whole text to create meaning. Examples included methodologies for helping students use the whole text to search for important themes and then support those themes through specific instances in the book.

Fourth, the chapter described approaches that help students improve their comprehension. Specific lesson plans and descriptions of methodologies were developed for semantic mapping or webbing, modeling, and questioning strategies.

Fifth, the chapter described approaches designed to encourage students to make predictions. Predictable books for younger students were discussed. Teaching students to identify foreshadowing was described as a way to help students find clues in the text and predict plot development, characterization, and theme. The chapter concluded with a unit on teaching biography.

ADDITIONAL READING AND LITERATURE ACTIVITIES

1. *Read an article in a current journal such as **The Reading Teacher, The Journal of Reading,** or **Language Arts** that describes a reading program developed around literature or approaches for using literature in the reading curriculum. Report your findings to the class.*
2. *Develop an introduction to a book that helps students identify prior knowledge or provides them with important background information before they read the book.*

3. *Try to analyze how you use your own prior knowledge of texts, your knowledge of the world, and the clues supplied by the text to create meaning (schema theory). To do this task read several pieces of unfamiliar literature chosen from different genres such as folktales, poetry, modern fantasy, and nonfictional informational books. Keep a journal account of how you created meaning during your reading.*

4. *Conduct a recreational reading group with a group of students or a group of your peers.*

5. *Develop a lesson plan in which students must use the whole text to find and support the theme or to trace the developing conflict in the story.*

6. *Develop a file of semantic maps or questioning strategies that could accompany specific books. Try to emphasize the development of higher comprehension skills.*

7. *Develop a file of books and accompanying activities that encourage students to make predictions.*

8. *Develop a thematic unit around reading and literature.*

BIBLIOGRAPHY

Ambrulevich, A. K. "An Analysis of the Levels of Thinking Required By Questions in Selected Literature Anthologies for Grades Eight, Nine, and Ten." Doctoral Dissertation, Univ. of Bridgeport, 1986. *Dissertation Abstracts International* 47,03A. (University Microfilms No. 86-13,043).

Anderson, Richard C. "Role of the Reader's Schema in Comprehension, Learning, and Memory." In *Theoretical Models and Processes of Reading.* 3d ed., edited by Harry Singer and Robert Ruddell. Newark, Del.: International Reading Association, 1985.

Barrett, Thomas. "Taxonomy of Reading Comprehension." In *Reading 360 Monograph.* Lexington, Mass.: Ginn, 1972.

Early, Margaret, and Ericson, Bonnie O. "The Act of Reading." In *Literature in the Classroom: Readers, Texts, and Contexts,* edited by Ben F. Nelms, pp. 31–44. Urbana, Ill.: National Council of Teachers of English, 1988.

Feitelson, Dina; Kita, Bracha; and Goldstein, Zahava. "Effects of Listening to Series Stories on First Graders' Comprehension and Use of Language." *Research in the Teaching of English* 20 (December 1986): 339–55.

Fleming, Margaret, and McGinnis, Jo, eds. *Portraits: Biography and Autobiography in the Secondary School.* Urbana, Ill.: National Council of Teachers of English, 1985.

Gordon, Christine J. "Modeling Inference Awareness Across the Curriculum." *Journal of Reading* 28 (February 1985): 444–47.

Guthrie, John T. "Story Comprehension and Fables." In *Theoretical Models and Processes of Reading.* 3d ed., edited by Harry Singer and Robert Ruddell. Newark, Del.: International Reading Association, 1985.

Hechinger, Fred M. "Study Blames Boring Texts for Students' Poor Comprehension." Austin, Texas: *Austin American-Statesman.* Saturday, April 9, 1988, p. E16.

MacCann, Donnarae. "Racism in Prize-Winning Biographical Works," In *The Black American In Books For Children: Readings in Racism.* Edited by Donnarae MacCann and Gloria Woodard, pp. 169–79. Metuchen, N.J.: Scarecrow Press, 1985.

Matthews, Dorothy, and Committee to Revise High Interest-Easy Reading. *High Interest-Easy Reading.* Urbana, Ill.: National Council of Teachers of English, 1988.

May, Jill P. "Creating A School Wide Literature Program: A Case Study." *Children's Literature Association Quarterly* 12 (Fall 1987): 135–37.

Meeks, Margaret. *Achieving Literacy: Longitudinal Case Studies of Adolescents Learning to Read.* London: Routledge & Kegan Paul, 1983.

Miller, Marilyn, and Luskay, Jack. "School Libraries and Reading Programs: Establishing Closer Ties." *Reading Today* 5 (January 1988): 1, 18.

Moss, Joy F. *Focus Units in Literature: A Handbook for Elementary Teachers.* Urbana, Ill.: National Council of Teachers of English, 1984.

National Assessment of Educational Progress. *Three National Assessments of Reading: Changes in Performance, 1970–1980.* Report 11-R-01. Denver: Education Commission of the States, 1981.

Norton, Donna E. "An Evaluation of the BISD/TAMU Multiethnic Reading Program."

Research Report, College Station: Texas A&M University, 1987.

Pearson, P. David, and Camperell, Kay. "Comprehension of Text Structures." In *Theoretical Models and Processes of Reading,* edited by Harry Singer and Robert Ruddell, pp. 323–42. Newark, Del.: International Reading Association, 1985.

Roehler, Laura, and Duffy, Gerald G. "Direct Explanation of Comprehension Processes." In *Comprehension Instruction,* edited by Gerald G. Duffy, Laura R. Roehler, and Jana Mason, pp. 265–80. New York: Longman, 1984.

Sawyer, Wayne. "Literature and Literacy: A Review of Research." *Language Arts* 64 (January 1987): 33–39.

Sutherland, Zena, and Arbuthnot, May Hill. *Children and Books.* Glenview, Ill.: Scott, Foresman, 1986.

CHILDREN'S LITERATURE REFERENCES

Adler, David. *Martin Luther King, Jr.: Free At Last.* Illustrated by Robert Casilla, New York: Holiday, 1987.

Asbjornsen, Peter Christian, and Moe, Jorgen E. *The Man Who Kept House.* Retold by Kathleen and Michael Hague. San Diego: Harcourt Brace Jovanovich, 1981.

Atwater, Richard, and Atwater, Florence. *Mr. Popper's Penguins.* Boston: Little, Brown & Co., 1938.

Brooks, Polly. *Queen Eleanor: Independent Spirit of the Medieval World.* Philadelphia: J. B. Lippincott Co., 1983.

Cendrars, Blaise. *Shadow.* Illustrated by Marcia Brown. New York: Charles Scribner's Sons, 1982.

Cleary, Beverly. *Dear Mr. Henshaw.* Illustrated by Paul O. Zelinsky. New York: William Morrow & Co. 1983.

_____. *Ramona and Her Father.* New York: William Morrow & Co., 1977.

Coerr, Eleanor. *Sadako and the Thousand Paper Cranes.* New York: G.P. Putnam's Sons, 1977.

Cook, Scott. *The Gingerbread Boy.* New York: Alfred A. Knopf, 1987.

Cresswell, Helen. "The Bagthorpe Saga" *Ordinary Jack* (1977), *Absolute Zero* (1978), *Bagthorpes Unlimited* (1978), and *Bagthorpes Abroad* (1984) New York: Macmillan Co.

Dalgliesh, Alice. *The Columbus Story.* New York: Charles Scribner's Sons, 1955.

Daniels, Guy. *The Peasant's Pea Patch.* New York: Delacorte Press, 1971.

D'Aulaire, Ingri and D'Aulaire, Edgar. *Columbus.* New York: Doubleday & Co., 1957.

DePaola, Tomie. *The Legend of the Bluebonnet.* New York: G.P. Putnam's Sons, 1983.

Fido, Martin. *Oscar Wilde: An Illustrated Biography.* New York: Harper & Row, 1987

_____. *Rudyard Kipling: An Illustrated Biography.* New York: Harper & Row, 1987.

Fleischman, Sid. *The Whipping Boy.* New York: Greenwillow, 1986.

Fox, Mem. *Hattie and the Fox.* New York: Bradbury, 1987.

Freedman, Russell. *Lincoln: A Photobiography.* New York: Clarion, 1987.

Fritz, Jean. *Where Do You Think You're Going, Christopher Columbus?* Illustrated by Margot Tomes. New York: Putnam, 1980.

Galdone, Paul. *The Three Wishes.* New York: McGraw-Hill Book Co., 1967.

Gauch, Patricia Lee. *Christina Katerina and the Time She Quit the Family.* New York: G.P. Putnam's Sons, 1987.

Goodnough, David. *Christopher Columbus.* Mahwah, N.J.: Troll, 1979.

Goodsell, Jane. *Eleanor Roosevelt.* New York: Thomas Y. Crowell Co., 1970.

Grifalconi, Ann. *Darkness and the Butterfly.* Boston: Little, Brown & Co., 1987.

Grimble, Ion. *Robert Burns: An Illustrated Biography.* New York: Harper & Row, 1987.

Hamilton, Virginia. *W.E.B. DuBois: A Biography.* New York: Thomas Y. Crowell Co., 1972.

_____, *Writings of W.E.B. DuBois.* New York: Thomas Y. Crowell Co., 1975.

Hastings, Selina. *Sir Gawain and the Loathly Lady.* New York: Lothrop, Lee & Shepard, 1985.

Hodges, Margaret. *Saint George and the Dragon.* Boston: Little, Brown & Co., 1984.

Hunt, Irene. *Across Five Aprils.* Chicago: Follett, 1964.

Ivimey, John. *The Complete Story of the Three Blind Mice.* Illustrated by Paul Galdone. New York: Clarion, 1987.

Keller, Beverly. *No Beasts! No Children!* New York: Lothrop, Lee & Shepard, 1983.

Lindgren, Astrid. *Pippi Longstocking.* New York: Viking Press, 1950.

Lowry, Lois. *Anastasia on Her Own.* Boston: Houghton Mifflin Co., 1985.

MacLachlan, Patricia. *Sarah, Plain and Tall.* New York: Harper & Row, 1985.

Magnus, Erica. *Old Lars.* Minneapolis: Carolrhoda, 1984.

Marshall, James. *Red Riding Hood.* New York: Dial Press, 1987.

O'Neill, Catharine. *Mrs. Dunphy's Dog.* New York: Viking Press, 1987.

Noyes, Alfred. *The Highwayman.* Illustrated by Charles Keeping. New York: Oxford, 1981.

Perrault, Charles. *Cinderella.* Illustrated by Marcia Brown. New York: Charles Scribner's Sons, 1954.

Polacco, Patricia. *Meteor!* New York: Dodd, Mead & Co., 1987.

Roosevelt, Elliot. *Eleanor Roosevelt, With Love.* New York: E.P. Dutton, 1984.

Sandburg, Carl. *Rootabaga Stories.* San Diego: Harcourt Brace Jovanovich, 1922.

Seuss, Dr. *Horton Hears A Who* New York: Random House, 1954.

Stevenson, James. *Could Be Worse!* New York: Greenwillow, 1977.

Ventura, Piero. *Christopher Columbus.* New York: Random House, 1978.

Voigt, Cynthia. *Dicey's Song.* New York: Atheneum Pubs. 1982.

_____. *A Solitary Blue.* New York: Atheneum Pubs., 1983.

Whitney, Sharon. *Eleanor Roosevelt.* New York: Watts, 1982.

Wood, Audrey. *Heckedy Peg.* Illustrated by Don Wood. San Diego: Harcourt Brace Jovanovich, 1987.

_____. *King Bidgood's in the Bathtub.* Illustrated by Don Wood. San Diego: Harcourt Brace Jovanovich, 1985.

Yates, Elizabeth. *Amos Fortune, Free Man.* New York: E.P. Dutton, 1950.

_____. *My Diary, My World.* Philadelphia: Westminister, 1981.

_____. *My Widening World.* Philadelphia: Westminster, 1983.

Yolen, Jane. *Owl Moon.* Illustrated by John Schoenherr. New York: Philomel, 1987.

Yorinks, Arthur. *Hey, Al.* Illustrated by Richard Egielski. New York: Farrar, Straus, & Giroux, 1986.

Zemach, Margot. *It Could Always Be Worse.* New York: Farrar, Straus, & Giroux, 1977.

Chapter Twelve

After completing this chapter on language arts, the media, and computer-assisted instruction, you will be able to:

1. List several language arts objectives that utilize different media.
2. Describe the use of films and filmstrips in the language arts class.
3. Describe or demonstrate the procedures for making slides.
4. Describe or demonstrate the procedures for filmmaking with children.
5. Develop a unit on filmmaking without using a camera.
6. Describe how television can motivate reading.
7. Develop a lesson or series of lessons for teaching critical reading and listening with television, radio, and newspaper advertisements.
8. Develop a learning center around the newspaper.
9. Describe computer-assisted instruction applications in the language arts.
10. List evaluative criteria for computer-assisted instruction.
11. Identify concerns and issues related to computer-assisted instruction.
12. Identify potential applications for computer-assisted instruction.

Media and Computer-Assisted Instruction

*B*oth educators and the general public express mixed reactions to the issue of using the media in education. The positive viewpoint stresses the many exciting new films, television programs, and other media forms that can deepen a child's understanding of the ideas presented in the classroom. On the other hand, many responsible people claim that children watch so much television that their achievement in reading and other academic subjects declines. Rather than debating these issues, we simply look for ways the various media and technology—which seem to be with us, for better or worse—can be used productively in the classroom.

LANGUAGE ARTS OBJECTIVES AND THE MEDIA

In a discussion of nonprint media, Iris Tiedt (1976) noted at least four ways media can stimulate teaching and learning. First, listening centers promote children's ability to read by allowing them to follow a printed story on a recorded cassette. Second, a literature film adds visual interpretation to a book, stimulates interest in reading the book, stimulates oral discussion, and extends children's imaginations. Third, art films stimulate oral, dramatic, and art expression. Fourth, television can be used constructively in a number of ways: a class can discuss reactions to certain shows; or assignments can treat television viewing positively by having children study the medium itself, analyzing its appeal, calculating viewing time, analyzing program content, and critically evaluating what they see and hear. Critical evaluation of what children view is one of the major objectives of media study.

A study of advertising in various media, for example, can develop the language arts objectives of critical evaluation, vocabulary enrichment, spelling improvement, oral language development, development of appropriate levels of language usage, reading improvement, and creative writing development. In this chapter, we consider ways of developing creativity, visual literacy, and critical thinking, listening, and reading skills through showing films, producing films and slides, using television as an instructional medium, and critical studies of propaganda techniques used in commercials and advertising.

American schools are rapidly entering the computer era. There are, however, both potentials for using computer-assisted instruction and issues connected with computer applications in school settings. In this chapter, we consider ways that computers can stimulate intellectual development, provide resources for an enriched curriculum, foster problem solving, and develop writing and editing skills.

FILMS AND FILMSTRIPS

Films and filmstrips add a visual dimension to the study of a book, or stimulate interest in it. Films or filmstrips can be shown either before or after reading a story. If they read the book or story before viewing the film, children can compare their own visualizations of the characters and settings with those in the film. They can also compare the unique qualities of the two media, to see how one relies strongly on pictures to develop setting and characters, while the other does so with the written word. The teacher can show a film before a book is read to motivate interest or provide background information. One method to stimulate interest in a book as well as to provide interesting background is to study the author. For example, the *Meet the Newbery Author* (Miller-Brody Productions) series offers a sound and color filmstrip on Marguerite Henry, author of *King of the Wind*. The filmstrip shows pictures of horses the author has used in her many books, and of experiences Ms. Henry has had in her travels, which she describes in her books.

There are several sources of information about available filmed media. Lists of notable films are published in *The Booklist* (American Library Association) and *School Library Journal*.

Art films also stimulate oral and dramatic involvement and interpretation of a subject. *The Loon's Necklace* (Britannica) uses Indian masks to develop the story. The award-winning film, *Pulcinella* (Connecticut Films), portrays a dream world in which Pulcinella runs, flies, and dances through a circus, a ballet, and a town, with Rossini's music as an exciting background for Pulcinella's humorous adventures. Another imaginative film is *Tchou Tchou* (Britannica), in which the principal characters are blocks and the two children who manipulate them. The blocks eventually become a block dragon, and the only way the two children can escape the dragon is to turn it into a block train. These films stimulate creative writing, as well as oral discussion and creative drama. Art films can also provide the springboard for an interest in making films, filmstrips, and slides.

REINFORCEMENT ACTIVITY

View several films or filmstrips that interpret a literature selection, provide background for a selection, or are categorized as art films. Which would you use with children? State your reasons. Describe several ways you might use one of the films with a group of children.

MAKING AND USING SLIDES

Slides, either hand-drawn or produced with a camera, present many exciting possibilities in the language arts. Slides may illustrate a folktale or book talk, provide background for literature, or accompany a research project in almost any content area. Slides can also motivate discussion in all content areas.

Equipment

To use slides, you will need a slide projector and screen. Many elementary schools have this equipment, and it is usually available from a university media center. Many parents also own equipment and are often willing to lend it to their child's classroom for a short time. Slides can be narrated "live," while they are shown, or the narration can be taped, and played simultaneously with the slide presentation.

To take slides, you will need a 35mm camera, a copy lens, if you are taking slides of a book or picture, and high-speed film. For taking pictures of a book, you will need a tripod or copy stand so the camera doesn't move while you take the picture. This equipment is also frequently available from a university media center.

Photographic Slides

Photographic slides can include copies of pictures in books or magazines, and pictures of scenery, historic points of interest, people, and so forth. One of my student teachers made slides of pictures in magazines as part of an outstanding unit on values clarification with sixth-grade students. She chose colored pictures from sources such as old *Life* magazines relating to major issues of the sixties and seventies. She found excellent pictures pertaining to ecology, integration, the Vietnam War, equal rights, and so forth. She used a copy stand, took a separate slide of each picture, and organized the slides according to issues. (Both sides of an issue were shown, along with well-known proponents of each view.) Next, the students looked at the slides and discussed the various issues, concentrating on one issue at a time. They researched the issue, looking for all viewpoints, and discussed what they found. They also investigated music, and associated protest songs with the particular issues. Finally, the students arranged all the slides in what they considered appropriate order, and taped music to accompany their slide presentation. The class viewed their final project, and showed it to other classes. The children had constructed a highly visual reenactment of the important issues in our recent past.

The teacher might also prepare photographic slides of items or locations relating to particular books or periods in history. For example, slides of the Caddie Woodlawn Memorial Park, near Menomonie, Wisconsin, would be an interesting introduction to *Caddie Woodlawn* by Carol Ryrie Brink. The slides could show her original home, some of the farm buildings, a rail fence, the site of the schoolhouse, the Red Cedar River, the Indian Burial Grounds, and pictures of the woods and fields in which Caddie and her family had their adventures.

If photography is a hobby, take pictures when you visit historical sites and restored colonial villages. Greenfield Village and the Henry Ford Museum in Dearborn, Michigan; Williamsburg, Virginia; and Sturbridge Village, Massachusetts offer opportunities to photograph a smithy working at a forge, a sawmill, a person in colonial dress dipping candles, hand weaving, horse-drawn carriages, and so forth. Such features of our past are often mentioned in books, but some students may never see them.

Handmade Slides

Children enjoy making 2 × 2-inch slides that can be shown with the same slide projector used with professionally developed slides. Handmade slides are quite inexpensive, but because they do require the ability to draw in a small space, they may not work with very young children. The easiest and most successful way to help children draw their own slides is to give them a ditto sheet divided into 2 × 2-inch boxes. (Use a slide mount as a stencil to determine the size of the opening.) Now the children can draw their pictures on the ditto sheet, which prevents wasting transparencies. Have them use a lead pencil, which transfers onto a transparency via thermofax. After they draw their illustrations just the way they want them, place the ditto and a transparency into the thermofax machine. The lead-pencil lines will reproduce onto the transparency, and the children can color their drawings with permanent, colored transparency markers. Cut these transparency drawings into the required 2 × 2-inch size and secure them in the cardboard slide mounts. You may find that asking parents to donate their unsuccessful slides will produce so many slide mounts you won't have to buy new ones. Simply take the used slides apart, remove the old slide, and tape the new transparencies into the old mount.

These handmade slides might accompany a creative storytelling experience, illustrate a poem or story as a child reads it, or accompany an oral book report. The handrawn slides can also illustrate a report in science, social studies, or math. The following sources are helpful for making and using slides and other types of photography:

- Eastman Kodak. *Classroom Projects Using Photography,* Part I: Elementary; Part II; Secondary. Eastman Kodak, 1975.
- Laybourne, K., and Cianciolo, Patricia. *Doing the Media: A Portfolio of Activities, Ideas and Resources.* Chicago: American Library Association, 1978.

REINFORCEMENT
ACTIVITY

---

1. *Select an area of study that could be stimulated by the use of photographic slides. This study might use pictures from books or magazines, or slides of historical points of interest. Begin such a slide collection.*
2. *Following the directions for making handmade slides, prepare a series of slides to use for instruction; or, if you are now teaching, instruct a group of children in the procedures for making their own slides.*

FILMMAKING WITH CHILDREN

What do you think a group of children could accomplish with felt-tip pens, paints, watercolors, inks, and clear leader tape? With these simple tools, they can make film without a camera. Making films with children is based on the idea that students learn a great deal through audiovisual media, and should learn to communicate in these media as well as in writing and reading. Cox (1985) stated that filmmaking helps children conceptualize, express, and retrieve meaning.

There are strong similarities between films and literature: both express feelings and ideas; both promote development of conceptual and sequential organization; and both must consider audience.

Film is unique, however, because it is visual and concrete. Filmmakers create an array of images with color, movement, and space. Let's see how children can create images with the fairly simple techniques of drawing or scratching designs on film.

Hand-drawn Films

When lines and other designs of bright, vivid colors are drawn directly on a film surface, they produce a film that dances with color and animation. To make such a film, you will need magic markers, acrylic paints, and a 16mm reel of white or clear leader tape. Permanent markers with fine points, such as those available for overhead transparencies, produce thin lines and colorful designs. The cheapest way to obtain the 16mm film is to ask for discarded film from a motion-picture processing laboratory, and bleach it in a strong solution of household bleach. After rinsing and drying it, the film is ready for children to use to experiment with filmmaking.

Usually, the children should practice on paper before they try drawing on the film, where the artwork must be confined to a very small space. They should also understand something about how a movie works before they try to produce one. They should realize the film goes through the projector quickly, that it takes four frames for your eyes to even see an image, and that eighteen frames go by during every second of viewing. To help students understand these facts before they draw directly on the film, give them a ditto sheet similar to the one show in figure 12–1.

Students can practice drawing small pictures and designs on similar ditto sheets. The ditto sheet can even be cut into strips the same size as the film, and the students can measure to see how much film they need for one second of viewing. This activity helps them understand the need for repetition and gradual change to show a sequenced action.

When students are ready to experiment with real film, they should draw their designs on the inside surface of the film. A good working procedure is to cover the floor or work tables, unroll the film into the long, covered area, mark four- to eight-foot sections per child (paper clips work well for this), and have the students work in groups until each has had a turn to draw designs in their alloted spaces. Finally, show the finished film on a 16mm projector. The children might want to play music to add interest to their animated designs.

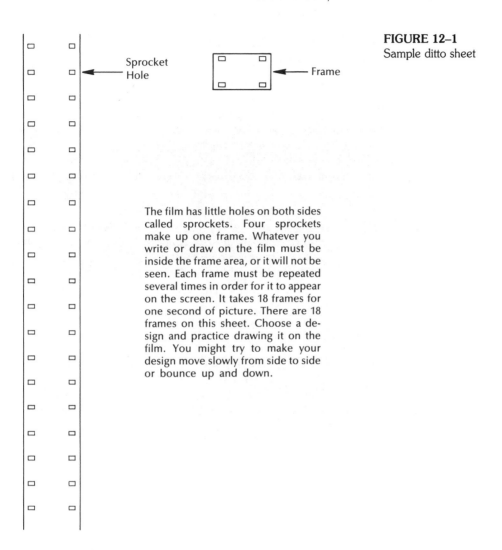

FIGURE 12–1
Sample ditto sheet

The film has little holes on both sides called sprockets. Four sprockets make up one frame. Whatever you write or draw on the film must be inside the frame area, or it will not be seen. Each frame must be repeated several times in order for it to appear on the screen. It takes 18 frames for one second of picture. There are 18 frames on this sheet. Choose a design and practice drawing it on the film. You might try to make your design move slowly from side to side or bounce up and down.

Scratch Film

Another technique for producing films is to scratch designs on underexposed or black film. Once again, you can often obtain discarded film from motion-picture processing laboratories. For this technique, however, do not bleach the film. To produce a design on the black film, use a knife, scissors, or other sharp object to scratch patterns into the emulsion side of the film. To find out which is the emulsion side, dampen you finger; the surface that sticks to the wet finger is the emulsion side. You can also project this film on a 16mm machine, and accompany it with music.

Units Developed around Filmmaking

Filmmaking can be modified for use with any age, and filmmaking units can also include listening, speaking, writing, and reading activities.

Introducing filmmaking   The film *Begone Dull Care* (McLaren, available through BAVI) provides a motivating introduction to filmmaking. McLaren, an artist, has made several films by drawing or scratching directly on the film. *Begone Dull Care* is an abstract composition synchronized to a jazz piano accompaniment.

    After viewing this film, students are usually excited about a new type of film, and want to learn more about this technique. They can discuss their reactions to the film and how they think it was produced. After they present their ideas, they can examine the film by unwinding a section for viewing. They will be able to see the colorful designs, how many frames are used just for the title, and how many repetitions are necessary to make the lines and designs come to life. After the children discuss and examine the film, show it again so they can see what happens when it is projected.

    The film will also motivate interest in filmmaking and generate questions about films, so other sources of information should be available so the students can read about this film technique and compare it to other techniques. Here are some sources for this activity:

- Andersen, Yvonne. *Make Your Own Animated Movies: Yellow Ball Workshop Film Techniques.* Boston: Little, Brown & Co., 1970.
- Cox, Carole. "Filmmaking as a Composing Process." *Language Arts.* 62 (January 1985): 60–69.
- Gaffney, M., and Laybourne, G. *What to Do When the Lights Go On: A Comprehensive Guide to 16mm Films and Related Activities for Children.* Phoenix: Oryx Press, 1981.

    Language experience charts can be kept for each activity the class develops. The introductory charts might include reactions to the motivational film, and other experience charts could show information about various other filming techniques. Creative writing stories might also accompany the film, or creative movement activities could pantomime the movement shown in the film.

REINFORCEMENT ACTIVITY

1. If you are now teaching children, develop a filmmaking unit with your class. Include an introduction to filmmaking, discussion activities, production of the film, and a culminating activity. Share the film your students produce with a group of your peers.
2. With a group of your peers, experiment with the various methods described in this section for producing a film without a camera. When you are finished, add musical accompaniment to your film.

Producing the film   Discussion is the first step in producing a film. This is a group production, consequently the students should discuss and plan what type of film they want to work on. Moving designs are easiest for young children, although older students may want to try other techniques, such as animated cartoons.

Once students decide what kind of film they want to make, they must go through the preparatory activities discussed earlier. Before drawing on the film, the class should review techniques and requirements. They might make a chart of these requirements. Finally, they will make and view the film.

Culminating activity   A film festival is a rewarding culminating activity for a film unit. The children might invite another class, or their parents, to the festival, and show slides, the film they have produced, creative writing that developed from the unit, and any other related information or creative activities that resulted from their study of film. The children can write their invitations and act as ticket takers, announcers, ushers, projectionists, light controllers, and welcomers.

USING TELEVISION TO MOTIVATE READING

The National Assessment of Educational Progress surveyed the viewing habits of nine-, thirteen-, and seventeen-year-olds and related these habits to reading achievement. Searls, Mead, and Ward (1985) concluded that the "highest reading levels occur among groups that combine 1 to 2 hours of reading with what appears to be the optimal amount of TV for their age group—3 to 4 hours for 9 year olds, 1 to 2 hours for 13 year olds, and under 1 hour for 17 year olds" (p. 162). It should be noted that as students become older, those who watch television the least are the best readers. In contrast, the survey found that students who watch television extensively, and who also report spending a great deal of spare time reading or doing homework are the poorest readers. The researchers concluded that this relationship is probably caused by students who try to read or study while watching television. Effective study habits and selective television viewing are apparently essential for reading achievement. Television is frequently criticized for diminishing reading as a leisure activity. By the time most children finish school, they have spent more time watching television than in the classroom! There are ways, however, for a creative teacher to use viewing interests to motivate reading.

The interest inventory in chapter 10 requires children to indicate their favorite television programs, and how frequently they watch television. Many television programs are based on books, or treat themes also found in literature. Hamilton (1976) allowed seventh-grade students to read books related to their favorite television programs. After a six-week period of instruction, he found the average number of television-related books preferred, per pupil, was 7.54, compared to 3.27 nonrelated books. He concluded that pupils who spend a great deal of time watching television are inclined to read television-related books. These books promote an interest in reading among students in lower socioeconomic groups and with lower IQs. No pupils felt that television-related books were difficult to read. Thus, a knowledge of television viewing habits, together with a supply of related literature, might be one way to get the more reluctant reader, who generally watches television, to read.

Another way to motivate reading is through selective television viewing.

Besides suggesting and providing television-related books for children's reading, teachers can also use television for developing discussion skills. Charles Silberman (1970) suggested that instructional television often fails because "television cannot do the entire teaching job; it can carry lectures and demonstrations, but it cannot direct a class discussion" (p. 65). Popular educational programs such as *Sesame Street* and *The Electric Company* have been criticized for their lack of interaction through discussion. A study completed in the early 1970s, "Rural Appalachia Program" (Stevenson, 1971), used a program format similar to Sesame Street's. This study, however, added an interaction component for an experimental group of three to five year olds through home visits by a paraprofessional and a weekly 1½-hour mobile-classroom session. In the study, a control group watched the television shows but did not take part in any interaction. Significant verbal improvement was shown only by the group of children who also received the opportunity for interaction. These results emphasize the need to provide discussion following instructional television if children are to gain maximum benefits from the programs and not be merely passive observers.

Another way to reduce the passive-observer role in television viewing and, at the same time, improve reading skills, is through use of scripts and videotapes of popular television programs. Bernard Soloman described a "Television Reading Program" that includes script scanning, viewing portions of a program while reading the script, developing an appropriate reading skill for each script, acting out parts of the script, and writing original scripts. Soloman stated, "Teacher and student interviews pointed out that the motivation to read television (scripts) and act out a favorite star's role . . . brought about instant movement towards reading instruction" (1976, p. 135).

Television can, apparently, provide both motivation to read and verbal improvement for children, but not without interaction for parents and teachers. In fact, children who are allowed to watch television indiscriminately will probably not develop the critical evaluation skills necessary to any instructional program that uses television.

In addition, there is a danger that television viewing may lower the literary quality of children's reading choices. Barbara Elleman (1982) compared the Chosen by Children booklist compiled during a national survey conducted by librarians and teachers with the 1982 juvenile best-seller list issued by Dalton Bookstores. Although the Chosen by Children list included books by highly respected authors, the Dalton list includes spin-offs of Disney, Sesame Street, Strawberry Shortcake, and the Smurfs. Elleman concluded that Dalton books are purchased without the guidance of teachers and librarians.

TEACHING CRITICAL READING AND LISTENING SKILLS WITH TELEVISION AND NEWSPAPERS

Critical reading or listening demands that a student evaluate carefully and exactly what she reads or listens to. To determine what and whom to believe requires considerable judgment of the source. Critical reading and listening go beyond factual comprehension; they require weighing the validity of facts, identifying the real problem, making judgments, interpreting implied ideas, interpreting character traits, distinguishing fact from opinion, drawing conclusions, and determining the adequacy of a source of information.

Television, radio, and newspaper stories, editorials, and commercials or advertisements provide abundant materials for teaching critical reading and listening skills. According to Arnold Cheyney (1984), students can learn many of these critical reading skills by carefully scrutinizing and questioning sources in newspapers or on television. Cheyney provided a list of questions for teachers to put on charts for student referral when they read or discuss news items:

Questions about News Items

The Writer's Competency and Integrity
  1. Is the writer an authority?
  2. How does this writer know?
  3. Does the writer make sense?

The Writer's Use of Sources and Evidence
  4. What evidence is presented to document the assertions?
  5. Is this fact or opinion?
  6. Is anything missing?
  7. What is the writer's purpose?
  8. Does the writer have a hidden motive?

The Reader's Ability to Form, Revise, and Test Opinions
  9. Are the premises valid?
 10. Why are these facts important to me?
 11. Do the conclusions necessarily follow?
 12. What have others said about this topic?
 13. Who stands to gain if I accept this without question?
 14. Does my lack of knowledge keep me from accepting this?
 15. Does my background make me intolerant of this point of view?
 16. Is the information as true today as when it was written?
 17. What more do I need to know before I come to my own conclusions?[1]

Fact versus Opinion in the Media

To answer these critical questions about news items, students need assignments that require them to read and listen to a variety of news sources. One of the first skills they will need is the ability to separate fact from opinion. The teacher can begin to help students understand the difference by reading them several paragraphs written around either fact or opinions; for example:

1. The grizzly bear lives in Alaska, Canada, and in the northern part of the United States. A big grizzly may weigh up to 1,000 pounds. The grizzly has a large hump above his shoulders; this makes him recognizable even at a distance. The grizzly likes to eat fish, fruit, and honey.

2. The grizzly bear is a cruel killer and should be outlawed in all parts of the United States. This killer not only attacks cattle, he also attacks and injures tourists in the national parks. The parks should be made safe for visitors and they will not be safe for campers and hikers until this animal is removed.

3. The grizzly bear has lived freely in the western part of the United States for hundreds of years. Man has so infringed on the grizzly's hunting grounds that he now is found only in secluded areas of Alaska and Canada and only infrequently in the high country in the United States. The U.S. government must protect the grizzly and keep the tourists and ranchers out of the grizzly's territory. This powerful, beautiful animal deserves our best protection.

After reading paragraphs similar to these, the teacher should ask students if they see any differences among the paragraphs. The discussion should produce the conclusion that the first paragraph is based on facts, and the other two are based on opinions; it should also note that the two opinion paragraphs are trying to sway the reader's opinion.

Next, the students should formulate their own definitions of fact and opinion writing; their definitions might resemble the following example:

Fact = Statements that can be checked and proven to be true are facts.
Opinion = Statements that tell what someone thinks or believes to be true, but cannot be proven, are called opinions.

---

[1]From Arnold B. Cheyney, *Teaching Reading Skills through the Newspaper.* 2d ed. (Newark, Del.: International Reading Association, 1984). Reprinted with permission of Arnold B. Cheyney and the International Reading Association.

Allow students to apply their knowledge of fact versus opinion and, subsequently, make judgments about new information, by having them search a newspaper for examples of factual writing and opinion writing. Make displays of the two kinds of writing and their characteristics. Students will find factual items in a newspaper in the form of obituaries, birth notices, court announcements, and most news stories, although the latter may also include opinion. They will find personal opinion writing in editorials, advertisements, letters to the editor, and advice columns. Students can reinforce their skills by writing factual and opinion paragraphs to share with the class; their classmates can then make judgments about the paragraphs.

Comparing Television and Newspaper Coverage of News

Students can compare how television, radio, newspapers, and newsmagazines handle news. I have had some stimulating discussions and learning experiences with children after they were assigned to report and analyze the various media presentations of an item we knew would be reported in the news. For example, if your city has coverage on the three commercial television networks and an educational station, several students might be assigned to listen to the evening news on each channel. Other students could search newspapers for news stories, editorials, and editorial cartoons about the same subject. Still other students can listen to the news reports on the area radio stations. If possible, the television and radio segments should be taped for classroom listening and critical analysis. The newspaper items would be brought to the classroom for sharing and discussion. (Controversial news items that may result in editorial comment are especially interesting for this investigation.)

The following day, students share their reports of the news presentations; discuss differences resulting from media requirements; discuss any bias noted in the various media or between networks; compare editorial comment in a paper with its news coverage; and discover words that might cause a biased reaction in the reader or viewer. The students should refer to Cheyney's list for evaluating the various news items. When they discuss media requirements, they should begin to see that television will be primarily visual, and present brief coverage of a story; radio relies entirely on words to develop an image; and newspapers may use both words and pictures, and also have more room to discuss the who, what, when, where, why, and how of a story.

Advertising as a Teaching Medium

We are bombarded with advertising, visually and auditorially, during most of our waking hours. The morning newspaper often appears to be predominately advertising; the same is true for many magazines. Radio and television sometimes seem to be one long series of commercials, interrupted occasionally by programming. Presumably, adults have acquired enough critical listening and reading skills to enable them to judge the value, worth, honesty, and acceptability of these ads that try to persuade them to invest in a product or vote a certain way in an election. There is a great deal of concern, however, for the preschool and school-age child, who may be unduly influenced by the propaganda techniques used in advertising. Elizabeth Gratz and J. E.

Gratz stated that "those who sell products are influencing the cognitive and the affective development of children as well as contributing to changes in society" (1984, p. 81).

Although good commercials can raise the standards of civilization, bad commercials can distort reality and lead young people to think that what they see in television ads portrays real life, leading them to desire and expect the wrong things for the wrong reasons. As teachers, we need to help our students look closely at both the good and bad in advertising, so they can critically evaluate what they see and hear.

Critical Evaluation of Propaganda Techniques in Advertising

To develop their powers of critical evaluation, students might begin by studying the various propaganda techniques frequently used to sway public opinion. A fourth-grade teacher used such a study to help children listen to and read advertisements more critically. This study was so successful that the students afterward did not believe anything in an ad unless they had investigated its claims and evaluated how authorities in a field would feel toward the advertisement.

The teacher introduced this particular study by asking a father who worked for an advertising agency to visit the class and talk about advertising. He discussed the purpose of advertising and the money spent on it, and showed the students some of the ways ads try to influence opinion about their products. Children saw examples of ads that used snob appeal, testimonials, transfer, glittering generalities, plain folks, bandwagon, and name-calling techniques to sway public opinion. They discussed the various methods, and asked numerous questions. This guest speaker gave a highly motivating introduction to advertising, and enabled the students to realize that companies spend millions of dollars trying to win prospective buyers.

Next, the students studied in-depth the various propaganda techniques presented by the guest speaker. They listed characteristics of each technique, found oral and written examples of them, and compiled table 12–1.

The students created several bulletin-board displays to illustrate the characteristics of the various propaganda techniques, with examples of newspaper, magazines, television, and radio ads and commercials. (They found that some ads relied on more than one approach.) Next, the teacher taped commercials from television and radio. The students listened to them carefully and identified the different techniques and key words related to each selling approach.

After listening, identifying, and analyzing, the students chose a particular ad or commercial to research. They listed questions to answer before making an evaluative decision; for example, if the ad was a testimonial, the students tried to answer the following questions:

Our Testimonial Question List

1. Who is the ad trying to persuade?
2. Who will benefit if I agree with the ad?
3. Who would be hurt if I agree with the ad?
4. Who is making the testimonial?

5. What is his or her credibility in relation to this product or issue?
6. Who could make an honest appraisal of this product or issue?
7. What facts support the ad?
8. What facts disagree with the ad?

They asked similar questions for the remaining categories of ads. The students presented their findings orally to the rest of the class, and put their lists of answered questions next to the example of the advertisement or commercial on the bulletin board.

This advertising study concluded with an activity in which students divided into groups, wrote their own commercials, and presented them to the rest of the class. The

TABLE 12–1

Characteristics and examples of propaganda techniques

Name	Characteristics	Example
1. Snob Appeal	1. Makes us want a product because superior or wealthy people have it. Association with a small, exclusive group.	1. The jet set dine at _____, you should too. Drive a _____ like _____.
2. Name-calling	2. Creates a feeling of dislike for a person or product by associating it with something disliked or undesirable. May appeal to hate and fear, and asks people to form a judgment without first examining evidence.	2. Congressman _____ associates with _____. The mayor has dishonest friends.
3. Glittering Generalities	3. Uses broad general statements whose exact meanings are not clear. appeals to our emotions of love, generosity, brotherhood. Words suggest ideals such as truth, freedom, and the American way.	3. Your wildest dreams for adventure will come true if you sail on the _____. Senator _____ will defend the American way.
4. Plain Folks	4. Identifies the person or product with the average man. Politicians may try to win our confidence by appearing to be like the good folks in your town.	4. For an honest used-car deal with your friendly neighborhood _____ dealer. Senator _____ is a hometown boy who will vote for your interests.

TABLE 12–1
continued

Name	Characteristics	Example
5. Transfer	5. Creates a good or bad impression by associating a product or a person with something we respect or do not respect. The feeling for the symbol is supposed to transfer to the new product or person.	5. _____ is as strong as Atlas.
6. Testimonial	6. A respected person, such as a movie star or athlete, recommends a product and suggests that we buy it because he uses it.	6. Lovely Lettie says, "I use Bright Hair shampoo because it leaves my hair silky and shiny."
7. Bandwagon	7. Follow the crowd and buy what everyone else is buying, or vote for the winner.	7. Everyone is rushing to _____ to buy the new _____. If you want to belong to the "In" generation, drink _____.

class decided to write commercials on opposing issues. They used the following four topics: (1) a commercial sponsored by an oil company that favored expansion and sale of oil and gasoline products; (2) an ecology commercial that opposed the expanded use of oil and favored greater conservation of natural resources; (3) a commercial to sell big cars; and (4) a commercial to sell small cars.

The students used any selling and propaganda techniques they wished, along with visual props, sound effects, and musical backgrounds. After each presentation, the remainder of the class identified the techniques in the commercial and discussed their truthfulness. (If the school has the equipment, the teacher can videotape these presentations and play them back as if they were commercials on a real television.)

FOR YOUR PLAN BOOK
An Instructional Unit Developed around Advertising

Susan Watson and Sarah Wells developed a unit to promote intelligent decision making among today's young consumers. Their three-week unit was designed for fourth-grade students. They incorporated art, consumer education, language arts, math, reading, and social studies activities around the theme of advertising. Their ideas should stimulate your own thinking about teaching through the medium of advertising.

Advertising

Idea One: American audiences are continually exposed to advertising.

1. *Students will view and listen to a one-hour program of their choice on television or radio. They will record the time devoted to commercials and compare the total commercial time to total program time by constructing a bar graph.*

2. *Each student will tabulate the number of advertising billboards, posters, and signs he sees from Monday through Friday.*

Idea Two: Advertising influences the American public's purchases.

1. *The teacher will facilitate discussion about the influences of advertising by comparing two products—one that is advertised a great deal and one that is not advertised. Questions include:*

   (a) *Which product would you buy?*

   (b) *Have you ever switched brands after seeing an advertisement that compared the old and new brands?*

   (c) *What advertisements can you name that might influence your decision about buying food, clothing, toys, and so forth?*

   (d) *How do you feel about an advertisement that influences your decision to vote, purchase a product, or contribute to a charity fund?*

Idea Three: Advertising is a multimillion dollar business in America.

1. *The students will choose a facet of the advertising industry that interests them. Possible choices include annual expenditure on advertising, central locations of the advertising industry, or an in-depth exploration of advertising spending for a particular type of product. The students will present their findings orally.*

Idea Four: Advertisements appear in many of the media; among them, television, radio, newspapers, magazines, and billboards.

1. *The teacher will utilize a guided discovery approach in a discussion of advertising media suitability. Topics will include advertising cost, coverage, applicable regulations, using jingles and catchy phrases, company logos, and visual appeal.*

2. *Each student will choose a product. The student will decide whether the product is suitable for television or radio advertising and will write a script to sell that product. The student will tape his advertisement. The presentation should exemplify characteristics of suitability to television or radio as discussed in the previous activity.*

3. The students will use their imaginations to write an original classified advertisement to buy or sell an item or to gain employment.

4. Each student will compute the cost of placing his ad in a newspaper.

Idea Five: Professional advertising opens many career options for today's young people.

1. The teacher will present an expository lesson plan on a career in advertising. The lesson plan will cover such topics as jobs, training required for each job, related-interest fields, and salaries.

2. The students will select a field in which they are interested or have special ability. Each student will research his chosen field and present a product exemplifying that field:

   (a) Drama—dramatic presentation for selling a product.

   (b) Speech—oral presentation of product, as on the radio with a microphone (disc jockey).

   (c) Art—drawing, sketching, painting.

   (d) Marketing—design package of product or catchy presentation of product.

   (e) Model—model clothes.

   (f) Journalism—written script and layout.

   (g) Photography—pictures of product.

   (h) Producer/director—produce skits or slide show.

Idea Six: Some advertisements are designed especially to appeal to specific consumer groups.

1. The teacher will videotape television commercials that attempt to appeal to specific groups. The students will discuss and categorize the intended groups.

2. Each student will choose one commercial appealing to a specific group and write an adaptation of this commercial to appeal to another group, identifying the old and new groups.

Idea Seven: Television advertisements attempt to appeal to different audiences during different times of the day and/or week.

1. The teacher will conduct a guided discovery lesson on the influence that time of day and/or week has on the types of products that will be advertised. The discussion will include such questions as these:

   (a) Who is most likely to be watching television during the day?

   (b) What types of products do you think would be advertised then?

   (c) How many of you watch the cartoons on Saturday mornings?

   (d) What kind of consumers watch television on Saturday morning?

   (e) What kinds of products are advertised then? Why?

Idea Eight: Advertisements are used by geographical areas to attract tourists.

1. The students will be told they have been chosen by the local Chamber of Commerce to write a description of their city so people will be encouraged to visit. They will be told to make their city come alive with vivid verbs and descriptive adjectives.

Idea Nine: Professional advertisers attempt to appeal to the five senses: touch, taste, sight, smell, and hearing.

1. The students will choose, cut out, and categorize magazine advertisements into the above categories, underlining key words.
2. Each student will create one original product appealing to at least two of the senses and will write an advertisement for the product.

Idea Ten: Much advertising is propaganda, the purpose of which is to influence the audience to buy something.

1. The students will bring three newspaper or magazine ads and analyze each by answering the following questions:
    (a) What product is being advertised?
    (b) Has the advertiser used any tricks to attract you to the product, to make you think it is better or more popular than it actually is? Explain.
    (c) Are there any definite claims cited by the manufacturers? If so, what are they?
    (d) Do you believe the ad? Why or why not?

2. The students will orally present a sales pitch to their group, who will role play the board of directors of an advertising firm. The salesman will persuade the board of directors to sponsor his product. Visual aids will be useful, and the student should be as persuasive as possible.

Idea Eleven: Testimonial advertising quotes a favorable statement made about the product by a famous person.

1. Students will identify and describe in writing at least one testimonial advertisement seen during afternoon or evening television viewing.
2. Students will list famous persons who have endorsed a product or participated in a testimonial advertisement, relying on past viewing experience.

Idea Twelve: Transfer advertising pictures a famous person or symbol with a product, although the person makes no statement about the product.

1. The students will search magazines and view television advertisements at home to find examples of the transfer propaganda technique.
2. The teacher will ask the students to think of a famous person in history whom they admire. The students will then write a transfer advertisement using that person with a product of their choice.

Culminating Activities:

1. The students will advertise an upcoming school or community event. The event might be a book fair, carnival, Christmas program, band or choir presentation, or any other type of assembly. The students will utilize as many forms of media as possible. They will be divided into groups according to the medium they prefer. Their advertisements should reflect the many forms and techniques studied during the unit.

2. *The students are to be divided into several small groups. The groups will take turns presenting several short skits with puppets. The group not presenting a skit at that time will present commercials between scenes and skits, so that each student will participate in both the commercials and puppet presentations. The commercials will be of the students' own creation, and should advertise imaginary products. The commercials, reviewed as a class effort, should reflect all the advertising techniques learned throughout the unit.*

Used with permission of Susan Watson and Sarah Wells, undergraduate elementary education majors, Texas A & M University.

REINFORCEMENT ACTIVITY

1. *Look for examples of the propaganda techniques described in this section. How could you use these examples to develop critical evaluation skills with children? What questions should be asked for each example?*
2. *Develop an instructional lesson or series of lessons for improving children's critical thinking, listening, and reading skills. Build your lessons around television, radio, and newspapers. State your objectives, motivation, instructional procedures, and reinforcement or enrichment activities you would use.*

Additional Reasons to Use the Newspaper in Language Arts

The newspaper can be used for many purposes in the language arts curriculum. The newspaper offers dynamic, up-to-date instructional materials. The skills connected with a newspaper are functional, because adults as well as young people need to be able to read a newspaper to gain specific information about a wide variety of subjects—news, sports, weather, editorials, business, television, crossword puzzles, fashions, society, books, entertainment, travel, art, cooking, and so forth. The newspaper has something for almost anyone.

The newspaper can also be used with all grade levels. Young children can use it to find letters of the alphabet, words and pictures that begin with certain sounds, or to make up their own stories about the comics, and tell stories about or answer oral questions about pictures. Primary-grade children can use the newspaper to motivate creative writing and develop oral language skills through discussion of various articles. They can write language experience stories around interesting news stories and pictures, and locate vocabulary words. The newspaper brings current problems into social studies and science classes. Older students with sufficient reading skills can use the newspaper to improve reading comprehension; to improve reading speed by practicing skimming for quick information; to improve rapid reading by looking for general ideas; and for careful and evaluative reading by analyzing ideas.

Many teachers use a study of the newspaper as an introduction to a classwritten newspaper, or to introduce editing and proofreading of a child's own writing. The writing activities that can be motivated by a newspaper are almost limitless. Newspapers are also useful for developing learning center activities. (See a discussion of learning centers in chapter 15.)

FOR YOUR PLAN BOOK

Learning Center Activities Based on the Newspaper

Meet the Press

Purpose: To increase the student's reading and writing ability through use of the newspaper.

Grade level: Intermediate upper elementary. Appropriate for individual, small-group and large-group activity. Suggested limit: 30

General materials: newspapers, paper, pencil, manila folders, typewriter (optional), picture file. More specific materials are listed for each activity.

Plan to progress: The students will receive a mimeographed handout with their assignments, determining level placement (e.g., Reporter Stacy). Space will be allotted for "Extra, Extra" activities to be assigned by the teacher to meet students' individual needs. This handout indicates the activities the students will complete as they progress through the learning center.

General instructions should be posted and explained orally to the students. They read as follows:

> *Dear Reporter,*
> *I have given you some assignments for our next newspaper. You may do the assignments in any order you wish. When you complete each assignment, check your answers. Then place your finished work in the "Ready For Print" box. When all the assignments are completed, sign up for an activity in the school newspaper. (The following categories are suggested: Lifestyle, Dennis the Menace, Editorials, Sports, News, Dear Abby, and Wedding.) In order to write your activity, interview a person in your field of interest. Place the article in the "Ready For Print" box. When all of your work is completed, you may begin work on the "Extra, Extra" box.*
>
> *Good Luck,*
> *The Editor*

Instructions: Besides the general instructions, each activity has specific instructions written on its manila envelope.

Evaluation: The activities are checked in many different ways. Some are self-checked; others need to be checked by the teacher. An answer box or envelope should be provided so the students can easily locate the answer keys they need. They will be kept in a class file. The students will indicate by a checkmark the activities they have completed. The teacher will observe the students' work and initial each student's handout upon approval.

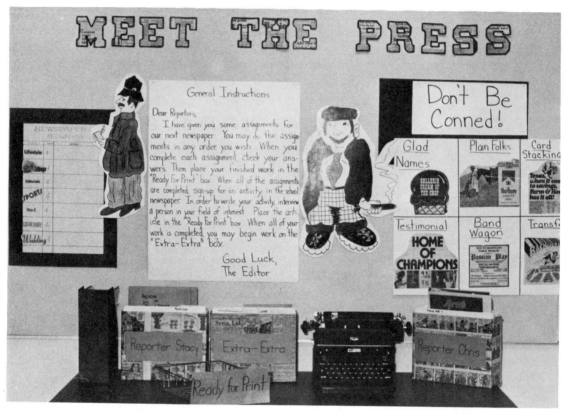

This "Meet the Press" learning center includes general instructions, activities designed for three ability levels, and evaluation checklists.

Activities:

Reporter Chris (low-level)

DEAR ABBY

Objectives: The students will match the advice column with the correct solution.

Materials: "Dear Abby" advice columns are glued to a manila envelope, with the problem separated from the solution. These are to be lettered and numbered.

Activity: The students will match the correct solution with the problem by writing their answers on a sheet of paper.

Evaluation: Answer key.

CAPTURE THE CAPTION

Objective: The students will match the pictures with the correct caption.

Materials: Manila folders, each containing several pictures and captions. The pictures will be lettered and the captions numbered.

Activities: *The students will choose one of the manila folders. They will match the picture with the caption and write their answers on paper.*
Evaluation: *Answer key.*

HEADLINE SCRAMBLE
Objective: *The students will match each article with the correct headline.*
Materials: *Manila folders, each containing several articles with matching headlines. The articles will be placed separately from the headlines in the manila folder. The articles will be numbered and the captions lettered.*
Activity: *The students will choose one of the folders. They will place the words in a line on their desk. They will then classify each picture under the appropriate word.*
Evaluation: *Student self-check.*

FEELINGS, FEELINGS
Objective: *In one word, the student will describe how each person feels.*
Materials: *A manila folder with several pieces of construction paper. Each piece of construction paper has pictures of people glued to it. Each picture is numbered. Each piece of construction paper is lettered.*
Activity: *The students will choose three of the sheets of construction paper with the pictures. In one word, they will describe how each person feels. They must use a different word for each person.*
Evaluation: *Pupil self-check.*

GETTING THE MEANING
Objective: *The students will match the article with the main idea.*
Materials: *A manila folder with numerous low-level, high interest articles. Each article will contain strips of paper with three possible main ideas.*
Activity: *The students will choose four articles from the manila folder. They will match the strip of paper that contains the main idea with the appropriate article.*
Evaluation: *Answer key.*

SEQUENCING CARTOONS
Objective: *The students will sequence cartoon frames.*
Materials: *Five multiframed cartoons, each backed with different-colored construction paper. These will be cut into separate frames, and numbered in the correct order on the back.*
Activity: *The students will group the cartoon frames according to color. They will then sequence them.*
Evaluation: *The students will check their answers by turning them over. The correct order is indicated by the numbers on the back.*

Reporter Stacy

HEADLINES
Objectives: *The students will write an article to go with a specific headline.*
Materials: *Headlines placed in a manila folder.*
Activity: *The students will choose one headline from the manila folder. They will write an article in journalistic form.*
Evaluation: *Rating scale (provided at the end of this unit).*

MAIN IDEAS

Objective: *The students will write out the main idea of an article they have read.*
Materials: *Articles without the headlines placed in a manila folder. They will read them and write out the main idea of each.*
Evaluation: *Rating scale.*

ARTICLE ARTISTRY

Objective: *The students will write an article and a caption from a photograph.*
Materials: *A picture file.*
Activity: *The students will choose a picture from the picture file. They will then write an article for the picture, and a caption for the article.*
Evaluation: *Rating scale.*

CATCHY CARTOONS

Objectives: *The students will write one caption for each of several pictures.*
Materials: *A picture file.*
Activity: *The students will choose several pictures from the picture file. They will then write one caption for each of the pictures.*
Evaluation: *Rating scale.*

DON'T BE CONNED

Objective: *The students will identify propaganda techniques contained in an advertisement.*
Materials: *Several advertisements illustrating propaganda techniques.*
Activity: *The students will read the advertisements and identify in writing the propaganda techniques used.*
Evaluation: *Answer key.*

CARTOON TALK

Objective: *The students will fill in cartoon balloons with appropriate words.*
Materials: *Liquid paper; several cartoons, each mounted on construction paper with their ballooned message opaqued out.*
Activity: *The students will fill in the cartoon balloons with appropriate dialogue.*
Evaluation: *Rating scale.*

ARTICLE SEQUENCING

Objective: *The students will sequence an article.*
Materials: *Several articles cut apart into sections and mounted on colored construction paper.*
Activity: *The students will group article pieces according to color, then sequence them.*
Evaluation: *Self-checked; answer on back.*

ADVICE EDITOR

Objective: *The students will write their advice to a given problem.*
Materials: *A variety of advice columns mounted on tagboard.*
Activity: *The students will choose one problem and write an original solution.*
Evaluation: *Rating scale.*

Extra Extra

WRITING CARTOONS

Objective: *After drawing a cartoon, the students will write a script to accompany it.*
Materials: *Blank cartoon frames, colored pencils.*
Activity: *The students will create a cartoon sequence using their favorite cartoon character. They will write a script to go along with their drawings.*
Evaluation: *Rating scale.*

JUNIOR JUMBLE

Objective: *The students will unscramble words.*
Materials: *Several Junior Jumbles mounted on poster board.*
Activity: *The students will choose three Junior Jumbles, unscramble the words, and solve the puzzle.*
Evaluation: *Answer key.*

TV QUIZ

Objective: *Using the* **TV Guide***, the students will answer several questions.*
Materials: **TV Guide***, typed questions mounted on tagboard.*
Activities: *The students will read the questions on the tagboard and skim the magazine to answer them.*
Evaluation: *Answer key.*

BRAIN TWISTERS

Objective: *The students will solve the problem in the brain twister.*
Materials: *Several brain twisters mounted on construction paper.*
Activity: *The students will read the brain twister, then solve the problem.*
Evaluation: *Self-checked; answers on the back.*

ADVERTISING

Objective: *The students will write an original advertisement using some propaganda technique.*
Materials: *Various pictures of products mounted on tagboard.*
Activity: *The students will choose at least one of the pictures and write an advertisement for it using one of the propaganda techniques.*
Evaluation: *Rating scale.*

FACT OR OPINION GAME

Objective: *The students will classify headlines into fact or opinion.*
Materials: *Game board made from construction paper; on it, a definition of fact and a definition of opinion. Under each of these definitions, a pocket is stapled to the construction paper. Twenty headlines, some of which are facts, the remainder of which are opinions.*
Activity: *The students will choose a headline and read it. They will then decide whether it is a fact or an opinion and place the headline into the correct pocket on the game board.*
Evaluation: *Self-checked; the answers are on the back.*

FUN FACTS FOR BROWSING

Objective: *The students will read fun facts for their enjoyment.*

Materials: *An envelope that contains fun facts items, such as Ripley's Believe It or Not,* Wrigley's *"Fun Fact" and* **Guinness Book of World Records** *excerpts.*

Activity: *This activity is merely for the enjoyment of reading; students can read these during free time.*

Evaluation: *None*

Rating Scale

Student's name: _____

		Low	. . .	. . .	. . .	High	
1.	Originality	1	2	3	4	5	NA*
2.	Creativity	1	2	3	4	5	NA
3.	Expressive	1	2	3	4	5	NA
4.	Form or Style-(Newspaper)	1	2	3	4	5	NA

    a. who _____

    b. what _____

    c. when _____

    d. where _____

    e. why _____

5.	Sentence Structure	1	2	3	4	5	NA
6.	Paragraph Structure	1	2	3	4	5	NA
7.	Consistency in theme	1	2	3	4	5	NA
8.	Grammatical structure	1	2	3	4	5	NA

    a. Spelling

    b. Punctuation

    c. Capitalization

Notes _____

_____

*NA means nonapplicable for this assignment.

_____

Used with permission of Nancy Mangano, Debra Penkert, and Jacqueline Thurston, graduate students at Texas A & M University.

COMPUTER TECHNOLOGY

World leaders stress the importance of computer technology in our rapidly changing world. Thatcher's introductory comments in the British publication *Micros in School Scheme* emphasize the importance of training students in the abilities and skills needed to design systems, write software, and develop new businesses and products (Department of Industry, 1981). She stated, "We must start in our schools. The microcomputer is the basic tool of information technology. The sooner children become familiar with its enormous potential the better. At present only some schools have microcomputers. That is why the Department of Industry has introduced its 'Micros in Schools' scheme. This scheme is closely linked with the Education Department's first in a series of initiatives which the government is taking to ensure that Britain stays with the leaders in the rapidly growing information technology market" (p. 2).

Computers are becoming common items in American schools. There have been numerous changes in computer technology and advances in computer applications in education since the second edition of this text in 1985. For example, Langer (1986) emphasized that the number of computer-based research studies submitted to the journal *Research in the Teaching of English* rose 300 percent between 1985 and 1986. Likewise, the third edition of Blanchard, Mason, and Daniel's *Computer Applications in Reading* (1987) contains almost 300 entries dated either 1985 or 1986. The types of computer applications are also increasing rapidly. Although there are still numerous applications in drill and practice activities and in word processing, there are also advances in administering assessment instruments, managing educational record keeping, providing tutorial programs, simulating real-life experiences such as role playing, accessing information through computer telecommunications, creating instructional programs and interactive stories, and teaching problem solving.

Bullough and Beatty (1987) stated that the computer has several unique attributes that make it ideal in educational settings. Computers encourage students to try new things without the fear of making mistakes, provide self-paced instruction, bring more interactive learning into classrooms, encourage cooperative learning, and monitor students' progress. The potential of the computer for stimulating intellectual development, developing mastery of a piece of modern technology, providing resources for an enriched curriculum, fostering problem solving, and developing writing and editing skills is not disputed. Many educators and computer specialists, however, criticize computer applications that are merely high-tech flash cards for content areas. Others warn that the transfer from computer training to other areas is not necessarily automatic. A final concern focuses on allocation of educational funding. Papert from MIT estimated that it will require another three million computers at an estimated $4.5 billion to reach the "threshold of seriousness" in computer applications (McGrath, 1983). In addition, he maintained that each dollar spent for computers requires two additional dollars for teaching teachers how to use them.

If teachers are to understand, evaluate, and use computers, colleges of education must provide training. This need is emphasized by Fiske, the *New York Times* education editor, who stated that American schools "are in the grip of computer mania. Ordinarily conservative and slow to change, schools are embracing new technology and new educational methods more rapidly than they can learn to use them. During the past few years, the influx of microcomputers in education has grown from a trickle to a torrent, engulfing teachers and administrators in a flood of confused expectations and unfulfilled promises" (1983, p. 86).

The semantic web in figure 12–2 illustrates current applications, evaluations, concerns, research examples, and potential applications in computer-assisted instruction. We consider each of these areas in the discussion of computer applications in language arts.

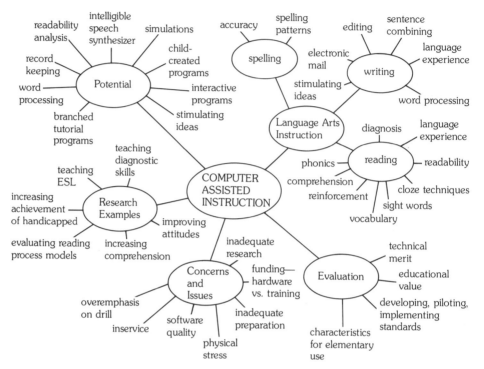

FIGURE 12–2
Semantic web of computer-based instruction

Computers and Language Arts Instruction

Computer-assisted programs may be designed to help students organize their ideas before writing, compose their stories, and revise and edit their selections. In addition, programs provide tutorial assistance, drill and practice, and instructional management. Writing, spelling, and reading programs are common in language arts.

Writing   Computers with various software programs for writing and word processing may encourage and motivate children to write because they can focus on content rather than penmanship, easily edit their written ideas, interact with peers during writing and editing, see what the finished product looks like before it is printed, and produce attractive final drafts. There are specific software programs and recommendations that may be used with each phase of the writing process.

At the prewriting stage, Gomez (1987) recommended three prewriting modules published by Milliken that allow students to brainstorm and organize their ideas on disks, to print these results, and to proceed to either write with pen and paper or to merge ideas with their files on the word processor. Other programs that have prewriting capabilities include *The Writers Helper* published by Conduit, *Bank Street Writer Activity Files* published by Scholastic, and *Quill* published by D. C. Heath.

At the composing level, various word processor programs encourage students to compose first drafts and then revise without recopying. Some programs have unique motivating capabilities: *That's My Story,* published by Learning Well, includes ideas for story starters; *Kidwriter,* published by Spinnaker, includes pictures that children may choose to write about; and *Story Tree,* published by Scholastic, has a choose-your-own adventure format.

The advantages created by student interactions during the composing process are emphasized by several studies. For example, Dickinson (1986) used computer teams and encouraged children to talk with each other about their writing plans, to discuss their actual writing, and to consider their teammates' reactions and feedback when they made their revisions. The capabilities of the *Quill* program mentioned earlier extend to the composing level. Bruce, Michaels, and Watson-Gegeo (1985) described students' interactions while using this program. They stated, "Students' writing is public and available to be read as it is entered into the computer (looking over the writer's shoulder as it appears on the screen). Later, using an information storage and retrieval system, students can retrieve their own or someone else's writing stored in the computer. Writing comes off the printer typed and formatted like published print (newspapers, magazine ads). It can then be seen on the wall (where its neatly typed format makes it easier to read and hence more accessible to classmates and outside visitors)" (p. 148). Student interactions during this observation occurred as they wrote on the computer, as they read and discussed other students' writing and revisions, and as they used the computers to send messages to each other.

At the revising and editing levels, word processing programs allow other students to respond to the text, encourage spelling checks, and make it easier for students to revise and edit their writing. Texts by Selfe (1986) and Wresch (1984) provided detailed directions for using computers in all stages of the writing process.

Educators also emphasize the use of computer applications to teach sentence combining and to increase and improve the writing skills of children with special needs. Bradley (1982) used the computer to help students improve their sentence structures by teaching them sentence combining through a computer program. Rubenstein and Rollins (1978) developed a letter-writing project with deaf children. During the year, forty previously reluctant writers used the computers to generate more than 1,500 letters to each other. The authors concluded that electronic mail is a powerful motivator for children's writing. Likewise, Kleiman and Humphrey (1982) found that learning disabled children ranging in age from seven to sixteen enthusiastically increased their writing when allowed and encouraged to use word processors.

Many educators are very enthusiastic about using various computer programs for writing. There are, however, several disadvantages and issues related to using computers as part of the writing process. If children are to make full use of the capabilities of computers during the writing process, they must have numerous opportunities to use computers. This means that they must have frequent access to computers. In addition, they must learn how to type on the computer keyboard, how to use a word processor, how to use various software packages, and how to use a printer. The greatest need may be for students and teachers to understand that word processing programs will not magically solve writing problems and provide instruction in all phases of writing.

This final point is emphasized by Dauite (1986) who studied the quality and the writing capabilities of seventh- through ninth-grade students. She concluded, "This study suggests that word processing features for revising are most useful for writing development when combined closely with cognitive and instructional aids that draw students into reading their texts and developing revising strategies such as self-questioning. Experienced writers who have benefited from the basic processing features already know what good writing looks like and how to evaluate their texts critically. If our students are to benefit from automatic recopying and reformatting, we have to continue to help them learn what good writing is and how to improve their own draft texts" (p. 158).

Spelling   Computer programs may be especially useful for helping children improve their spelling accuracy and increase their knowledge of spelling patterns. Gustafson (1982) compared spelling achievement scores and spelling retention ability of students who learned to spell with a computer spelling program with the scores of children who learned to spell with a teacher-directed approach. Both achievement and retention were higher in the computer-based group.

Mason (1983) maintained that computer programs are excellent accuracy trainers. When students print in words incorrectly, they must find their errors before continuing. He stated, "Teachers using this approach report dramatic improvement in spelling in addition to greater accuracy in reading words. They also report incidents of student jubilation when pesky bugs are found and corrected and programs finally run as they should. This joy in accomplishment is so profound that it often shocks teachers who had thought nothing in school could excite these children" (p. 506).

The ease of changing words in a word processor prior to printing can motivate the proofreading for spelling and mechanical errors as well as the improvement of other aspects of composition. This potential is not yet reached. Becker, director of the Johns Hopkins classroom-technology study, stated that one of the best uses for computers with older children is the teaching of writing (McGrath, 1983). He estimated, however, that less than 7 percent of computer time in high school programs involves word processing activities.

Balajthy (1986) provided the following guidelines to use when selecting spelling software: It should allow teachers to add new words, contain colorful and interesting programs, use voice synthesis to present words, provide feedback, review misspelled words, and allow for recordkeeping.

Reading and literature   Computer applications in reading receive mixed evaluations. Kinzer, Sherwood, and Bransford (1986) pointed out "Reading appears to be a difficult instructional area to computerize. Much reading software is little more than a high-technology workbook, although software which merely imitates a workbook format is generally a poor use of resources. Much of the software available and being used in schools is of the drill-and-practice variety. Although there is much that reading researchers still do not know about reading, most would agree that no one learns to read from an instructional diet of drill and practice alone" (p. 215). In contrast to drill and practice activities, these computer experts identify promising reading approaches including simulation activities that encourage planning and decision making; language experience approaches that allow teachers to type in the story, quickly print and distribute the story, develop lessons from the story, and keep records of students' vocabularies for individualized word banks; tutorial programs; and data management programs that allow teachers to record test scores and other vital information.

Likewise, other computer experts identify both poor and good types of computer assisted reading programs. Attitudes toward computer-assisted reading programs are also changing. In 1983, Mason, Blanchard, and Daniel were extremely critical of reading software because the programs over-emphasized drill and practice activities. In their latest text, Blanchard, Mason, and Daniel (1987) were more optimistic. They stated, "the microcomputer reading software of the late eighties promises to fulfill many of the unmet promises made about earlier computer-based instruction programs. Reading software is improving and will continue to improve, but we have a long way to go" (p. 23).

Many of the same advantages cited for using computers during the writing process are also cited for computer-assisted instruction in reading. Bradley (1982), for example, used a microcomputer to develop language experience lessons with first graders. She types the children's stories into the computer rather than printing them on a chart. She believes this approach has the following advantages: (1) it is highly motivating, (2) it is easily transcribed, (3) the story is easily changed, and (4) the printed copy is ready for immediate distribution. Disadvantages of this approach relate to limitations in various programs, such as a computer's inability to display lower-case letters on the screen or printed lines that are easily read. Likewise, both Grabe and Grabe (1985) and Smith (1985) stated that computers aid teachers and students during language experience activities.

Numerous software programs are available to reinforce sight words, expand vocabularies, provide drill in phonics, analyze words according to structural analysis, and test comprehension. Software reading programs are not the only choices for language arts teachers. Some computer games are recommended by reading clinicians who teach children with reading and/or motivational problems. If games require reading of words or printed descriptions, comprehending written text, and responding to questions or directions, reading may be enhanced.

Children with special educational needs seem to benefit from computer-assisted instruction in reading just as they do from using computers to aid writing. Harper and Ewing (1986) found that using the computer to have children read passages and answer questions was more effective than having them complete similar tasks in workbook format.

Some educators are using children's interests in technology to encourage reading. For example, Sharp (1985) used graphics programs such as Logo to have children create their own designs for a memory quilt as depicted in Ann Jonas's literature selection, *The Quilt* (1984). She used word processing programs and various books that contain specific language patterns such as Margaret Wise Brown's *The Important Book* (1949). For this activity she typed the recurring pattern into the word processor leaving blanks for the main ideas. In this way children created their own stories. The format for this activity with *The Important Book* looks like this:

The important thing about _____ is that it is _____ . It is _____ and _____ , and _____ . But the most important thing about _____ is that it is _____ .

Other ideas include creating original concrete poems after reading concrete poems in poetry anthologies and creating flow charts illustrating major decisions or actions in a story. The software program *Story Tree* published by Scholastic allows students to develop cognitive maps or flow charts related to literature selections.

Evaluation of Computer-Assisted Instruction

Evaluation of computer-assisted instruction focuses on several areas in program development and implementation. Educational groups frequently focus on the evaluation of generic computer programs in classroom settings or the assessment of training programs and classroom management. Evaluation considerations may also focus on desirable characteristics of word processors purchased for instructional programs in elementary grades or guidelines for software developed for educational applications.

The Association for Teacher Training in Europe (1982) evaluated computer-assisted instruction in elementary schools and reported the following positive and negative aspects of computer-assisted instruction:

1. Computer-assisted instruction frees the teachers from extensive participation in skills/drills exercises.
2. Computers, whether used in individualized or group-oriented projects, are patient and highly stimulating vehicles for learning and practicing elementary number and literary skills.
3. Computers have strong motivational attributes that are not temporary phenomena; these motivational attributes are especially pertinent to slow learners.
4. Computers aid teachers in controlling learning experiences and they have potential as learning resources.
5. On the negative side, student work cannot be undertaken without considerable teacher involvement.

Another evaluative study was carried out by the Dundee College of Education and the Committee on Primary Education (Lorick, 1983). This project studied the impact of computers in Scottish primary education. The following research findings and recommendations resulted from the committee's examination of teacher training, appropriate curriculum areas, classroom management, and pupil reaction:

1. The initial training of teachers should be intensive in the area of machine handling; the provision of a simple manual plus one hour of instruction is not enough.
2. Enthusiastic teachers can be taught how to manipulate data within a program, but they cannot become skilled programmers without a high degree of interest and the devotion of considerable time to the task.

3. Teachers must be given considerable advice on how to use computers.
4. Programs should not be issued without documentation. This documentation should include a description of the program, with suggestions for usage and indications of prerequisite learning.
5. Advanced planning is essential for optimal usage. A regularly scheduled slot in the day for use of the machine is not sufficient to achieve maximum gain for children.
6. Group teaching may be advantageous, provided proper management techniques are implemented.
7. The classroom layout may dictate the placement of the computer; conversely, group seating arrangements will facilitate the creation of a computer section in the classroom.
8. Group discussion is a prominent feature of pupil reactions in most programs; this implies that a self-contained computer area must be located suitably within the classroom.
9. Classroom management must include appropriate provisions for ancillary material such as dictionaries, workbooks, and folders.
10. Staff attitudes toward computers range from cautious interest to complete distrust. Staff members frequently display a false notion about the complexity of handing a microcomputer.

These findings and recommendations suggest both potentials for and concerns related to teacher training and installation of computer-assisted programs. Guidelines developed by Bradley (1982) and Jelden (1981) provided recommendations for computers and software. Bradley recommended that the following characteristics be considered when selecting computers and word processors in elementary schools:

1. A screen editor is vital so that children can see major portions of their work.
2. The screen should display upper and lowercase letters.
3. The text should be displayed in a double-spaced format.
4. The capability to display large, primary letters or normal-sized ones by giving a command would be an asset.
5. Deletion and retrieval functions are accepted by children.
6. Mnemonic commands that relate to children's vocabularies are easy to learn and to remember.
7. A dictionary that checks for spelling and other structural errors is a useful component.
8. The smallest number of commands should be used.
9. A quiet printer is important to avoid disturbing others in the classroom.

Finally, guidelines such as those developed by Jelden are useful when selecting software with educational value (1981, pp. 212–13). According to Jelden, educational software should include (1) course and performance objectives, (2) pace control by students, (3) variations for individual differences, (4) extensive feedback, (5) branching and looping potential, (6) built-in student evaluation, (7) requirements for correct answers before continuing to more complex instructions, (8) recycling potential, (9) 85 percent probable success, (10) instructional assistance, (11) practical, hands-on experiences, (12) verbal integration with practical experiences, (13) liberal use of graphics, and (14) no more than three frames before child responds.

Bullough and Beatty (1987) developed an evaluation form for software that includes many of these concerns. Figure 12–3 includes questions pertaining to program characteristics, content, running the program, strengths and weaknesses, and recommendations.

REINFORCEMENT ACTIVITY

Select several language arts programs that teach language arts subjects through computer-assisted instruction. Evaluate the software using the Software Evaluation Form shown in figure 12–3.

As computer programs increase, the demand for software and computers that have both technical merit and educational value will also increase. The next decade will certainly emphasize the development, the piloting, and the implementation of standards.

Concerns and Issues

The rapidly increasing use of computers in educational settings and the corresponding expansion of available software create numerous concerns and issues. Moursund stated, "No one has stopped to resolve the basic issue, what and how much students should learn about computers" (in McGrath, 1983, p. 64). Watts reaffirmed this concern. He stated, "Computer literacy has become a political football. No one can tell you exactly what it is, but everyone feels it's good for you" (1983, p. 68).

These two quotes suggest major concerns about inadequate research and field testing. Considerable research is required before educators identify effective instructional strategies and software programs. Research should also focus on developing, piloting, and implementing standards for evaluating technical merit and educational value of hardware and software.

Preparation at the university and classroom level is a major concern. Inadequate university preparation means that beginning teachers graduate without computer capabilities. At the classroom level, intensive inservice is essential.

Adequate preparation, however, creates another issue. There may be inadequate funds to buy hardware, purchase software, and provide teacher training. Without adequate teacher training in computer applications, computer evaluation, and computer capability, programs are ineffective.

As discussed earlier, software quality is an issue. Even though gains have been made, most of the software programs overemphasize drill. Some language arts software programs are programmed workbook lessons placed in a new format. There is, however, considerable potential for developing higher level skills, stimulating ideas, and creating branched tutorial programs. A review of research studies suggests classroom potentials and educational interests in computer-assisted instruction.

SOFTWARE EVALUATION FORM

Program title _____ Version _____

Producer _____ Cost _____ Copyright date _____

Required hardware _____ Required software _____
   (Include mirocomputer brand, memory, other)

Storage medium: ____ 3″ disk ____ 5″ disk ____ 8″ disk ____ cassette tape ____ cartridge

Name of reviewer _____ Date _____

Address or school _____

Backup policy _____

I. PROGRAM CHARACTERISTICS

1. Subject matter area _____ Specific topic _____

2. Grade level(s)   Pre   K   1   2   3   4   5   6   7   8   9   10   11   12   Adult   College

3. Objectives: Clearly stated? ____ Yes ____ No   If stated, list them. If not stated, describe what you perceive them to be: _____
_____

4. Is documentation provided? ____ Yes ____ No   If provided, describe briefly: _____
_____

5. What prerequisite skills should the student have? _____

6. Appropriate number of users: ____ individual ____ pairs ____ small group ____ entire class

7. Nature of the program (check as many as apply)

   ____ Drill and practice   ____ Demonstration
   ____ Game   ____ Problem solving
   ____ Simulation   ____ Tool (i.e., word processing, graphics)
   ____ Testing   ____ Computer-managed instruction
   ____ Tutorial   ____ Other (specify)

II. DESCRIPTION: In your own words, describe the program. Tell what it is about, how it is structured, etc.

FIGURE 12–3

Software Evaluation Form (from Robert V. Bullough and Lamond F. Beatty, *Classroom Applications of Microcomputers*. Copyright 1987 by Merrill Publishing Co. Used with permission.)

III. CONTENT

Key: Y = Yes   ? = Not Sure   N = No   NA = Not Applicable

Y ? N NA 1. The content of the program is accurate.
Y ? N NA 2. The content is appropriate for the objectives.
Y ? N NA 3. The content is consistent with expectations of school, district.
Y ? N NA 4. The level of sophistication is appropriate.
Y ? N NA 5. The content is free of bias.

IV. RUNNING THE PROGRAM

Y ? N NA   1. The instructions are clear and easy to understand.
Y ? N NA   2. The screen display is well designed.
Y ? N NA   3. The program is free of bugs.
Y ? N NA   4. The material is well organized and presented effectively.
Y ? N NA   5. Various ability levels are provided for.
Y ? N NA   6. Graphics and sound are used to enhance the program rather than as embellishments.
Y ? N NA   7. The student engages in ongoing interaction with the computer.
Y ? N NA   8. Feedback, both negative and positive, is effective and not demeaning.
Y ? N NA   9. The student is assisted through the program with appropriate cues and prompts.
Y ? N NA 10. Pacing and sequencing can be controlled.
Y ? N NA 11. Instructions can be skipped if desired.
Y ? N NA 12. Instructions and help screens can be accessed at any time.
Y ? N NA 13. A tutorial or sample program is provided.
Y ? N NA 14. The program will tolerate inappropriate input without malfunctioning.
Y ? N NA 15. The program represents an appropriate use of the computer.
Y ? N NA 16. The program achieves the stated objectives.

V. MAJOR STRENGTHS AND WEAKNESSES

Identify the major strengths of this program.

Identify the major weaknesses of this program.

VI. RECOMMENDATION (check one only)

_____ Excellent program; recommend purchase
_____ Good program; consider purchase
_____ Fair program; might wait
_____ Poor program; would not purchase

FIGURE 12–3
(continued)

498

Research in Computer Applications

Educational journals are now publishing studies that compare pupils' progress following computer-assisted instruction with pupils' achievement following traditional educational procedures. Some of these studies refer to published software programs. Many, however, are one of a kind research projects such as doctoral dissertations or university research and development projects. These studies are valuable because they focus attention on computer potential and careful evaluation of computer-assisted instruction. Some studies show dramatic gains for students engaged in computer-assisted instruction; others, however, favor traditional approaches.

Computer Applications in Reading 3d ed. (Blanchard, Mason, and Daniel, 1987) provides an annotated bibliography of computer research. The text also includes a list of journals that specialize in computer-related articles. Another source for educational applications for computer-assisted instruction is *The Reading Teacher.* This educational journal contains a monthly column devoted to computers in educational settings. *The Computing Teacher* provides many practical applications for computers in the classroom.

Computer Potential and Language Arts

Improved technology, expanded computer programs, and increased software quality will certainly influence education in the next decade. Record keeping and management capabilities may free teachers and administrators from time-consuming paperwork. Computerized libraries will probably replace card catalogs and change circulation desks. Software sections will increase the multimedia capabilities of libraries.

As programs become more sophisticated, learning potential will improve. Interactive programs will probably stimulate cognitive development and problem solving. Branched tutorial programs should increase individualized instruction and foster higher-level comprehension abilities.

The readability potential of computer programs should allow teachers to check readability levels of texts and rewrite materials or lessons at appropriate levels. The contextual analysis potential of computer programs will help teachers interact with children during the writing process. The word processing capability will hopefully increase children's writing abilities as they discover the ease of proofreading and changing text before printing. Computer programs may stimulate and expand children's ideas and knowledge base.

A rich potential exists for computer-assisted instruction. Research and practice in the next decade, however, will answer many questions about effective instruction, evaluative standards, and teacher training.

SUMMARY

Many feel the media can stimulate understanding in the classroom, whereas others attribute declining academic achievement to overexposure to the media, especially television. We discussed some ways the media can be used productively in the classroom.

Language arts objectives that can be met through media study include the development of listening, oral language, and discussion skills, imagination, dramatic expression, critical evaluation, spelling, vocabulary, language usage, and creative writing.

Both films and filmstrips add visual interpretation to, stimulate interest in, or provide background information on a book. Art films motivate creative writing, and often lead to an interest in producing student-made films and slides. Both photographic and handmade slides are valuable for instruction. We described a method for students to make their own slides by drawing on transparency film. They can make films and filmstrips by drawing on clear leader tape. Film units that include listening, speaking, writing, and reading activities provide an exciting means of instructing in creative imagination.

Television is an influential medium and can be useful in instruction if certain guidelines are met. Instruction must overcome the passive observer role, and encourage interaction among parents, teachers, and peers. Television viewing can stimulate the reading of television-related books, and promote discussion skills, reading skills, and critical reading and listening skills.

Television, radio, and newspaper commercials and advertisements as well as news items provide abundant materials for the development of critical reading and listening skills. Students should learn to question a writer's competence and integrity, use of sources and evidence, and use of facts versus opinions. Critical evaluation of advertisements can be approached through a study of the various propaganda techniques used to sway public opinion. We have suggested several instructional approaches and learning activities for stimulating critical evaluation of advertisements and news sources.

Newspapers offer a chance for other dynamic instruction in the classroom, because they provide a variety of information that appeals to unlimited interests. We concluded with an example of a learning center that uses the newspaper to develop functional and creative skills.

Computer-assisted instruction is a rapidly expanding area in education. Tutorial programs, drill and practice, and dialogue programs are found in language arts programs using computer-assisted instruction. Evaluative criteria include guidelines for computer education, hardware, and software. Concerns and issues include funding, teacher preparation, inadequate research, software quality, and overemphasis on drill practice. Potentials for future development include more interactive programs, stimulation of ideas, branched tutorial programs, word processing, contextual analysis, record keeping, readability analysis, and speech synthesizers.

ADDITIONAL MEDIA AND COMPUTER-ASSISTED INSTRUCTIONAL ACTIVITIES

1. *Identify a television program that could motivate reading or research. Develop a lesson plan that would stimulate children's interests either before or following the production.*

2. *Read Herbert London's article "What TV Drama Is Teaching Our Children" in* **The New York Times,** *Sunday, August 23, 1987, pp. H23, 30. Do you agree with his findings that television viewing is molding the ethical standards of young people? If you agree with his position, how would you develop a curriculum that would encourage students to be more selective and more critical of what they view on television? If you disagree with his position, provide a rationale for your position.*

3. *Identify a language arts area that could be stimulated or reinforced through the newspaper. Develop a lesson plan that focuses on your objective. Share the lesson plan with your language arts class.*

4. *Collect newspaper examples that develop differing viewpoints in the editorials and/or editorial cartoons. Develop a critical evaluation lesson to accompany the newspaper articles.*

5. *Read Shirley Baechtold, Terrell O. Culross, and Gwendolyn Gray's article "The News Magazine in the College Reading Classroom" in* **Journal of Reading** *29 (January 1986): 304–10. Adapt one of their suggestions for the elementary classroom.*

6. *Visit a computer center. Evaluate the various types of computers according to their potentials for elementary language arts.*

7. *Visit an elementary school that uses computers. What areas of the curriculum use computers? Interview the teachers about their attitudes toward the computers and their beliefs in the potentials for computer-assisted instruction.*

8. *Interview several school children who use a computer for instruction. What do they like or dislike about the computer?*

9. *Read a computer article in an educational journal. Share a summary of the article with your language arts class.*

10. *If possible, evaluate several software programs in reading, writing, or spelling. Share your evaluation with your language arts class.*

BIBLIOGRAPHY

Association for Teacher Training in Europe. *Microelectronics Education and Teacher Training—A Compendium.* Manchester: Microelectronics Education Program, 1982.

Baechtold, Shirley; Culross, Terrell O.; and Gray, Gwendolyn. "The News Magazine in the College Reading Classroom." *Journal of Reading* 29 (January 1986): 304–10.

Balajthy, E. "Using Microcomputers to Teach Spelling." *The Reading Teacher* 39 (1986): 438–43.

Blanchard, Jay S.; Mason, George E.; and Daniel, Dan. *Computer Applications in Reading.* 3d ed. Newark, Del.: International Reading Association, 1987.

Bradley, Virginia N. "Improving Students' Writing with Microcomputers." *Language Arts* 59 (October 1982): 732–43.

Bruce, Bertram; Michaels, Sarah; Watson-Gegeo, Karen. "How Computers Can Change the Writing Process." *Language Arts* 62 (February 1985): 143–49.

Bullough, Robert V., and Beatty, Lamond F. *Classroom Applications of Microcomputers.* Columbus, Oh.: Merrill Publishing Co., 1987.

Cheyney, Arnold B. *Teaching Reading Skills through the Newspaper.* 2d ed. Newark, Del.: International Reading Association, 1984.

Cox, Carole. "Filmmaking as a Composing Process." *Language Arts* 62 (January 1985): 60–69.

Daiute, Colette. "Physical and Cognitive Factors in Revising: Insights from Studies With Computers." *Research in the Teaching of English* 20 (May 1986): 141–59.

Department of Industry. *Micros in Schools Scheme.* London: Industry/Education Unit, 1981.

Dickinson, David K. "Cooperation, Collaboration, and a Computer: Integrating a Computer into a First-Second Grade Writing Program." *Research in the Teaching of English* 20 (December 1986): 357–78.

Elleman, Barbara. "Chosen by Children." *Booklist* 79 (December 1, 1982): 507–9.

Fiske, Edward. "Computer Education: Update '83." *Popular Computing* 2 (August 1983): 83–86.

Gomez, Mary Louise. "Using Microcomputers in the Language Arts." In *Language Arts Instruction and the Beginning Teacher,* edited by Carl Personke and Dale Johnson. Englewood Cliffs, N.J.: Prentice-Hall, 1987.

Grabe, M., and Grabe, C. "The Microcomputer and the Language Experience Approach." *The Reading Teacher* 38 (1985): 508–11.

Gratz, Elizabeth, and Gratz, J. E. "Who Is Controlling Children's Learning? Subliminals?" *Review Journal of Philosophy and Social Science* 9 (1984): 81–96.

Gustafson, B. "An Individualized Teacher-Directed Spelling Program Compared with A Computer-Based Spelling Program." Doctoral Dissertation, Iowa State University, 1982. *Dissertation Abstracts International* 991A–992A. (University Microfilms No. DA 8221191)

Hamilton, Harlan. "TV Tie-Ins as A Bridge to Books." *Language Arts* 53 (February 1976): 129–30.

Harper, J., and Ewing, N. "A Comparison of the Effectiveness of Microcomputer and Workbook Instruction on Reading Performance of High Incidence Handicapped Children." *Educational Technology* 26 (1986): 40–45.

Jelden, D. L. "The Microcomputer as a Multi-User Interactive Instructional System." *AEDS Journal* 4 (1981): 208–17.

Kinzer, Charles K.; Sherwood, Robert D.; and Bransford, John D. *Computer Strategies for Education: Foundations and Content-Area Applications.* Columbus, Oh.: Merrill Publishing Co., 1986.

Kleiman, Glenn, and Humphrey, Mary. "Word Processing in the Classroom." *Compute* 22 (March 1982): 96–99.

Langer, Judith A. "Musings . . . Computers and Conversation." *Research in the Teaching of English* 20 (May 1986): 117–19.

Lorick, Brenda A. *The United Kingdom Microelectronics Program: A Success or Failure for CAI.* Research paper prepared during a comparative education, Study Abroad Program. College Station, Tex.: Texas A&M University, 1983.

Luzzati, Emanuele, and Gianini, Giulio. *Pulcinella.* Westport, Ct.: Connecticut Films, 1974.

Mason, George E. "The Computer in the Reading Clinic." *The Reading Teacher* 36 (February 1983): 504–7.

_____; Blanchard, Jay S.; and Daniel, Danny B. *Computer Applications in Reading.* Newark, Del.: International Reading Association, 1983.

McGrath, Ellie. "Education: The Bold Quest for Quality." *Time,* October 10, 1983, 58–66.

Meet the Newbery Author. New York: Miller-Brody Productions, 1974.

National Film Board of Canada. *Tchou Tchou.* Chicago: Encyclopedia Britannica Educational Corp., 1974.

Parry, James D., and Macfarlane, Christine A. "Computer Based Instruction (CBI): The Transition from Research Findings to Teaching Strategies." *Educational Research Quarterly* 10 (1986): 30–39.

Rubenstein, R., and Rollins, A. *Demonstration of the Use of Computer Assisted Instruction with Handicapped Children: Final Report.* Cambridge, Mass.: Bolt Beranek & Newman, 1978.

Searls, Donald T.; Mead, Nancy A.; and Ward, Barbara. "The Relationship of Students' Reading Skills to TV Watching, Leisure Time Reading, and Homework." *Journal of Reading* 29 (December 1985): 158–62.

Selfe, Cynthia. *Computer-Assisted Instruction In Composition: Create Your Own.* Urbana, Ill.: National Council of Teachers of English, 1986.

Sharp, Peggy Agostino. "Children's Books and Computers—A Perfect Team." *The Computer Teacher*. June 1985, pp. 9–11.

Silberman, Charles. *Crisis in the Classroom*. New York: Random House, 1970.

Smith, N. "The Word Processing Approach to Language Experience." *The Reading Teacher* 38 (1985): 556–59.

Soloman, Bernard. "The Television Reading Program." *Language Arts* 53 (February 1976): 135.

Stevenson, H. W. "Television and the Behavior of Preschool Children," vol. 2. Washington, D.C.: Government Printing Office, 1971.

Tiedt, Iris. "Input, Media Special." *Language Arts* 53 (February 1976): 119.

Wresch, William, ed. *The Computer in Composition Instruction: A Writer's Tool*. Urbana, Ill.: National Council of Teachers of English, 1984.

Chapter Thirteen

After completing this chapter about the school media center and reference skills, you will be able to:

1. Describe the composition of print and non-print items recommended for the school library.

2. List the objectives for reference-skill instruction.

3. Evaluate a student's use of reference skills.

4. Describe the library reference skills taught in the primary, intermediate, and upper-elementary grades.

5. Develop a learning activity for teaching library reference skills in the primary, intermediate, and upper-elementary grades.

6. Describe several methods for acquiring minimal-cost materials for the classroom or school library.

The School Library
Media Center and
Reference Skills

*T*he school library no longer contains only books and other printed references. In fact, John Freeman (1976) described the school library as "a supermarket of communication, with variety as its key feature." If you visit a well-developed media center, you will find children working on projects that require them to use encyclopedias, dictionaries, literature collections, newspapers, magazines, films, filmstrips, and recordings.

To provide adequately for learning possibilities and activities, the library must contain an assortment of materials covering a number of reading levels and subjects. In 1975, the American Library Association published guidelines and recommendations for use by local school districts in balancing their media collections and supplying adequate materials and library staff. The ALA recommended that a library designed for 500 or fewer students have a minimum of 20,000 books and other references. The ALA suggests the following proportions of print and nonprint media: (1) 8,000 to 12,000 books; (2) 50 to 175 titles for periodicals; (3) 2,000 to 6,000 slides and transparencies; (4) 1,500 to 2,000 tapes and other recordings; (5) 500 to 2,000 filmstrips; (6) access to 3,000 16mm film titles and 500 to 1,000 8mm films; (7) 400 to 750 games; (8) 200 to 500 models and sculptures; and (9) 200 to 400 specimens. To manage these materials, the ALA recommended a full-time library specialist, a media technician, and one aide for every 250 students.

Research shows that children who have access to a well-staffed, well-stocked school library read more books and have higher verbal scores (Marchant, Broadway, Robinson, and Shields, 1984). Children whose school has a good library read more than twice as many books, of greater variety, than children who have access to only a classroom library or a centralized, school-district library. Both a school library and classroom library are essential for easy and frequent access to materials. We have already enumerated, in our discussions of literature, reading, and the media, the materials that should be included in the school and classroom libraries.

OBJECTIVES OF REFERENCE SKILL INSTRUCTION

Why is it so important for students to develop effective reference skills? Think for a moment about the last time you were required to write a factual composition or a

Reference skills help children find information in the media center.

research paper. Where did you look for information? What sources did you use? Did you read an entire reference book? How did you know where to find what you needed in the reference book? How did you pull together the ideas from several sources to write your final report? As a college student, the task was probably fairly easy for you. If you can remember the first time you had to write a report, however, you probably found the assignment anything but easy—you may even have been totally confused. You probably did not know where to locate information or what to do with the reference books after you found them.

Reference skills are essential for finding information throughout one's lifetime. Teaching reference skills helps students develop independent study skills and positive attitudes toward investigation; it stimulates an interest in functional and recreational reading and creates an interest in and knowledge about the library that will enrich their lives.

Instruction in library reference skills cannot be left to incidental learning. Usually, the librarian and classroom teacher teach reference skills cooperatively. In fact, Hart (1985) maintained that research evidence shows that "instruction in the use of library media resources needs to be correlated with classroom instruction" (p. 1). Certain skills are taught as early as kindergarten, when the kindergarten teacher introduces the children to the proper care of books: how to hold a book, how to turn pages, how to use a bookmark, how to put a book back on the shelf, how to place a book down on a desk or table, and how to care for a book when taking it home. The kindergarten teacher also introduces children to the school and classroom libraries. The children should be shown where to find the picture books and easier-to-read books as well as other media. They also need to learn how to check out books and return them.

During the primary grades, the library skills are extended as the children learn where to find specific types of materials, to refer to books by author and title, to manage the classroom library, and to use some of the reference materials. To teach effective use of reference materials, the teacher introduces certain prerequisite skills, such as alphabetizing, using an index, using a table of contents, and categorizing items according to subject, title, or author. Children become acquainted with picture dictionaries, primary school dictionaries, thesauruses, and encyclopedias, so they will know the value of each.

In the intermediate and upper-elementary grades, library reference skills are taught more formally, and children become competent in using the dictionary for finding word meanings, spellings, parts of speech, synonyms, and antonyms; locating books on the shelves from information in the card catalogue entries; preparing new books for the shelves; and creating bulletin boards and displays of colorful books.

REINFORCEMENT ACTIVITY

1. Visit an elementary school library. Observe the types of materials included in the library or media center. Do the print and nonprint materials meet the guidelines of the American Library Association? Observe what learning activities are going on in the library. Are these activities a vital extension of the classroom?

2. Interview a school librarian. Find out what instructional tasks she performs. What does she feel is the most important aspect of the job? How does this librarian cooperate with classroom teachers?

Evaluation of Reference Skills

Formal instruction in library skills does not usually begin until the intermediate grades, although the readiness and prerequisite skills are taught in the primary grades. Attainment of these skills is evaluated by observing students while they use the library and its various print and nonprint reference materials. Therefore, instead of suggesting formal and informal paper-and-pencil tests, we offer some checklists the teacher can use to evaluate the student's reference skills, and discuss some learning activities that are prerequisites for assignments requiring reference skills.

Primary grades   In the primary grades, the teacher wants to develop positive attitudes toward books, teach the children where to find and how to use specific types

TABLE 13–1
Checklist of library reference skills—primary grades

Skill	Yes	Some-times	No
1. Care of Books—Is able to			
a. hold book correctly	————	————	————
b. turn pages correctly	————	————	————
c. care for book when taking it home	————	————	————
2. Library Orientation—Is able to			
a. find specific types of books and other media in the library	————	————	————
b. refer to books by author and title	————	————	————
c. check out books	————	————	————
3. Reference Skills—Is able to			
a. manage the classroom library	————	————	————
b. make simple author-card for classroom library	————	————	————
c. make simple title-card for classroom library	————	————	————
d. make simple subject-card for the classroom library	————	————	————
e. alphabetize a short list of words by first letter	————	————	————
f. use primary picture dictionary	————	————	————
g. make and use class telephone directory	————	————	————
h. use a class-made encyclopedia	————	————	————

of books, and introduce other prerequisites for using a library to its fullest. The checklist in table 13–1 illustrates the reference skills that might be evaluated for a primary grade child.

Intermediate grades Reference-skill instruction becomes more formal in the intermediate grades, consequently there are more opportunities to evaluate the students' use and attainment of certain skills. The checklist in table 13–2 illustrates the abilities to evaluate in the intermediate grades.

Upper-elementary and middle school As the various subject areas begin to demand greater involvement with library references, the upper-elementary and middle-school child must be able to find and use library reference materials efficiently and effectively. These skills should be at least reviewed every year, so students feel comfortable and secure in their library searches. We have found especially crucial times for teaching and diagnosing library skills are when students change from elementary to middle school, at the end of fifth grade, or when they change from elementary to junior high school, at the end of sixth grade. The student not only changes schools, but changes from a more self-contained teaching environment to

TABLE 13–2

Checklist of library reference skills—intermediate grades

Skill	Evaluation		
	Yes	Some-times	No
1. Library Orientation—Is able to			
a. find specific books and media	_____	_____	_____
b. check out materials independently	_____	_____	_____
2. Library Use—Is able to			
a. understand and demonstrate appropriate uses for library at different times	_____	_____	_____
b. use librarian for appropriate reasons	_____	_____	_____
c. explain how books are arranged	_____	_____	_____
d. select materials at an appropriate level of difficulty	_____	_____	_____
3. Dictionary Skills—Is able to			
a. alphabetize words according to 2nd, 3rd, and 4th letters	_____	_____	_____
b. use guide words	_____	_____	_____
c. locate correct meaning for a word from multiple definitions	_____	_____	_____
d. locate synonyms and antonyms	_____	_____	_____
4. Other Reference Materials—Is able to			
a. use a children's encyclopedia	_____	_____	_____
b. use a telephone directory	_____	_____	_____
c. understand importance of various reference materials	_____	_____	_____
d. use card catalogue to find books on shelves	_____	_____	_____
5. Parts of the Book—Is able to			
a. use table of contents and index to find information in reference book	_____	_____	_____
b. refer to copyright date, and use to evaluate date of information	_____	_____	_____

one in which classes are completely departmentalized. In addition, the new campus may be quite distant from the elementary school in both geographic location and teacher expectancies. Although the elementary teacher usually provides considerable guidance in a library search, this is not necessarily the case in the middle-school content areas.

In the middle school, any language arts skills, including use of reference skills, are usually the responsibility of the language arts or English teacher. But often, students lack the reference skills they need to do reports in the content areas, so a major instructional shift may be required. Students need to be introduced to the appropriate skill before and during the time it is needed. Teachers should not assume students have mastered adequate library reference skills before coming to class, and need to find out what the students do or do not know. Table 13–3 illustrates which reference skills to evaluate for an upper-elementary or middle-school student.

TABLE 13–3

Checklist of library reference skills—upper-elementary or middle school

Skill	Evaluation		
	Yes	Some-times	No
1. Dictionary Usage—Is able to			
a. alphabetize words by 5th or 6th letters, if necessary	____	____	____
b. use guide words	____	____	____
c. find spellings for phonetically regular words	____	____	____
d. find spellings for phonetically irregular words	____	____	____
e. find the correct meaning from multiple meanings	____	____	____
f. find synonyms and antonyms	____	____	____
g. use pronunciation symbols to find correct pronunciation	____	____	____
2. Library Orientation—Is able to			
a. locate specific items in the library (e.g., check-out desk, card catalogue)	____	____	____
b. locate different categories of books	____	____	____
3. Card Catalogue—Is able to			
a. demonstrate how and when to refer to a subject card	____	____	____
b. demonstrate how and when to refer to an author card	____	____	____
c. demonstrate how and when to refer to a title card	____	____	____
d. understand the need to use, and uses, a cross-reference card	____	____	____
4. Classification Systems—Is able to			
a. use the Dewey Decimal or Library of Congress classification system	____	____	____
b. locate books on the shelf by call numbers	____	____	____
5. Encyclopedia—Is able to			
a. find and use the index of an encyclopedia to locate related articles on a subject	____	____	____
b. use the cross reference in an encyclopedia	____	____	____
c. use guide words in an encyclopedia	____	____	____
d. check copyright date of encyclopedia	____	____	____
e. choose an encyclopedia on an appropriate level of reading difficulty	____	____	____
f. use more than one encyclopedia to compare information	____	____	____
6. Other Reference Tools—Is able to			
a. understand the purposes for, and use, other reference materials biographical dictionaries	____	____	____
atlases	____	____	____
almanacs	____	____	____
specialized encyclopedias (e.g., science and social studies)	____	____	____
7. Finding Information in Periodicals—Is able to			
a. understand that periodicals can provide most recent information on a subject	____	____	____
b. identify subject headings that might provide references	____	____	____
c. use the *Reader's Guide to Periodical Literature*	____	____	____
d. understand abbreviations in the *Reader's Guide*	____	____	____
e. find the correct periodicals after locating references in the *Reader's Guide*	____	____	____

REINFORCEMENT ACTIVITY

1. *If you are now teaching students, use the appropriate checklist to evaluate a student's use of reference skills. What skills have reached the level of independence? What skills require further instruction?*
2. *Look at the language arts textbooks used in the elementary or middle schools. Choose a specific series and grade level, and design a library reference skill checklist to correspond with that series.*

TEACHING LIBRARY SKILLS

Instruction in the use of library media centers is receiving increased interest. Hart (1985) stated that "since 1978, twice as many items have developed for instructing students in the use of a library media center as had been developed previously" (p. 1). To use the library successfully, students need to know which references to use, where they are located, and how to use them effectively. There are also essential prerequisite skills to using a dictionary, an encyclopedia, or a card catalogue. Have you ever watched children grope through a dictionary when they do not have sufficient skills in alphabetizing? Alphabetizing is equally important for locating information in encyclopedias or card catalogues. Equally important for finding information in encyclopedias, card catalogues, and reader's guides is the ability to choose promising subjects or categories under which the appropriate information may be found. Whereas formal dictionary, encyclopedia, and card catalogue instruction may not begin until about fourth grade, readiness for these sophisticated skills begins in the primary grades.

Instruction in the Primary Grades

Even first-grade children can use the easy picture dictionaries published for them. These dictionaries are especially useful with young children because a picture is used to define each word entry. These books can be introduced when children are discussing the meaning of a new word in a reading class or in another content area, or during early writing activities, to help children see how they can use a dictionary to find the proper spelling of an unknown word. The pictures and words are alphabetized, so the teacher can also introduce the concept and importance of alphabetizing.

Language experience word banks   Several activities normally used in the primary grades lend themselves to the development of alphabetizing skills. The language experience approach described in chapter 14 mentions the use of word cards and word banks. Each of these cards contains a word the child knows, and as a result they form a convenient collection of words for individualized instruction. As the number of words increases, and the child has more difficulty finding a specific word for sentence construction or some other writing activity, the need for a more efficient way of grouping the words becomes apparent. The words can first be alphabetized according to first letters, then according to second and third letters. For example:

I
and

II
basket

ball

belt

car

big

Class-made picture dictionaries   Another activity that allows children to understand and practice alphabetizing skills is constructing individual and class picture dictionaries. Picture dictionaries might center around specific categories, such as animals or famous people. The children might start with one page for each letter (e.g., ant, bee, cat) and, later, add pages that require arrangement according to second- (e.g., ladybug, lion, lobster, lynx) and third-letter alphabetical order (e.g., ladybug, lamb). They can draw the accompanying pictures, or cut them from newspapers and magazines. Real photographs also add interest to the picture dictionaries.

Class telephone books   The telephone directory is another reference source that requires understanding of alphabetical order. Students can examine real telephone directories to see how they are organized, then assemble a class telephone directory that contains the same kind of information.

Using manipulative objects   Alphabet blocks allow children to practice the sequential order of the alphabet. Letters may also be printed on wooden beads, and the beads strung in alphabetical order. An alphabet tray also offers practice in sequential order. To make an inexpensive alphabet tray, fill an old cookie sheet with damp sand or plaster of paris (this would make a permanent tray). The teacher carves the letters into the sand or plaster of paris with a blunt object, such as a tongue depressor. Students use the tray to trace the letters with their fingers as they recite and observe the order of the alphabet.

Developing an alphabetical order learning center   Learning centers can be constructed with manipulative activities, games, and other activities that require placing items in alphabetical order. Sandra Kaplan's *Change for Children* (1973) illustrates one such learning center and its activities. The activities range from manipulating wooden alphabet blocks to developing individual dictionaries and even making up a new alphabet.

Introduction to the encyclopedia   Harris and Smith (1976) recommended that primary-grade children be introduced to the scope and use of an encyclopedia. They suggested that teachers may find it worthwhile to develop their own encyclopedia. This class-made primary encyclopedia is actually a picture file arranged in alphabetical order according to index or subject words. Subject categories such as animals, city, families, farm, holidays, houses, seasons, transportation, and weather are listed on dividers. Pictures and other information relevant to each subject are filed under the appropriate subject. As the encyclopedia increases in size, students will realize the necessity for developing a cross-reference index, because many materials may fall into more than one category.

Introduction to the card catalogue   To use a card catalogue effectively, children must eventually understand three kinds of classifications: title, author, and subject. Primary children can begin to understand the card catalogue by constructing a simple catalogue for the books in their classroom library. Class discussion will lead to several suggestions for arranging the books; someone usually mentions alphabetical order by author's name or by title. The children should first concentrate on one kind of classification, and put the necessary information for each book on a 3-inch × 5-inch index card, then arrange the finished cards by alphabetical order in a file box. Whichever method they begin with, the children will soon realize that more than one arrangement of categorizing and filing can be useful. If the library is using a computer system for cataloging, students will require instruction in the use of the computer.

An Integrated Approach to Reference Skills

Keel (1985) identified the following practices to improve the instruction of library media skills:

1. Teach children only the skills they need, at the time they need them.
2. Make instruction specific, simple, practical, brief, and entertaining.
3. Involve as many senses as possible in the presentation.
4. Lead children to discover conclusions for themselves.
5. Give them opportunities to use new skills as soon as possible.
6. Give each child praise, encouragement, and guidance. (p. 51)

Keel teaches library skills lessons that coincide with the beginning of various social studies units. To meet a multimedia goal and to involve as many senses as possible, she uses filmstrips, pictures, charts, posters, transparencies, tape recordings, records, and slides. For example, when introducing the reference skills needed for a third-grade study of Mexico, she uses Mexican folk music and develops a narrated slide presentation showing children how to locate the subject heading "Mexico," where to find these books in the library, and how to find books that would be most useful in answering specific questions about Mexico. Later, she teaches children how to find and use other materials about Mexico such as those found in filmstrips, picture files, vertical files, and recordings.

Instruction in the Intermediate Grades

Formal instruction in reference skills usually begins at about fourth grade. At this point the children have developed enough reading skills to use the easier-to-read dictionaries and encyclopedias. Some children may already be quite proficient, whereas others will require considerable formal instruction.

Alphabetizing   The intermediate-grade teacher should have children progress from simple to more demanding alphabetizing skills, so she can discover which students are ready to handle other dictionary skills and which students still require assistance in alphabetizing. They might begin by alphabetizing a list of words by initial letters: frog, count, vine, round, light, antler, water, peace. Students who have difficulty with this task may need to work on the readiness activities suggested for primary grades, especially the alphabet tray and other manipulative devices.

The next activity asks students to alphabetize a list of words according to second letters: after, avenue, apple, air, able, amount, acrobat, attic. Then they can learn to alphabetize according to the third letter of a word: dew, death, deck, detail, debate, defend, design, delight. Finally, they must alphabetize according to the fourth letter of a word: stress, strong, stray, struggle, stride. Similar lessons can require students to alphabetize in different combinations of difficulty. The children can use envelopes filled with word cards of different combinations for alphabetizing according to first letter, second letter, and so forth, for individualized instruction and additional practice.

Dictionary   To find words quickly in a dictionary, a student must understand the function of and be able to use guide words. Even when students know how to alphabetize, they often ignore the guide words at the top of the dictionary page. The teacher can approach this understanding inductively. She might, for example, have the students open their dictionaries to a specific page and ask them to examine it carefully to see how it is organized. After they have discovered the organization, they can look at the tops of the pages to see what is printed at the left and the right sides of each. Next, they can try to decide how the two words at the top of the page were chosen. They will usually deduce quickly that the word on the left is the first word on the page and the word on the right is the last word on the page. Then they can look at other pages, to see if this generalization is true for all pages and also for other dictionaries. The teacher can now tell them these words are called guide words, and are there to help us locate a word quickly, because all words on the page must come after the left-hand guide word and before the right-hand guide word.

Gillingham and Stillman (1977) recommended the use of a skeleton dictionary to focus attention on guide words. They suggested the skeleton dictionary lessons come after an introduction to the dictionary and demonstration of its organization. The skeleton dictionary consists of several pages containing only guide words and page numbers; the pages should correspond to a classroom dictionary that will be used for instruction. Figure 13–1 is an example of a skeleton dictionary page. The students compare the skeleton pages with the real dictionary, then indicate where certain words would appear in the skeleton dictionary, explain why they would be on a certain page, and verify the location with the real dictionary. Finally, from another list of words, the students would locate the guide words that should appear on the page containing a specific word and verify that the word is actually on that page.

Another dictionary skill that needs practice is locating the correct definition for a word from the multiple definitions presented. Some students have a tendency to focus on only the first definition after a word, and ignore the remaining information, which can give them some very odd ideas. For example, one student interpreted the

FIGURE 13–1
A skeleton dictionary page

ground	318	guard

Are these words on this dictionary page? grumble, green, guess, gruff, gypsy, great, growl. Why or why not?

sentence, "John will fast all day Friday," as "John will be held firmly to some object all day Friday, because fast means to fasten firmly." This problem results from dictionary practice in which students find the meanings of words presented in isolation. To stress multiple meanings, the teacher can list several sentences using the same word with different meanings:

1. The clock is *fast;* it shows 5:00.
2. The boat was tied *fast* to the dock.
3. The Olympic runner was very *fast.*
4. It was a holy day, so the monk had to *fast.*

The teacher could first ask for a definition of the word *fast*, then discuss with the students whether that definition makes sense in each sentence. Next, the students can look up *fast* in the dictionary, notice how many meanings are given, and find the correct definition as it applies to each sentence.

Dictionary races give the children practice in finding words. For a dictionary race, the teacher writes a sentence on the board with one word underlined. Points are allotted to each team that finds the word first, gives the correct guide words on the page, gives the correct meaning, pronounces the word correctly, gives the correct part of speech, and gives correct synonyms and antonyms.

Encyclopedias in the library and classroom   In the intermediate grades, children begin using the encyclopedia as a reference source. The encyclopedias available in libraries are written for different levels of reading ability, thus it will be helpful to both teacher and students to know what reading level each encyclopedia requires. The following encyclopedias are identified for different ability groups:

1. Upper-Elementary (Fourth Grade and Higher):
   Britannica Junior (Chicago: Encyclopedia Britannica). *The World Book Encyclopedia* (Chicago: Field Enterprises Educational Corp.).
2. Upper-Elementary and Middle School:
   Compton's Encyclopedia (Chicago: F. E. Compton & Co.).
3. Middle School and High School:
   Collier's Encyclopedia (New York: Crowell Collier & Macmillan). *Encyclopedia International* (New York: Grolier).
4. High School, College, and Adult:
   Encyclopedia Americana (New York: Americana Corp.). *The Encyclopedia Britannica* (Chicago: Encyclopedia Britannica).

Card catalogue   If intermediate-grade children have made classroom card catalogues, they already understand the function of author, title, and subject cards. Some intermediate children may have been using a library card catalogue since early primary grades, whereas other children have no understanding of the system.

Harris and Smith (1976) identified three things to teach intermediate-grade children about the card catalogue: (1) the abbreviations used on the cards (e.g., *c 1988* means copyright, 1988; *127 p. illus.* means the book has 127 pages and contains illustrations); (2) cross-reference cards direct the user to other sources of information; and (3) call numbers are needed to find a book in the library.

Instruction in these three areas may take only a few lessons for some children, but it is extremely important that they digest the information thoroughly, however long it takes. One fourth-grade teacher, for example, introduces her students to the card catalogue by having them put on a play about the card catalogue, adapted from a play printed in *Instructor* (Sister Mary Denis Tompkins, 1971). The play introduces Charlie Catalog, Art Author, Terry Title, Sam Subject, Dewey Decimal, Ruthie Reference, Dottie Dictionary, Al Atlas, Amy Almanac, and Ed Encyclopedia.

After presenting the play, the students make an author card, a title card, and a subject card for the same book. Next, they use the card catalogue in the library to search

for answers to questions such as the following: Who is the author of _____ ?
What books has _____ written? How many pages does the book
_____ have? What books might help us find out more information about
Mars? This teacher gives the children many opportunities to use the card catalogue,
and guides them while they use the references.

Instruction in Upper-Elementary and Middle School

In the upper-elementary and middle-school grades, reference skills are reviewed and
refined through instruction requiring students to use encyclopedias, card catalogues,
and other reader's guides and references designed for specific content areas.

FOR YOUR PLAN BOOK

A Unit for Teaching Library Reference Skills in the Upper-Elementary and Middle School

*An effective way to supplement initial orientation visits to the library is to prepare a se-
ries of tape presentations for the classroom. The tapes can be used:*

1. *to introduce students to library facilities at the beginning of the year;*
2. *to reinforce the location of materials and facilitate their use when they are
   needed;*
3. *to introduce the library to students who transfer into class later in the year;*
4. *to organize an effective learning center on library skills; and*
5. *to provide reinforcement and additional study for those students who need
   more time to assimilate the necessary information.*

*In addition, the tapes can be "made-to-order" to fit the available library facilities, and to
present necessary information in a sequence the school's teachers and libraries consider
most appropriate.*

Lesson One: Introduction to the Library
Overview of Script: *The narrator leads the students on a walking tour of the library that
includes:*

1. *the location and a brief explanation of the most important physical facilities;*
2. *a discussion of the rationale behind the usual behavior expected in libraries;
   and*
3. *an outline of regulations specific to the library the students will use.*

Activities to Accompany Lesson One
*To encourage students to reinforce the introductory information they have received
from the tape, you will probably want to include some application activities. These activities could be included on the tape, or put on cards and incorporated into a learning
center.*

Students show their ability to use the card catalog.

1. Draw a sketch of our school library, locating the following:
   a. circulation desk
   b. card catalog
   c. shelves
      (1) Fiction
      (2) Nonfiction
      (3) Reference
      (4) Periodicals
      (5) Audiovisual materials
2. Using your own materials and imagination, make a bookmark suitable for use in any book.
3. Using your own materials and imagination, make a cartoon or drawing that shows an important library regulation or encourages use of the library. You may use pictures cut from magazines, and so forth. This can be notebook size.
4. Using your own materials and imagination, make a poster suitable for display in the library or a classroom illustrating a library regulation or encouraging use of the library.

5. *Interview one of the librarians to find out what kinds of tasks librarians do besides checking out books, and how one prepares to become a librarian. Here are some questions you might ask (you can probably think of others):*
   a. *Who chooses books for the library and how are they chosen?*
   b. *What steps does the librarian go through to get a book on the shelf?*
   c. *What other tasks does a librarian do?*
   d. *How does a librarian learn what to do?*
   e. *What are the advantages and disadvantages of being a librarian?*

Lesson Two: Arrangement of Books in the Library
Overview of Script: *This lesson is designed to:*

1. *enable students to become proficient in locating books of fiction and biography;*
2. *introduce students to the general arrangement of nonfiction books; and*
3. *introduce students to library classification systems.*

Script Outline:
By following this general outline, you can provide your students with the necessary information to become proficient in finding books in the library.

I. *Books of fiction*
   A. *General information*
      1. *Definition of fiction*
      2. *Distinction between full-length books of fiction and story collections*
      3. *Symbols often used on the spine to indicate a book of fiction and those often used to indicate a story collection*
   B. *How fiction books are arranged on shelves*
      1. *Alphabetically by author's name*
      2. *Alphabetically by title when several books are available by the same author (**a, an,** and **the** are disregarded in alphabetizing)*
   C. *How to use the card catalog to locate a book of fiction*
      1. *Use of the title card to find out the author*
      2. *Use of the subject index to select a book of interest*
II. *Biography and autobiography*
   A. *General information*
      1. *Definitions of biography and autobiography*
      2. *Distinction between individual biographies and collective biographies*
   B. *How biographies are arranged on shelves*
      1. *Biographies and autobiographies*
      2. *Collective biography*
   C. *How to use the card catalog to locate a biography*
III. *Nonfiction books*
   A. *Distinction between fiction and nonfiction*
      1. *Definition of nonfiction*
      2. *Contrast with fiction*
         Often, teachers falsely assume that students have a clear concept of the distinction between fiction and nonfiction. Actually, many students have

translated these two terms into "false" and "true." As a result, their concept of fiction is limited to books of fantasy, while their concept of nonfiction includes all books about "real life."

    To have a common reference base, you might use popular television shows as examples to explore your students' concept of these two terms.

    3. *Provide examples of nonfiction books*

IV. Library classification systems
   A. Rationale behind library classification systems
     1. Need for order to provide easy access
     2. Desire to organize knowledge
   B. Two major classification systems used
     1. Library of Congress—formerly used with large libraries, such as universities, but now becoming popular in public libraries
     2. Dewey Decimal System—still the most popular in school libraries
   C. General divisions of the Dewey Decimal System

    Avoid any temptation to get too detailed here. Students are to be users of a library, not librarians. However, many students are fascinated by learning what the numbers all mean.

    You may wish to include an explanation of the interrelationship of the broad classifications of the Dewey system, as well as prepare a handout for the students.

Activities to Accompany Lesson Two

1. Put the following books of fiction in the order they would appear on the shelf. Put author's name first.

The Call of the Wild—London
The Witch of Blackbird Pond—Speare
The Bronze Bow—Speare
The Secret Garden—Burnett
The Voyages of Dr. Dolittle—Lofting
Charlotte's Web—White

Mary Poppins Comes Back—Travers
And Now Miguel—Krumgold
It's Like This Cat—Neville
Mary Poppins—Travers
The Big Road—Clarke

2. Find the answers to these questions about our school library and write them on notebook paper.
   a. How are fiction books marked?
   b. How are biographies marked?
   c. How are collective biographies marked?
   d. How are story collections marked?

3. On the plan of our library you made for Lesson I, locate the biography shelves.
Using the card catalog, get the title, author, and mark whether biography or autobiography, about the following:
   a. a President of the United States
   b. a sports figure
   c. a famous woman

4.   Survey the biography shelves in the library to locate:
     a.   three books about one famous person.
     b.   a collective biography that contains an article on that person.
     c.   any three biographies.
             List the title and author of each book and write a short identification of the subject of each book. Arrange your list in the order these should be found on the shelves.

5.   Making the Dewey System work for you:
     a.   From the tape, make a chart of the 10 main classifications. List two types of books you would find in that category.
             Example: 000–009   General Works
                                 encyclopedias, dictionaries
     b.   Using your chart, give the **general numbers** (e.g., 500) for the following types of books (all are nonfiction):
             (1)   American history   *(5)*   How to build a house
             (2)   Basketball   *(6)*   Teenage problems
             (3)   Poetry   *(7)*   Greek gods
             (4)   Chemistry
     c.   Give the general numbers (e.g., 500) of sections where you would expect to find the following titles:

     John Paul Jones   **Stars for Sam**
     Alligators and Crocodiles   **Pioneer Art in America**
     History of South America   **Poems Every Child Should Know**
     The Yellow Fairy Book   **Myths of Greece and Rome**
     Mathematics Made Easy   **America Sings**

6.   Use the card catalog to discover a biography or autobiography of each of the following persons, and list the titles and authors of the books. Look through the book to answer these questions:
     a.   Approximately when did this person live?
     b.   What nation claims this person?
     c.   What was this person's main occupation?
             Louisa May Alcott   Richard Byrd
             Hans Christian Andersen   Cleopatra
             Louis Armstrong   Coronado
             John James Audubon   Bob Cousy
             Benedict Arnold   Althea Gibson
             Johann Sebastian Bach   G. Marconi
             P.T. Barnum   Jim Thorpe
             Clara Barton

7.   Make a set of posters illustrating the classifications of the Dewey Decimal System.

Used with permission of Mary Russell, an English professor who teaches courses in English and language arts methods.

Further Library Activities

Once your students have been introduced to the library through taped presentations or orientation programs, you may want to encourage them to use library facilities to find information. One of the most useful tools is the card catalog. The following activities and exercises may prove helpful in designing instruction on the card catalog.

1. On manila folders or cardboard notebook covers, prepare sample author, title, and subject cards that contain only essential information so the students can learn the location of the information they will use.

Sample Subject Card

781.5 [a]

MUSICAL FORM [b]
Tovey, Sir Donald Francis, [c] 1875-1940.
The forms of music. [d] New York, [e]
Meridian Books, [f] 1956. [g]
251 p. [h] illus. [i]

a. Call Number
b. Subject Heading
c. Author
d. Title
e. Place of Publication

f. Publisher
g. Date of Publication
h. Number of Pages
i. Illustrations

Sample Author Card

781.5　Tovey, Sir Donald Francis, 1875–1940.
The forms of music. New York,
Meridian Books, 1956.
251 p. illus.

Sample Title Card

781.5　　　　The forms of music.
Tovey, Sir Donald Francis, 1875–1940.
The forms of music. New York,
Meridian Books, 1956.
251 p. illus.

2. Prepare exercises similar to the following that will give students practice in interpreting the information on cards.

```
┌─────────────────────────────────────────────┐
│                                             │
│          LEWIS AND CLARK EXPEDITION         │
│                                             │
│   917.8            Daugherty, James         │
│              Of courage undaunted. New York, Viking │
│                     Press, 1951.            │
│                     168 p. illus.           │
│                                             │
└─────────────────────────────────────────────┘
```

Using the sample card shown, answer the following questions:

(1) What is the title?_____

(2) What is the call number?_____

(3) Who is the author?_____

(4) When was the book published?_____

(5) Where was it published?_____

(6) Who published it?_____

(7) How many pages?_____

(8) What does *illus.* mean?_____

(9) What kind of card is this one?_____

(10) Under what letter of the alphabet would this card be filed?_____

Using the information on the sample card (Lewis and Clark Expedition), make a sample of the other two types of cards. Be sure to label the kind of card under your drawing.

```
┌─────────────────────────────────────┐   This card would
│                                     │   be filed under
│                                     │   the letter _____
│                                     │
│                                     │
│                                     │
│                                     │
│                                     │
│                                     │
└─────────────────────────────────────┘
```

Kind of card _____

Develop an instructional lesson plan to teach a library reference or readiness skill at the primary, intermediate, upper-elementary, or middle-school level. If you are now teaching, present the lesson to a group of children; if you are not teaching, present the lesson to a group of your peers.

ADDING MATERIALS TO THE CLASSROOM OR SCHOOL LIBRARY

If you are teaching in a school that does not yet have an outstanding library, or if you wish to increase your classroom library, there are a number of ways to enlarge the collections at minimal cost. Schools that are instituting new language arts or reading programs must often employ rather creative methods for procuring additional recreational reading and content reference materials.

In one school, for example, a new program was intended to teach communication and language arts skills to fifth- and sixth-grade students of lower reading ability. The teachers needed motivating reading materials in the classrooms and recreational reading materials a student could keep. The teachers used two methods to acquire the paperback books many of the students preferred. First, they sent a letter to parents of children in the school district, explaining the new educational program and their hope of obtaining appropriate paperbacks. They asked the parents to donate children's books, especially paperbacks, that were not being used, so that other children could enjoy them. The response was excellent. An attractive book corner was set up in the room, and there were enough books left over to start a paperback reading center in the school library, as well. The students could use the books in the classroom, or check them out. They also established a paperback book exchange in the classroom. If a student found a book he especially wanted to keep, he had only to bring in another paperback book to trade.

The teachers used several methods to increase the amount of nonfiction reference materials in the classroom. The letter asking for paperbacks also asked for donations of reference periodicals, such as *National Geographic, Smithsonian,* old *Life Magazines,* hobby magazines, *Popular Science, Popular Mechanics,* sports magazines, and so forth. The periodicals served as reading materials for special interests and provided materials for a picture-information file. The students compiled the file, and thus learned, inductively, some of the library reference skills, such as classifying materials, alphabetizing subjects, and developing a card index of their own materials.

A final means for collecting additional reference information was to have the students write for free and inexpensive materials. This activity required them to use letter-writing skills, and they found many colorful, up-to-date materials this way. When the materials arrived, of course, they had to be filed for easy retrieval, thus reinforcing library skills. If the materials contained advertising, the children critically evaluated them for bias and other propaganda techniques.

After this experience, the teachers concluded that the original lack of materials was compensated by the learning experiences that took place during the search for new and interesting materials. These ideas may be used with any classroom, and will provide inductive instructional methods for introducing library reference skills.

SUMMARY

The school library no longer houses only books; it has become a media center, emphasizing a student's ability to use both nonprint and print media. Access to a good school library increases both the number and variety of books students read. We described several methods for acquiring free or inexpensive materials for the classroom, including (1) asking parents for donations of used children's books; (2) a paperback book exchange; (3) requesting donations of used reference periodicals; and (4) writing letters to request free and inexpensive materials.

Reference skills are essential for finding materials efficiently in the library. These skills are necessary to the development of independent study skills, positive attitudes toward investigation, interest in functional and recreational reading, and knowledge about the library that will last a lifetime. Instruction in library reference skills should not be left to incidental learning. We have included three checklists of Library Reference Skills to use in evaluating the reference skills of primary, intermediate, and upper-elementary and middle-school students.

Students need to know which references to use, where the references are located, and how to use them efficiently. Instruction in the primary grades emphasizes library readiness and prerequisite skills, such as familiarization with the library, picture dictionaries, alphabetizing, class-made dictionaries, class-made telephone directories, class-made encyclopedias, and class-made card catalogues. Formal instruction in reference skills usually begin in the intermediate grades, and includes dictionary skills such as alphabetizing, using guide words, locating correct definitions for multiple definitions, pronunciation, and finding synonyms, antonyms, and parts of speech. Students are also taught to use the card catalogue and other library references. These library skills are reviewed and refined in the upper-elementary and middle-school grades, where they are frequently used in the content areas.

ADDITIONAL MEDIA CENTER AND REFERENCE SKILLS ACTIVITIES

1. *Visit several school, public, and/or university libraries. Interview the librarians. What reference skills do the librarians believe students and adults require?*

2. *Look at an elementary language arts scope and sequence. What library and reference skills are included? At what grade levels is instruction recommended?*

3. *Look at several journals such as **Book List** and **School Library Journal** that review literature. If possible, find references to nonprint media. Compile a current list of recommended nonprint media and share it with your language arts class.*

4. *Interview a media center specialist at your university. What changes does the specialist believe will occur during the next ten years? How will these changes affect the language arts program?*

5. *Select several dictionaries or encyclopedias designed to be used with early elementary grades, middle-elementary grades, and upper-elementary grades. Compare the readability levels, content, and illustrations.*

6. *Hart's **Introduction in School Media Center Use (K–12)** (1985) provides descriptions of numerous instructional activities and includes references to a wide range of approaches for teaching various library-related skills. Read about several of these approaches and describe the ones that you believe are the most interesting or beneficial. Share your findings with your class.*

BIBLIOGRAPHY

Douglas, Mary Peacock. *The Teacher-Librarian's Handbook*. Chicago: American Library Association, 1949.

Freeman, John. "School Library: See Media Center." *Language Arts* 53 (October 1976): 803–4.

Gillingham, Anna, and Stillman, Bessie W. *Remedial Training for Children with Specific Disability in Reading, Spelling, and Penmanship*. Cambridge, Mass.: Educators Publishing Service. 1977.

Harris, Larry A., and Smith, Carl B. *Reading Instruction*. New York: Holt, Rinehart & Winston, 1976.

Hart, Thomas L. *Instruction in School Library Media Use (K–12)*. Chicago: American Library Association, 1985.

Kaplan, Sandra Nina; Kaplan, Jo Ann Butom; Madsen, Sheila Kunishima; Taylor, Bette K. *Change for Children*. Pacific Palisades, Calif.: Goodyear Publishing Co., 1973.

Keel, Helen. "Library Skills." In *Instruction in School Library Media Center Use (K–12),* edited by Thomas L. Hart. Chicago: American Library Association, 1985.

Merchant, Maurice P.; Broadway, Marsha D.; Robinson, Eileen; and Shields, Dorothy. "Research into Learning Resulting from Quality School Media Service." *School Library Journal* 30 (April 1984): 20–22.

Tompkins, Sister Mary Denis. "A Friend in Need." *Instructor* 81 (November 1971): 90–91.

Wisdom, Aline C. *Introduction to Library Services*. New York: McGraw-Hill Book Co., 1974.

Chapter Fourteen

After completing the chapter on the linguistically different child, you will be able to:

1. *Describe the differences between a linguistically different and a linguistically deficient concept of language.*

2. *Understand the problems associated with the use of standardized tests when evaluating children who do not speak standard English.*

3. *Demonstrate an informal method for evaluating language usage.*

4. *Describe several factors, such as teacher attitudes and environment, that may influence the performance of linguistically different children.*

5. *Describe how knowledge of the values and learning styles of Mexican-American children could be utilized for effective education.*

6. *Describe the educational alternatives suggested by linguists and educators and the rationale for each approach.*

7. *Demonstrate the teaching of a language experience approach and understand its philosophy, how it is used with linguistically different children, and how the approach integrates writing, oral discussion, and reading.*

8. *Describe several components of successful intervention programs for use with inner-city populations and preschoolers.*

9. *List criteria that could be used to evaluate multicultural literature.*

The Linguistically Different Child and Multicultural Education

A major issue facing educators today is the lower achievement of many inner-city, Black, Mexican-American, and poor rural white children. The literature of the 1960s and 1970s increasingly dealt with various aspects of children who are often called "linguistically different" or "linguistically diverse." Currently educators are concerned with educating students categorized as having limited English proficiency. Educators in the 1980s emphasized the need for educational excellence for minority students. For example, Kronkosky (1982), executive director of the Southwest Educational Development Laboratory, emphasized reading and writing literacy, educational equity, computer literacy, and training and retraining effective teachers. Likewise, Corrigan (1983) anticipated a minority school population that will shortly exceed 50 percent in several southwestern states. High minority populations are also in large urban school districts such as New York, Detroit, Washington, D.C., Chicago, and Los Angeles. This concern in metropolitan Los Angeles was highlighted in a recent study analyzing achievement of high-school students. The study conducted by the University of Chicago (Daniels, 1987) found that reading, writing, and mathematics scores were lower for low income Black and Hispanic students.

Educators who are concerned with developing a quality educational system for all students face issues related to linguistically different speech patterns, non-English speech, and cultural diversity. A heightened sensitivity to the needs of all people in American society has led to the realization that the language arts program should heighten self-esteem and create a respect for the individuals, the contributions, and the values of all cultures. This chapter considers the selection of positive literature that presents the values, diversity, and contributions of Blacks, native Americans, Hispanics, and Asians.

LINGUISTICALLY DIFFERENT POPULATIONS

The term *dialect* is often used to describe linguistically different speech patterns, although everyone speaks some sort of dialect. Malmstrom defined dialect as "a distinct variety of a language used by a group of people who share interests, values, goals, and communication" (1977, p. 67). Dialects differ from one another in pronunciation, grammar, and vocabulary. We recognize a variety of regional dialects in the United States. We can often tell by listening to people whether they come from New England, the Midwest, the deep South, the Far West, and so forth.

Beginning in the 1960s, most American dialect studies began to focus on the speech of urban centers, such as New York, Chicago, and Washington, D.C., investigating influences on speech of social class, urbanization, race, economic levels, and group values. Because the majority of people living in inner cities are Black, this type of nonstandard English came to be referred to most commonly as "Black English." We can appreciate the educational dilemma faced by those in the inner cities when we realize that the English spoken and taught in most American schools, and recognized by most American teachers, is standard English.

Several studies have focused on the low achievement of inner-city children. In 1969, Cohen reported that 83 percent of disadvantaged Black children and 45 percent of disadvantaged white children in New York City were one to three years retarded in reading by third grade. Eisenberg's (1974) comparative study of metropolitan, commuter county, suburban, and independent schools showed that metropolitan schools had failure rates two-thirds higher than the commuter county, three times higher than suburban, and more than fifty times higher than independent schools. Although the achievement of Black students is still lower than the achievement of white students, there is a steady decline in the achievement differences

[handwritten margin note: black dialect is a lang system w/ a set of rules. Will repeat bad in their own dialect]

A good multicultural literature program includes selections of literature about people from around the world.

between Blacks and whites. For example, Burton and Jones (1982) analyzed data based on the National Assessment of Educational Progress in reading, writing, mathematics, science, and social studies. They concluded that achievement differences had lessened between 1970 and 1980. Increased understanding of linguistically different speech patterns may account for educational improvements.

Two quite different theories have been proposed to explain the causes of low achievement by many Black inner-city children. In the early 1960s, psychologists discovered that on standardized language tests, these children were unable to reach the standardized norms. Thus, psychologists such as Bereiter and Englemann (1966), Deutsch (1967), and Bernstein (1961) argued that many Black inner-city children have a language deficit characterized by the lack of a fully developed language system and illogical grammatical structures in their dialect. They argued that Black children often came to school with a deficient language system that had to be remediated if the children were to progress in school. Intensive language remediation programs such as Englemann's and Osborn's DISTAR reflect this view.

Other educators oppose this language deficit theory. In the late sixties and seventies, much of the research pertaining to speakers of Black English was conducted by linguists who analyzed speech samples under various conditions. Linguists studied Black dialects and examined in great detail how those dialects differed from standard English dialects in terms of phonology (speech sounds), morphology (word forming, including inflection, derivation, and compounding), and syntax (the way words are put together to form phrases, clauses, and sentences). Extensive research by Labov (1969, 1970, 1971, and 1973) and by Baratz and Shuy (1969) demonstrated that children who speak Black English have a fully developed, but different, language system from that of standard English. These investigators disagree with the deficit theory that stressed illogical grammatical structures in Black English. In fact, these linguists maintain that Black English is just as logical and consistent as standard English. Labov found that structures of Black English function in the same way that similar structures do in other languages. For example, the use of the double negative to signify negation, which appears in Black English, also appears in both French and German.

The syntactic consistency of Black-English-speaking children is suggested by Baratz (1969). In a study, she asked Black-English-speaking children and standard-English-speaking children to repeat sentences read to them in both standard and nonstandard English, and found that Black-English-speaking students had consistent verbal responses. For example, 97 percent of the Black-English-speaking children responded in the same way to the sentence: "I asked Tom if he wanted to go to the picture that was playing at the Howard." These Black-English-speaking children responded with: "I aks Tom did he wanna go to the picture at the Howard." Sixty percent of the Black-English-speaking children responded to the sentence: "Does Deborah like to play with the girl that sits next to her in school?" by repeating: "Do Deborah like to play wif the girl what sit next to her in school?"

The white standard-English-speaking children also followed a definite pattern by translating Black English back into standard English when they were asked to repeat sentences. Seventy-eight percent of the standard-English-speaking children repeated the sentence: "I aks Tom do he wanna go to the picture that be playing at the Howard" as "I asked Tom if he wanted to go to the picture that was playing at the Howard." Likewise, 68 percent of the standard-English-speaking children repeated the sentence: "Do Deborah like to play wif the girl what sit next to her at school" as "Does Deborah like to play with the girl that sits next to her at school?"

This study not only demonstrated consistency of speech patterns; it also, according to Baratz, demonstrated that the dialect of Black-English-speaking children interferes when the children attempt to use standard English. The problems related to how much interference results and how to provide an effective education for speakers of nonstandard English have not as yet been resolved. Research such as Baratz's shows that Black-English-speaking children often translate sentences into their own dialects. However, according to Harber and Beatty (1978), research does not show whether this translation negatively affects comprehension of standard English materials.

According to linguist Guy Bailey, there are many unanswered questions related to how or if oral language interferes with writing (personal communication, 1983). Bailey stated, "Over the last decade an increasing number of scholars have suggested that students' oral language interferes with their writing: the syntactic structures and coherence devices characteristic of the oral code are used in lieu of those characteristics of the written code. This interference apparently involves both the intrusion of oral styles in writing (for example, failure to include explicit reference and coherence devices in compositions) and the influence of dialect (the absence of tense markers, for instance) or in some cases the student's first language. However, linguists have yet to specify the exact relationships between speech and writing because they lack comparable data from both modes" (p. 1).

Many linguists and educators believe that Black English is systematically structured; therefore, it is different and not deficient. But even among those who agree on "different" versus "deficient," there is little agreement on the best educational approach to use to improve the school achievement of Black-English-speaking children. Suggested educational alternatives include (1) contrastive analysis of Black-English and standard-English sentence structures, (2) language experience approaches, and (3) dialect reading and speaking of conventional instructional materials. We examine each of these instructional approaches.

Evidence suggests that other factors, such as testing bias, teacher attitudes, early identification, parent involvement, early intervention, and long-term intervention programs can influence the achievement of the linguistically different child. We discuss these factors, and review several inner-city and early-intervention programs that appear to be successful.

TESTING

Educators are concerned about the criticism that many commonly used standardized tests may be both culturally and linguistically biased in favor of middle-class, standard-English-speaking children. In fact, Bartel, Grill and Bryen (1973) went so far as to suggest that there were no standardized tests appropriate for Black-English-speaking children, and that use of available tests may result in extreme errors in student placement. These two authors also concluded, in 1977, that the Grammatic Closure subtest of the Illinois Test of Psycholinguistic Ability (ITPA) should not be used as a diagnostic instrument with children who speak Black English. If Bartel and Grill were right, many teachers are probably wasting a great deal of instructional time trying to remediate problems with solutions that do not address the real difficulty at all.

Many tests may also be biased because items on the test are outside the child's experience. Even pictured items on such tests as the Peabody Picture Vocabulary Test may be totally foreign to an inner-city child or to a poor rural child. Hall and Freedle (1975) stressed the problems in finding culturally fair tests. Each teacher will need to evaluate the tests he is thinking of using with specific children's backgrounds in mind.

Some educators suggest the use of nontraditional tests to measure cognitive ability and language. Adler (1973) recommended such tests for evaluating linguistically different children. He believed the language test should consist of analysis of spontaneous speech in a naturalistic setting. In fact, the best information about verbal capacity may be obtained by asking children questions that are interesting to them; for this reason, tests should encourage involvement of the student. For example, the Picture Story Language Test by Myklebust is designed to encourage the telling of a story. When administering any test to a linguistically different child, Adler urges the examiner to repeat directions and use practice questions to ascertain whether the child actually understands what is expected of her. The examiner will probably need to use more oral directions and examples than are provided with the test manual, and must be able to understand the child's nonstandard speech.

Mantell (1974) made several suggestions for obtaining language samples and evaluating linguistically different children's language patterns by nonstandardized methods. One device she recommended is using photographs of people or places familiar to the children. In this situation, the teacher would ask open-ended questions about the pictures, such as, "What is the person in the photograph doing? Why is he doing that? How does the person look? How would he talk about himself? What would the person in the photograph do if . . . ?" By using familiar photographs, children are motivated to talk about something they know and feel comfortable talking about.

A second device for obtaining language samples is using what Mantell described as "The Faceless X." For this informal test, the teacher draws a circular face on a transparency. Students are asked to fill in expressions using eyebrows, mouth, hairstyle, skin color, headdresses, and so forth. The teacher then asks open-ended questions about the face. This activity can also provide insights into students' attitudes and biases.

As a third testing device, Mantell suggested the use of role playing in order to analyze students' abilities to alter their language according to the requirements of a specific situation. Role-playing situations can be formal or informal. A student who role plays an interview with the school principal for an article in the school paper or an interview with an employer for a part-time job would be displaying an ability or inability to use more formal language. In contrast, role playing an argument between a brother and sister over whose turn it is to do the dinner dishes would allow analysis of the student's ability to use casual language. During role-playing situations, other students are asked to watch for specific language usage. The teacher can note their ability to recognize various usage situations.

Several educators recommend modifying scoring procedures for reading tests so that errors attributable to dialect differences will not be counted as reading errors. Hunt (1974) found that Black, inner-city third- and fourth-graders scored significantly higher on the Gray Oral Reading Test when responses reflecting Black English variations in language were not counted as errors than they did when the tests were scored according to the test manual directions. The Goodman Miscue Analysis also stresses the importance of recognizing dialect differences in reading. It suggests that dialect errors in oral reading not be counted as comparable to errors that change the meaning of the selection. In order to do this, of course, the teacher must be knowledgeable about the dialect of the children being tested.

Another researcher (Hutchinson, 1972) found that inclusion or deletion of dialect-prejudiced items in a word discrimination test also influenced the grade level scores of students taking the test. When dialect-prejudiced items were included in the test, 40 percent of the Black, inner-city, higher-ability third-graders scored below grade level; when the dialect-prejudiced items were removed, only 2.6 percent of these students scored below grade level. It is important, then, to examine tests for possible unintentional bias before administering them to linguistically different children.

Linguists have provided us with a great deal of knowledge about the differences between standard English and Black English. They recommend that teachers become aware of these differences, and utilize their knowledge to more fairly assess a Black-English-speaking child's achievement. Teachers need to become familiar with differences in both pronunciation and syntax. It is very important that teachers distinguish between differences in dialect pronunciation and an erroneous interpretation of test material. Only a teacher familiar with the differences between Black and standard English can accurately evaluate a Black-English-speaking student's language patterns and establish priorities for effective instruction in oral language, writing, and reading.

Guidelines for analysis of the major differences between pronunciation and syntax are provided in table 14–1 and table 14–2.

The type of informal testing recommended by many educators requires that teachers understand whatever linguistically different speech patterns the children use. Knowledge of these speech patterns is also necessary for effective instruction. The guides in this chapter will help you familiarize yourself with differences between black English and standard English; the following resources will also be useful:

- "Major Differences Between Standard English and Black English," in *Reading and the Black English Speaking Child* (Harber and Beatty, 1978), pp. 46–47.
- "Black English: A Descriptive Guide for the Teacher," in *Black Dialects and Reading* (Fryburg, 1974), pp. 190–96.

TABLE 14–1
Phonological differences between standard English and Black English

Feature	Standard English	Black English
Simplification of consonant	test	tes
clusters	past	pas
th sounds		
voiceless *th* in initial position	think	tink
voiced *th* in initial position	the	de
voiceless *th* in medial		
position	nothing	nofin
th in final position	tooth	toof
r and *l*		
in postvocalic position	sister	sistah
	nickel	nickuh
in final position	Saul	saw
Devoicing of final *b*, *d*, and *g*	cab	cap
	bud	but
	pig	pik
Nasalization		
ing suffix	doing	doin
i and *e* before a nasal	pen	pin
Stress—absence of the first	about	'bout
syllable of a multisyllabic		
word when the first		
syllable is unstressed		
Plural marker	three birds	three bird
	the books	de book
Possessive marker	the boy's hat	de boy hat
Third person singular marker	He works here	He work here
Past tense—simplification of fi-	passed	pass
nal consonant clusters	loaned	loan

SOURCE: Reprinted with permission of Jean R. Harber and Jane N. Beatty, and the International Reading Association, from *Reading and the Black English Speaking Child.* Newark, Del.: International Reading Associaiton, 1978.

REINFORCEMENT ACTIVITY

1. *Review an often-used standardized test. Look at both pictures and vocabulary, and find items you feel might be biased because of a child's experience or English usage. Share your findings with your language arts class.*

2. *If you are teaching children, try one of the nonstandardized testing techniques suggested in this chapter. If you are not teaching at this time, develop a series of pictures and questions or a role-playing strategy, and present the informal techniques to a group of your peers.*

TABLE 14–2
Syntactic differences between standard English and Black English

Feature	Standard English	Black English
Linking verb	He is going.	He goin' He is goin'
Pronomial apposition	That teacher yells at the kids.	Dat teachah, she yell at de kid.
Agreement of subject and third-person singular verb	She runs home. She has a bike.	She run home. She have a bike.
Irregular verb forms	They rode their bikes.	Dey rided der bike.
Future form	I will go home.	I'm a go home.
"If" construction	I asked if he did it.	I aks did he do it.
Indefinite article	I want an apple.	I want a apple.
Negation	I don't have any.	I don't got none.
Pronoun form	We have to do it.	Us got to do it.
Copula (verb "to be")	He is here all the time.	He be here.
Prepositions	Put the cat out of the house.	Put de cat out de house.
	The dress is made of wool.	De dress is made outta wool.

SOURCE: Reprinted with permission of Jean R. Harber and Jane N. Beatty, and the International Reading Association, from *Reading and the Black English Speaking Child*. Newark, Del.: International Reading Association, 1978.

FACTORS INFLUENCING PERFORMANCE

We have seen that a number of educators have investigated the problems arising from differences between standard English and Black English. Educators are also interested in the roles of such factors as teacher attitudes, intelligence, environment, and self-concepts.

Teacher Attitudes

The literature and research on teacher attitudes toward the linguistically different child are quite startling. Teachers often not only react negatively toward students who speak nonstandard English, but also rate them as lower class, less intelligent, and less able to achieve academically than their standard-English-speaking peers. For example, Crowl and MacGinitie (1974) audiotaped both Black and white ninth-grade boys responding with identically worded answers to school questions. Experienced white teachers were then asked to evaluate the appropriateness of the answers. The teachers assigned significantly higher grades to the answers given by the white students than to the identical answers given by Black students.

Similar attitudes are found in elementary teachers. Watson-Thompson (1985) found that elementary teachers expressed negative attitudes toward the use of Black English. Negative attitudes were also uncovered in a questionnaire sent to both Black and white teachers of elementary school children. Blodgett and Cooper (1973) found that 53 percent of the white teachers and 26 percent of the Black teachers viewed nonstandard-English-speaking children as less intelligent than standard-English-speaking children. Undergraduate students in elementary education not only expected Black, lower-socioeconomic-class children to do less well in school than middle-class children, but also felt that Black children would be responsible for their own failures (Cooper, Baron, and Love, 1975). Linguists Shuy and Williams (1973) analyzed ratings of speech, and found that Black respondents rated Black English more positively than white respondents. Byrd and Williams (1981) also reported significant relationships among a teacher's definition of Black dialect, attitudes toward Black dialect, and race of the teacher.

There is no doubt that negative teacher attitudes affect the teaching that takes place in the classroom. The author has observed classrooms in which negative and positive attitudes toward linguistically different children have been openly displayed. One teacher of mostly Black, lower-socioeconomic-class children, for example, mentioned that she taught her 25 students as one group because testing indicated they were all at a very low level of ability. She implied that one could not expect much from a class made up of mostly linguistically different students. In contrast, another teacher divided a similar group of linguistically different students into ability groups and taught highly motivating lessons, because she felt the students had a great deal of potential that required exceptional teaching. She not only accepted nonstandard reading renditions from her Black-English-speaking students, but also spoke Spanish for the benefit of her Mexican-American students. The atmosphere in the room was so positive that you could almost feel the learning that was taking place.

Several studies indicate that a Black teacher or Black test examiner may achieve better responses from Black-English-speaking children than white standard-English-speaking teachers. Gantt, Wilson, and Dayton (1975) studied the language and listening comprehension of Black, Title I third-graders, and found that a Black examiner elicited more words and more divergent English syntactical characters than a white examiner. They also concluded that the teacher's race and teaching style may affect achievement. Murnane (1975) also stressed teachers' effectiveness in relation to achievement. Following a study of the achievement of Black, inner-city school children, he concluded that Black teachers, male teachers, and teachers with several years experience are more effective than other teachers with inner-city children.

Teacher attitudes about linguistically different speech have also been found to affect student oral reading results. A popular method of checking oral reading ability is to have the student read a selection to the teacher. In fact, this activity is duplicated millions of times each day in classrooms across the nation. During oral reading activities, teachers are not usually as concerned with correcting errors that do not change the meaning of the selection as they are with correcting errors that do cause a meaning change. But Cunningham (1977) found that 78 percent of the teachers in her study would correct dialect errors that did not change meaning, whereas only 27 percent indicated they would correct nondialect errors that did not change meaning. Consequently, Cunningham recommended that teacher training stress the meaning equivalence between standard and Black English, and also the grammatical nature of Black English. If a teacher spends a great deal of reading-instruction time correcting dialect differences, it is questionable that the children are receiving effective reading instruction.

It is clear that teacher attitudes are important to students' achievement. Following a review of attitudinal research, Harber and Beatty (1978) concluded, "To improve the self-image and academic achievement among black, lower socioeconomic status children, teachers need to be trained to expect more from them, to judge their capabilities independent of race and socio-economic status, and to understand and respect their dialect and cultural background."

Overcoming long-standing attitudes and the effects of racism is a complex process. William Raspberry, a syndicated columnist with the *Washington Post,* stated that "the reasonable course for dealing with effects of racism is the course we are taking with AIDS: to learn as much as we can about how to cure it, how to immunize ourselves against it, and how to halt its tragic spread" (1988, p. 6A).

REINFORCEMENT ACTIVITY

1. *Observe an elementary classroom that includes linguistically different children. Record in writing the questions asked by the teacher, the responses made by the students, and the responses made to the students by the teacher. Analyze the questions and responses. Does the teacher respond differently towards children who speak Black English as opposed to those who speak standard English?*

2. *Interview the children's teacher. Ask the teacher to identify which children he expects will do well in the class, and those he expects not to do well. Is there a relationship between the answers and the speech of the children? Discuss your findings in class. If negative attitudes exist, what is the implication for academic achievement? How will you make sure that all children in your classes receive a quality education?*

Environment

The suggestion has been made that children's achievement and intellectual development is negatively affected by poor environment. Lack of medical care may result in malnutrition, birth handicaps, and poor health, all of which hinder a child's progress. Common sense tells us, for example, that a child who is absent from school because of illness will have more academic problems than a child who attends consistently.

The current emphasis on day-care centers and Head Start programs was instituted in response to an apparent need for early stimulation of children. Studies such as those reported by Bradley and Caldwell (1976) found that a mother's emotional and verbal responses, appropriate play materials, and maternal involvement with the child were all influential in developing the child's intelligence. Cravioto and De Licardie (1975) stressed that many lower-socioeconomic-class parents are so busy trying to acquire the minimum basic needs for life that they cannot help their small children discover and manipulate auditory, visual, and tactile stimuli in the environment. Many educational authorities recommend both well-run day-care centers and parental education to help overcome such environmental deficiencies.

An interesting study by Greenberg and Davidson (1972) compared home environments of higher-academic-achieving and low-achieving Black, urban-ghetto children. They found that the high achievers had a more structured home life, had more room, and had more books than low achievers. In addition, their parents showed more respect and concern for their children's education, used less punishment, offered more responses to their children, and demonstrated broader social concerns than parents of low-achieving children.

The literature on linguistically different children often mentions that Black, lower-socioeconomic-class children do not feel they have as many opportunities as middle-class white or Black children. It is often suggested that the educational content for linguistically different children must build the child's self-concept and utilize instructional materials that show minorities as contributing members of the society. We look at the development of self-concept when we investigate the instructional approaches that may be used with linguistically different children, and when we look at suggested materials for providing relevant content in the literature program.

BILINGUAL EDUCATION AND STUDENTS WITH LIMITED ENGLISH PROFICIENCY

Teachers are concerned with educating the diverse populations in American schools. Education, however, has not been as effective as it should be for children who have limited English proficiency. (Journal articles and researchers in this area frequently abbreviate this description to LEP.) Because there is considerable research associated with Mexican-American students we use this population to illustrate why there is concern for these children, to suggest how much still needs to be accomplished, and to show that there are successful instructional programs for Mexican-American students.

According to Litsinger (1973), several generalizations can be made about the Mexican-American student's achievement: (1) Mexican-American students have one of the highest dropout rates of any minority group; (2) they fall progressively farther behind other students as they move through the educational system; and (3) few Mexican-American students go to college. Litsinger pointed out that these generalizations do not represent shortcomings within students of Mexican-American heritage, but they do represent failures within the structure of educational institutions.

The differences identified by Litsinger in the 1970s are still apparent in the 1980s. Maestas (1981) examined the effects of rural and urban backgrounds and ethnic differences on the achievement of high school seniors. He found that Mexican-American achievement, as measured by the Comprehensive Test of Basic Skills and the Iowa Test of Basic Skills, was lower than that of non-Mexican Americans. This was found again in 1987 in a study that analyzed achievement in Los Angeles (Daniels, 1987).

Interaction with teachers and in small groups is important for students with limited English proficiency.

If the Mexican-American child has been raised during preschool years in the traditional culture of the Mexican-American community, the child will be required, upon entering school, to adjust to cultural values, language, and teaching styles that differ from anything previously experienced. These children must also continue to function effectively in their culture, and the adjustment problems for Mexican-American children are prodigious. This bicultural existence may be especially difficult for young children when they begin the educational process.

It is important to understand how the culture shapes the individual's responses; consequently Purves (1986) contended that we must investigate the social values placed on reading, literature, and writing by the culture. These beliefs, which may have deep historical and religious roots, are so important that Purves concluded, "When an individual is transplanted from one culture to another culture, the individual has a great deal both to unlearn and to learn if he or she is to be accepted as a writer in that culture. To some extent, it may be impossible for an individual thoroughly acculturated in one culture to become indistinguishable as a writer from members of another culture" (p. 194).

Critics of the educational system for Mexican-American children maintain that educators have created a barrier between the child's culture and the school, and have often forced the child to choose between home and school. It is probably also true that few educators have been aware they were doing this. Cultural influences on the learning of the Mexican-American child began to receive considerable attention in the

1970s. The literature now provides us with information about the Mexican-American's value systems, learning styles, and language that the teacher can use effectively to improve the child's learning environment.

Values

Four major Mexican-American value clusters have been identified by Ramirez and Castañeda (1974). First, there are strong family ties in the Mexican-American culture; the needs and interests of the individual are considered secondary to those of the family. Because of these ties, the community and the total ethnic group become an extension of the family. This strong family and community identification contrasts with the stress on a child's separate identity and the development of self-awareness emphasized in most American schools.

Second, interpersonal relationships in Mexican-American communities are characterized by openness, warmth, and commitment to mutual dependence. Emphasis is on the development of sensitivity to the feelings and needs of others. Achievements are considered the result of cooperative rather than individual effort. Ramirez and Price-Williams (1971) found that Mexican-American children in Houston scored higher than Anglo-American children on their need to cooperate with other children as well as on their need for sympathetic aid from another person. They also expressed greater need for guidance, direction, and support from authority figures. Consequently, Mexican-American children may achieve better in a cooperative rather than a competitive atmosphere. This cultural characteristic of mutual dependence and cooperative achievement contrasts with the individual competition encouraged by American schools.

The third value cluster is role and status definition within the family and the community. Older people hold greater status, and children are expected to model themselves after their parents and other respected members of the community. The Mexican-American child may produce a higher level of achievement if the teacher takes on the role of this status figure and provides a model for the child.

The final value cluster is the support and reinforcement of these values provided by the Mexican-Catholic ideology. Emphasis on family ties is reinforced through religious commitment. The individual is thus encouraged to be respectful of adults and social conventions and to be open to guidance.

Ramirez and Castañeda contended that development of teachers' understanding of Mexican-American values is one of the primary concerns of educators. Educators have recently been taking this advice; many school districts with predominately Mexican-American populations are providing inservices to teachers to help make the teacher aware of both cultural characteristics and learning styles of Mexican-American students.

Learning Styles

Children develop their own styles of learning and means of assimilating information about their environments. A culture's values and socialization styles affect the development of a specific learning style in children. Cultural values are instrumental in developing preferences for certain types of rewards, and for the type of learning environment that will be most rewarding to the child. If these preferences are recognized, schools may utilize them in designing an educational climate that is rewarding and that will produce higher achievement.

The motivational styles of Mexican-American children appear to be quite different from the motivational styles of Anglo Americans. Kagan and Madsen (1971) found, as would be expected from Mexican-American values, that Mexican-American children were more highly motivated in a cooperative setting than in a competitive setting. In contrast, Anglo-American children were found to be more competitive.

Mexican-American children have also shown greater desire to interact with others and to belong to a social group (Ramirez and Price-Williams, 1974). Children are socialized to become sensitive to the feelings of others so they will be able to respond acceptably to others. Due to their humanistic orientations, Mexican-American children may do better when the curriculum has a human content and is closely related to the needs of others as well as to the family and community.

Language

The Mexican-American children in American schools also face problems beyond the differences in cultural values and learning styles; they frequently face differences in language. Bilingual programs are attempting to meet the needs of the Spanish-speaking child in the English-speaking school environment. The Bilingual Education Act of 1968, Title VII, provided funds for bilingual education programs. The programs are found throughout the United States, with the greatest numbers in California and Texas.

Several components essential to sound bilingual programs are identified by Litsinger (1973):

1. Utilization of the native language in the educational process to teach basic concepts and skills necessary for future learning.
2. Continued language development in the native language.
3. Development of subject matter in the native language.
4. Development of subject matter in the second language.
5. Development of a positive self-image and cultural identity.
6. Continued language development in the second language.

Examples of bilingual education programs may be found in *The Challenge of Teaching Mexican American Students* by Dolores Litsinger. The text includes exemplary programs from Arizona, California, Colorado, Florida, Massachusetts, Michigan, New Mexico, Nevada, New York, and Texas, as well as several examples of units for instruction written in both Spanish and English.

There has not yet been enough experience and research in bilingual education to produce agreement on the issue of whether to teach reading in the child's native language, then in English, or to teach reading only in English.

EDUCATIONAL IMPLICATIONS FOR LIMITED ENGLISH PROFICIENCY STUDENTS

Learning a second language causes additional academic challenges for students with limited English proficiency. Wong-Fillmore's (1986) four-year study of the effects of instructional practices on the learning of English by students from Chinese and Hispanic backgrounds provides several educational implications. First, the research shows that students vary greatly in their ability to learn a second language. Whereas some students learn in just a few years, most students require from four to six years to achieve the proficiency needed for full participation in school.

Furthermore, the research supports certain effective instructional practices. For example, Hispanic students gain in both production and comprehension skills if they have opportunities to use the new language while interacting with their peers. In contrast, interacting with peers has little value for Chinese students until they have enough proficiency in the new language to feel confident when using English with their classmates. For the Chinese students, close interactions with teachers and opportunities to use English during teacher-directed instructional activities relates to gains in oral language. Although both Chinese and Hispanic students gain from high quality instruction and clear language usage, Hispanic students (mostly Mexican-Americans) do especially well when the instruction is clearly presented and well-organized.

Wong-Fillmore is especially critical of the quality of much of the instruction. She stated,

> I came to realize that what these LEP children generally get in school does not add up to a real education at all. Much of what they are being taught can be described as 'basic skills' rather than as 'content.' Instruction in reading, for example, is mostly focused on developing accuracy in reading rather than on understanding or appreciating textual materials. Writing focuses on accuracy in spelling, punctuation, the niceties of grammar rather than on communicating ideas in written form or on the development of sustained reasoning and the use of evidence in supporting written arguments. (p. 478)

Canadian educators provide additional guidelines and implications for effective education. For example, Piper (1986) recommended using the traditional stories and fables from various cultural sources and focusing on students' awareness of each others different languages and different cultural backgrounds. He suggested that teachers compile lists of words that are the same across many languages, construct signs in schools that are written in several representative languages, explore emigration patterns of families within the classroom by using maps and family histories, and involve parents in the design and implementation of curricular materials. Sealey (1984) emphasized that instruction should encourage students to accept and be sensitive to cultural diversity, to understand that similar values frequently underlie different customs, to have quality contact with people from other cultures, and to role play experiences that allow students to be involved with other cultures. The literature recommended later in this chapter and the activities that teach children how to identify values in the literature encourage cultural sensitivity.

Educators who have investigated the Mexican-American child's cultural values and learning styles emphasize significant educational implications. The following suggestions for curriculum and teaching strategies that match a Mexican-American child's learning style are by Dixon (1976):

1. Activities that emphasize improving skills of all members of a group may be more successful than activities that emphasize only individual improvement.
2. Activities that require cooperation rather than competition may be successful. The school environment should encourage cooperation by assigning tasks for small groups of children to work on together, and should encourage a cooperative attitude about classroom behavior.
3. Mexican-American children will profit from educational experiences that allow them to interact with the teacher or other students.
4. Seeing other children in authority roles in the classroom may be natural and desirable as a result of the generalization from the home culture's use of other children as authority figures. Peer teaching is recommended for bilingual classes because the peer approach makes use of other children in authority roles and strengthens human involvement and interpersonal relationships within the classroom. In peer teaching, a student who has a superior skill "teaches" or assists a child who needs help. The language experience approach (see pp. 549–557) is considered very good for bilingual children because the children can use both their own language and cultural values.

Other implications for education are found in the research of Ramirez and Castañeda (1974). According to these authorities on Mexican-American education, the teacher should use many nonverbal indications of acceptance, including smiling, touching the child, allowing the child to work next to the teacher, and sharing laughter and other warm experiences with the child. These actions are reminiscent of the close family ties developed in the traditional Mexican-American culture. The teacher should also try to know the children as individuals and to relate lesson material to their experiences whenever possible.

The close ties of the traditional Mexican-American family make it appropriate for the teacher to work closely and cooperatively with the children's family. Send notes to the family telling them how pleased you are with their child's progress. Notes accompanying excellent examples of student work are especially appropriate. The students may also be rewarded with frequent reminders of how proud their families will be about their progress. Meeting personally with and talking frequently to the student's family also help the teacher develop a closer relationship.

As we have noted, researchers suggest that traditional Mexican-American children look up to an authority figure and model themselves after older, revered members of the community. Teachers may provide that model by having children repeat words after them in language instruction or by providing a model of how to write, paint, or do any other classroom activity.

When you are trying to provide a positive teacher model, you should also show respect for the children's language by using Spanish frequently during the school day. If the highly respected authority model uses the child's native language, it conveys the message that the language is to be respected. The teacher can also convey respect for the language by teaching Spanish songs and rhymes in addition to addressing Spanish-speaking children and adults in Spanish. Respect for the culture can also be developed by studying famous Mexican-Americans and Mexican holidays and other celebrations.

Educators who use second-language techniques provide guidance for working with Mexican-American children who have limited skills in the English language. Current approaches in English as a second language (ESL) classes are described in Carter's (1982) *Non-Native and Nonstandard Dialect Students: Classroom Practices in Teaching English, 1982–1983*. For example, de Félix's (1982) language arts approach is based on the following basic principles: (1) fluency precedes accuracy; (2) students learn concepts and structures when they are ready to use them; (3) teachers organize successively more challenging tasks; and (4) acquisition of the linguistic function precedes the acquisition of form. The ESL methods in the approach require teachers to slow the pace for students with limited English proficiency, to introduce phonics after students have increased their vocabulary, to build on children's native language, to understand the nature of language strategies or function, and to keep informed of new strategies. The following five-step approach for working with ESL students is recommended by de Félix:

1. *Develop Comprehension.* Teachers use a variety of resources such as pictures, concrete materials, native language materials, community resources, pantomime, and peer tutors to help students understand the concepts behind the language to be taught.

2. *Select and Analyze for Vocabulary, Grammar, and Linguistic Function.* Teachers analyze the materials to be used to help students integrate knowledge. For example, they select vocabulary words that can be taught through concrete examples and experiences. When possible, expand this concrete understanding through categories. (If the word is *meat,* teach different kinds of meat or sample foods from different nutritional groups.) Next, select two or three types of sentences found in the lesson and extend the students' use of grammar and the target vocabulary. Construct sentences showing how the vocabulary fits into different sentence patterns. ("I like meat," "I like beef and ham," "Beef and ham are types of meat.") Finally, select linguistic functions that may be taught through the lesson. Examples are as follows: phrases that justify actions or identify needs; phrases that direct students' actions; phrases or words that help them interpret materials by recognizing sequences, relationships, and causes; phases that allow them to hypothesize, anticipate consequences, recognize problems, and predict solutions; experiences such as puppetry that allow students to explore language; and experiences such as songs and games that encourage children to have fun with language.

3. *Plan Activities to Meet Students' Needs.* Teachers plan activities that consider individual needs such as age, cultural background, and English proficiency.

4. *Present Lesson Proceeding from Known to Unknown.* Teachers present new information on the basis of prior knowledge and understanding of concepts to be acquired.

5. *Integrate and Reaffirm Skills.* Teachers help students integrate concepts into their personal lives and school content areas and help them reaffirm skills in various contexts.

REINFORCEMENT
ACTIVITY

If you have a bilingual class near your university or college, visit a classroom. Observe the children and describe some of the effective instructional techniques you see. (You may also observe some that are not effective. If this happens, see if you can decide why they are not effective). If you do not have access to bilingual programs, read about an exemplary educational program for Mexican-American students. Is the program utilizing knowledge about learning styles as justified in the research?

INSTRUCTIONAL APPROACHES

Thus far in our discussion of linguistically diverse children, we have found a difference of opinion as to the nature of the language spoken by many inner-city children. One group of investigators believes that the nonstandard English spoken by many Black students is deficient and lacks logical grammatical structures. Another group of investigators analyzed the speech structures of many inner-city students and found that Black-English-speaking children actually have a fully developed but different language system from that of standard English. In addition to language difficulties, factors such as environment, teacher attitudes, and test bias have also been found to influence the low achievement of many linguistically different children. Linguists and educators suggest a number of educational alternatives.

Oral Language Expansion through Contrastive Analysis

The aim of many educators who try to expand a child's language is the child's improved oral and written communication in the classroom. Some programs stress teaching standard English before children learn to read; other programs teach standard English as an alternative dialect and as a useful tool for school and society. This second approach usually teaches standard English usage while maintaining and respecting the student's original dialect. Thus, students theoretically have a choice: they may use whichever form of language is most useful for the specific occasion. In this section we look at a program designed for early primary language expansion and another designed for the middle grades.

This first language expansion program for early primary grades was developed by Cullinan, Jaggar, and Strickland (1974) for kindergarten through third-grade children attending predominately Black public schools. The program was designed to expand language experience and the children's control over language through a literature-based oral language program. The teachers in the program were trained to recognize the contrasts between standard English and Black English so they could help children perceive the differences between the two forms of language. The program utilized literature selections that were read to the children daily. After hearing the selections, the children took part in creative activities that provided opportunities to expand oral language as well as to practice specific patterns of standard English.

The program's daily lessons were designed to include one of six different types of creative language-expansion activities. The first activity utilized dramatization and puppetry. Students discussed the story, prepared the scenes to be dramatized, and presented the scenes to the class. The second activity was built on storytelling by the children. Students listened to a story, such as "Curious George," talked about the main character or the parts of the story that were most enjoyable, then told stories about the character in another situation. (For example, what would happen if George found himself locked overnight in a supermarket?) The third activity used book discussions, stressing various listening comprehension skills as well as clear expression of ideas. The fourth activity was built on role playing. Children talked about the characters in a story and how the characters felt. Specific lines were chosen from the stories to illustrate a particular language pattern that needed practice. Students then did role playing using the specific sentence patterns from the story. The fifth activity used even more specific language patterns. Following the story and the discussion, a puppet named "Peter Parrot" asked the children to repeat sentences from the story. Children repeated the sentences in unison and individually. The final activity used after a literature selection was choral speaking. After a story or poem was read to the children, they were divided into groups to perform the story or poem as a choral-speaking activity.

A language expansion program built on contrastive analysis (comparing the points of differences between two forms of language) has been designed by Mantell (1974) for middle-grade Black-English-speaking students. Mantell's program is designed to identify the most frequently used grammatical features of Black English and then to focus specifically on those features that most sharply separate one social group from other social groups.

The list of features on which Mantell says it is most important to concentrate is based on findings by Wolfram (1970). Wolfram plotted features of Black English against the following five criteria: (1) sharpness of social stratification; (2) generality of application of the rule; (3) whether the rule is grammatical or phonological; (4) whether the significance is general or regional; and (5) the frequency of occurrence of the feature. According to his findings, there are seven syntactic features of Black English that are important for students to control in respect to social prestige. These features, in the order ranked by Wolfram, include the following categories and examples:

1. *s* third person singular—he go
2. Multiple negation—didn't do nothing
3. *s* possessive—man hat
4. invariant *be*—he be home
5. copula absence—he nice
6. *been* auxiliary in active sentence—he been ate the food
7. existential *it*—it is a whole lot of people

The purpose of Mantell's program is to help middle-school children become aware of the syntactic differences and similarities between Black English and standard English, as well as to test the appropriateness of each form of language. Language expansion activities are developed through role-playing and game situations. Students are made aware of differences by listening to recorded samples of Black English and standard English, and are then asked to isolate the differences on mimeographed copies of the recordings. Students are also asked to collect language samples showing how various people express possession, plurality, time, and negation.

Other language expansion activities include the following: providing short sentences as appropriate captions to photographs; storytelling; eyewitness reports of current or fantasy events for a news commentary; and role playing. In role-playing activities, the appropriateness of language is rated by a panel of students.

A final instructional approach in this program uses an incomplete sentence technique. The teacher provides the beginning of a sentence and asks students to complete it; for example, I . . . ; We . . . ; They . . . . This program does not use published materials, but utilizes drama, literature, and topics of interest to the students.

REINFORCEMENT ACTIVITY

Thus far in this section, we have discussed instructional materials and approaches that stress oral language expansion. Look at an instructional program that uses this approach. What do you feel is the basic philosophy of the approach? What are the instructional goals and how are these goals developed? What do you feel are some strengths or weaknesses of the material?

Language Experience

Following a review of the literature, Harber and Beatty (1978) maintained that allowing students to use their own dialect when reading conventional materials, along with a multiple approach to instruction, is currently the acceptable educational alternative for linguistically different children. The multiple approach includes the use of language experience, literature, oral language development, development of abstract thinking, listening, structured sequential programs, and multiethnic materials.

Educators such as Batty and Batty (1985) recommended using the language experience approach because it offers social and psychological benefits to readers, eliminates the possible mismatch between language and print, allows children to incorporate their own experiences and thoughts into the reading program, personalizes instruction, and fosters positive attitudes toward reading. We deal with the language experience approach in detail, because it incorporates so many language arts skills and is useful not only with the linguistically different, but with every child.

The growth of the language experience approach in our nation's schools has been quite dramatic during the late 1960s and 1970s. Hoover (1971) surveyed teacher education colleges and universities in the United States and reported that, prior to 1960, there was little teaching about a language experience approach. It was only in the later 1960s that professors began to include this approach in college classes. I still find many elementary teachers who do not know how to develop a language experience approach. This topic is frequently requested for inservice programs, especially from remedial teachers and other special education teachers.

Philosophy of the approach   The title "language experience" provides a clue to the nature of the approach. All the communication skills, listening, speaking, reading, and writing are included in this language-arts-oriented approach. The philosophy of the approach is best described by Lee and Allen's (1963) rationale:

1. What a child thinks about, she can talk about.
2. What one talks about can be expressed in writing.
3. Anything the child or teacher writes, can be read.
4. One can read what she writes and what other people write.
5. What a child has to say is as important to her as what other people have written for her to say.

The language experience approach thus emphasizes that the child's own ideas are worth expressing. They are not only worth saying, they are also worth writing down and being read by the child, the teacher, and other classmates. This approach stresses the natural flow of language. First, the child's oral expression is stimulated by art, literature, or other experiences; then her written expression is developed from her oral expression; her motivation for reading is developed because she has seen her own language in print; and, finally, after reading her own language, she is moved naturally into reading the language of other authors found in published books.

Language experience programs   As early as 1933, the use of a language experience approach was advocated with Mexican-American children whose language and cultural background were different from the language of the school. The students learned to read materials they dictated orally, so there were no problems with cultural or language interference.

The literature contains numerous references to programs that have used language experience approaches with other linguistically or culturally diverse groups. Hall (1965) developed a language experience approach for culturally disadvantaged Black children in Washington, D.C. Her students made greater gains in reading readiness, in word recognition, and in sentence reading than a group of students who were taught with traditional basal materials. Calvert (1973) used a language experience approach with middle-school Mexican-American students and found that his students made significant gains in writing, vocabulary, comprehension, and attitudes toward teachers.

The language experience method The language experience stimulates oral language and writing by providing opportunities for discussion, exploration of ideas, and expression of feelings. The language experiences in turn provide the content for group and individual stories composed by the children and recorded by the teacher. Many teachers introduce students to the language experience approach through development of a group chart story. This can be done with any age group, but it is a frequent readiness- or early-reading activity in kindergarten or first grade.

The chart story is usually written by an entire group, guided by the teacher, following a shared motivational experience. The motivational experience may be any activity that involves the group and encourages oral language as the children are drawn into a discussion about the activity. For example, one of my students brought a baby chick into the classroom. The first graders looked at and listened to the chick, carefully touched it, and fed and watered it. They discussed what they saw, heard, felt, and cared for. An excellent chart story resulted from each child's contributions to the story. A third-grade remedial-reading teacher involved the children in making pudding before they wrote an experience story. A seventh- and eighth-grade learning-disabilities resource teacher read literature selections to her students, then had the class develop chart stories about the main characters. Examples of other motivational materials and activities include trips, films, building projects, pictures, guest speakers, science experiments, puppet shows, music, and poetry. A room that offers a rich environment of activities will provide many motivational experiences for the chart story.

After the motivational experience and the oral discussion, the teacher records the story exactly as it is said by each child. The teacher may record the story on a large piece of newsprint, a poster board chart, or on the chalkboard, from which it will later be reprinted onto a chart. Whichever means of recording is used, it is essential that children sit so they can see each word as it is written in the proper left-to-right sequence. They will thus see that sentences follow a progressive pattern from the top of the page to the bottom, and that sentences begin with capital letters and end with periods.

As the children dictate the story, the teacher writes each word, repeating the word aloud as he writes. Following each sentence, the teacher reads the whole sentence to the group, using his hand to underline the word being read. The teacher asks the child who has dictated the sentence whether the sentence is correct.

After the chart story is completed, the teacher reads the whole story. Then, the students reread the story with the teacher. If a title has not already been given, the class may discuss an appropriate title. Some students may feel confident enough to try to read the selection individually, whereas other students may read only the sentence they dictated.

I recently demonstrated the development of a chart story that could be used for oral expression and reading with a group of learning-disabled third-graders. The following lesson plan was used during the demonstration.

Language Experience Chart Story

The objectives of the lesson are to (1) involve the students in a motivational activity that will result in oral discussion and a group-dictated chart story, (2) improve the students' ability to comprehend the sequential order of a selection, and (3) help the students increase the number of sight words they recognize.

As a motivational activity, the teacher can begin the discussion by asking the students to tell the group about their favorite desserts. Show a picture of a chocolate pie made from chocolate pudding and a graham-cracker crust. Ask the students how many of them have helped make a pie like the one in the picture. Discuss how they think it would be accomplished. Ask them what they think they would need to make the pie in the picture. As various items are mentioned, remove them from a box and place them on the table (include instant chocolate pudding mix, milk, measuring cup, bowl, egg beater, spoon, and graham cracker crust). Explain to the students that we are not only going to make a pie, but we are also going to eat the pie, then write a story about our experience so we can share it with the rest of the class and with our parents.

The procedures are as follows:

1. Lead an oral discussion about desserts, as described in the motivational activity.
2. Place the ingredients and utensils on the table. Clarify terms such as "instant chocolate pudding mix," "glass bowl," "measuring cup," "eggbeater," and "graham cracker crust" as each item is placed on the work table. Label each item so the students can visualize the printed term and associate it with each item.
3. Place the ingredients in the order the students believe is appropriate. Ask them how we can verify whether we have them in the proper order. A student may respond that you have to read a recipe. If no one does, tell them what a recipe is.
4. Read the recipe to the group. As each step is read, have the students check to see if the ingredients and utensils are in the correct order.
5. Orally review the whole sequence using the concrete items as reminders for the procedure.
6. Make the pie according to the recipe directions. Allow each child to take part in some portion of the pie-making activity.
7. Cut the pie and allow all the children to have some.
8. Have the students dictate a chart story about the pie-making experience. Print the story on a large chart. (Follow the procedure for writing the dictation described earlier.)

The following story was dictated as a result of this lesson plan:

We Made a Chocolate Pie

We put chocolate pudding into a glass bowl. We poured milk into the bowl. We beat the milk and pudding with an eggbeater. The pudding got thick. We poured the pudding into a graham cracker crust. We ate the pie. It was good.

REINFORCEMENT
ACTIVITY

1.   Have the students match the labels on the concrete examples with the same terms in the chart story.

2.   Cut a duplicate chart story into sentences. Mix the sentences and have the students put them into the correct sequential order.

3.   Duplicate the story. During the next class period, have the students read the story, underlining words they know. Test these words in isolation. If the words are recognized the following day, place the words in the student's word bank.

The preceding lesson was an example of a group chart story; the language experience approach is often used as an individualized approach in which each student dictates her own story to the teacher. Motivation is also essential for individual stories. A room that is rich with manipulative objects, pictures, animals, and learning centers also provides motivation for story dictation. I have found that a student's own art work is often good motivation for a language experience. The artwork is highly personal, and requires complete involvement on the part of the child. Stockler (1971) investigated motivational topics for Black, culturally different, inner-city students and found that topics related to "self" were of overwhelming concern, with family, friends, school, recreation, and the Black image often preferred. She concluded that Black inner-city children could be motivated with language experience materials that expressed concerns about themselves, their cultures, and their lives. A language experience story developed by a student teacher, Nancy Matula, is included at the close of this section and illustrates how to motivate children to write a story using the concept of autobiographies.

To develop an individual language experience lesson, the children take turns dictating individual stories to the teacher or a teacher's aide. The teacher records the story on lined paper, on newsprint, or with a primer-size typewriter. The story is written exactly as the child dictates it. The dictation procedure is similar to the group chart story; as the child dictates the story, the teacher reads back the words and sentences to the child to make sure they are properly recorded and to reinforce the story's development. It is usually preferable that a beginning story be only a few sentences long, because reading a long story may be discouraging.

Several linguistic authorities discuss the issue of precisely how a child's linguistically different language should be recorded. Jaggar (1974) answered this question by suggesting that the teacher record pronunciation differences in standard spelling (e.g., "des" is written "desk"), but record the different syntactic patterns just as they are represented by the child. She believed that if the child says, / "d principal, he stei In d afis" /, the teacher writes, "The principal, he stay in the office," recognizing that subject redundancy and deletion of third-person verb marker are regular grammatical features of Black English and should be recorded to best represent the child's meaning. To do this accurately, of course, the teacher will need knowledge of the language of the children who are being instructed.

After the dictation, you may have the child read the selection with your help. Have the child underline words that she knows in the story. For example, the following story was written about a child's drawing of an owl. The girl who dictated the story reread the story with the teacher's help and underlined each word she knew.

This owl <u>sitting</u> under <u>the</u> moon. <u>He</u> sitting on <u>a</u> branch. <u>He</u> sleep in <u>the</u> morning. <u>He</u> hunt at <u>night</u>.

The underlined words form a beginning for the child's word bank (words that are instantly recognized as sight words). The teacher gradually reduces the number of visible cues in order to help the child recognize the words not underlined. According to Stauffer (1970), this reduction consists of the following four steps:

1. Use the total story context to help the child read her story.
2. Point to individual words and ask the child to identify the words.
3. Use a window-card to isolate words in the story sequence and ask the child to identify the words.
4. Use a window-card to isolate words in the story in a random order and ask the child to identify these words.

The window-card is easily constructed: cut an opening large enough to show a word in a 3 × 5-inch file card. The teacher moves the card across the story and stops at individual words, testing whether the child can identify the words without the help of the total sentence.

come

Every word that is still recognized the second and third day after dictation is printed on a small card. Stauffer recommended the card size by ⅜-inch by 1½-inches. These cards are placed in an individual file box labeled with the child's name. The cards are used for frequent review of words, sentence-building activities, word-recognition activities, development of alphabetizing skills, and creative writing.

Most children are motivated for further learning as they see their word banks grow. The child can construct many new sentences with the cards in the word bank. He can also experiment with sentence expansion. For example, the word cards:

| Chris | went | to | the |

| store |

can become:

| Chris | and | his | Father |

| went | to | the | grocery |

| store |

or they might form this sentence:

| A | hungry | Chris | went |

| to | the | grocery | store |

| to | buy | a | candy |

| bar |

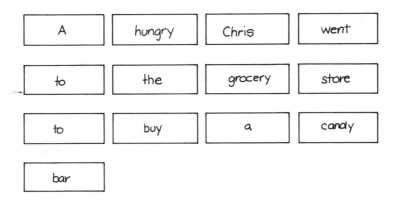

Language experience stories and word banks may also be used for teaching phonics and structural analysis skills. Children can find all the words that begin like

☾★ or all the words that end with "at," like "cat." They can find all the words that have a certain word ending, such as "ed" in "walked." After a number of words are in the word bank, the children may discover they are difficult to find. This occasion provides a good opportunity to introduce alphabetizing. You may wish to have students place small envelopes, one for each letter, inside their word boxes, to facilitate the beginning of alphabetical instruction. This is also a logical time to introduce the use of a dictionary. (This use of word banks was described in the chapter on reference skills.) Finally, the word banks provide a readily available source of words for creative writing.

A LANGUAGE EXPERIENCE ACTIVITY FOR LINGUISTICALLY DIFFERENT CHILDREN

The following activity was developed by one of my students, Nancy Matula, during a practicum experience with fifth-grade students.[1] The group consisted primarily of Black and Mexican-American students. (As you read the activity, try to see how she motivated the students to write and how she involved discussion as part of the experience. When you read about her results, note any examples of Black dialect.) This is Nancy's activity described in her own words:

The Language Experience activity I used with the fifth-graders dealt with biographies and autobiographies. To begin the activity, I read the students a relatively easy biography about Kit Carson. Following the reading, we discussed the book and talked about the specific type of book. The term *biography* was introduced, and students located the meaning of the word in the dictionary. Next, we discussed several characteristics of a biography. We discussed the fact that every one of us could have a biography written about us.

After the discussion of biographies, I introduced the word *autobiography* to the class. We looked at several autobiographies and discussed the differences and similarities between biography and autobiography. We talked about and listed some of the things that could be included in an autobiography. Such items as birthplace, favorite colors, foods, sports, and pets, among other items, made up our list.

At this point, I explained that we were going to write our own book. Because the book (constructed out of poster board held together by rings) had blank white pages, we needed to fill up the pages. Through discussion, it was suggested that we needed both words and pictures. We took both group pictures and individual pictures. It was decided that the group pictures would decorate the cover of our book and that each page would contain one student's picture. Beneath each person's picture would be his or her autobiography. Because we could not write our autobiographies on this same day, I asked the students to continue thinking about what they wanted to include in their autobiographies.

During the following session, we reviewed what could go into our autobiographies. These items were written on the board for further reference. Next, we reviewed writing rules such as beginning sentences with capital letters and ending them with the correct punctuation. The students were off and running (or should I say some were off and walking). Many of the students had difficulty in beginning their writing. They needed assistance with spelling, and a few needed to dictate their stories orally.

Each student completed his or her autobiography and used the story for reading instruction. I also found they were highly motivated to read each other's autobiographies.

---

[1]Used with permission of Nancy Matula, elementary teacher.

The book is to be left in the classroom until the end of the school year, because everyone wants to read the whole book. At the end of the school year, each student can remove and keep his page from the book.

The students were extremely excited and happy with this writing experience. We all learned a great deal while writing about a subject we each know better than any other— ourselves."

The following autobiography was written by one fifth-grade boy:

> *I want to be a singer and a movie star. I have black and white eyes a black pupil, brown hair and afro. My favorite sport is basketball. My favorites singing group is the Jackson five. My favorites foods is hamburger, frenchfries, hot dog and green, green peas. My favorite music is hotline. My color is black.*

REINFORCEMENT ACTIVITY

Choose a motivational activity that might foster both discussion and writing. If you are teaching children, develop a language-experience chart story with a group. If you are not teaching, demonstrate your language experience activity with a group of your peers. Include a motivational activity, the writing of a dictated story, and a reinforcement activity similar to the language experience activity described on page 552.

INNER-CITY AND EARLY-INTERVENTION MODELS

We have looked at several instructional approaches often used for the linguistically different child. Academic achievement of linguistically different children is often lower than that of other children in the schools, thus you may be wondering if any schools have been able to successfully meet the needs of the linguistically different child. Some schools use early preschool intervention programs in an effort to increase language skills and other aspects of school readiness. (As you read the following description of several successful intervention programs and the descriptions of four successful inner-city schools, remember the factors influencing performance of linguistically different children and see if you can discover why these programs may be more successful than others.)

Successful Compensatory-Education Models

The literature describing the long-range achievement of children in some of the early intervention programs has not been encouraging. One reason for the poor results of some of these programs may be the brevity of the intervention program. Eisenberg (1974) suggested characteristics of an intervention program that could succeed. Eisenberg's intervention program would include mental and child health programs for both mother and child, health and education programs for the preschool child, and revised curricula and classroom conditions throughout the years of public schooling.

The mental and child health portion of the program would emphasize the reduction of complications related to malnutrition, poor hygiene, and inadequate medical care by creating special programs to provide medical care and supplementary education for high-risk mothers (e.g., those who are very young, lower-socioeconomic-class). Next, preschool programs would stress early cognitive development. Eisenberg also stressed that early nursery programs must provide parallel educational programs for parents; these educational programs must make sense to inner-city parents. Finally, Eisenberg felt the preschool enrichment program must be augmented by revisions in school programs. These revisions would include selecting the best teachers for inner-city schools; reducing class sizes to private-school levels of fifteen to twenty students per class; curriculum modification; and extension of school programs to include after-school tutoring as well as recreational activities. This early-intervention program would thus begin in the pediatric clinic, continue into early-nursery intervention programs, and conclude with an educational program to meet the unique demands of inner-city children throughout their education.

A compensatory program for inner-city children closely resembling the Eisenberg model has been developed by the Chicago Public Schools. The Child Parent Center program enrolls disadvantaged children, ages three through nine, in a six-year program that includes two years of preschool education, one year of kindergarten, and the first, second, and third grades (Stenner, 1973).

The instructional program is unique at each center, and is tailored to the individual needs of the students in the area it serves. Evaluators of the Chicago program identified the following four contributors to the program's success: (1) early student involvement; (2) heavy parent involvement, especially in the preschool years; (3) continuity of the program into the elementary grades; and (4) structured language/basic skills orientation. The evaluators concluded, "When these four ingredients are present, the traditional gap between disadvantaged students and their more advantaged counterparts can be systematically and substantially eliminated. Likewise, hopes that an isolated preschool experience can increase the achievement potential for disadvantaged children will go unfulfilled unless the receiving school programs are radically altered up through at least grade 3" (Stenner, 1973, p. 248).

The goals of this program are apparently being met because 82 percent of the students in the program demonstrated readiness for kindergarten, and their reading and math scores in first, second, and third grades were above the national norms. The strong language component in the program also stresses the necessity for language arts instruction. Some of the centers provided language arts instruction by using tightly structured linguistic programs, whereas other centers used language experience programs. We have described methods used for a structured language approach as well as methods that utilize the language experience approach. All of the centers stressed growth in the students' use of language, and all programs used praise or other rewards when students completed their tasks successfully.

Successful Inner-City Schools

Achievement in the elementary grades in inner-city schools is usually quite low. Four exceptional inner-city schools, however, were found in a study by Weber (1972).

These schools met the study's inner-city criteria because they were nonselective public schools in the central areas of large cities and were attended by very poor children. The schools met a high achievement criterion because their reading achievement medians equalled or exceeded the national norms and their percentages of nonreaders were exceptionally low for inner-city schools. In fact, the achievement levels of these four successful inner-city schools were found to be approximately that of typical average-income schools. The four schools that met these criteria were P.S. 11 in Manhattan, New York; the John H. Finley School in Manhattan, New York; the Woodland School in Kansas City, Missouri; and the Ann Street School in Los Angeles.

What makes these schools so different from other schools in inner-city areas? Weber maintained that these four successful schools showed characteristics usually absent from unsuccessful inner-city schools.

First, the successful schools showed strong leadership. Either the principals or the area superintendent inaugurated effective programs and thoroughly evaluated their progress. Second, both teachers and administrators had high expectations in regard to the ability of inner-city students. All four schools had teachers with positive attitudes toward their students. Third, the atmosphere in the schools was purposeful, conveying a sense of pleasure in learning. The atmosphere was not characterized by disorder, tension, or confusion. Fourth, there was strong emphasis on reading in all of the successful schools. No single approach was common to the four schools; various materials were used, including basals, *Bank Street* readers, the *We Are Black* series by Science Research Associates, and the Scholastic Library of Paperbacks. One school started instruction with word charts and experience stories. Many library books were utilized in the programs; one school used library books for individualized instruction. Two schools also used programmed reading materials. Fifth, the schools had reading specialists to work with the primary grades and to assist teachers in planning instruction and diagnosing needs. Sixth, the reading program was supplemented with extensive phonics instruction. Seventh, there was a concern for each child's progress, and work was modified to take into account the child's learning problems. Finally, there was careful evaluation of each pupil's progress. The evaluation was continuous and included both formal and informal methods.

The majority of the children in these successful inner-city schools could be classified as linguistically different. As we investigated the linguistically different child, we saw how important teacher attitudes, a motivating environment, oral language activities, and respect for the student's language are to effective instruction. It is clear from Weber's description of these four schools that these institutions have been careful to consider all these factors, and that their effect on student achievement is not just theoretical, but very real.

MULTICULTURAL EDUCATION

America is a multicultural nation, including people with European backgrounds, native Americans, Black Americans, Hispanics, and Asians. A heightened sensitivity to the needs of all people in American society has led to the realization that the language arts program should heighten self-esteem and create a respect for the individuals, the contributions, and the values of all cultures. Unfortunately, literature and other language arts resources frequently ignore the needs of minority ethnic groups or present stereotypic images of ethnic groups.

Contemporary education, at both the university and public school levels, is beginning to stress the need for multicultural education. In this section we focus on Blacks, Native Americans, Hispanics and Asians. Lewis and Doorlag (1987) presented the following reasons for developing multicultural education:

1. Commonalities among people cannot be recognized unless differences are acknowledged.
2. A society that interweaves the best of all of its cultures reflects a truly mosaic image.
3. Multicultural education can restore cultural rights by emphasizing cultural equality and respect.
4. Students can be taught basic skills while also learning to respect cultures; multicultural education need not detract from basic education.
5. Multicultural education should enhance the self-concepts of all students because it provides a more balanced view of American society.
6. Students must learn to respect others.

Positive multicultural literature provides one of the best sources for the language arts teacher. In addition to the values already identified, literature helps children identify a cultural heritage, understand sociological change, respect the values of a minority group, raise aspirations, and expand an understanding of the products of imagination and creativity. My own research (1981, 1984, 1987) also shows that positive multicultural literature and language art activities related to the literature can improve reading scores and improve attitudes toward Black Americans, native Americans, and Hispanic Americans if multicultural literature and literature-related activities are encouraged by placing them in the curriculum and if teachers are provided instructions in selecting this literature and developing teaching strategies that can be used to accompany it. In contrast, this research also indicates that merely placing the literature in the classroom, without subsequent interaction, does not change children's attitudes.

Selecting Multicultural Literature

Two selection problems face the language arts teacher. First, although the number of books about Blacks, native Americans, Hispanics, and Asians is increasing, there is still a disproportionate number that deal in any way with minorities. Second, the selection of books that do not contain stereotypic views of minority cultures and individuals requires careful evaluation. A brief review of children's literature written prior to the late 1960s suggests stereotypes for each minority. For example, books about Blacks frequently characterized them as physically unattractive, musical, dependent on whites, religious with superstitious beliefs, and required to select life goals that benefit Black people. Native Americans were often characterized as savage, depraved, and cruel or noble, proud, silent, and close to nature. Their culture was frequently shown as inferior and not worth retaining. Hispanic literature was criticized for recurring

Literature is an important part of a multicultural program.

themes related to poverty, intervention of Anglos in problem-solving situations, and superficial treatment of problems. Finally, Asian-American literature frequently suggested that Asians looked alike, lived in quaint communities in the midst of large cities, and clung to outworn, alien customs.

Excellent criteria for evaluating books involving Black people are found in Latimer's *Starting Out Right—Choosing Books about Black People for Young Children* (1972, pp. 7–12).[2] Latimer recommended the following criteria:

1. Has a Black perspective been taken into consideration? There should be no stigma attached to being Black, and the characters should not conform to old stereotypes. The dignity of the characters should be preserved.
2. How responsible is the author in dealing with issues and problems? Is it an honest presentation, with problems presented clearly but not oversimplified? Must a Black character exercise all the understanding and forgiveness?
3. Do the Black characters look natural? This is especially crucial in picture books; characters should not have exaggerated features.

[2]Used with permission of Wisconsin Department of Public Instruction, Madison, Wisconsin.

4. Will the young reader realize he is looking at a Black person, or do the characters look "grey" in appearance? ("Grey" refers to a merely darkened version of Caucasian-featured characters.)
5. Is the Black character a unique individual, or is he a representative of a group?
6. Does clothing or behavior perpetuate stereotypes of Blacks as primitive or submissive?
7. Is the character glamorized or glorified, especially in a biography? Some situations may be glorified while others are ignored, resulting in an unreal and unbalanced presentation.
8. Is the setting authentic, so the child can recognize it as an urban, suburban, rural, or fantasized situation?
9. Does the author have a patronizing tone?
10. Is a Black character used as a vehicle to get a point across so that he becomes a tool of literary exploitation and acts artificial rather than real?
11. How are Black characters shown in relationship to white characters? Is either placed in a submissive or inferior role without justification? Is the white person always shown as the benefactor?
12. If dialect is used, does it have a purpose? Does it ring true and blend naturally with the story, or is it used as an example of "substandard" English?
13. If the story deals with historical or factual events, how accurate is it?
14. If the book is a biography, are both the personality and accomplishments of the main character shown?

Well-written literary selections are excellent sources of materials for balancing with multicultural content of the curriculum. They may also be used with any content area. For example, books on African art or American Black art may be used in the art curriculum content. African myths may be added to a literature study of mythology. Mythology provides an excellent source of materials for adding knowledge about a people's culture. Books of poetry from the specific culture can also be added to the study of literature. Biographies of great people from the culture may be used to increase self-esteem and awareness. At the end of this chapter, you will find a list of books that will increase your teaching effectiveness and will also meet Latimer's criteria.

Native American literature includes traditional tales, poetry, historical fiction, contemporary realistic fiction, and informational books. The following criteria may be useful when evaluating native American literature (Norton, 1983):

1. Are the Indian characters portrayed as individuals with their own thoughts, emotions, and philosophies? The characters should not conform to stereotypes or be dehumanized.
2. Do the Indian characters belong to a specific tribe, or are they grouped together under one category referred to as Indian?
3. Does the author recognize the diversity of Indian cultures? Are the customs and ceremonies authentic for the Indian tribes?

4. Is the Indian culture respected, or is it presented as inferior to the white culture? Does the author believe the culture is worthy of preservation or that it should be abandoned? Must the Indian fit into an image acceptable to white characters in the story?
5. Is offensive and degrading vocabulary used to describe the characters, their actions, their customs, or their lifestyles?
6. Are the illustrations realistic and authentic, or do they reinforce stereotypes or devalue the culture?
7. If the story has a contemporary setting, does the author accurately describe the life and situation of the native American in today's world?

Award-winning Hispanic literature for children tends to be about a small segment of the Spanish-American population, the sheepherders of Spanish Basque heritage, whose ancestors emigrated to America before the regions became part of the United States. Award-winning picture storybooks tend to be about Christmas celebrations. Schon (1981) was critical because the "overwhelming majority of recent books incessantly repeat the same stereotypes, misconceptions, and insensibilities that were prevalent in the books published in the 1960s and the early 1970s" (p. 79). Schon supported this contention by reviewing books published in 1980 and 1981 that develop the stereotypes of poverty, children's embarrassment about their backgrounds, distorted and negative narratives about pre-Columbian history and simplistic discussions of serious Latin American problems.

The following criteria may be useful when evaluating Hispanic literature (Norton, 1983):

1. Does the book suggest that poverty is a natural condition for all Hispanics? This is a negative stereotype suggested in some literature.
2. Are problems handled individually, allowing the main characters to use their own efforts to solve their problems? Or are all problems solved through the intervention of an Anglo American?
3. Are problems handled realistically or superficially? Is a character's major problem solved by learning to speak English?
4. Is the cultural information accurate? Are Mexican-American, Mexican, or Puerto Rican cultures realistically pictured? Is the culture treated with respect?
5. Do the illustrations depict individuals, not stereotypes?
6. Is the language free from derogatory terms or descriptions?
7. If dialects are portrayed, are they a natural part of the story and not used to suggest a stereotype?
8. Does the book have literary merit?
9. If the Spanish language is used, are the words spelled and used correctly?

There are few highly recommended books written from an Asian-American perspective. The books by Lawrence Yep, who writes with a sensitivity about Chinese-Americans, have characters who overcome stereotypes associated with Asian-American literature. In addition, his stories integrate information about the cultural heritage into the everyday lives of the people involved. Other authors are writing stories with settings in Vietnam and America. These stories frequently highlight difficulties associated with characters who move into a different culture. Historical fiction set in World War II tells about Japanese-American families who are placed in internment centers. There are also excellent traditional Asian tales that stress the traditional values.

The Council on Interracial Books for Children (1977, pp. 87–90) recommends the following criteria for evaluating Asian-American literature:

1. The book should reflect the realities and way of life of the Asian-American people. Is the story accurate for the historical period and cultural context of the story? Are the characters from a variety of social and economic levels? Does the plot exaggerate the "exoticism" or "mysticism" of the customs and festivals of the Asian-American culture? Are festivals put into the perspective of everyday activities?

2. The book should transcend stereotypes. Are problems handled by the Asian-American characters or is benevolent intervention from a white person required? Does the character have to make a definite choice between two cultures or is there an alternative in which the two cultures can mingle? Do the characters portray a range of human emotions or are they docile and uncomplaining? Is there an obvious occupational stereotype in which all Asian-Americans work in laundries or restaurants?

3. The literature should seek to rectify historical distortions and omissions.

4. The characters in the book should avoid the "model" minority and "super" minority syndromes. Are characters respected for themselves, or must they display outstanding abilities to gain approval?

5. The literature should reflect an awareness of the changing status of women in society. Does the author provide role models for girls other than as subservient females?

6. The illustrations should reflect the racial diversity of Asian-Americans. Are the characters all look-alikes, with the same skin tone and exaggerated features such as slanted eyes? Are clothing and settings appropriate to the culture depicted?

Books that meet the criteria for excellent multicultural literature are listed in the annotated bibliography at the end of this chapter. The books include an interest level designated by age (I:5–8) and a readability level designated by grade (R:3).

Identifying Traditional Values in Multicultural Literature

Educators referenced earlier in this chapter emphasized the desirability of teaching content and developing reasoning ability (Wong-Fillmore, 1986), the advantages of using traditional stories from various cultural sources (Piper, 1986), and the benefits of encouraging students to accept cultural diversity and to understand that similar values frequently underlie different customs (Sealey, 1984). Reading, analyzing, and discussing multicultural literature is one of the best ways to develop these goals. Traditional literature (folktales, myths, and legends) provides excellent sources for discovering the traditional values of a culture. Students gain insights and hone their reasoning abilities as they listen for values, analyze possible values reflected in the stories, compare values across cultures, and even consider if those values are still found in the contemporary literature.

I have used this type of activity with African, Native American, Hispanic, and Asian literature. It is especially beneficial with students in the middle- and upper-elementary grades. The following lesson plan identifies and compares values found in a native American folklore selection and in a realistic fiction selection. Examples of similar activities for African and Hispanic literature are located in Norton's *Language Arts Activities for Children* (1985).

FOR YOUR PLAN BOOK

Identification and Analysis of Traditional Native American Values

Purpose

1. To develop an appreciation for a culture that places importance on oral tradition; respect for nature; understanding between animals and humans; knowledge of the elderly; folklore as a means of passing on culture and tribal beliefs; and diversity of Native American folktales, cultures, and customs.
2. To listen for and to identify traditional values found in native American folklore.
3. To compare the traditional values in folklore and contemporary literature.

Sources

1. Tomie de Paola's **The Legend of the Bluebonnet** (1983)
2. White Deer of Autumn's **Ceremony-In the Circle of Life** (1983)

Procedures

1. Discuss ways in which students can identify traditional values found in folklore. For example, they can read the tales to discover answers to the following questions and then analyze what those answers may reveal about the values and the beliefs of the people:
   a. What is the problem faced by the characters?
   b. What is desired by the characters?
   c. What actions or values are rewarded or respected?
   d. What actions are punished or not respected?
   e. What rewards are given to the heroes, the heroines, or the great people in the stories?
   f. What are the personal characteristics of the heroes, the heroines, or the great people in the stories?
2. Print each question on a chart. Allow room to include several tales. Use the chart to identify the answers to the questions and then to consider what these

answers might reveal about the beliefs and the values of the people. (Using the same chart form for different cultures helps students compare values.)

Comparisons	Folklore The Legend of the Bluebonnet	Realistic Fiction Ceremony-In the Circle of Life
What is the problem?	Drought and famine killed the Comanche	Humans are destroying the land
What is desired?	To end drought and famine	To comfort and honor Mother Earth
	To save land and people	To teach humans about Earth
What actions or values are rewarded or respected?	Sacrifice to save land and tribe	Honoring and caring for Mother Earth
	Belief in Great Spirit	Living in harmony Knowledge, truth, belief
What actions are punished or not respected?	Selfishness Taking from Earth without giving back	Pollution and destruction of Earth
What rewards are given?	Bluebonnets: sign of forgiveness	A living pipe: Symbolizes vision of Earth
	Rain, end of drought Name change	New strength and knowledge Understanding
What are the personal characteristics of heroes or heroines?	Unselfishly loves people and land Respects Great Spirit	Loves animals and land Respects Star Spirit and ways of the People
	Willing to sacrifice for benefits for all	Wants knowledge and truth

3. *Introduce de Paola's* **The Legend of the Bluebonnet**. *Ask the students to listen for the answers to the questions printed on the chart.*

4. *After reading* **The Legend of the Bluebonnet** *aloud, have students identify the answers and place them on the chart. Discuss the implications for the various answers. How do these answers relate to possible values and beliefs of the people?*

5. *During another class time introduce White Deer of Autumn's* **Ceremony-In the Circle of Life**. *Tell the students that this is a contemporary story written by a native American author who wants to share his vision for the earth and his concerns with native American children and other children of the world. Ask students to listen for the answers to the same questions, consider the possible implications of those answers, and compare the values reflected in the two books. Ask them to provide possible reasons for the similarities in values expressed in the two books.*

6. *During other class periods read other traditional literature from various native American tribes and other groups. Ask the children to consider what information they learn from these sources. Additional books for this activity are found in Norton's* **Through the Eyes of a Child: An Introduction to Children's Literature**, *chapter 11 (1987).*

REINFORCEMENT ACTIVITY

Using Latimer's criteria for evaluating multiethnic books (or one of the other criteria for evaluating native American, Hispanic, or Asian-American literature), find several books that you feel are excellent examples of literature showing a Black, native American, Hispanic, or Asian perspective. Find several other books that violate these guidelines. Refer to specific examples in the books in order to substantiate your judgment.

SUMMARY

We have seen that issues relating to instruction of the linguistically different child are complex. This subject is one of great concern, because the majority of children classified as linguistically different have consistently made smaller academic gains compared to children who speak standard English. If linguistically different children are to compete with standard-English speakers, teachers must be specifically prepared to work with the linguistically different child.

Although the current viewpoint stresses the "linguistically different" rather than the "linguistically deficient" philosophy, there is not as yet agreement about the best instructional procedures to use. This chapter reviewed the educational alternatives suggested by linguists and educators. Alternatives include oral language expansion programs, reading conventional materials in dialect, and the language experience approach.

Investigators are concerned not only with the interference of Black English in instruction with standard English materials, but also with the problems of testing linguistically different children using tests written in standard English. Some educators suggest that if standardized tests are used, scoring procedures should be modified; for example, oral reading tests should be scored so that mistakes attributable to dialect differences are not counted as reading errors. Other educators recommend the use of nonstandardized tests. This chapter reviewed some informal testing procedures, such as using pictures to elicit an oral language sample, and role-playing techniques. We noted that it would be difficult to provide evaluation or instruction unless the teacher is familiar with differences between Black English and standard English. In addition to bias within the test, teacher attitudes and environment also influence testing and academic results.

Several successful inner-city and early-intervention models were reviewed. There appear to be a number of contributing factors to successful programs, including

ADDITIONAL MULTICULTURAL EDUCATION ACTIVITIES

1. *Informally interview a child who uses Black English or listen to conversations on a playground (tape conversations). Can you identify any of the phonological and syntactic features discussed in this chapter?*

2. *Read the contrasting viewpoints expressed by S. I. Hayakawa in "Why the English Language Amendment?" and Victor Villanueva, Jr. in "Whose Voice Is It Anyway? Rodriguez's Speech in Retrospect." Both articles are in the **English Journal** 76 (December 1987): 14–21. Discuss each of these viewpoints and then consider how the adoption of each one would influence language arts instruction.*

3. *Read and summarize the findings in three current journal articles that provide information on (1) teaching linguistically different children, (2) teaching bilingual classes, and (3) teaching English as a second language.*

4. *Compare the characteristics of native Americans in a book published before 1965 (for example, Walter Edmond's **The Matchlock Gun**) with the characteristics of native Americans in a book published after 1975 (for example, Brent Ashabranner's **Morning Star, Black Sun: The Northern Cheyenne Indians and America's Energy Crisis**). Find quotations that show stereotypic or nonstereotypic characterizations.*

5. *Read one of the books awarded The Coretta Scott King Award for portraying "people, places, things, and events in a manner sensitive to the true worth and value of things." Read one of the books that the Children's Literature Review Board does not recommend because of stereotypes, unacceptable values, or terms used. For example, David Arkin's **Black and White**; Florine Robinson's **Ed and Ted**; Shirley Burden's **I Wonder Why**; Anco Surany's **Monsieur Jolicoeur's Umbrella**; May Justus's **New Boy in School**; or William Pappas's **No Mules**. Develop a rationale for the recommendations for the King award and for the negative responses for the stereotypic books.*

strong leadership; early student involvement; parent involvement; continuity of the program in elementary grades; language/basic skills orientation; high expectations; positive attitudes on the part of teachers; purposeful atmosphere characterized by pleasure in learning; and careful evaluation of student progress.

This chapter concluded with recommendations for developing course content relevant for linguistically different students. The recommendation was made that teachers evaluate multiethnic literature and balance the educational program by selecting literature that both recognizes that Americans take pride in their race, religion, and social background, and helps students in their search for identity.

ANNOTATED BIBLIOGRAPHY OF MULTICULTURAL CHILDREN'S LITERATURE

The following books are examples of multicultural literature that meet the criteria listed in this chapter. The books include an interest level designated by age (I:5–8) and a readability level designated by grade (R:3).

Black Literature

Traditional Tales

Aardema, Verna. *Bringing the Rain to Kapiti Plain: A Nandi Tale*. Illustrated by Beatriz Vidal. New York: Dial Press, 1981. (I:5–8). A cumulative tale from Kenya tells how a herdsman pierces a cloud with his arrow and brings rain to the parched land.

_____. *What's so Funny, Ketu? A Nuer Tale*. Illustrated by Marc Brown. New York: Dial Press, 1982. (I:5–8, R:6). The repetitive language and African words add authenticity.

_____. *Who's in Rabbit's House?* Illustrated by Leo and Diane Dillon. New York: Dial Press, 1977. (I:7 +, R:3). A Masai folktale illustrated as a play performed by villagers wearing masks.

Bryan, Ashley. *Beat the Story-Drum, Pum-Pum*. New York: Atheneum Pubs., 1980. (I:7 +, R:6). Text includes five Nigerian folktales.

Grifalconi, Ann. *The Village of Round and Square Houses*. Boston: Little, Brown & Co., 1986. (I:4–9, R:6). A folktale from Cameroon.

Hamilton, Virginia. *The People Could Fly: American Black Folktales*. Illustrated by Leo and Diane Dillon. New York: Alfred A. Knopf, 1985. (I:9 +, R:6). Animal, fanciful, supernatural, and slave tales.

Harris, Joel Chandler. *Jump Again! More Adventures of Brer Rabbit*. Adapted by Van Dyke Parks, illustrated by Barry Moser. New York: Harcourt Brace Jovanovich, 1987. (I:8 +, R:6). Five stories in an illustrated version.

Faulkner, William. *The Days when the Animals Talked*. Chicago: Follett, 1977. An excellent adult reference includes Black American folktales and interpretations about how they developed.

Jaquith, Priscilla. *Bo Rabbit Smart for True: Folktales from the Gullah*. Illustrated by Ed Young. New York: Philomel, 1981. (I:all, R:6). Four tales from the islands off the Georgia coast.

Lester, Julius. *The Tales of Uncle Remus: The Adventures of Brer Rabbit*. Illustrated by Jerry Pinkney. New York: Dial Press, 1987, (I:8 +, R:5). A new retelling of the Joel Chandler Harris tales.

Poetry and Songs

Adoff, Arnold. *All the Colors of the Race*. Illustrated by John Steptoe. New York: Lothrop, Lee & Shepard, 1982. (I:all). Poems written from the point of view of a child who has a Black mother and a white father.

_____. *Black Is Warm Is Tan*. New York: Harper & Row, 1973. (I:all). A story in poetic form about an integrated family.

Brown, Marcia. *Shadow* (from the French of Blaise Cendrars). New York: Charles Scribner's Sons, 1982. (I:8 +). Poetry and collage paintings evoke the image of African storytellers and shamans.

Bryan, Ashley. *I'm Going to Sing: Black American Spirituals, Volume Two.* New York: Atheneum Pubs., 1982. (I:all). Words and music are complemented by woodcuts.

———. *Walk Together Children: Black American Spirituals.* New York: Atheneum Pubs., 1974. (I:all). Includes twenty-four spirituals.

Davis, Ossie. *Langston: A Play.* New York: Delacorte Press, 1982. (I:10+). Langston Hughes's poetry used as part of the dialogue in a play about the famous poet.

Feelings, Tom. *Something on My Mind.* New York: Dial Press, 1978. (I:all). Poems and illustrations express hopes and fears associated with growing up.

Contemporary Realistic Fiction

Flournoy, Valerie. *The Patchwork Quilt.* Illustrated by Jerry Pinkney. New York: Dial Press, 1985. (I:5–8, R:4). Constructing a quilt brings a family together.

Greene, Bette. *Phillip Hall Likes Me. I Reckon Maybe.* Illustrated by Charles Lilly. New York: Dial Press, 1974. (I:10+, R:4). A humorous book about a girl's first crush. Set in the Arkansas mountains.

Greenfield, Eloise. *Sister.* Illustrated by Moneta Barnett. New York: Thomas Y. Crowell Co., 1974. (I:8–12, R:5). A thirteen-year-old reviews the memories written in her book.

Grifalconi, Ann. *Darkness and the Butterfly.* Boston: Little, Brown, & Co., 1987. (I:5–8, R:6). An African child overcomes her fear of the dark.

Hamilton, Virginia. *The House of Dies Drear.* Illustrated by Eros Keith. New York: Macmillan Co., 1968. (I:10+, R:4). A contemporary suspenseful story about a family living in a home that was on the Underground Railroad.

———. *M. C. Higgins, the Great.* New York: Macmillan Co., 1974. (I:10+, R:4). A boy dreams of leaving his home and the spoil heap that threatens his security; instead, he builds a wall to protect his home.

———. *The Planet of Junior Brown.* New York: Macmillan Co., 1971. (I:10+, R:6). Three outcasts create their own world in a secret basement room.

———. *Zeely.* Illustrated by Symeon Shimin. New York: Macmillan Co., 1967. (I:8–12, R:4). A girl discovers Zeely's identity and also discovers her own.

Mathis, Sharon Bell. *The Hundred Penny Box.* Illustrated by Diane Dillon. New York: Viking Press, 1975. (I:6–9, R:3). A penny represents each of the important years in the life of a 100-year-old woman.

Steptoe, Joe. *Stevie.* New York: Harper & Row, 1969. (I:3–7, R:3). A boy is at first jealous, but then discovers the importance of friendship.

Historical Fiction

Brenner, Barbara. *Wagon Wheels.* Illustrated by Don Bolognese. New York: Harper & Row, 1978. (I:6–9, R:1). An easy-to-read history book based on the story of a family who moves from Kentucky to Kansas in 1878.

Collier, James, and Collier, Christopher. *Jump Ship to Freedom.* New York: Delacorte Press, 1981. (I:10+, R:7). A slave obtains his and his mother's freedom.

Monjo, F. N. *The Drinking Gourd.* Illustrated by Fred Brenner. New York: Harper & Row, 1970. (I:7–9, R:2). An "I can read" history book tells about a family on the Underground Railroad.

Petry, Ann. *Tituba of Salem Village.* New York: Thomas Y. Crowell Co., 1964. (I:11+, R:6). A talented, sensitive slave becomes part of the Salem witch trials in the 1690s.

Taylor, Mildred D. *Let the Circle Be Unbroken.* New York: Dial Press, 1981. (I:10+, R:6). The author continues the story of the family in *Roll of Thunder Hear My Cry.*

———. *Roll of Thunder, Hear My Cry.* Illustrated by Jerry Pickney. New York: Dial Press, 1976. (I:10+, R:6). A Mississippi family in 1933 experiences night riders and humiliating experiences but retains their independence.

Nonfiction

Adler, David A. *Martin Luther King, Jr.: Free at Last.* Illustrated by Robert Casilla. New York: Holiday, 1986. (I:7–10, R:5). Stresses magnitude of King's work and the reasons he fought against injustice.

Adoff, Arnold. *Malcolm X.* New York: Thomas Y. Crowell Co., 1970. (I:7–12, R:5). A biography of the Black leader, highlighting the changes that took place in his life.

Greenfield, Eloise, and Little, Lessie Jones. *Childtimes: A Three Generation Memoir.* New York: Thomas Y. Crowell Co., 1979. (I:10+, R:5). Three Black women tell about their childhood experiences.

Haskins, James. *Black Theater in America.* Thomas Y. Crowell Co., 1982. (I:10+, R:6). The history of the theater proceeds from minstrel shows to contemporary theater.

Patterson, Lillie. *Frederick Douglas: Freedom Fighter.* Champaign, Il.: Garrard, 1965. (I:6–9, R:3). Biography of the great Black leader.

_____. *Sure Hands, Strong Heart, the Life of Daniel Hale Williams.* Nashville: Tenn.: Abingdon Press, 1980. (I:10+, R:5). Biography of the Black physician.

Tobias, Tobi. *Arthur Mitchell.* Illustrated by Carol Byard. New York: Thomas Y. Crowell Co., 1975. (I:7–9, R:5). Describes the life of the founder of the Dance Theatre of Harlem.

Asian-American Literature

Traditional Tales

Asian Cultural Center for UNESCO. *Folk Tales from Asia for Children Everywhere,* Book Three. Weatherhill, 1976. (I:8–12, R:6). A collection of folktales from many Asian nations.

Carrison, Muriel Paskin, retold by. *Cambodian Folk Stories.* Rutland, Vt.: Tuttle, 1987. (I:8+, R:6). Tales of scoundrels and rascals, kings and lords, and foolishness and fun.

Clark, Ann Nolan. *In the Land of Small Dragon.* Illustrated by Tony Chen. New York: Viking Press, 1979. (I:7–12, R:7). A Vietnamese "Cinderella" story.

Haviland, Virginia. *Favorite Fairy Tales Told in Japan.* Illustrated by George Suyeoka. Boston: Little, Brown, & Co., 1967. (I:8–10, R:5). A collection of Japanese folktales.

Ike, Jane, and Zimmerman, Baruch. *A Japanese Fairy Tale.* New York: Warne, 1982. (I:5–8, R:5). A hunchback takes the disfiguration of his future wife, allowing her to be beautiful.

Laurin, Anne. *The Perfect Crane.* Illustrated by Charles Mikolaycak. New York: Harper & Row, 1981. (I:5–9, R:6). A Japanese folktale about friendship between a magician and the crane he creates from paper.

Lee, Jeanne M. *Legend of the Milky Way.* New York: Holt, Rinehart & Winston, 1982. (I:5–8, R:4). A Chinese folktale tells about the origin of the Milky Way.

Louie, Ai-Lang. *Yeh Shen: A Cinderella Story from China.* Illustrated by Ed Young. New York: Philomel, 1982. (I:7–9, R:6). A Chinese variation of the "Cinderella" story.

Newton, Patricia Montgomery. *The Five Sparrows: A Japanese Folktale.* New York: Atheneum Pubs., 1982. (I:5–8, R:6). Kindness is rewarded and greed is punished in this Japanese folktale.

Philip, Neil, ed. *The Spring of Butterflies and Other Folktales of China's Minority Peoples.* Translated by He Liyi, illustrated by Pan Aiqing and Li Zhao. New York: Lothrop, 1986. (I:9+, R:6). Tales from northwestern China.

Sadler, Catherine Edwards. *Treasure Mountain: Folktales from Southern China.* Illustrated by Chen Mung Yun. New York: Atheneum Pubs., 1982. (I:8+, R:6). Six folktales depict values such as kindness and humor and disliked characteristics such as greed.

Yagawa, Sumiko. *The Crane Wife.* Illustrated by Suekicki Akaba. New York: William Morrow & Co., 1981. (I:7–10, R:6). A wife returns to her animal form when her husband breaks his promise.

Contemporary and Historical Literature

Clark, Ann Nolan. *To Stand against the Wind.* New York: Viking Press, 1978. (I:10+, R:4). A Vietnamese boy now living in America prepares for the traditional Day of the Ancestors.

Davis, Daniel S. *Behind Barbed Wire: The Imprisonment of Japanese Americans during World War II.* New York: E. P. Dutton, 1982. (I:10+, R:7). The author explores the actions taken against Japanese Americans following the declaration of war.

Friedman, Ina R. *How My Parents Learned to Eat.* Illustrated by Allen Say. Boston: Houghton Mifflin Co., 1984. (I:6–8, R:3). A humorous story about trying to eat with chopsticks or with knives and forks.

Lord, Bette Bao. *In the Year of the Boar and Jackie Robinson.* Illustrated by Marc Simont. New York: Harper & Row, 1984. (I:8–10, R:4). Developing a love for baseball helps a Chinese girl make friends in America.

Nhuong, Huynh Quang. *The Land I Lost: Adventures of a Boy in Vietnam.* Illustrated by Vo-Dinh Mai. New York: Harper & Row, 1982. (I:8–12, R:6). The author tells about his boyhood experiences.

Uchida, Yoshiko. *Journey Home.* Illustrated by Charles Robinson. New York: Atheneum Pubs., 1978. (I:10+, R:5). In a sequel to *Journey to Topaz,* Yuki and her parents return to California after World War II.

Wallace, Ian. *Chin Chiang and the Dragon's Dance.* New York: Atheneum Pubs., 1984. (I:6–9, R:6). A young boy dreams of dancing on the first day of the Year of the Dragon.

Yep, Lawrence. *Child of the Owl.* New York: Harper & Row, 1977. (I:10+, R:7). A Chinese-American girl learns about her heritage when she lives with her grandmother.

———. *Dragonwings.* New York: Harper & Row, 1975. (I:10+, R:6). A historical book set in San Francisco, 1903.

———. *Sea Glass.* New York: Harper & Row, 1979. (I:10+, R:6). A boy faces problems as he tries to make his father understand his desires.

Native-American Literature

Traditional Literature

Baker, Betty. *Rat Is Dead and Ant Is Sad.* Illustrated by Mamoru Funai. New York: Harper & Row, 1981. (I:6–8, R:2). A cumulative Pueblo Indian tale stresses the consequences of reaching wrong conclusions.

Baker, Olaf. *Where the Buffaloes Begin.* Illustrated by Stephen Gammell. New York: Warne, 1981. (I:8+, R:7). A Prairie Indian legend about the lake in which buffaloes are created and how the buffaloes help a boy who believes.

Baylor, Byrd. *And It Is Still That Way: Legends Told by Arizona Indian Children.* New York: Charles

Scribner's Sons, 1976. (I:all, R:3). A collection of tales told by children.

———. *God on Every Mountain.* Illustrated by Carol Brown. New York: Charles Scribner's Sons, 1981. (I:6–10, R:5). Southwest Indian tales about sacred mountains.

Cleaver, Elizabeth. *The Enchanted Caribou.* New York: Atheneum Pubs., 1985. (I:6–10, R:6). An Inuit tale of transformation.

Coatsworth, Emerson, and Coatsworth, David. *The Adventures of Nana Bush: Ojibway Indian Stories.* Illustrated by Frances Kagige. New York: Atheneum Pubs., 1980. (I:8+, R:6). Sixteen tales about a powerful spirit.

de Paola, Tomie. *The Legend of the Bluebonnet.* New York: G. P. Putnam's Sons, 1983. (I:all, R:6). A Comanche tale in which unselfish actions are rewarded.

Goble, Paul. *The Girl Who Loved Wild Horses.* Scarsdale, N.Y.: Bradbury, 1978. (I:6–10, R:5). An American Indian girl, who loves wild horses, joins them in a flight during a storm.

Grinnell, George Bird. *The Whistling Skeleton: American Indian Tales of the Supernatural.* Edited by John Bierhorst. Illustrated by Robert Andrew Parker. New York: Four Winds, 1982. (I:10+, R:6). Nine mystery tales told by nineteenth century storytellers from the Cheyenne, Pawnee, and Blackfoot tribes.

Haviland, Virginia. *North American Legends.* Illustrated by Ann Strugnell. New York: Collins, 1979. (I:8+, R:6). An anthology of North American tales.

Highwater, Jamake. *Anpao: An American Indian Odyssey.* Illustrated by Fritz Scholder. Philadelphia: J. B. Lippincott Co., 1977. (I:12+, R:5). A native American travels across the history of traditional tales in order to search for his destiny.

Robinson, Gail. *Raven the Trickster. Legends of the North American Indians.* Illustrated by Joanna Troughton. New York: Atheneum Pubs., 1982. (I:8–12, R:6). Nine tales from the Northwest.

Spencer, Paula Underwood. *Who Speaks for Wolf.* Illustrated by Frank Howell. Austin: Tribe of Two Press, 1983. (I:all). A native American learning story that emphasizes the need to consider the animals.

Poetry and Songs

Baylor, Byrd. *Before You Came This Way.* Illustrated by Tom Bahti. New York: E. P. Dutton, 1969. (I:all). The Indian petroglyphs of the Southwest are described in poetic form.

_____. *The Desert Is Theirs.* Illustrated by Peter Parnall. New York: Charles Scribner's Sons, 1975. (I:all). The poetic form captures the life of the Papago Indians.

_____. *Hawk, I'm Your Brother.* Illustrated by Peter Parnall. New York: Charles Scribner's Sons, 1976. (I:all). A young boy longs to glide through the air like a hawk.

Belting, Natalia. *Whirlwind Is a Ghost Dancing.* Illustrated by Leo and Diane Dillon. New York: E. P. Dutton, 1974. (I:all). The lore of numerous tribes is depicted in poetic form.

Bierhorst, John. *A Cry from the Earth: Music of the North American Indians.* New York: Four Winds, 1979. (I:all). A collection of native American songs.

Contemporary and Historical Literature

Dodge, Nanabah Chee. *Morning Arrow.* Illustrated by Jeffrey Lunge. (New York: Lothrop, Lee, and Shepard, 1975. (I:7–10, R:3). A ten-year-old Navaho boy lives with and helps his partially blind grandmother.)

Hudson, Jan. *Sweetgrass.* Edmonton: Tree Frog, 1984. (I:10+, R:4). A Blackfeet girl grows up during the winter of a smallpox epidemic in 1837.

Martin, Bill, and Archambault, John. *Knots on a Counting Rope.* Illustrated by Ted Rand. New York: Holt, Rinehart & Winston, 1987. (I:7+) Rhythmic tale of a blind native American boy who rides in a horse race.

Miles, Miska. *Annie and the Old One.* Illustrated by Peter Parnall. Boston: Little, Brown, & Co., 1971. (I:6–8, R:3). Annie's love for her grandmother causes her to interfere with the completion of a rug she associates with the probable death of her grandmother.

O'Dell, Scott. *Sing Down the Moon.* Boston: Houghton Mifflin Co., 1970. (I:10+, R:6). The 1864 forced march of the Navaho is told through the viewpoint of a young girl.

Rockwood, Joyce. *Groundhog's Horse.* Illustrated by Victor Kalin. New York: Holt, Rinehart & Winston, 1978. (I:7–12, R:4). A humorous, warm story tells of a Cherokee boy, who in 1750, rescues his horse.

Sneve, Virginia Driving Hawk. *High Elk's Treasure.* Illustrated by Oren Lyons. New York: Holiday, 1972. (I:8–12, R:6). A dream beginning in 1876 is renewed in the late 1970s.

_____. *Jimmy Yellow Hawk.* Illustrated by Oren Lyons. New York: Holiday, 1972. (I:6–10, R:5). The story of a contemporary boy who lives on an Indian reservation in South Dakota.

White Deer of Autumn. *Ceremony-In the Circle of Life.* Illustrations by Daniel San Souci. Milwaukee: Raintree, 1983. (I:all). A contemporary boy learns about his heritage and the importance of living in harmony with nature.

Nonfiction

Ashabranner, Brent. *Morning Star, Black Sun: The Northern Cheyenne Indians and America's Energy Crisis.* Photographs by Paul Conklin. New York: Dodd, Mead & Co., 1982. (I:10+, R:7). The text traces the history of the Northern Cheyenne Indians and discusses the tribe's fight to save their lands from power companies and strip mining.

Fall, Thomas. *Jim Thorpe.* Illustrated by John Gretzer. New York: Thomas Y. Crowell Co., 1970. (I:7–9, R:2). A simple biography about the great athlete.

Freedman, Russell. *Indian Chiefs.* New York: Holiday, 1987. (I:8+, R:7). Biographies of six native American chiefs, photographs add authenticity.

Hirschfelder, Arlene. *Happily May I Walk: American Indians and Alaska Natives Today.* New York: Charles Scribner's Sons, 1986. (I:10+, R:6). An excellent resource book about contemporary life.

McGraw, Jessie Brewer. *Chief Red Horse Tells about Custer: The Battle of Little Bighorn: An Eyewitness Account Told in Indian Sign Language.* New York: Elsevier/Nelson, 1981. (I:8+). The historical background and glossary of Indian terms add to the story based on pictographs.

Tobias, Tobi. *Maria Tallchief*. Illustrated by Michael Hampshire. New York: Thomas Y. Crowell Co., 1970. (I:7–12, R:4). A biography of the ballerina who was a member of the Osage Indian tribe.

Weiss, Malcolm. *Sky Watchers of Ages Past*. Illustrated by Eliza McFadden. Boston: Houghton Mifflin Co., 1982. (I:10+, R:6). The author introduces readers to some of the astronomers of the past such as the Anasazi Indians and the Mayans.

Hispanic Literature

Traditional Tales

Aardema, Verna. *The Riddle of the Drum: A Tale from Tizapán, Mexico*. Illustrated by Tony Chen. New York: Four Winds, 1979. (I:6–10, R:3). The man who marries the king's daughter must guess the kind of leather in a drum.

Belpré, Pura. *The Rainbow-Colored Horse*. Illustrated by Antonio Martorell. New York: Warne, 1978. (I:6–10, R:6). Three favors granted by a horse allow Pio to win the hand of Don Nicanoro's daughter.

Bierhorst, John, ed. *Black Rainbow: Legends of the Incas and Myths of Ancient Peru*. New York: Farrar, Straus & Giroux, 1976. (I:10+, R:7). Twenty traditional tales.

_____. *Spirit Child: A Story of the Nativity*. Illustrated by Barbara Cooney. New York: William Morrow & Co., 1984. (I:8–12, R:6). Pre-Columbian style illustrations accompany an Aztec story.

dePaola, Tomie. *The Lady of Guadalupe*. New York: Holiday, 1980. (I:8+, R:6). A Mexican tale about the patron saint of Mexico who appeared to an Indian in 1531.

Hinojosa, Francisco, adapted by. *The Old Lady Who Ate People*. Illustrated by Leonel Maciel. Boston: Little, Brown & Co., 1984. (I:all, R:6). Four frightening tales from Mexico.

Jagendorf, M. A., and Boggs, R. W. *The King of the Mountains: A Treasury of Latin American Folk Stories*. New York: Vanguard, 1960. (I:9+, R:6). A collection of tales from twenty-six countries.

Contemporary and Historical Literature

Clark, Ann Nolan. *Year Walk*. New York: Viking Press, 1975. (I:10+, R:7). A Spanish Basque sheepherder faces loneliness and the challenges of becoming a man when he crosses the desert into the high country.

Ets, Marie Hall, and Labastida, Aurora. *Nine Days to Christmas, A Story of Mexico*. New York: Viking Press, 1959. (I:5–8, R:3). The book includes lovely illustrations of a girl preparing for a Mexican Christmas holiday.

Krumgold, Joseph. *And Now Miguel*. Illustrated by Jean Charlot. New York: Thomas Y. Crowell Co., 1953. (I:10+, R:3). Miguel, a member of a proud sheep-raising family, wishes to go with the sheepherders to the Sangre de Cristo Mountains.

Mohr, Nicholasa. *Felita*. Illustrated by Ray Cruz. New York: Dial Press, 1979. (I:9–12, R:2). A family tries to adjust to a new neighborhood.

O'Dell, Scott. *The Captive*. Boston: Houghton Mifflin Co., 1977. (I:10+, R:6). A Spanish seminarian witnesses the exploitation of the Mayas during the 1500s.

_____. *Carlota*. Boston: Houghton Mifflin Co., 1977. (I:9+, R:4). A girl fights beside her father in California during the Mexican War.

_____. *The Feathered Serpent*. Boston: Houghton Mifflin Co., 1981. (I:10+, R:6). The seminarian in *The Captive* witnesses the arrival of Cortés.

Politi, Leo. *The Nicest Gift*. New York: Charles Scribner's Sons, 1973. (I:5–8, R:6). There are many Spanish words in the text and illustrations from the barrio in East Los Angeles.

_____. *Song of the Swallows*. New York: Charles Scribner's Sons, 1949. (I:5–8, R:4). Illustrations of Spanish architecture in a story set in Capistrano.

Taha, Karen T. *A Gift for Tia Rosa*. Illustrated by Dee deRosa. Minneapolis: Dillon, 1986. (I:5–8, R:3). A young girl faces a neighbor's death and learns about love.

Nonfiction

Ashabranner, Brent. *Children of the Maya: A Guatemalan Indian Odyssey*. Photographs by Paul Conklin. New York: Dodd, Mead & Co., 1986. (I:9–12, R:6). A contemporary report about Mayans who settled in Florida after escaping from Guatemala.

Brown, Tricia. *Hello, Amigos!* Photographs by Fran Ortiz. New York: Holt, Rinehart & Winston, 1986.

(I:6–9). A photographic essay about a Mexican-American boy who lives in San Francisco.

Franchere, Ruth. *Cesar Chavez.* Illustrated by Earl Thollander. New York: Thomas Y. Crowell Co., 1970. (I:7–9, R:4). A biography depicting Chavez's struggles to improve the pay and living conditions for migrant workers.

Martinello, Marian L., and Nesmith, Samuel P. *With Domingo Leal in San Antonio 1734.* The University of Texas, Institute of Texas Cultures at San Antonio, 1979. (I:8+, R:4). The text is based on research on the lives of Spanish settlers who arrived in Texas during the 1730s.

Meltzer, Milton. *The Hispanic Americans.* Photographs by Morrie Camhi and Catherine Noren. New York: Thomas Y. Crowell Co., 1982. (I:9–12, R:6). The author discusses the influence in America of Puerto Ricans, Chicanos, and Cubans.

Wolf, Bernard. *In This Proud Land: The Story of a Mexican American Family.* Philadelphia: J. B. Lippincott Co., 1978. (I:all, R:4). Photographs and text follow a family from the Rio Grande Valley to Minnesota for summer employment.

BIBLIOGRAPHY

Adler, Sol. "Data Gathering: The Reliability and Validity of Test Data from Culturally Different Children." *Journal of Learning Disabilities* 6 (August/September 1973): 429–34.

Allen, Roach Van. *Language Experiences in Communication.* Boston: Houghton Mifflin Co., 1976.

Baratz, Joan C. "A Bi-Dialectical Task for Determining Language Proficiency in Economically Disadvantaged Children." *Child Development* 40 (December 1969): 889–901.

———, and Shuy, Roger. *Teaching Black Children to Read.* Washington, D.C.: Center for Applied Linguistics, 1969.

Bartel, Nettie R.; Grill, Jeffrey, J.; and Bryen, Diane N. "Language Characteristics of Black Children: Implications for Assessment." *Journal of School Psychology* 11 (Winter 1973): 351–64.

Batty, Constance J., and Batty, Beauford R. "Teaching Minority Children to Read in Elementary School." In *Tapping Potential: English and Language Arts for the Black Learner,* edited by Charlotte K. Brooks. Urbana, Ill.: National Council of Teachers of English, 1985.

Bereiter, Carol, and Engelmann, Siegfried. *Teaching Disadvantaged Children in the Preschool.* Englewood Cliffs, N.J.: Prentice-Hall, 1966.

Bernstein, B. "Social Structure, Language, and Learning." *Educational Research* 3 (1961): 163–76.

Blodgett, Elizabeth G., and Cooper, Eugene B., "Attitudes of Elementary Teachers toward Black Dialect." *Journal of Communication Disorders* 6 (June 1973): 121–33.

Bradley, Robert H., and Caldwell, Bettye M., "The Relation of Infants' Home Environments to Mental Test Performance at Fifty-Four Months: A Follow-Up Study." *Child Development* 47 (December 1976): 1172–74.

Brooks, Charlotte K. *Tapping Potential: English and Language Arts for the Black Learner.* Urbana, Ill.: National Council of Teachers of English, 1985.

Burton, Nancy W., and Jones, Lyle V. "Recent Trends in Achievement Levels of Black and White Youth." *Educational Research* (April 1982): 10–14, 17.

Byrd, M. L., and Williams, H. S. *Language Attitudes and Black Dialect: An Assessment.* (1) Language Attitudes in the Classroom. (2) A Reliable Measure of Language Attitudes. Paper presented at the annual meeting of the Speech Communication Association, Anaheim, California, November 1981 (ERIC Document Reproduction Service No. ED213062).

Calvert, John D. *An Exploratory Study to Adapt the Language Experience Approach to Remedial Seventh and Tenth Grade Mexican American Students.* Doctoral dissertation, Arizona State University, 1973.

Carter, Candy, ed. *Non-Native and Nonstandard Dialect Students: Classroom Practices in Teaching English, 1982–1983.* Urbana, Ill.: National Council of Teachers of English, 1982.

Cohen, S. A. *Teach Them All to Read.* New York: Random House, 1969.

Cooper, Harris M.; Baron, Reuben M.; and Love, Charles A. "The Importance of Race and Social Class Information in the Formation of Expectancies about Academic Performance." *Journal of Educational Psychology* 67 (April 1975): 312–19.

Corrigan, Dean. *Teaching Excellence.* Paper presented at the meeting of American Association of University Women, College Station, Texas, October, 1983.

Council on Interracial Books for Children. "Criteria for Analyzing Books on Asian Americans." In *Cultural Conformity in Books for Children,* edited by Donnarae MacCann and Gloria Woodard. Metuchen, N.J.: Scarecrow, 1977.

Cravioto, Joaquin, and DeLicardie, Elsa R. "Environmental and Learning Deprivation in Children with Learning Disabilities." In *Perceptual and Learning Disabilities in Children, Research and Theory,* vol. 2, edited by William R. Cruickshank and Daniel P. Hallahan. Syracuse, N.Y.: Syracuse University Press, 1975.

Crowl, Thomas K., and MacGinitie, Walter H. "The Influence of Students' Speech Characteristics on Teachers' Evaluations of Oral Answers." *Journal of Educational Psychology* 66 (June 1974): 304–8.

Cullinan, Bernice E.; Jaggar, Angela M.; and Strickland, Dorothy. "Oral Language Expansion in the Primary Grades." In *Black Dialects and Reading,* edited by Bernice E. Cullinan. Urbana, Ill.: National Council of Teachers of English, 1974.

Cunningham, Patricia M. "Teachers' Correction Responses to Black Dialect Miscues Which Are Non-Meaning Changing." *Reading Research Quarterly* 12 (Summer 1977): 637–53.

Daniels, Lee A. "Study Links Academic Skills to Race and Family Income." *New York Times,* Sunday, October 25, 1987, p. 15.

de Felix, Judith Walker. *"Steps to Second Language Development in the Regular Classroom."* In Candy Carter, editor, *Non-Native and Nonstandard Dialect Students: Classroom Practices in Teaching English, 1982–1983.* Urbana, Ill.: National Council of Teachers of English, 1982.

Deutsch, M.; Bloom, R. D.; Brown, B. R.; Deutsch, C. P.; Goldstein, L. S.; John, V. P.; Katz, P. A.; Levinson, A.; Peisach, E. C.; and Whiteman, M. *The Disadvantaged Child.* New York: Basic Books, 1967.

Dixon, Carol N. "Teaching Strategies for the Mexican-American Child." *The Reading Teacher* 30 (November 1976): 141–45.

Eisenberg, Leon. "The Epidemiology of Reading Retardation and a Program for Preventive Intervention." In *The Disabled Reader,* edited by John Money. Baltimore: Johns Hopkins University Press, 1974.

Gantt, Walter N.; Wilson, Robert M.; and Dayton, Mitchell C. "An Initial Investigation of the Relationship between Syntactical Divergency and the Listening Comprehension of Black Children." *Reading Research Quarterly* 10 (Winter 1974–75): 193–211.

Greenberg, Judith W., and Davison, Helen H. "Home Background and School Achievement of Black, Urban Ghetto Children." *American Journal of Orthopsychiatry* 42 (October 1972): 803–10.

Grill, J. Jeffrey, and Bartel, Nettie R. "Language Bias in Tests: ITPA Grammatic Closure." *Journal of Learning Disabilities* 4 (April 1977): 229–35.

Hall, MaryAnne. *The Development of Evaluation of a Language Experience Approach to Reading with First-Grade Culturally Disadvantaged Children.* Doctoral dissertation, University of Maryland, 1965.

Hall, William S., and Freedle, Roy O. *Culture and Language: The Black American Experience.* New York: John Wiley & Sons, 1975.

Harber, Jean R., and Beatty, Jane N. *Reading and the Black English Speaking Child.* Newark, Del.: International Reading Association, 1978.

Hoover, Irene. *Historical and Theoretical Development of a Language Experience Approach to Teaching Reading in Selected Teacher Education Institutions.* Doctoral dissertation, University of Arizona, 1971.

Hunt, Barbara Carey. "Black Dialect and Third and Fourth Graders' Performance on the Gray Oral Reading Test." *Reading Research Quarterly* 10 (Fall 1974): 103–23.

Hutchinson, June O'Shields. "Reading Tests and Nonstandard Language." *Reading Teacher* 25 (February 1972): 430–37.

Jaggar, Angela M. "Beginning Reading: Let's Make It a Language Experience for Black English Speakers." In *Black Dialects and Reading,* edited by Bernice E. Cullinan. Urbana, Ill.: National Council of Teachers of English, 1974.

Kagan, S., and Madsen, M. "Cooperation and Competition of Mexican, Mexican American, and Anglo American Children of Two Ages under Four Instructional Sets." *Developmental Psychology* 5 (1971): 32–39.

Kronkosky, Preston C. "Testimony Sponsored by the U.S. Department of Education's Public Hearings on Excellence in Education." Speech presented at Dallas, Texas, October 1982.

Labov, William. *Language in the Inner City: Studies in the Black English Vernacular.* Philadelphia: University of Pennsylvania Press, 1973.

_____. *The Logic of Non-Standard English.* Washington, D.C.: Georgetown Monograph Series on Language and Linguistics No. 22, 1969.

_____. "The Logic of Non-Standard English." In *Language and Poverty,* edited by Frederick Williams. Chicago: Markham Publishing Co. 1971.

_____. *Study of Non-Standard English.* Urbana, Ill.: National Council of Teachers of English, 1970.

Latimer, Bettye. *Starting Out Right—Choosing Books about Black People for Young Children.* Madison, Wis.: Wisconsin Department of Public Instruction Bulletin Number 2314, 1972.

Lee, Doris, and Allen, Roach Van. *Learning to Read through Experience.* New York: Appleton Century Crofts, 1963.

Lewis, Rena B., and Doorlag, Donald H. *Teaching Special Students in the Mainstream.* 2d ed. Columbus, Oh.: Merrill Publishing Co., 1987.

Litsinger, Dolores Escobar. *The Challenge of Teaching Mexican-American Students.* New York: American Book Co., 1973.

Maestas, Leo Carlos. "Ethnicity and High School Student Achievement across Rural and Urban District." *Educational Research Quarterly* (Fall 1981): 33–42.

Malmstrom, Jean. *Understanding Language.* New York: St. Martin's Press, 1977.

Mantell, Arlene. "Strategies for Language Expansion in the Middle Grades." In *Black Dialects and Reading,* edited by Bernice E. Cullinan. Urbana, Ill.: National Council of Teachers of English, 1974.

Murnane, Richard J. *The Impact of School Resources on the Learning of Inner City Children.* Cambridge, Mass.: Ballinger, 1975.

Norton, Donna E. "Changing Attitudes toward Minorities: Children's Literature Shapes Attitudes." *Review Journal of Philosophy and Social Science* 9 (1984): 97–113.

_____. "The Development, Dissemination, and Evaluation of a Multi-Ethnic Curricular Model for Preservice Teachers, Inservice Teachers, and Elementary Children." New Orleans: International Reading Association, National Conference, April 1981.

_____. *Language Arts Activities for Children.* 2d ed. Columbus, Oh.: Merrill Publishing Co., 1985.

_____. *Through the Eyes of a Child: An Introduction to Children's Literature.* 2d ed. Columbus, Oh.: Merrill Publishing Co., 1987.

Piper, David. "Language Growth in the Multiethnic Classroom." *Language Arts* 63 (January 1986): 23–36.

Purves, Alan, and Purves, William. "Viewpoints: Cultures, Text Models, and the Activity of Writing." *Research in the Teaching of English* 20 (May 1986): 174–97.

Ramírez III, Manuel, and Castañeda, Alfredo. *Cultural Democracy, BiCognitive Development, and Education.* New York: Academic Press, 1974.

Ramírez III, Manuel, and Price-Williams, D. R. "Cognitive Styles of Children of Three Ethnic Groups in the United States." *Journal of Cross-Cultural Psychology* 5 (1974): 212–19.

_____. *The Relationship of Culture to Educational Attainment.* Center for Research in Social Change and Economic Development, Houston, Tex.: Rice University, 1971.

Raspberry, William. "Black America's House Fire." Bryan, Texas: *Eagle.* Friday, February 19, 1988, p. 6A.

Schon, Isabel. "Recent Detrimental and Distinguished Books about Hispanic People and Cultures." *Top of the News* 38 (Fall 1981): 79–85.

Sealey, D. Bruce. "Measuring the Multicultural Quotient of a School." *TESL Canada Journal/Revue TESL Du Canada* 1 (March 1984): 21–28.

Shuy, Roger W., and Williams, Frederick. "Stereotyped Attitudes of Selected English Dialect Communities." In *Language Attitudes: Current Trends and Prospects,* edited by Roger W. Shuy and Ralph W. Fasold, Washington, D.C.: Georgetown University Press, 1973.

Stauffer, Russell. *The Language-Experience Approach to the Teaching of Reading.* New York: Harper & Row, 1970.

Stenner, A. Jackson, and Mueller, Siegfried. "A Successful Compensatory Education Model." *Phi Delta Kappan* (December 1973): 246–48.

Stockler, Dolores S. *Responses to the Language Experience Approach by Black Culturally Different, Inner-City Students Experiencing Reading Disability in Grades Five and Eleven.* Doctoral dissertation, Indiana University, 1971.

Vick, Marian. "Relevant Content for the Black Elementary School Pupil." In *Literacy for Diverse Learners,* edited by Jerry L. Johns. Newark, Del.: International Reading Association, 1974.

Watson-Thompson, O. B. "An Investigation of Elementary School Teachers' Attitudes Toward the Use of Black English in West Alabama." Doctoral Dissertation, Univ. of Alabama, 1985. *Dissertation Abstracts International* 46, 11A. (University Microfilms No. DA 8600778)

Weber, George. *Inner-City Children Can Be Taught to Read: Four Successful Schools.* Washington, D.C.: Council for Basic Education, Occasional Papers No. 18, 1972.

Wolfram, Walt. "Sociolinguistic Implications for Educational Sequencing." In *Teaching Standard English in the Inner City,* edited by Ralph Fasold and Roger W. Shuy. Washington, D.C.: Center for Applied Linguistics, 1970.

Wong-Fillmore, Lily. "Research Currents: Equity or Excellence? *Language Arts* 63 (September 1986): 474–81.

Chapter Fifteen

EFFICIENT LANGUAGE ARTS INSTRUCTION
Diagnosing Student Needs ▪ Instructional Time
▪ Flexible Grouping ▪ Integrating Language Arts
Instruction ▪ Flexible Room Arrangements

LEARNING CENTERS
Characteristics of Centers ▪ Components of
Centers ▪ Creating the Center

ORGANIZING INSTRUCTION
Steps in Blocking Time Periods ▪ Examples of
Instructional Planning Blocks

SUMMARY

*After completing the chapter on classroom
organization and management, you will be able to:*

1. *List and describe the factors influencing effi-
    cient language arts instruction.*
2. *Suggest ways to group children according to
    recommended diagnostic techniques.*
3. *Describe the use of flexible grouping practices
    that utilize whole-class instruction, smaller-
    group instruction, and individual instruction
    and activities for independent work.*
4. *Describe flexible room arrangements that
    would be appropriate for whole-class instruc-
    tion, various types of group instruction, and
    individualized and independent work.*
5. *Describe the development of effective assign-
    ments.*
6. *Describe how you would integrate language
    arts instruction.*
7. *Develop a language arts learning center where
    children can work independently.*
8. *Organize classroom instruction using an in-
    structional blocking technique.*

Classroom Organization and Management

In this chapter, we discuss organizing the classroom for effective language arts instruction, which occupies a major portion of the elementary classroom day. We have already discussed effective assessment and instruction in oral language, listening, handwriting, spelling, writing, grammar, reading, literature, media, and reference skills. For instruction in all these areas to be maximally effective, the teacher must be able to manage classroom organization efficiently.

Solving the problems associated with classroom organization and management is not unique to beginning teachers. They are problems of major concern to experienced teachers, who frequently request discussions on classroom organization and management during inservice training programs. At these sessions, teachers ask: (1) How can I effectively divide my class into groups? (2) How much time should be spent on various language arts? (3) How can I provide flexible classroom groupings? (4) If I divide my class into groups, how can I provide profitable assignments for the groups who are not working directly with me? (5) Do all language arts skills need to be taught in small groups or individualized? (6) Will interest centers improve the organization and management of the classroom? (7) How can I interrelate language arts with other subject areas?

EFFICIENT LANGUAGE ARTS INSTRUCTION

Answers to these questions reflect many of the factors that influence efficient management and organization in the language arts. Petty, Petty, and Backing (1973) suggested that effective instruction should provide for individual needs. They believed that one way to provide for these individual needs is to use evaluative and diagnostic procedures followed by instruction that uses a variety of grouping practices. Dorothy Hennings (1978) identified three ways to organize instruction so as to include in the curriculum a literature, oral language, individual writing, and sharing component. These grouping procedures include full-class instruction, small-group instruction, and individual activity. In addition to discovering individual needs and providing instruction through appropriate groupings, Kean and Personke (1976) suggested the need for flexibility in the curriculum; balance in the subjects studied, so that instructional time is provided for all important aspects of communication; balance in subjects studied, so activities such as creative writing, creative use of literature and media, speaking, and listening are not de-emphasized; balance between formal and informal learning; balance between oral and written work; and balance in cultural experiences.

A practical list of management and organizational skills is provided by Harris and Sipay (1985). They concluded that effective planning and organization require instructional time, motivation, grouping for various purposes, manageable numbers of groups, manageable numbers of students within groups, flexible grouping practices, flexible classroom arrangements that allow desired groupings, adequate materials, assignments for students to perform independently while the teacher is working with other groups or individual students, pupil or aide leaders who can act as resource persons, whole-class activities, group activities, and individual activities.

Diagnosing Student Needs

Inservice teachers ask, "How can I effectively divide my class into groups?" We have suggested numerous ways to evaluate student abilities and interests in the various curriculum areas of language arts. The more information the teacher has, the more instruction will meet the students' individual needs. For example, the information gained from evaluating students' spelling instructional levels should be used to divide children into appropriate groups for spelling instruction, or to provide individualized remediation. Each of the language arts areas has diagnostic techniques that permit the teacher to evaluate abilities, needs, or interests. The teacher must seriously consider each evaluation before grouping students. It becomes apparent that these groupings should be flexible, so children can change groups as the need arises.

REINFORCEMENT ACTIVITY

Review the various diagnostic techniques we have covered in this book. Choose one area of the language arts that could be taught effectively through grouping. Stipulate how you would use diagnostic techniques to group children in that area. Share your suggestions with your language arts class or instructor.

Instructional Time

A review of an elementary teacher's weekly lesson plans will show a large portion of instructional time spent either in direct teaching of the language arts or in language arts-related activities in other subject areas (e.g., discussing a topic, writing a report, reading a reference). The amount of direct teaching depends on the children's age level, their ability level, and whether the specific skill is being taught for the first time or is being reviewed or reinforced.

The greatest block of direct teaching time in language arts is usually devoted to reading. Harris and Sipay (1985) recommended that instructional time for reading in first grade should allow about ninety minutes per day for developmental reading [groups using a basal or other reading approach, Brophy (1986) recommended that twenty-five to thirty minutes of this time be used for each reading group instruction], and sixty minutes per day for other reading-related activities, such as language experience stories, special-need groups, individualized reading, recreational reading, storytelling, choral reading, and dramatizations of stories. Many of the activities Harris and Sipay suggested for reading instruction also allow instructional or reinforcement time for the other language arts.

As children progress through the grades, the total time spent on reading instruction remains approximately the same; however, the proportion spent on developmental reading and other reading activities changes. Because the fourth-grader is becoming a more independent reader, Harris and Sipay suggested decreasing developmental reading time to approximately sixty minutes per day and increasing related activities to about eighty minutes per day. This adjustment allows more time for functional reading, research, special needs, literature, and recreational reading.

Upper-elementary students are usually able to work with considerable independence. The amount of time spent in developmental reading groups may again decrease slightly, with approximately fifty minutes for developmental instruction and about ninety minutes for related instruction. In many upper-elementary grades, instruction may be departmentalized, with one period for reading and another period for English or language arts. Reading in the content areas and other language arts-related skills are also included in science and social studies.

Research does not indicate the amount of instruction time required for mastery of various language arts skills. (The reading times are recommendations by reading specialists.) Individual children have quite different needs and many of the language arts are interrelated, so it would be quite difficult to estimate direct instructional or reinforcement time for language arts skills. Perhaps the most one can do in this area is to take note of Kean and Personke's (1976) warning about the need for balance in the language arts curriculum, and to avoid teaching one skill to the exclusion of others.

Flexible Grouping

The effective language arts curriculum should include opportunities for children to work individually, in pairs, in small groups, and with the whole class because there are advantages and disadvantages with each type of grouping. Although individual work is necessary for many diagnostic procedures and independent study, it would be impossible to develop effective discussion strategies without the use of larger grouping practices that encourage student and teacher interaction. Without flexible groupings, the teacher would be forced, at one extreme, to utilize only whole-class instruction, or, at the other extreme, complete individualization or small groups. Neither of these alternatives allows maximum efficiency of student or teacher time and instruction. For a better understanding of grouping in the language arts curriculum, we examine whole-class, small-group, and individual instruction to see how they can be used to advantage.

Whole-class activities   Many language arts skills call for whole-class instruction, which has several obvious advantages. First, the teacher works with all the children at one time, and need not provide independent activities to occupy a few children while she works with the rest. Second, the teacher is immediately available to answer questions or to clarify misconceptions. Third, there are no stigmas attached to belonging to a lower-ability group. Fourth, some students may be motivated by others who have a greater interest in the subject. Fifth, students hear or see a model demonstrating desired behaviors (e.g., the teacher orally reading a literature selection to the class).

Whole-class instruction is appropriate for choral reading; motivating a language experience chart or individual story; introducing new ideas and skills; appreciative listening; and oral reading or storytelling. Table 15–1 lists some whole-class instructional activities.

Smaller-group instruction   Grouping children has several advantages. The grouping most often used is the ability grouping, in which children of approximately equal abilities are given instruction designed to meet their specific needs. All instructional

TABLE 15–1
Whole-class instructional activities

Language Arts Area	Activities
Oral language development	1. Show and tell 2. Listing effective speaking characteristics 3. Discussion and questioning centered around a common need or interest 4. Choral reading and speaking 5. Enrichment experiences for oral vocabulary development 6. Pantomime 7. Creative presentations 8. Buzz sessions 9. Role playing
Listening	1. Informal assessment of attentive listening in the classroom 2. Listening to rhymes to develop auditory perception 3. Setting conditions for listening in the classroom 4. Listening to a guest speaker 5. Listening to an educational television program 6. Questioning prior to a listening experience 7. Appreciative listening to a story, recording, drama, or choral speaking 8. Listening to a group presentation 9. Critically listening to a commercial, news report, television presentation, or speaker 10. Listening to a presentation for predetermined purposes, such as recognizing the main idea, important details, sequence of events

TABLE 15–1
continued

Language Arts Area	Activities
Creative writing and composition	1. Motivating a creative writing activity 2. Teacher and peer interaction to develop and expand ideas 3. Motivating a language experience chart story 4. Providing stimulating experience for creative writing 5. Enrichment activities for vocabulary development
Handwriting	1. Informal assessment of normal writing, fast writing, and best writing 2. Readiness activities 3. Introduction of individual manuscript letters 4. Introduction of cursive writing
Grammar, usage, and mechanics	1. Role playing 2. Introducing sentence expansion activities 3. Introducing new punctuation or capitalization skills
Literature	1. Reading and telling stories 2. Creative interpretation
Media	1. Producing a film 2. Putting together a class newspaper 3. Critical evaluation of TV commercials and newspaper advertisements, discussion of propaganda techniques 4. Critical evaluation of TV, radio, or film presentations 5. Enrichment activities for extending knowledge about literature, folklore, authors, etc.
References	1. Introducing a new reference skill 2. Introducing reference materials 3. Instruction in library usage

materials are on a suitable level of difficulty. Time is not wasted by providing instruction that might be too difficult, or by forcing children to repeat skills they have already mastered. Diagnosis is extremely important in forming ability groupings.

Ability groupings vary in longevity. Ability groups for reading instruction may continue throughout the school year, although children should be able to move from group to group as their abilities change. On the other hand, an ability group for corrective instruction in punctuation, for example, may stay together only as long as it takes to correct the problem. Corrective groupings are efficient because only the students who need extra instruction are placed in the group. The corrective group may remain together for only one class period, or for several weeks. Corrective ability groupings are based on efficient diagnosis; a group is formed as soon as a need arises and before excessive remediation is necessary.

Conclusions from teacher effectiveness research reported by Brophy (1986) provide guidelines for developing effective small-group instruction in reading. First, the programming should emphasize continuous progress by providing twenty-five to thirty minutes of small-group instruction each day, include effective management of students

The teacher groups her students during reading instruction.

who work independently as well as for students within the reading group, move through the curriculum in a way that is brisk enough to produce continuous progress, and continue with practice and review until smooth and correct performance is achieved.

The group should also be organized so that the students will start lessons quickly and the teacher can work with the group and still monitor the rest of the class.

In addition, lessons and activities should be introduced with an overview to provide students with a mental set and help them anticipate what they will be learning. Students should demonstrate that they know what to do and how to do it before they are asked to do activities independently.

Moreover, encourage everyone's participation by asking students questions about key concepts or meanings, by using a system that insures that all students have opportunities to read and answer questions while still emphasizing that students must wait their turns and respect the turns of others. Also monitor individuals to assure that each receives feedback and achieves mastery of the lesson.

Furthermore, concentrate questions on academic content, include word attack questions, wait for answers if students are still thinking about the question, give help if you think the students cannot reason out the correct answer by rephrasing or simplifying the question or by giving clues, give the answer when necessary, and explain answers by providing the steps one would go through in order to successfully solve the problem and answer the question.

Also remember to acknowledge when students respond correctly, provide feedback that emphasizes the methods used to get the answers, and, when appropriate, ask follow-up questions that help students integrate relevant information.

Finally, praise in moderation, identify specifically what is being praised, and if criticism is necessary, focus on the academic content and include corrective feedback.

In addition to ability groupings, language arts instruction also utilizes research and interest groupings. These groupings allow students with diverse abilities to work together until they complete a research project, or satisfy an area of mutual interest.

The size of each group depends on its purpose and the students' abilities. Students who need a great deal of help should work in smaller groups, so they will receive individual attention and be able to participate frequently. Research and interest groups should be of a convenient size to allow discussion and interaction.

The number of groups formed in any classroom depends not only on abilities and interests, but also on the teacher's management skills. The language arts teacher should use different types of groupings, and not rely solely on ability groupings. Table 15–2 shows some language arts activities to use with small groups.

Individual instruction and independent work   The term *individual* may refer to completely individualized instruction, characterized by self-pacing and self-selection, or to activities children complete independently while the teacher works with another group. Both types of individual instruction make specific demands on the teacher's classroom management capability.

Individualized instruction, in which children establish their own pace and sequence of instruction according to individual needs and differences, is especially

TABLE 15–2

Activities for research, interest, and ability groupings

Area of Language Arts	Type of Grouping	Activities
Oral language development	Research	1. Panel discussions 2. Round table discussions 3. Buzz sessions 4. Interviewing to investigate research topic 5. Presentation of research findings to class
	Interest	1. Puppet theater presentations 2. Creative dramatizations
	Ability	1. Readers' Theatre (enrichment for accelerated readers) 2. Choral reading (reinforcement of reading and language skills for remedial students)
Listening	Research	1. Listening to an interview 2. Listening to auditory references
	Interest	1. Listening to group-chosen recording, story, or book
	Ability	1. Listening to improve auditory perception 2. Listening to improve attentiveness 3. Listening to improve comprehension of main ideas, details, sequence, etc. 4. Listening to improve critical evaluation
Creative writing and composition	Research	1. Outlining research questions, needs, and findings 2. Writing report of research findings
	Interest	1. Creative writing stimulated by a topic of mutual interest 2. Group chart story
	Ability	1. High-ability group interpreting mythology and relating mythology to themselves 2. Improving paragraph organization 3. Vocabulary development 4. Language experience stories to remediate specific skill areas—developing a main idea, developing a logical sequence, following directions
Handwriting	Research	1. Producing legible reports, outlines, charts, letters

TABLE 15–2
continued

Area of Language Arts	Type of Grouping	Activities
	Interest	1. Producing legible stories for an audience
	Ability	1. Remediation of printing or cursive writing difficulties
Spelling	Research	1. Instruction using spelling words associated with research topic 2. Proofreading research outlines, charts, reports, letters
	Interest	1. Instruction using spelling words associated with interest area 2. Proofreading stories, creative writings, etc.
	Ability	1. Fernald approach to spelling for disabled spellers
Grammar, usage, and mechanics	Research	1. Proofreading for grammar, punctuation, and capitalization 2. Instruction in skills needed for research reporting
	Interest	1. Proofreading for grammar, punctuation, and capitalization 2. Instruction in skills needed for interest reporting
	Ability	1. Sentence building activities 2. Sentence transformation activities 3. Role playing 4. Cloze activities to develop understanding of language functions 5. Instruction in needed punctuation and capitalization skills
Literature	Research	1. Reading historical fiction to learn more about a time period studied in social studies 2. Investigating techniques used to illustrate picture books 3. Investigating the significance of a country's folk literature and music 4. Investigating an author 5. Investigating the significance of Mother Goose rhymes and their historical purpose
	Interest	1. Reading and discussing literature associated with a subject of mutual interest 2. Book discussions during and after

TABLE 15–2

continued

Area of Language Arts	Type of Grouping	Activities
		reading a book chosen by an interest group
		3. Enrichment activities following the literature reading (e.g., a group creates a puppet theater, another does creative writing, a third assembles a bulletin board)
		4. Recreational reading groups
	Ability	1. Reading high interest, lower reading level literature (lower ability and remedial groups)
		2. Readers' Theatre created around a literature selection (higher ability groups)
		3. Choral reading of poetry, rhymes, and easy-to-read books (e.g., *Green Eggs and Ham* or *Cat in the Hat*) to reinforce language and reading skills
		4. Literature club discussion (higher ability groups)
Reading	Research	1. Form a group to research a subject area; dividing into research groups after the development of a web of interest
		2. Reading information needed for a research project in any content area
	Interest	1. Recreational reading groups
		2. Reading stories or topics about a mutual interest
	Ability	1. Developmental reading groups (high, average, low)
		2. Remedial reading classes
		3. Corrective skill groups
Media	Research	1. Investigation of film production, making your own film
		2. Investigation of propaganda techniques found in media
		3. Investigation of newspaper production, editing, writing, etc.
		4. Investigation of TV production
	Interest	1. Stimulating reading before or after a TV or movie presentation

TABLE 15–2
continued

Area of Language Arts	Type of Grouping	Activities
	Ability	2. Creating a TV show, radio production, film or newspaper
		1. Functional reading skills needed to read the newspaper
		2. Developing listening comprehension skills
		3. Stimulating interest, developing background information
Reference	Research	1. Introduction to library skills
		2. Instruction in various skills needed for research project
	Interest	1. Instruction in reference skills needed for topic studied by interest group
		2. Instruction in locating books and other media about a specific interest or hobby
	Ability	1. Instruction and corrective skill grouping in dictionary usage
		2. Instruction in using encyclopedias, making a group encyclopedia
		3. Instruction in using a card file for locating references

demanding. Lapp and Flood (1986) identified six teacher-knowledge areas essential for directing an individualized program. The teacher must be knowledgeable in (1) the reading process (language process, writing process, etc.): (2) organizational skills; (3) sequential skill development; (4) assessment techniques; (5) instructional materials for numerous ability levels; and (6) classroom management procedures. Students must have several specific characteristics to succeed in this type of program because they do much of their work independently. They must be able to work independently for longer periods of time and be able to complete work successfully at a realistic pace without constant teacher or group interaction for successful work.

Reading is the most frequently individualized area of the language arts. This individualization usually occurs after children have acquired a background in reading skills that enables them to work effectively at their own pace and instructional levels. You will find directions for individualizing the reading program from several sources, among them books by Burns, Roe, and Ross (1988).

The individualized reading program is characterized by one-to-one pupil–teacher conferences. We have previously mentioned other areas of the language arts that also benefit from individual pupil–teacher conferences; for example, the writing process. Leonard Sealey, Sealey, and Millmore made a strong point for teacher–pupil interaction during writing: "Finding a time for writing is not just a question of how to find time for children to write when the curriculum is already filled, but involves a serious assessment of how much time the teacher herself can devote to becoming involved" (1979, p. 15). The primary teacher must be prepared for children to make continuous demands on her or another resource person. Providing frequent opportunities for individual teacher–pupil interaction and conferences puts heavy demands on teacher management. According to Spache and Spache, "The number of conferences and small-group sessions that a teacher can manage per week is an individual matter. The recommendation of two to four conferences per week suggested by some writers may be ideal but it is almost impossible for some teachers" (1977, p. 327).

The second type of independent assignment is that given to children for completion while the teacher works with another group or confers individually with another child. Although this assignment may be the same for each member of the group, the teacher must still plan the assignment carefully and give clear directions so the students can work successfully without assistance. The teacher must also plan independent activities that supply meaningful supplemental work for students who have completed their other assignments.

Effective, individual, language arts activities range from individual pupil assessment to interest centers in reading, literature, listening, or writing, where children can learn new skills, reinforce previously learned skills, or participate in enrichment activities. Table 15–3 lists individual instructional activities.

REINFORCEMENT ACTIVITY

We have discussed both the time requirements for language arts and the various flexible grouping practices that might be used in language arts. If you are a preservice teacher, visit a classroom and observe the time spent in various language arts and the types of instruction that use whole-class, small-groups, or individual activities. Keep a time log of each language art you see and the grouping practices used. If you are an inservice teacher, keep a weekly log of the amount of time you spend in language arts instruction and the grouping practices you use during that week.

Integrating Language Arts Instruction

Many of the activities presented in this text emphasize an integrated approach to the language arts. For example, the composition approach to the study of biography outlined on pages 353 to 357 integrates literature through an analysis of literary elements in biographical writing, oral language through discussion of biographies and interviewing needed to obtain information prior to writing biographies, listening through critical evaluation of biographies read by the teacher and peers, writing through composing biographies, grammar and mechanics of language through writing biographies, reading through critical evaluation of biographies, and library skills through research conducted prior to writing biographies or to evaluate the authenticity of published biographies. In addition, the affective domain is heightened by developing students' appreciation for biographical literature. In this example, the central focus for the integration of language arts is literature.

In self-contained classes, teachers have many opportunities to integrate not only the language arts but other content areas such as social studies and science as well.

TABLE 15–3
Individual activities

Area of the Language Arts	Activities
Oral language development	1. Individual assessment of oral language–language experience, storytelling, story retelling, picture interpretation
Listening	1. Individual assessment of hearing acuity, auditory perception, listening comprehension 2. Listening interest center 3. Listening to tapes or records for a specific purpose
Creative writing and composition	1. Individual pupil-teacher interaction during a writing assignment 2. Pupil-teacher writing conferences 3. Individual writing of a creative story following whole-class or group stimulation 4. Individual writing of paragraphs, compositions 5. Writing interest centers 6. Individual writing of a language experience story following group motivation or individual motivation 7. Vocabulary enrichment using an audio-visual tape
Handwriting	1. Individualized instruction for handwriting difficulties 2. Assistance during pupil-teacher writing conference 3. Handwriting learning center with models of manuscript or cursive writing
Grammar, usage, and mechanics	1. Informal assessment of usage during oral language assessment 2. Sentence expansion activities 3. Instruction in sentence structure, punctuation, and capitalization during pupil-teacher writing conference or pupil-

TABLE 15–3

continued

Area of the Language Arts	Activities
	teacher interaction during writing assignment 4. Learning center on letter writing
Spelling	1. Individual instruction to overcome spelling problem unique to one or two children 2. Learning spelling words of particular need and interest 3. Individual or small-group instruction using Fernald techniques
Literature	1. Recreational reading 2. Independent work associated with a literature selection 3. Preparing oral, written, or art book reports 4. Individual pupil-teacher conference about literature 5. Interest centers developed around literature themes and selections
Reading	1. Administering an informal reading inventory or other type of individual reading test 2. Individual pupil-teacher conferences 3. Individualized reading 4. Reading group activities completed independently while teacher works with another reading group 5. Corrective instruction when needed 6. Individual work as part of research or interest groups 7. Recreational reading 8. Reading learning centers—decoding skills, comprehension, basic study skills, rate building, critical reading, etc.
Media	1. Independent activities developed around newspapers, filmstrips, etc. 2. Interest centers 3. Using media such as television or film to motivate individual creative writing
Reference-library	1. Individual instruction in library skills when needed 2. Library skill assignment to group to be accomplished independently: dictionary, encyclopedia, card catalogue, study sheets 3. Library skill learning centers: alphabetizing, dictionary skills, etc.

Again, literature may provide the focus for such integration. For example, during a focus on science fiction (Norton, 1987) the students develop an understanding of and appreciation for literature by identifying and analyzing techniques authors use to develop credible science fiction, enhance oral language by discussing possibilities of living in a space colony, reinforce science concepts and reference skills by evaluating if the content of the science fiction is based on scientific principles, and enhance writing skills by writing their own science fiction or speculating about what they might see from NASA's proposed space telescope.

An extended language arts block provides many opportunities for integrating language arts. Think of ways to focus on literature and integrate the language arts in the lower-elementary grades. My own graduate and undergraduate students are analyzing basal readers used in many school districts, identifying literature-related skills that will be taught in the basals, identifying literature selections that develop those same skills, and then developing curricular plans that integrate the language arts through the literature and literature-related activities. This approach allows teachers to place a greater emphasis on primary literature sources and other language arts areas. For example, let us consider how we might use an integrated approach to teach one of the literary elements: personification. First, after identifying a logical place in the basal to provide broader applications, we would identify literature selections that develop personification. For example, there is personification of objects in Virginia Lee Burton's *The Little House* (1942) and Anthony Browne's *Gorilla* (1983). Personification of animals appears to Emily Arnold McCully's *Picnic* (1984) and in Dr. Seuss's *The Cat in the Hat* (1957). Personification of nature is developed in Susan Jeffers's illustrated version of Longfellow's *Hiawatha* (1983).

Next, we would introduce the concept of personification by modeling a personification lesson with Burton's *The Little House*. This listening and discussing lesson would introduce examples of personification, tell children the advantages of learning about personification, and explore ways that they can use this knowledge in reading, writing, and oral language. We would then go through the steps in modeling; the teacher first reads the text orally to the first question about personification, answers the question, cites the evidence, and explores the reasoning process needed to solve the problem. Then we would continue the lesson while gradually encouraging the students to enter into the process. Additional language arts skills would be integrated and understanding of personification expanded by having students both pantomime the feelings expressed by the house and create conversations that might occur between the house and her city or country neighbors. Writing is added as children consider other characters in the story and think about how they might personify the lives of other objects such as the horseless carriages, the subway, the apartment houses, and the moving truck. Mechanics of writing are taught or reinforced as children write dialogues for their personified characters, which requires correct punctuation.

Additional lessons would use the other books to develop the understanding that animals and nature may also be personified. Considerable oral discussion and writing would accompany the wordless book *Picnic* as students practice their ability to personify mice as the mice respond to different settings, increasing conflict, and different characters. Reading skills would be enhanced as children read independently additional books containing personification.

Analytical skills may be expanded by asking students to compare characteristics of the personified objects, animals, and nature with characteristics of objects, animals, and nature in realistic fiction as well as nonfiction. For example, students could analyze and compare the characters, setting, story content, and illustrations in the personified *Picnic* with the nonfictional depiction of mice in Oxford Scientific Film's *Harvest Mouse* (1982).

These opportunities to integrate the language arts through primary sources of literature enhance children's appreciation of literature and provide many opportunities to develop and use numerous language arts skills. Many teachers discover the enjoyment that can be had by expanding such possibilities.

Flexible Room Arrangements

The flexible grouping practices and integration of language arts make certain demands on classroom arrangement; each type of grouping, to be effective, has certain requirements.

Room arrangements for whole-class instruction   All the activities suggested for whole-class instruction do not require the same room arrangements. Show and Tell and appreciative listening when the teacher reads or tells a story may be best conducted by grouping the children in a semicircle in front of the teacher. This arrangement allows eye contact, easy viewing of pictures, and involves everyone closely in the listening activity. Children might also sit on a colorful rug in the reading center while they listen to stories.

Other whole-group activities require different arrangements. Large, uninterrupted spaces are required for movement and pantomime activities. The classroom may be too restricted for some of the activities; another location, such as a gymnasium, may be more appropriate. Some whole-class activities require extensive work space in the classroom; you can move tables together for creating films or murals. Other activities require individual work space, so students will sit at desks or tables. If the teacher is presenting a handwriting readiness lesson to the whole class, for example, each child needs his own working space. Desks or tables should be arranged so the teacher can move easily from child to child. If the whole-class activity involves listening to a speaker, viewing a film, or watching the teacher present material on the chalkboard, the room should be arranged so children can easily focus their attention on the subject. Movable chairs, desks, and tables are necessary for such a variety of whole-class activities.

Room arrangements for group activities   The multitude of potential group activities in language arts instruction also calls for flexible classroom arrangements. The circle, semicircle, or chairs around a rectangular work-table arrangement work very well for panel discussions, round-table discussions, and reading or other language arts research, interest, and ability groups. Some group activities, such as the reading group, are teacher-directed and need only one area for group instruction, because the remainder of the class can work independently at their seats or interest centers. At other times, however, the classroom must have work and discussion space for several groups to work concurrently. Several research or interest groups can group together tables or desks, while another interest group works at a puppet theater.

Room arrangements for individualized and independent work   If the teacher is assessing an individual student or conducting a pupil–teacher conference, she needs a location with minimal distractions. This location must usually be in the classroom, so the teacher can observe the activities of the rest of the class while she confers with one student. Some teachers overcome the management problems connected with conferences by scheduling conferences for specific blocks of time. While the teacher confers with one student at a table, the next student on the schedule waits his turn in a chair near the conference area. When that student's turn for a conference comes up, the next student on the schedule takes his place in the waiting chair. This procedure saves considerable time in a busy teaching day.

When children work independently, the room arrangement must minimize contact with other students. There are times in the day when conversation and group work are undesirable, and separate work or desk areas are preferable. Attentive studying for a spelling test, reading a basal reader, reading a reference to learn specific information, or doing an individual assignment to reinforce a skill taught during a group activity, require concentration and few distractions.

Learning and interest centers also permit children to work individually or in small groups. Many elementary teachers arrange interest centers around the perimeter of the room. For example, an early-elementary class might have a library corner, where children can select recreational reading books or read; a creative writing center, where children can find individual stimulation for writing, write a short story, and display their results; an art center, where a child can manipulate clay, paints, and other materials; a listening center; a game center; a science center; and/or a creative drama center, complete with puppet theater. Centers are also appropriate for middle- and upper-elementary grades, although the content of the centers changes. An upper-elementary classroom might have a literature center, a creative writing center; a study skills center; a science center; a social studies center; a listening center; and/or a media center. A classroom arranged for flexible grouping and learning centers is shown in figure 15–1.

Making assignments   We have noted that, if teachers are to effectively manage the whole-class, group, and individual activities, they must make appropriate assignments that students can work on while the teacher works with other groups and individuals. Teachers must plan these assignments so the children can complete them without assistance. The children's ability level influences the selection of assignments; for example, many first graders require assignments that do not call for reading ability. To overcome this problem, teachers might match a child with reading or writing ability with a child who has not acquired this ability. Language arts assignments in the middle and upper grades are not quite so restricted, because children have usually developed independent work habits, are able to read independently, and can often work effectively for longer periods without direct teacher supervision. For all assignments,

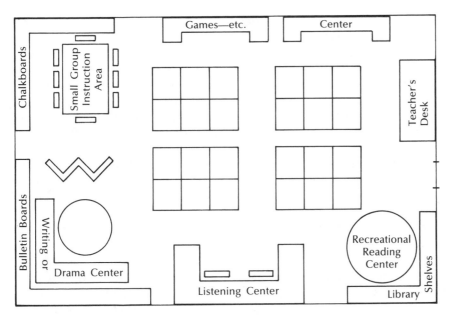

FIGURE 15–1
A classroom arrangement for flexible grouping

regardless of grade level, the teacher must give clear and explicit directions and complete examples to the children so they know what is expected of them. (These principles apply to homework as well as in-class assignments; parents become frustrated when children bring home assignments they apparently do not know how to do, or lack the prerequisite skills necessary to complete the work.)

Classroom management must also take into account the necessity for activities for children to do when they have completed their other assignments. If children have interesting activities to turn to, they are less apt to waste time or become behavior problems. Some teachers make a list of activities the children can do after they finish their group work. A list for a first-grade, average-ability group might look like this:

Mon., Jan. ——————
After I Do My Class Work I Can

1. Read a book from the library corner.
2. Listen to the tape of "Winnie the Pooh and the Blustery Day" at the listening corner.
3. Make a clay model of my favorite pet at the art center.
4. Work on the Picture Dictionary with someone from (*high ability*) group.
5. Finish my "Where the Wild Things Are" mobile.
6. Put on a puppet show with someone from (*low ability*) group.

Teachers can more easily master classroom management if they designate a classroom helper who can answer routine questions about assignments, help children who are late, or review or give assignments to children who are returning to the classroom from a resource room or other special educational class. The classroom helper should be a mature child who understands the assignments and is able to give some assistance or answer questions correctly. The position should rotate, so many of the children will have a chance to function in this capacity; otherwise, the few children who are always asked to help others will never to able to complete their own enrichment activities.

Some classrooms are fortunate to have an adult teacher's aide to answer children's questions about assignments, or provide tutorial help to students who require more one-to-one instruction. Other teachers use adult volunteers when they need extra help. These volunteers might even teach minicourses in fields where they have outstanding expertise. A parent or other adult from the community can be especially helpful with creative writing, creative drama, or use of the media.

Another way to offer help for especially demanding assignments (e.g., writing individual language experience stories) is to use cross-grade or cross-ability-level groupings. In cross-grade grouping, students in a lower grade are grouped with students in an upper grade. A first-grade teacher who wants her students to dictate individual language experience stories or other creative writing could group the children with fourth- or fifth-graders who have the necessary handwriting and spelling skills. The older students benefit from the experience of taking dictation, and reinforce their own writing skills.

Teaming students of different ability levels also increases the capability of some students to complete assignments or to do reinforcement activities without constant teacher supervision. Many games designed to reinforce reading skills, for example, cannot be played properly unless at least one of the players can verify a correct answer. Many interest-center activities are also more effective when the children work in pairs or small-group arrangements; the teacher can team children who have diverse abilities but common interests.

LEARNING CENTERS

The term *learning* or *interest center* has been used frequently in this chapter on classroom organization and management. A learning center gives students a chance to work independently or in small groups, thus freeing the teacher to work with other students. (There are examples of learning centers that provide individual and small-group activities in chapters 4 and 12.) A center occupies a defined area in the room, and relates to a specific theme, topic, or skill. The learning activities instruct, reinforce, or enrich a child's knowledge in that particular area. Several learning centers help attain a desired balance in the language arts curriculum.

Characteristics of Centers

A good learning center is easy for children and teacher to manage. It is especially true of the learning center that students be able to follow assignments easily. If the teacher

A well-managed learning center allows students to work independently.

expects children to use the center independently, it must be well-organized, with clearly written directions, so students know what to do and where to find materials. Second, the center must be geared to the children's needs and interests. Teachers can create a learning center to meet the needs of a specific ability group (e.g., a center for handwriting remediation, or one for literature enrichment for the gifted child), or they can develop a multilevel center with activities for the whole class (e.g., the "Meet the Press" learning center in chapter 12 specifies activities for several ability levels). Third, a well-balanced learning center has activities that can be accomplished individually, in pairs, and in small groups. Fourth, because many of the activities in a learning center are accomplished without direct teacher involvement, they must be motivating to the children. Finally, the well-planned learning center includes several learning modalities—activities that call for listening, viewing media, reading, writing, oral expression, creative manipulation, and art expression.

Components of Centers

A learning center also requires specific components for effective management. General directions and directions for each activity in the center are both essential. The general directions tell the purposes for the center and how to use it. Specific directions for activities require enough information for students to complete tasks successfully. Examples are always useful for clarifying directions, and not all directions need to be written; some are more effective if recorded on tape. The learning center should use multimedia: films, books, slides, filmstrips, records, and tapes; and should call for written, oral, and artistic interpretations.

The center requires some method for evaluating answers to specific activities. The activities might be self-evaluated, evaluated by another student, or teacher-evaluated. Many of the creative language arts activities, of course, will not have one correct response. Some of these will be shared with the group, displayed on a bulletin board, or discussed with the teacher or a small group.

There must be a defined place for students to put completed work. Individual folders, in-and-out boxes, or bulletin boards will do for some activities. Management is simplified if each activity gives directions for what to do with the finished product. Children like to share their work, and it is motivating to know who the audience will be before they start the activity.

Three types of record forms are often specified in connection with learning centers: (1) records used to schedule students for work in the centers; (2) records in notebooks or checklists kept by the teacher; and (3) records kept by the student. The first type improves the center's usefulness by organizing the times and numbers of students who will work there. Scheduling depends on the purpose for the center, the number of children who can work in it at one time, and the number of activities that can be performed away from the learning center. If teachers want a whole group to work in a specific center, they might use the schedule shown in figure 15–2. The name of the group that is to work in each center is hung on the hook or attached to the

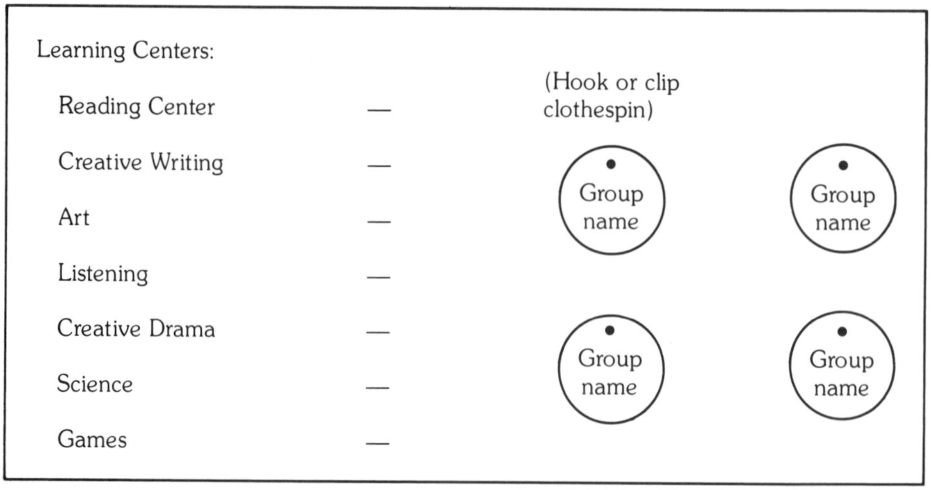

FIGURE 15–2
A schedule for working in learning centers

clothespin. This method can also be used for assigning individual children to a center. A less permanent scheduling system involves duplicating sheets with the name of each center. The name of the center is at the top of the sheet, followed by the number of spaces the center can accommodate.

The second learning-center record is the teacher checklist or notebook to record what the children do. After children complete an activity, their names are checked with the appropriate key on the checklist. A checklist for "Meet the Press" is shown in figure 15–3.

The teacher can keep more detailed records in notebook format. Some teachers keep anecdotal records in a loose-leaf notebook with a separate page for comments on each child's activities and progress in the various learning centers. Teachers can use this information during pupil–teacher conferences, to plan further learning experiences, and during parent–teacher conferences.

The final type of learning-center record is kept by each child. These individual records can use different formats depending on the children's ages. For example, an older child might keep a straightforward list of the books he reads for recreation (figure 15–4), and a younger child might create a "book train" (figure 15–5).

Activity:	Mary		Jack		John				
Dear Abby									
Capture the Caption									
Headline Scramble									
Feelings, Feelings									
Getting the Meaning									
Sequencing Cartoons									
Headlines									
Main Ideas									
Article Artistry									
Catchy Cartoons									
Don't Be Conned									
Cartoon Talk									

Key: Satisfactorily Completed X
Redo the Activity   RD
Conference is Required   C

FIGURE 15–3
A checklist for "Meet the Press" learning center

Jimmy Johnson's Reading Record			
Date started	Name of the book	Date completed	My thoughts about the book

FIGURE 15–4
An individual record of books read

FIGURE 15–5
A record of books read by a younger child

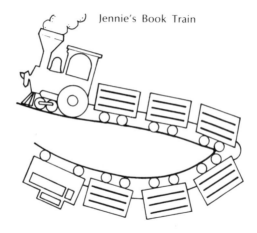

Jennie's Book Train

Creating the Center

Kaplan, Kaplan, Madsen, and Taylor (1973) mentioned six steps for creating a learning center:

1. Select a subject area.
2. Determine the skill or concept to be taught, reinforced, or enriched.
3. Develop the skill or concept into a learning activity—manipulating, experimenting, listening, or viewing.
4. Prepare the skill or concept into an applying activity—filling in, arranging in order, putting together, taking apart, listing, classifying, matching, tracing, writing, locating, or labeling.
5. Incorporate the skill or concept into an extending activity—comparing, developing your own, researching, reconstructing, finding *what other,* or deciding *what if.*
6. Put all the games, worksheets, and charts together in one area of the room for children to use in a self-selected manner.

When the center is ready, the teacher must introduce it. Even the most attractive center will not self-motivate all the children. The teacher needs to motivate the children to use the center, explain its purpose, describe the procedures for using it, and direct several activities generated from the center. (An activity to introduce a learning center is presented in the last section of this chapter.)

REINFORCEMENT ACTIVITY

With a group of your peers, select a subject area around which to create a learning center. (1) Follow Kaplan's six steps to create the learning center. (2) Include in your center the characteristics and components of a well-developed learning center. (3) Create the learning center and share it with your language arts class.

ORGANIZING INSTRUCTION

We have reviewed some of the elements for effective management of language arts instruction. But the teacher must still combine these elements—adequate time, flexible grouping practices, flexible room arrangements, and appropriate assignments—into an organized day, week, and year. If teachers combine their management skills with effective diagnosis, effective instruction, and adequate instructional materials, the children will learn and enjoy the language arts skills.

One method of organization effectively used by some teachers involves blocking out large time periods; identifying whole-group, small-group, and individual activities to include in those periods; and identifying teacher responsibilities and location during each segment of the time block. Blocking allows teachers to visualize whole-class management for that period of time. They can ask themselves certain questions:

1. Can the groups or individuals that I am not with perform their tasks without my constant supervision?
2. Have I included whole-group, small-group, and individual activities? (These groupings need not all occur daily, but should appear in the weekly and monthly plans).
3. Does the physical organization of the classroom permit each type of group or independent activity?
4. When will I give instructions to each group so they can work independently on a task? (The ability level of the group influences this decision, because some children retain directions longer than others. All children, however, require at least a quick review of expected procedures.)
5. Have I chosen someone to be responsible for answering questions and giving directions while I work with a different group? Have I prepared that person for the task?
6. Have I included enrichment or reinforcement tasks for children to do after they complete their assigned tasks?

7. Do the enrichment and reinforcement tasks correspond with individual differences, needs, and interests?
8. Have I allowed enough flexible time so that I can form a small group when I identify a skill that calls for immediate correction?
9. Do my activities balance within the language arts? Are there oral language, writing, listening, and reading activities?

When blocking time periods, it is helpful if the teacher is able to quickly visualize the organization of that period. One approach we use frequently with inservice teachers is to have them color-code their instructional plan. For example, when blocking activities for ability-group instruction, activities that are teacher-directed might be written or underlined in red; individual assignments for the high-ability group in blue; individual assignments for the middle-ability group in green; and individual assignments for the low-ability group in black. Similar methods might be used for blocking research and interest grouping activities; in this case, the group the teacher is working with appears in one color and each research or interest group in a different color.

Steps in Blocking Time Periods

Using color-coding, the teacher first maps her locations on the instructional plan. The map for a second-grade, three reading ability grouping arrangement, might look like this:

Teacher's Location During Three Reading Ability Groups

10 min.	Teacher Directed	Review Directions for the groups	
25 min.	Teacher Directed—Low Ability Group	(Middle group) (Independent work)	(High group) (Independent work)
25 min.	(Low group) (Independent work)	Teacher Directed—Middle Ability Group	(High group) (Independent work)
25 min.	(Low group) (Independent work)	(Middle group) (Independent work)	Teacher Directed—High Ability Group
20 min.	Teacher-Directed Whole Class Activity—Introducing a new interest center. Doing a teacher-directed activity from the center.		

Now the teacher can visualize her location and which groups must work independently while she directs instructional activities with another group. Specific assignments should be placed in each slot, and some team assignments included. For example, looking at the above plan will show the teacher which groups are working independently. Are there some activities that might be accomplished by teaming the low- and middle-ability groups, the low and high groups, or the middle and high groups? The teacher can see quickly when these groups are available to work together. Are there activities that take more time? If the high-ability group is working on a puppet presentation, they may need an entire fifty-minute period to work independently. The lower-ability group, however, may require more nonreading activities, or may need to be teamed with other students from different groups, if they are to benefit fully from this longer period without teacher instruction. If the students are to work at interest centers after completing their reading assignments, are they adequately prepared for the activities? Finally, the teacher should block out longer-range time periods, so she can evaluate the plan for balance and manageability.

Examples of Instructional Planning Blocks

The color-coding and blocking technique can be used with any language arts activities, and is most helpful when instruction requires simultaneous group and individual activities. The following time block shows teacher locations and assignments while introducing the media learning center in chapter 12. (Times are approximate; they will differ according to group needs.)

Instructional Plan for Introducing Multilevel Media Interest Center, "Meet the Press"

15 min.   Teacher-Directed, Whole-Class Activity: Introduce Learning Center. Discuss importance of newspaper and information received from it.

Discuss the purpose of the center and materials in it; describe the "plan to progress" sequence; discuss general instructions, including correction of answers, "Ready For Print" box, and "Extra, Extra" box; have students repeat directions; discuss specific instructions for each activity; discuss evaluation of specific activities. (Pass out multiple copies of one activity from "Reporter Chris," "Reporter Stacy," and "Extra, Extra." Divide class into three groups and allow students to try the activities independently.)

	Reporter Chris Group	*Reporter Stacy Group*	*Extra, Extra Group*
15 min.	Teacher with group, sequencing cartoons—Students read and explain directions, and try example with teacher guidance.	Independently read directions and fill in cartoon balloons with appropriate script.	Independently read directions and draw a cartoon and write script to accompany it.
	Reporter Chris Group	*Reporter Stacy Group*	*Extra, Extra Group*
15 min.	Do several sequencing activities; independently check answers.	Teacher directed—verify if directions were followed, work independently with students.	Continue drawing and writing own cartoons.

| 15 min. | Independent work—Team with a member of Reporter Stacy Group; share sequencing activity, then cartoon balloons activity. Select another cartoon without captions and write new caption together. Place results on a bulletin board. | Teacher directed—verify if directions were followed, work independently with students. Group shares their own cartoons. Place cartoons on bulletin board. |

The teacher answers questions and clarifies any misconceptions about the center.

The teacher can also block weekly and monthly plans. Table 15–4 shows a weekly schedule for spelling instruction.

REINFORCEMENT ACTIVITY

Choose an area of language arts instruction. (1) Block out the instructional activities you would include for a one-week period. (2) Color-code the assignments according to teacher-directed, group, or individual activities. (3) Include specific assignments for each group, whole-class, or individual activity. (4) If applicable, include activities for children to do when they complete the class or group assignments.

SUMMARY

Several factors influence the efficiency of language arts instruction. First, teachers need appropriate diagnostic techniques for instructional grouping. Both informal and formal techniques can be used to assess language arts abilities. Second, adequate instructional time must be provided for all aspects of communication—oral language, writing, literature, reading, listening, and media study. Third, because language arts objectives require children to work individually, in pairs, in small groups, and as a whole class, the teacher should use flexible grouping practices. For example, whole-class activities are appropriate for choral reading, motivation for a language experience chart story, or storytelling, but smaller groupings are frequently advantageous. Children may be divided into ability, research, and interest groups. All of these groupings have advantages and disadvantages, so every child should have opportunities to work in each kind of group. Effective teaching practices also frequently require individualized instruction and activities. Fourth, the various grouping arrangements make specific demands on classroom arrangement, and because of the different requirements, a flexible room arrangement, with movable desks, chairs, and tables, is preferable. Fifth, the teacher must make appropriate assignments so students can work independently while the teacher meets with other groups and individual children. Library, game, and interest-center activities will give children things to do

TABLE 15–4

Weekly spelling schedule—third grade (three ability groups)

Second-grade level	Third-grade level	Fourth-grade level
Monday: Pretest		
Teacher-directed: Teacher presents pretest to whole class by dictating second-grade list to one group, third-grade list to next, and fourth-grade list to final group. Each student uses corrected-spelling approach.		
Teacher with lower group to help correct spelling and immediate study. May require Fernald method.	Corrected spelling approach and immediate study of misspelled words.	Corrected spelling approach and immediate study of misspelled words.
Tuesday: Self-study and Expansion		
Teacher directions for group activities.		Teacher introduces individualized activities.
Students use self-study	Teacher Directed: Inductive approach to a spelling generalization.	Self-study of missed words. Activity which stresses multiple meaning of words.
Teacher-directed: checks progress; meaning-expansion activity.	Self-study of missed words. Finding and using additional words using generalization.	Individualized spelling activities.
Wednesday: Midweek Test		
Teacher-directed: Teacher presents midweek test of all words given on pretest. (Children who spell words correctly are not required to take additional test on Friday.) Children who miss words use self-study procedure.		
Thursday: Self-study and Enrichment		
Teacher directions for group or individual activities.		
(For Some Children) Self-study of missed words.	(For Some Children) Self-study of missed words.	Teacher-directed; checks and directs individualized work, enrichment activity, learning center, individual conferences.
Enrichment activity—Second- and third-level spellers team for reinforcement activity, such as spelling game.		
Friday: Final Test for Some Students		
Teacher-directed: Final test for all students who missed words on Wednesday. Place any missed words in student's individual notebook for periodic review. (Students who do not take spelling test may do enrichment activities in a learning center or recreational reading.)		

when they finish other assignments. Finally, literature provides an excellent focus for an integrated approach to the language arts.

The teacher must combine the elements of adequate time, flexible grouping, flexible room arrangements, and appropriate assignments into the organized day, week and year. Instructional time can be organized by blocking out time periods and using color-coding to show teacher-directed activities and activities for various groups.

ADDITIONAL CLASSROOM ORGANIZATION AND MANAGEMENT ACTIVITIES

1. *Computer centers are located in many elementary classrooms. Visit a classroom with a computer center. What types of learning activities are reinforced through computer-assisted instruction? How does the teacher manage the computer center? Consider how the teacher selected children to use the computer, how the teacher observes the learning while other children are engaged in learning experiences, and how the teacher records student progress.*
2. *Analyze a lesson plan you wrote for one of the language arts subjects. What is the most effective grouping? Why is it the most effective? How would you organize the classroom for your grouping scheme?*
3. *Choose a specific grade level. With a group of your peers, develop a list of activities that children could do when they have completed their assignments. Share the activities with your language arts class. How do the activities differ from each grade level identified in your class?*
4. *Locate a learning center that is described in an educational journal or textbook. Evaluate the learning center according to language arts objectives, to quality of instruction, and to ease of management.*
5. *Select one of the series of activities developed in **Language Arts Activities for Children** (Norton, 1985). Develop the activities into a learning center.*
6. *Develop a proposal for integrating the language arts through a focus on literature.*

BIBLIOGRAPHY

Brophy, Jere. "Principles for Conducting First Grade Reading Group Instruction." In *Effective Teaching of Reading: Research and Practice,* edited by James V. Hoffman. Newark, Del.: International Reading Association, 1986.

Burns, Paul; Roe, Betty, and Ross, Elinor. *Teaching Reading in Today's Elementary Schools.* Boston: Houghton Mifflin Co., 1988.

Harris, Albert J., and Sipay, Edward R. *How to Increase Reading Ability.* 8th ed. New York: Longman, 1985.

———. *How to Teach Reading.* New York: Longman, 1979.

Hennings, Dorothy Grant. *Communications in Action.* Chicago: Rand McNally College Publishing Co., 1978.

Kaplan, Sandra; Kaplan, Jo Ann; Madsen, Sheila; and Taylor, Bette. *Change for Children.* Pacific Palisades, Calif.: Goodyear Publishing Co., 1973.

Kean, John M., and Personke, Carl. *The Language Arts.* New York: St. Martin's Press, 1976.

Lapp, Diane, and Flood, James. *Teaching Students to Read.* New York: Macmillan Co., 1986.

Norton, Donna E. *Language Arts Activities for Children.* 2d. ed. Columbus, Oh.: Merrill Publishing Co., 1985.

_____. *Through the Eyes of a Child: An Introduction to Children's Literature.* 2d ed. Columbus, Oh.: Merrill Publishing Co., 1987.

Petty, Walter T.; Petty, Dorothy C.; and Backing, Marjorie F. *Experiences in Languages.* Boston: Allyn & Bacon, 1973.

Sealey, Leonard; Sealey, Nancy; and Millmore, Marcia. *Children's Writing—An Approach for the Primary Grades.* Newark, Del.: International Reading Association, 1979.

Spache, George D., and Spache, Evelyn B. *Reading in the Elementary School.* Boston: Allyn & Bacon, 1977.

CHILDREN'S LITERATURE REFERENCES

Browne, Anthony. *Gorilla.* New York: Watts, 1983.

Burton, Virginia Lee. *The Little House.* Boston: Houghton Mifflin Co., 1942.

Longfellow, Henry Wadsworth. *Hiawatha.* Illustrated by Susan Jeffers. New York: E. P. Dutton, 1983.

McCully, Emily Arnold. *Picnic.* New York: Harper & Row, 1984.

Oxford Scientific Films. *Harvest Mouse.* Photographs by George Bernard, Sean Morris, and David Thompson. New York: Putnam, 1982.

Seuss, Dr. *The Cat in the Hat.* New York: Random House, 1957.

Name Index

612

Subject Index